FRACTIONAL PARTIAL DIFFERENTIAL EQUATIONS AND THEIR NUMERICAL SOLUTIONS

FRACTIONAL PARTIAL DIFFERENTIAL EQUATIONS AND THEIR NUMERICAL SOLUTIONS

Boling Guo

Institute of Applied Physics and Computational Mathematics, China

Xueke Pu

Chongqing University, China

Fenghui Huang

South China University of Technology, China

NEW JERSEY • LONDON • SINGAPORE • BEIJING • SHANGHAI • HONG KONG • TAIPEI • CHENNAI

Published by

World Scientific Publishing Co. Pte. Ltd.

5 Toh Tuck Link, Singapore 596224

USA office: 27 Warren Street, Suite 401-402, Hackensack, NJ 07601

UK office: 57 Shelton Street, Covent Garden, London WC2H 9HE

Library of Congress Cataloging-in-Publication Data
Guo, Boling.
    Fractional partial differential equations and their numerical solutions / by Boling Guo (Institute of Applied Physics and Computational Mathematics, China), Xueke Pu (Chongqing University, China), Fenghui Huang (South China University of Technology, China).
        pages cm
    Includes bibliographical references and index.
    ISBN 978-9814667043 (hardcover : alk. paper)
    1. Fractional differential equations. 2. Differential equations, Partial. 3. Differential equations--Numerical solutions. I. Pu, Xueke. II. Huang, Fenghui. III. Title.
    QA372.G94 2015
    515'.353--dc23
                                          2015002791

British Library Cataloguing-in-Publication Data
A catalogue record for this book is available from the British Library.

Fractional Partial Differential Equations and Their Numerical Solutions
© Guo Bolin, Pu Xueke, Huang Fenghui
The Work is originally published by Science Press in 2011.
This edition is published by World Scientific Publishing Company Pte Ltd by arrangement with Science Press, Beijing, China.
All rights reserved. No reproduction and distribution without permission.

Printed in Singapore

Preface

In recent years, fractional-order partial differential equation models have been proposed and investigated in many research fields, such as fluid mechanics, mechanics of materials, biology, plasma physics, finance, chemistry and so on. Fractional-order differential equations, such as fractional Fokker-Plank equation, fractional nonlinear Schrödinger equation, fractional Navier-Stokes equation, fractional quasi-geostrophic equation, fractional Ginzburg-Landau equation and fractional Landau-Lifshitz equation have clear physical background and opened up related new research fields. In fact, some mathematicians (such as L'Hôpital, Leibniz, Euler) began to consider how to define the fractional derivative as early as the end of the 17th century. In 1870s, Riemann and Liouville obtained the definition of fractional derivative for a given function by extending the Cauchy integral formula,

$$ {}_0\mathrm{D}_t^{-v} f(t) = \frac{1}{\Gamma(v)} \int_0^t (t-\tau)^{v-1} f(\tau) \mathrm{d}\tau $$

where $\mathrm{Re}\, v > 0$. Nowadays, the commonly used fractional derivative definitions include Riemann-Liouville definition, Caputo definition, Grünwald-Letnikov derivative and Weyl definition. Kohn and Nirenberg began the research on pseudo-differential operator in 1960s.

In recent years, we collected and summarized the researches on nonlinear fractional differential equations and their numerical methods for specific physical problems appearing in the fields of atmosphere-ocean dynamics and plasma physics, and studied the mathematical theory of these problems. This book introduces the latest research achievements in these areas, as well as some researches of the authors and our collaborators. To give a systematic understanding of fractional problems to our readers, here we also briefly introduce some basic concepts of the fractional calculus, algorithms and their basic properties. In particular, we give brief introductions of numerical methods for the fractional differential equations. The aim of this book is to give a basic understanding of recent developments in this research field for readers who are interested in this topic. Our expectation is that the readers, who want to engage in this field, can access to the frontier of this study based on reading this book, and thus promote a more vigorous development.

Due to the time and knowledge limited, errors and inadequacies of the book are inevitable. Any suggestions and comments are welcome. At last, we express our heartful thanks to the seminar members of Institute of Applied Physics and Computational Mathematics. We also thank Professor W. Chen and his team at Hohai University who translated the Chinese version into English of the first version, which greatly reduced our burden of translation. We also express our gratitude to all those unnamed here.

December 1, 2010

Contents

Chapter 1

Physics Background

Fractional differential equations have profound physical background and rich theory, and are particularly noticeable in recent years. They are equations containing fractional derivative or fractional integrals, which have received great interest across disciplines such as physics, biology and chemistry. More specifically, they are widely used in dynamical systems with chaotic dynamical behavior, quasi-chaotic dynamical systems, the dynamics of complex material or porous media and random walks with memory. The purpose of this chapter is to introduce the origin of the fractional derivative, then introduce some physical background of fractional differential equations. Due to space limitations, this chapter only gives some brief introductions, but these are sufficient to show that the fractional differential equations, including fractional partial differential equations and fractional integral equations, are widely employed in various applied fields. However, the mathematical theory and the numerical algorithms of fractional differential equations need to be further studied. Interested readers can refer to the monographs and literature.

1.1 Origin of the fractional derivative

The concepts of integer order derivative and integral are well known. The derivative $\mathrm{d}^n y/\mathrm{d}x^n$ describes the changes of variable y with respect to variable x, and has a profound physical background. The present problem is how to generalize n into a fraction, even a complex number.

This long-standing problem can be dated back to the letter from L'Hôpital to Leibniz in 1695, in which it is asked what the derivative $\mathrm{d}^n y/\mathrm{d}x^n$ is when $n = 1/2$. In the same year, the derivative of general order was mentioned in the letter from Leibniz to J. Bernoulli. The problem was also considered by Euler(1730), Lagrange(1849) et al, and gave some relevant insights. In 1812, by using the concept of integral, Laplace provided a definition of fractional derivative. When $y = x^m$, employing the gamma function

$$\frac{\mathrm{d}^n y}{\mathrm{d}x^n} = \frac{\Gamma(m+1)}{\Gamma(m-n+1)} x^{m-n}, \quad m \geqslant n, \tag{1.1.1}$$

was derived by Lacroix, which gives

$$\frac{\mathrm{d}^{1/2} y}{\mathrm{d}x^{1/2}} = \frac{2\sqrt{x}}{\sqrt{\pi}}. \tag{1.1.2}$$

When $y = x$ and $n = \dfrac{1}{2}$. This is consistent with the so-called Riemann-Liouville fractional derivative.

Soon later, Fourier (1822) gave the definition of fractional derivative through the so-called Fourier transform. Noting that the function $f(x)$ can be expressed as a double integral

$$f(x) = \frac{1}{2\pi} \int_{-\infty}^{\infty} \int_{-\infty}^{\infty} f(y) \cos \xi(x-y) \mathrm{d}\xi \mathrm{d}y,$$

and

$$\frac{\mathrm{d}^n}{\mathrm{d}x^n} \cos \xi(x-y) = \xi^n \cos\left(\xi(x-y) + \frac{1}{2}n\pi\right),$$

replacing n with a general ν, and calculating the derivative under the sign of integration, one then generalizes the integer order derivative into the fractional derivative

$$\frac{\mathrm{d}^\nu}{\mathrm{d}x^\nu} f(x) = \frac{1}{2\pi} \int_{-\infty}^{\infty} \int_{-\infty}^{\infty} f(y) \xi^\nu \cos\left(\xi(x-y) + \frac{1}{2}\nu\pi\right) \mathrm{d}\xi \mathrm{d}y.$$

Consider the Abel integral equation

$$k = \int_0^x (x-t)^{-1/2} f(t) \mathrm{d}t, \tag{1.1.3}$$

where f is to be determined. The right hand side defines a definite integral of fractional integral with order $1/2$. In Abel's research on the above integral equation, its right end was written as $\sqrt{\pi} \dfrac{\mathrm{d}^{-1/2}}{\mathrm{d}x^{-1/2}} f(x)$, then $\dfrac{\mathrm{d}^{1/2}}{\mathrm{d}x^{1/2}} k = \sqrt{\pi} f(x)$, which indicates that the fractional derivative of a constant is no longer zero.

In 1930s, Liouville, possibly inspired by Fourier and Abel, made a series of work in the field of fractional derivative, and successfully applied them into the potential theory. Since

$$D^m \mathrm{e}^{ax} = a^m \mathrm{e}^{ax},$$

the order of the derivative was generalized into an arbitrary order by Liouville (ν can be a rational number, irrational number, even a complex number)

$$D^\nu e^{ax} = a^\nu e^{ax}. \tag{1.1.4}$$

If the function f can be expanded into an infinite series

$$f(x) = \sum_{n=0}^{\infty} c_n e^{a_n x}, \qquad \text{Re}\, a_n > 0, \tag{1.1.5}$$

then its fractional derivative can be obtained as

$$D^\nu f(x) = \sum_{n=0}^{\infty} c_n a_n^\nu e^{a_n x}. \tag{1.1.6}$$

Which method can be employed to obtain the fractional derivative if f can not be written in the form of equation (1.1.5)? Maybe Liouville had noticed this problem, and he gave another expression by using the Gamma function. In order to take advantage of the basic assumptions (1.1.4), noting that

$$I = \int_0^\infty u^{a-1} e^{-xu} = x^{-a} \Gamma(a),$$

one then obtains

$$D^\nu x^{-a} = \frac{(-1)^\nu}{\Gamma(a)} \int_0^\infty u^{a+\nu-1} e^{-xu} du$$
$$= \frac{(-1)^\nu \Gamma(a+\nu)}{\Gamma(a)} x^{-a-\nu}, \quad a > 0. \tag{1.1.7}$$

So far, we have introduced two different definitions of fractional derivatives. One is the definition (1.1.1) with respect to $x^a (a > 0)$ given by Lacroix, the other one is the definition (1.1.7) with regard to $x^{-a} (a > 0)$ given by Liouville. It can be seen that, Lacroix's definition shows that the fractional derivative of a constant x^0 is no longer zero. For instance, when $m = 0, n = \dfrac{1}{2}$,

$$\frac{d^{1/2}}{dx^{1/2}} x^0 = \frac{\Gamma(1)}{\Gamma(1/2)} x^{-1/2} = \frac{1}{\sqrt{\pi x}}. \tag{1.1.8}$$

However, in Liouville's definition, since $\Gamma(0) = \infty$, the fractional derivative of a constant is zero (despite Liouville's assumption $a > 0$). As far as which is the correct form of fractional derivative between the two definitions, Willian Center pointed out that the whole problem can be attributed to how to determine $\mathrm{d}^\nu x^0/\mathrm{d}x^\nu$, and as De Morgan pointed out (1840), both of them may very possibly be parts of a more general system.

The present Riemann-Liouville's definition (R-L) of fractional derivative may be derived from N. Ya Sonin (1869). His starting point is the Cauchy integration formula, from which the nth derivative of f can be defined as

$$D^n f(z) = \frac{n!}{2\pi i} \int_C \frac{f(\xi)}{(\xi - z)^{n+1}} d\xi. \qquad (1.1.9)$$

Using contour integration, the following generalization can be obtained (in which, Laurent's work contributed!)

$$_c D_x^{-\nu} f(x) = \frac{1}{\Gamma(\nu)} \int_c^x (x - t)^{\nu-1} f(t) dt, \qquad \mathrm{Re}\,\nu > 0, \qquad (1.1.10)$$

where the constant $c = 0$ is commonly used, which is known to be the Riemann-Liouville fractional derivative, i.e.,

$$_0 D_x^{-\nu} f(x) = \frac{1}{\Gamma(\nu)} \int_0^x (x - t)^{\nu-1} f(t) dt, \qquad \mathrm{Re}\,\nu > 0. \qquad (1.1.11)$$

In order to make the integral convergent, a sufficient condition is $f(1/x) = O(x^{1-\varepsilon}), \varepsilon > 0$. An integrable function with this property is often referred to as belonging to the function of the Riemann class. When $c = -\infty$,

$$_{-\infty} D_x^{-\nu} f(x) = \frac{1}{\Gamma(\nu)} \int_{-\infty}^x (x - t)^{\nu-1} f(t) dt, \qquad \mathrm{Re}\,\nu > 0. \qquad (1.1.12)$$

In order to make the integral convergent, a sufficient condition is when $x \to \infty$, $f(-x) = O(x^{-\nu-\varepsilon})(\varepsilon > 0)$. An integrable function with this property is often referred to as belonging to the function of the Liouville class. This integral also satisfies the following exponential rule

$$_c D_x^{-\mu} {_c D_x^{-\nu}} f(x) = {_c D_x^{-\mu-\nu}} f(x).$$

When $f(x) = x^a (a > -1)$, $\nu > 0$, from the equation (1.1.11), it is easy to get

$$_0 D_x^{-\nu} x^a = \frac{\Gamma(a + 1)}{\Gamma(a + \nu + 1)} x^{a+\nu}.$$

By using the chain law, has $D[D^{-\nu} f(x)] = D^{1-\nu} f(x)$, then one can obtain

$$_0 D_x^\nu x^a = \frac{\Gamma(a + 1)}{\Gamma(a - \nu + 1)} x^{a-\nu}, \qquad 0 < \nu < 1, \ a > -1.$$

Specially, when $f(x) = x$, $\nu = \frac{1}{2}$, Lacroix's equation (1.1.2) can be recovered; when $f(x) = x^0 = 1$, $\nu = \frac{1}{2}$, then the equation (1.1.8) can be also recovered.

In addition, the Weyl's definition of fractional integral is frequently used now

$$_xW_\infty^{-\nu}f(x) = \frac{1}{\Gamma(\nu)} \int_x^\infty (t-x)^{\nu-1} f(t)\mathrm{d}t, \quad \mathrm{Re}\ \nu > 0. \qquad (1.1.13)$$

Using the R-L's definition of fractional derivative (1.1.12), and taking the transform $t = -\tau$, one obtains

$$_{-\infty}D_x^{-\nu}f(x) = -\frac{1}{\Gamma(\nu)} \int_\infty^{-x} (x+\tau)^{\nu-1} f(-\tau)\mathrm{d}\tau.$$

Then taking the transform $x = -\xi$, one derives the following equation

$$_{-\infty}D_{-\xi}^{-\nu}f(-\xi) = \frac{1}{\Gamma(\nu)} \int_\xi^\infty (\tau-\xi)^{\nu-1} f(-\tau)\mathrm{d}\tau.$$

Let $f(-\xi) = g(\xi)$, then the right end of Weyl's definition (1.1.13) can be recovered.

1.2   Anomalous diffusion and fractional advection-diffusion

Anomalous diffusion phenomena are ubiquitous in the natural sciences and social sciences. In fact, many complex dynamical systems often contain anomalous diffusion. Fractional kinetic equations are usually an effective method to describe these complex systems, including diffusion type, diffusive convection type and Fokker-Planck type of fractional differential equations. Complex systems typically have the following characteristics. First, the system typically contains a large diversity of elementary units. Secondly, strong interactions exist among these basic units. Thirdly, the anomalous evolution is non-predictable as time evolves. In general, the time evolution of, and within, such systems deviates from the corresponding standard laws. These systems now exist in a large number of practical problems across disciplines such as physics, chemistry, engineering, geology, biology, economics, meteorology, and atmospheric. We do not plan to give a systematic introduction of anomalous diffusion or fractional advection diffusion, but just introduce some fractional differential equations to describe complex systems. We refer the reader to some monographs mentioned below.

In the classical exponential Debye mode, the relaxation of the system usually satisfies the relation $\Phi(t) = \Phi_0 \exp(-t/\tau)$. However, in complex systems it often satisfies the exponential Kohlrausch-Williams-Watts relation $\Phi(t) = \Phi_0 \exp(-(t/\tau)^\alpha)$ for $0 < \alpha < 1$, or the following asymptotic power law $\Phi(t) = \Phi_0(1 + t/\tau)^{-n}$ for $n > 0$. In addition, the conversion from the exponential to power-law relationship can be observed in practical systems. Similarly, in many complex systems, the diffusion process no longer follows the Gauss statistics. Then, the Fick second law is not sufficient to describe the transport behavior. In the classical Brownian motion, linear dependence of the time-mean-square displacement can be observed

$$< x^2(t) > \sim K_1 t. \tag{1.2.1}$$

But in anomalous diffusion, the mean-square displacement is no longer a linear function of time. The power-law dependence is common, i.e., $< x^2(t) > \sim K_\alpha t^\alpha$. Based on the index α of the anomalous diffusion, different anomalous diffusion types can be defined. When $\alpha = 1$, it is the normal diffusion process. When $0 < \alpha < 1$, it is sub-diffusion process or dispersive, slow diffusion process with the anomalous diffusion index. When $\alpha > 1$, it is ultra-diffusion process or increased, fast diffusion process.

There have been extensive research results on anomalous diffusion process with or without an external force field situation, including:

(1) fractional Brownian motion, which can be dated to Benoît Mandelbrot [153, 154];

(2) continuous-time random walk model;

(3) generalized diffusion equation [28];

(4) Langevin equation;

(5) generalized Langevin equation;

...
Among them, (2) and (5) appropriately depict the memory behavior of the system, and the specific form of the probability distribution function [162], however, it is insufficient to directly consider the role of the external force field, boundary value problem or the dynamics in the phase space.

1.2.1   The random walk and fractional equations

The following is a brief description of the random walk and the fractional diffusion equation. Considering the one-dimensional random walk, the test particle is assumed to jump randomly to one of its nearest neighbour sites in discrete time steps of span Δt, with lattice constant Δx. Such a system can be described by the following equation

$$W_j(t + \Delta t) = \frac{1}{2}W_{j-1}(t) + \frac{1}{2}W_{j+1}(t),$$

where $W_j(t)$ represents the probability of the particle located at site j, at time t, the coefficient $\frac{1}{2}$ means the walks of the particle are isotropic, i.e, the probability of jumping to left or right is $\frac{1}{2}$. Consider the continuum limit $\Delta t \to 0$, $\Delta x \to 0$, and by the Talyor series expansion, we can get

$$W_j(t + \Delta t) = W_j(t) + \Delta t \frac{\partial W_j}{\partial t} + O((\Delta t)^2),$$

$$W_{j\pm1}(t) = W(x,t) \pm \Delta x \frac{\partial W}{\partial x} + \frac{(\Delta x)^2}{2} \frac{\partial^2 W}{\partial x^2} + O((\Delta x)^3),$$

which leads to the diffusion equation

$$\frac{\partial W}{\partial t} = K_1 \frac{\partial^2}{\partial x^2} W(x,t), \qquad K_1 = \lim_{\Delta x \to 0, \Delta t \to 0} \frac{(\Delta x)^2}{2\Delta t} < \infty. \qquad (1.2.2)$$

Based on simple knowledge of partial differential equations, the solution of the equation (1.2.2) can be expressed as

$$W(x,t) = \frac{1}{\sqrt{4\pi K_1 t}} \exp\left(-\frac{x^2}{4K_1 t}\right). \qquad (1.2.3)$$

The function (1.2.3) is often called the propagator, i.e. the solution of the equation (1.2.2) with initial data $W_0(x) = \delta(x)$. The solution of equation (1.2.2) satisfies the exponential decay law

$$W(k,t) = \exp(-K_1 k^2 t), \qquad (1.2.4)$$

for individual mode in Fourier phase space.

For anomalous diffusion, we first consider the continuous-time random walk model. It is mainly based on the idea: for a given jump, the jump length and waiting time between two adjacent jumps are determined by a probability density function $\psi(x,t)$. The respective probability density functions of the jump length and waiting time are

$$\lambda(x) = \int_0^\infty \psi(x,t)dt, \quad w(t) = \int_{-\infty}^\infty \psi(x,t)dx. \qquad (1.2.5)$$

Here $\lambda(x)dx$ can be understood as the probability of the jump length in the interval $(x, x+dx)$, and $w(t)dt$ is the probability of a jump waiting time in time slice $(t, t+dt)$. It is easy to see that if the jump time and jump length are independent, then $\psi(x,t) = w(t)\lambda(x)$. Different continuous-time random walk processes can be determined by the converging or diverging characteristics of the waiting time $T = \int_0^\infty w(t)t dt$, and the variance of the jump length $\Sigma^2 = \int_{-\infty}^\infty \lambda(x)x^2 dx$. Now, the following equation can depict a continuous-time random walk model

$$\eta(x,t) = \int_{-\infty}^\infty dx' \int_0^\infty dt' \eta(x',t')\psi(x-x',t-t') + \delta(x)\delta(t), \qquad (1.2.6)$$

which links the probability density function $\eta(x,t)$ of the particle arrived at the site x at time t and the event of the particle arrived at the site x' at time t'. The second item on the right hand side represents the initial condition. Thus, the probability density function $W(x,t)$ of the particle at the site x at time t can be expressed as

$$W(x,t) = \int_0^t dt' \eta(x,t') \Psi(t-t'), \quad \Psi(t) = 1 - \int_0^t dt' w(t'). \qquad (1.2.7)$$

The items of the equation (1.2.7) have the meanings: $\eta(x,t')$ means the probability density function of the particle at the site x at time t', and $\Psi(t-t')$ is the probability density function of the particle which does not leave before time t, thereby $W(x,t)$ is the probability density function of the particle at the site x at time t. By using the Fourier transforms and Laplace transform, $W(x,t)$ satisfies the following algebraic relation [126]

$$W(k,u) = \frac{1-w(u)}{u} \frac{W_0(k)}{1-\psi(k,u)}, \qquad (1.2.8)$$

where $W_0(k)$ represents the Fourier transform of the initial value $W_0(x)$.

When $w(t)$ and $\lambda(t)$ are independent, i.e. $\psi(x,t) = w(t)\lambda(x)$, and T and Σ^2 are finite, the continuous-time random walk model is asymptotically equivalent to the Brownian motion. Consider the probability density function of the Poisson waiting time $w(t) = \tau^{-1}\exp(-t/\tau)$, and $T = \tau$, and the Gauss probability density function of the jump length $\lambda(x) = (4\pi\sigma^2)^{-1/2}\exp(-x^2/(4\sigma^2))$, $\Sigma^2 = 2\sigma^2$. The Laplace transforms and the Fourier transform have the following forms, respectively $w(u) \sim 1-u\tau+O(\tau^2)$ and $\lambda(k) \sim 1 - \sigma^2 k^2 + O(k^4)$.

Consider a special case: fractional time random walk. This will lead to the fractional diffusion equation to describe the sub-diffusion process. In this model, the characteristic waiting time T is divergent and the variance Σ^2 of the jump length is finite [196]. Introduce the probability density function of the long-tail waiting time, whose asymptotic behavior and the Laplace transform satisfy, respectively, $w(t) \sim A_\alpha(\tau/t)^{1+\alpha}$ and $w(u) \sim 1 - (u\tau)^\alpha$, where the specific form of $w(t)$ is insignificant. Taking into account the above mentioned probability density function $\lambda(x)$ of Gauss jump length, we can obtain the probability density function

$$W(k,u) = \frac{[W_0(k)/u]}{1+K_\alpha u^{-\alpha} k^2}. \qquad (1.2.9)$$

Using the Laplace transform of the fractional integral [16, 69, 165, 175, 195]

$$\mathscr{L}\{{}_0D_t^{-p}W(x,t)\} = u^{-p}W(x,u), \quad p \geqslant 0,$$

and noticing $\mathscr{L}\{1\} = 1/u$, one obtains the following fractional integral equation from the equation (1.2.9),

$$W(x,t) - W_0(x) = {}_0D_t^{-\alpha}K_\alpha \frac{\partial^2}{\partial x^2}W(x,t). \qquad (1.2.10)$$

Introducing the operator $\dfrac{\partial}{\partial t}$ of the time derivative, then we can get the fractional derivative equation

$$\frac{\partial W}{\partial t} = {}_0D_t^{1-\alpha}K_\alpha \frac{\partial^2}{\partial x^2}W(x,t), \qquad (1.2.11)$$

where the Riemann-Liouville operator ${}_0D_t^{1-\alpha} = \dfrac{\partial}{\partial t}{}_0D_t^{-\alpha}(0 < \alpha < 1)$ is defined as (please refer to the next chapter)

$${}_0D_t^{1-\alpha}W(x,t) = \frac{1}{\Gamma(\alpha)}\frac{\partial}{\partial t}\int_0^t \frac{W(x,t')}{(t-t')^{1-\alpha}}dt'. \qquad (1.2.12)$$

Since the integral kernek $M(t) \propto t^{\alpha-1}$ in the definition, the sub-diffusion process defined in the equation (1.2.11) does not have the Markov properties. In fact, it can be shown that [162]

$$< x^2(t) >= \frac{2K_\alpha}{\Gamma(1+\alpha)}t^\alpha.$$

The equation (1.2.11) can also be transformed into its equivalent form

$${}_0D_t^\alpha W - \frac{t^{-\alpha}}{\Gamma(1-\alpha)}W_0(x) = K_\alpha \frac{\partial^2}{\partial x^2}W(x,t),$$

where, unlike the normal diffusion process, the initial value $W_0(x)$ no longer has the exponential decay property, but the power law decay instead [17] (compare with equation (1.2.4)).

Consider another special form: Levy flights. The characteristic waiting time T is finite and Σ^2 is divergent. This model possesses a Poisson waiting time and a Lévy distribution for the jump length, i.e.,

$$\lambda(k) = \exp(-\sigma^\mu|k|^\mu) \sim 1 - \sigma^\mu|k|^\mu, \quad 1 < \mu < 2, \qquad (1.2.13)$$

which asymptotically satisfies $\lambda(x) \sim A_\mu\sigma^{-\mu}|x|^{-1-\mu}$ for $|x| \gg \sigma$. Since T is finite, this process has the Markov property. Substituting the asymptotic expansion of $\lambda(k)$ in equation (1.2.13) into the equation (1.2.8) leads to $W(k,u) = 1/(u + K^\mu|k|^\mu)$. By Fourier and Laplace inverse transform, the following fractional derivative equation can be obtained

$$\frac{\partial W}{\partial t} = K^\mu {}_{-\infty}D_x^\mu W(x,t), \quad K^\mu \equiv \sigma^\mu/\tau. \qquad (1.2.14)$$

Here $_{-\infty}D_x^\mu$ is the Weyl operator (please refer to the next chapter), which is equivalent to the Riesz operator ∇^μ in the one-dimensional case. Use the Fourier transform, the propagator can be expressed as $W(k,t) = \exp(-K^\mu t|k|^\mu)$. If both of Σ^2 and T are divergent, then we can get the following fractional derivative equation [162]

$$\frac{\partial W}{\partial t} = {}_0D_t^{1-\alpha}K_\alpha^\mu\nabla^\mu W(x,t), \qquad K_\alpha^\mu \equiv \sigma^\mu/\tau^\alpha. \qquad (1.2.15)$$

1.2.2   Fractional advection-diffusion equation

Here, we consider the fractional advection-diffusion equation. In a Brownian motion, when a system has an additional velocity field v or under the influence of a constant external force field, it can be described by the following advection-diffusion equation

$$\frac{\partial W}{\partial t} + v\frac{\partial W}{\partial x} = K_1\frac{\partial^2}{\partial x^2}W(x,t). \qquad (1.2.16)$$

The equation will no longer well describe the anomalous diffusion. Some common generalizations are considered below.

First, note that the equation (1.2.16) is Galilean invariance, i.e. the problem is invariant under the transform $x \to x - vt$. Assume, when considered under the moving frame (reference frame) with homogeneous velocity field v, the jump function of the tested particle in random walk is $\psi(x,t)$, then the corresponding jump function of the particle to be tested under the laboratory frame is $\phi(x,t) = \psi(x - vt,t)$. Using the corresponding Fourier-Laplace transform, we get $\phi(k,u) = \psi(k,u+ivk)$. When T is divergent and Σ^2 is finite, the propagator can be obtained from the equation (1.2.8) that $W(k,u) = 1/(u + ivk + K_\alpha k^2 u^{1-\alpha})$. Then the fractional advection-diffusion equation can be deduced (compare with the equation (1.2.11))

$$\frac{\partial W}{\partial t} + v\frac{\partial W}{\partial x} = {}_0D_t^{1-\alpha}K_\alpha\frac{\partial^2}{\partial x^2}W(x,t), \qquad (1.2.17)$$

whose solution can be obtained through the Galilean transform of the equation (1.2.11), i.e.

$$W(x,t) = W_{v=0}(x - vt,t).$$

Some moment statistics of the equation (1.2.17) are

$$< x(t) >= vt, \quad < x^2(t) >= \frac{2K_\alpha}{\Gamma(1+\alpha)}t^\alpha + v^2t^2,$$

$$< (\Delta x(t))^2 >= \frac{2K_\alpha}{\Gamma(1+\alpha)}t^\alpha. \qquad (1.2.18)$$

It can be seen that, the mean square displacement $< (\Delta x(t))^2 >$ only contains the distribution information of the molecule. The first moment $< x(t) >$ explains the parallel translation along the velocity field v. This Galilean-invariant sub-diffusion can depict the motion of the particles in the flow field, where the liquid itself has sub-diffusion phenomenon.

If the velocity field $v = v(x)$ depends on the space variable [49-51], one assumes $\phi(x, t; x_0) = \psi(x - \tau_a v(x_0), t)$, then we can deduce the following fractional differential equation

$$\frac{\partial W}{\partial t} = {}_0 D_t^{1-\alpha} \left[-A_\alpha \frac{\partial}{\partial x} v(x) + K_\alpha \frac{\partial^2}{\partial x^2} \right] W(x, t). \tag{1.2.19}$$

For a homogeneous velocity field, the following fractional differential equation can be obtained:

$$\frac{\partial W}{\partial t} = {}_0 D_t^{1-\alpha} \left[-A_\alpha \frac{\partial}{\partial x} v + K_\alpha \frac{\partial^2}{\partial x^2} \right] W(x, t). \tag{1.2.20}$$

It can be proved that the solution of the fractional equation does not satisfy the Galilean transform of the form $W(x - v^* t^\alpha, t)$. Some statistics of the equation are

$$< x(t) > = \frac{A_\alpha v t^\alpha}{\Gamma(1+\alpha)}, \qquad < x^2(t) > = \frac{2 A_\alpha^2 v^2 t^{2\alpha}}{\Gamma(1+2\alpha)} + \frac{2 K_\alpha t^\alpha}{\Gamma(1+\alpha)}, \tag{1.2.21}$$

in which case, the first moment increases sub-linearly.

For the Lévy flight under an external velocity field v, i.e. T is finite and Σ^2 is divergent, the following fractional differential equation can be deduced

$$\frac{\partial W}{\partial t} + v \frac{\partial W}{\partial x} = K^\mu \nabla^\mu W(x, t), \tag{1.2.22}$$

which can be used to describe the Markov process with divergent mean square displacement.

1.2.3   Fractional Fokker-Planck equation

The Fokker-Planck equation (FPE) can be used to describe the classical diffusion process under an external force field [84, 162, 178, 191, 217]:

$$\frac{\partial W}{\partial t} = \left[\frac{\partial}{\partial x} \frac{V'(x)}{m \eta_1} + K_1 \frac{\partial^2}{\partial x^2} \right] W(x, t), \tag{1.2.23}$$

where m is the mass of the tested particle, η_1 is the friction coefficient between the tested particle and the environment, the external force can be expressed as $F(x) = -\dfrac{\mathrm{d}V}{\mathrm{d}x}$ using the external field. Its properties can be found in related literature. To compare it with the following fractional Fokker-Planck equation (FFPE), several important basic properties are given below.

(1) When the external force does not exist, the equation (1.2.23) degenerates into Fick's second law, and hence the mean square displacement satisfies the linear relation described in equation (1.2.1);

(2) Single-mode relaxation decay with time exponent:

$$T_n(t) = \exp(-\lambda_{n,1}t), \qquad (1.2.24)$$

where, $\lambda_{n,1}$ is the eigenvalue of the Fokker-Planck operator $L_{FP} = \dfrac{\partial}{\partial x}\dfrac{V'(x)}{m\eta_1} + K_1\dfrac{\partial^2}{\partial x^2}$;

(3) The steady state solution $W_{st}(x) = \lim_{t\to\infty} W(x,t)$ is given by the Gibbs-Boltzmann distribution

$$W_{st} = N\exp(-\beta V(x)), \qquad (1.2.25)$$

where N is the regularization constant, $\beta = (k_B T)^{-1}$ is the Boltzmann factor;

(4) FPE satisfies the Einstein-Stokes-Smoluchowski relationship.

$$K_1 = k_B T/m\eta_1;$$

(5) The second Einstein relationship is established.

$$< x(t) >_F = \frac{FK_1}{k_B T}t, \qquad (1.2.26)$$

which links the first moment under the constant external force F and the second moment $< x^2(t) >_0 = 2K_1 t$ without external force.

The FPE equation and its applications have been extensively studied. To describe the anomalous diffusion under an external field, the generalized FFPE is introduced [17, 159, 160, 162]

$$\frac{\partial W}{\partial t} = {}_0 D_t^{1-\alpha}\left[\frac{\partial}{\partial x}\frac{V'(x)}{m\eta_\alpha} + K_\alpha\frac{\partial^2}{\partial x^2}\right]W(x,t). \qquad (1.2.27)$$

This equation has the following properties.

(1) Without the external force field,

$$< x^2(t) >_0 = \frac{2K_\alpha}{\Gamma(1+\alpha)}t^\alpha,$$

which degenerates into the equation (1.2.11) when $V =$ is a constant.

(2) The single-mode relaxation is given by the Mittag-Leffler function (compare with the equation (1.2.24)).

Using the method of separating variables, we let

$$W_n(x,t) = T_n(t)\varphi_n(x),$$

then the equation (1.2.27) can be decomposed into the following equations

$$\frac{dT_n}{dt} = -\lambda_{n,\alpha 0} D_t^{1-\alpha} T_n(t), \tag{1.2.28}$$

$$L_{FP}\varphi_n(x) = -\lambda_{n,\alpha}\varphi_n(x). \tag{1.2.29}$$

When $T_n(0) = 1$, $T_n(t)$ is given by the Mittag-Leffler function.

$$T_n(t) = E_\alpha(-\lambda_{n,\alpha}t^\alpha) = \sum_{j=0}^{\infty} \frac{(-\lambda_{n,\alpha}t^\alpha)^j}{\Gamma(1+\alpha j)}.$$

(3) The steady state solution is given by the Gibbs-Boltzmann distribution.

The right end of the equation (1.2.27) as follows

$$-{_0}D_t^{1-\alpha}\frac{\partial S(x,t)}{\partial x}, \quad S(x,t) = \left[-\frac{\partial}{\partial x}\frac{V'(x)}{m\eta_\alpha} - K_\alpha\frac{\partial^2}{\partial x^2}\right]W(x,t), \tag{1.2.30}$$

where $S(x,t)$ represents the probability current. In the case of the steady-state solution, $S(x,t)$ is a constant, thus

$$\frac{V'(x)}{m\eta_\alpha}W_{st}(x) + K_\alpha\frac{d}{dx}W_{st}(x) = 0, \tag{1.2.31}$$

whose solution is given by

$$W_{st}(x) = N\exp(-\frac{V(x)}{m\eta_\alpha K_\alpha}).$$

Similar to the classical case, W_{st} is given by the Boltzmann distribution.

(4) The generalized Einstein-Stokes-Smoluchowski relationship: $K_\alpha = k_B T/m\eta_\alpha$.

(5) The second Einstein relationship still holds for FFPE.

$$< x(t) >_F = \frac{F}{m\eta_\alpha\Gamma(1+\alpha)}t^\alpha = \frac{FK_\alpha}{k_B T\Gamma(1+\alpha)}t^\alpha.$$

This relationship reduces to equation (1.2.26), since $\Gamma(2) = 1$.

Consider a special case $V(x) = \frac{1}{2}m\omega^2 x^2$, the system depicts the motion of sub-diffusive harmonic restrained particles. Now, the equation (1.2.27) is simplified as

$$\frac{\partial W}{\partial t} = {}_0D_t^{1-\alpha}\left[\frac{\partial}{\partial x}\frac{\omega^2 x}{\eta_\alpha} + K_\alpha\frac{\partial^2}{\partial x^2}\right]W(x,t).$$

By using the method of separating variables and the definition of the Hermite polynomial [2], the solution of this equation can be obtained [159]

$$W = \sqrt{\frac{m\omega^2}{2\pi k_B T}}\sum_0^\infty\frac{1}{2^n n!}E_\alpha\left(\frac{-n\omega^2 t^\alpha}{\eta_\alpha}\right)H_n\left(\frac{\sqrt{m}\omega x'}{\sqrt{2k_B T}}\right)$$

$$H_n\left(\frac{\sqrt{m}\omega x}{\sqrt{2k_B T}}\right)\exp\left(-\frac{m\omega^2 x^2}{2k_B T}\right),$$

where, H_n's are the Hermite polynomials. The steady state solution can be expressed as

$$W_{st}(x) = \sqrt{\frac{m\omega^2}{2\pi k_B T}}H_0\left(\frac{\sqrt{m}\omega x'}{\sqrt{2k_B T}}\right)H_0\left(\frac{\sqrt{m}\omega x}{\sqrt{2k_B T}}\right)\exp\left(-\frac{m\omega^2 x^2}{2k_B T}\right)$$

$$= \sqrt{\frac{m\omega^2}{2\pi k_B T}}\exp\left(-\frac{m\omega^2 x^2}{2k_B T}\right),$$

which is the Gibbs-Boltzmann distribution, as expected.

Using the Laplace transform with the same initial value $W_0(x) = \delta(x-x')$, the solution of the equation (1.2.27) satisfies

$$W_\alpha(x,u) = \frac{\eta_\alpha}{\eta_1}u^{\alpha-1}W_1\left(x,\frac{\eta_\alpha}{\eta_1}u^\alpha\right), \quad 0 < \alpha < 1, \tag{1.2.32}$$

where W_1 and W_α respectively represent the solutions of the equations (1.2.23) and (1.2.27). This shows that under the Laplace transform, the sub-diffusion system and the classical diffusion differ by a scale. Furthermore, W_α can be expressed in terms of W_1 through

$$W_\alpha(x,t) = \int_0^\infty dsA(s,t)W_1(x,s), \tag{1.2.33}$$

where $A(s,t)$ is given by the inverse Laplace transform

$$A(s,t) = \mathscr{L}^{-1}\left[\frac{\eta_\alpha}{\eta_1 u^{1-\alpha}}\exp\left(\frac{\eta_\alpha}{\eta_1}u^\alpha s\right)\right]. \tag{1.2.34}$$

If we consider the non-local jump process, i.e. assume that Σ^2 is divergent, then we can obtain the following FFPE [162]

$$\frac{\partial W}{\partial t} = {}_0D_t^{1-\alpha}\left[\frac{\partial}{\partial x}\frac{V'(x)}{m\eta_\alpha} + K^\mu\nabla^\mu\right]W(x,t). \tag{1.2.35}$$

When $\mu = 2$, i.e. Σ^2 is finite, the equation reduces to the sub-diffusive FFPE(1.2.27). Considering the opposite case, i.e. T is finite and Σ^2 is divergent, one obtains similar to (1.2.35)

$$\frac{\partial W}{\partial t} = \left[\frac{\partial}{\partial x} \frac{V'(x)}{m\eta_1} + K_1^\mu {}_{-\infty} D_x^\mu \right] W(x,t). \tag{1.2.36}$$

This is the Lévy flight with the external field $F(x)$.

1.2.4   Fractional Klein-Framers equation

Based on the continuous-time Chapamn-Fokker equation [107, 217, 218], and the Markov-Langevin equation describing the damped particles with an external fore field, the fractional Klein-Kramers(FKK) equation can be derived whose velocity averaged high-friction limit reproduces the fractional Fokker-Planck equation, and explains the occurrence of the generalised transport coeffcients K_α and η_α. The FKK equation is of the form [162]

$$\frac{\partial W}{\partial t} = {}_0 D_t^{1-\alpha} \left[-v^* \frac{\partial}{\partial x} + \frac{\partial}{\partial x} \left(\eta^* v - \frac{F^*(x)}{m} \right) + \eta^* \frac{k_B T}{m} \frac{\partial^2}{\partial v^2} \right] W(x,v,t) \tag{1.2.37}$$

where, $v^* = v\vartheta$, $\eta^* = \eta\vartheta$, $F^*(x) = F(x)\vartheta$ and $\vartheta = \tau^*/\tau^\alpha$. Integrating this equation w.r.t. v, one obtains the following

$$\frac{\partial W}{\partial t} + {}_0 D_t^{1+\alpha} \frac{W}{\eta^*} = {}_0 D_t^{1-\alpha} \left[-\frac{\partial}{\partial x} \frac{F(x)}{m\eta_\alpha} + K_\alpha \frac{\partial^2}{\partial x^2} \right] W(x,t). \tag{1.2.38}$$

The equation (1.2.38) is of the type of the generalized Cattaneo equation, which reduces to the telegraph equation when $\alpha = 1$, in the limiting case of the Brownian motion. When considering the high-friction limit or the long time limit, one recovers the fractional Fokker-Planck equation (1.2.27). Integrating with respect to the position coordinates of the above equation and considering the undamped limit, one then obtains the fractional Rayleigh equation

$$\frac{\partial W}{\partial t} = {}_0 D_t^{1-\alpha} \eta^* \left[\frac{\partial}{\partial v} v + \frac{k_B T}{m} \frac{\partial^2}{\partial v^2} \right] W(v,t), \tag{1.2.39}$$

whose solution $W(v,t)$ of probability density distribution depicts the process tending to the stable Maxwell distribution

$$W_{st}(v) = \frac{\beta m}{2\pi} \exp\left(-\frac{\beta m}{2} v^2 \right).$$

1.3   Fractional quasi-geostrophic equation

Fractional quasigeostrophic equation (quasigeostrophic equation) has the following form [53]

$$\frac{D\theta}{Dt} = \frac{\partial\theta}{\partial t} + v \cdot \boldsymbol{\nabla}\theta = 0, \tag{1.3.1}$$

where, $v = (v_1, v_2)$ is a two-dimensional velocity field which is decided by the stream function

$$v_1 = -\frac{\partial\psi}{\partial x_2}, \qquad v_2 = \frac{\partial\psi}{\partial x_1}. \tag{1.3.2}$$

Here the current function ψ and θ has the relationship

$$(-\Delta)^{\frac{1}{2}}\psi = -\theta. \tag{1.3.3}$$

By using the Fourier transform, the fractional Laplace operator can be defined as

$$(-\Delta)^{\frac{1}{2}}\psi = \int e^{2\pi \mathrm{i} x \cdot k} 2\pi |k| \hat{\psi}(k) \mathrm{d}k.$$

Here, θ is the potential temperature, v is the current velocity, ψ can be regarded as the pressure. When the viscous term is considered, the following equation can be derived

$$\theta_t + \kappa(-\Delta)^{\alpha}\theta + v \cdot \boldsymbol{\nabla}\theta = 0,$$

where θ and v are still determined by the equation (1.3.2)-(1.3.3), $0 \leqslant \alpha \leqslant 1$ and $\kappa > 0$ is a real number. More generally, we can consider the fractional QG equation with an external force term

$$\theta_t + u \cdot \boldsymbol{\nabla}\theta + \kappa(-\Delta)^{\alpha}\theta = f.$$

For simplicity, f is usually assumed to be independent of time.

The fractional QG equation (1.3.1)-(1.3.2) and the three-dimensional incompressible Euler equation share many similarities in physics and mathematics. The three-dimensional vorticity equation has the following form

$$\frac{D\omega}{Dt} = (\boldsymbol{\nabla}v)\omega, \tag{1.3.4}$$

where $\dfrac{D}{Dt} = \dfrac{\partial}{\partial t} + v \cdot \boldsymbol{\nabla}$, $v = (v_1, v_2, v_3)$ is the three-dimensional vorticity vector, and $\mathrm{div}v = 0$ and $\omega = \mathrm{curl}v$ is the vorticity vector. Introduce the vector $\boldsymbol{\nabla}^{\perp}\theta =^t (-\theta_{x_2}, \theta_{x_1})$. One can find that the roles of the vector field $\boldsymbol{\nabla}^{\perp}\theta$ in the two-dimensional QG equation are similar to ω in the three-dimensional Euler equation. Differentiating the equation (1.3.1), one obtains

$$\frac{D\boldsymbol{\nabla}^{\perp}\theta}{Dt} = (\boldsymbol{\nabla}v)\boldsymbol{\nabla}^{\perp}\theta, \tag{1.3.5}$$

where $v = \boldsymbol{\nabla}^\perp \psi$ and hence $\mathrm{div}\, v = 0$. It can be seen that $\boldsymbol{\nabla}^\perp \psi$ in the equation (1.3.5) and the vorticity ω in the equation (1.3.4) satisfy the same equation.

Then we examine its analytic structure. For the three-dimensional Euler equation, the velocity v can be expressed by its vorticity, i.e. by the well known Biot-Savart law

$$v(x) = -\frac{1}{4\pi} \int_{\mathbf{R}^3} \left(\boldsymbol{\nabla}^\perp \frac{1}{|y|} \right) \times \omega(x+y) dy.$$

The matrix $\boldsymbol{\nabla} v = (v^i_{x_j})$ can be decomposed into the symmetric part and the antisymmetric part

$$\mathcal{D}^E = \frac{1}{2}[(\boldsymbol{\nabla} v) + (\boldsymbol{\nabla} v)^t], \quad \text{and} \quad \Omega^E = \frac{1}{2}[(\boldsymbol{\nabla} v) - (\boldsymbol{\nabla} v)^t],$$

where the symmetric part $\mathcal{D}^E$ can be expressed as a singular integral

$$\mathcal{D}^E(x) = \frac{3}{4\pi} P.V. \int_{\mathbf{R}^3} \frac{\boldsymbol{M}^E(\hat{y}, \omega(x+y))}{|y|^3} dy.$$

As the fluid is incompressible, $tr\mathcal{D}^E = \sum_i d_{ii} = 0$. Here the matrix $\boldsymbol{M}^E$ is given by

$$\boldsymbol{M}^E(\hat{y}, \omega) = \frac{1}{2}[\hat{y} \otimes (\hat{y} \times \omega) + (\hat{y} \times \omega) \otimes \hat{y}],$$

where $a \otimes b = (a_i b_j)$ is the tensor product of two vectors. Obviously, the Euler equation can be rewritten as

$$\frac{D\omega}{Dt} = \omega \cdot \boldsymbol{\nabla} v = \mathcal{D}^E \omega.$$

For the two-dimensional fractional QG equation

$$\psi(x) = -\int_{\mathbf{R}^2} \frac{1}{|y|} \theta(x+y) dy,$$

and hence

$$v = -\int_{\mathbf{R}^2} \frac{1}{|y|} \boldsymbol{\nabla}^\perp \theta(x+y) dy.$$

Now, the symmetric part $\mathcal{D}^{QG}(x) = \frac{1}{2}((\boldsymbol{\nabla} v) + (\boldsymbol{\nabla} v)^t)$ of the matrix of the velocity gradient can be written as the singular integral

$$\mathcal{D}^{QG} = P.V. \int_{\mathbf{R}^2} \frac{\boldsymbol{M}^{QG}(\hat{y}, (\boldsymbol{\nabla}^\perp \theta)(x+y))}{|y|^2} dy, \tag{1.3.6}$$

where, $\hat{y} = \dfrac{y}{|y|}$ and

$$M^{QG} = \frac{1}{2}(\hat{y}^\perp \otimes \omega^\perp + \omega^\perp \otimes \hat{y}^\perp).$$

For fixed ω, the mean of the function M^{QG} in the singular integral is zero over the unit circle. The velocity in the two-dimensional QG equation and three-dimensional Euler equation has similar expression

$$v = \int_{\mathbf{R}^d} K_d(y)\omega(x+y)\mathrm{d}y,$$

where $K_d(y)$ is a homogeneous kernel function of order $1 - d$. The symmetric parts $\mathcal{D}^E$ and $\mathcal{D}^{QG}$ can be represented by the singular integral of $\omega(x)$, whose kernel function is $-d$ order homogeneous function and has the standard cancellation property. From the above discussions, we know that the roles of $\nabla^\perp \theta$ in the two-dimensional QG equation and the vorticity ω in the three-dimensional incompressible Euler equation are equivalent .

Consider the vortex lines of the three-dimensional Euler equation. The smooth curve $C = \{y(s) \in \mathbf{R}^3 : 0 < s < 1\}$ is called the vortex line at fixed time t, if the curve and the vorticity ω are tangential at each point, i.e.

$$\frac{\mathrm{d}y}{\mathrm{d}s}(s) = \lambda(s)\omega(y(s), t), \quad \lambda(s) \neq 0.$$

Let $C = \{y(s) \in \mathbf{R}^3 : 0 < s < 1\}$ be the initial vortex line, as time evolves, it develops into $C(t) = \{X(y(s), t) \in \mathbf{R}^3 : 0 < s < 1\}$, where $X(\alpha, t)$ denotes the trajectory of the particle of α. Using the vorticity equation, one can show that $X(\alpha, t)$ satisfies the equation

$$\omega(X(\alpha, t), t) = \nabla_\alpha X(\alpha, t)\omega_0(\alpha).$$

From the definition of $C(t)$, we know that

$$\frac{\mathrm{d}X(y(s), t)}{\mathrm{d}s} = \nabla_\alpha X(y(s), t)\frac{\mathrm{d}y(s)}{\mathrm{d}s} = \nabla_\alpha X(y(s), t)\lambda(s)\omega_0(y(s)),$$

and hence

$$\frac{\mathrm{d}X}{\mathrm{d}s}(y(s), t) = \lambda(s)\omega(X(y(s), t), t).$$

It shows that in the ideal fluid, the vortex line moves with the fluid. Let LS^{QG} represent the level set of the two-dimensional QG equation, i.e. θ is a constant. From the equation (1.3.1), LS^{QG} moves with the fluid, and $\nabla^\perp \theta$ is tangent to the level set LS^{QG}. This shows that the level set for the two-dimensional QG equation plays similar roles as the vortex line does for the three-dimensional Euler equation. Furthermore, for the three-dimensional Euler equation,

$$\frac{D|\omega|}{Dt} = \alpha^E |\omega|,$$

where $\alpha^E(x,t) = \mathcal{D}^E(x,t)\xi \cdot \xi$, $\xi = \dfrac{\omega(x,t)}{|\omega(x,t)|}$. Analogously, for the two-dimensional QG equation, the development of $|\nabla^\perp \theta|$ satisfies the same equation

$$\frac{D|\nabla^\perp \theta|}{Dt} = \alpha|\nabla^\perp \theta|, \qquad (1.3.7)$$

where $\alpha^{QG} = \mathcal{D}^{QG}(x,t)\xi \cdot \xi$, $\mathcal{D}^{QG}$ is defined in the equation (1.3.6), and $\xi = \dfrac{\nabla^\perp \theta}{|\nabla^\perp \theta|}$ is the direction vector of $\nabla^\perp \theta$.

Now we investigate the conserved quantity of the equation. Using the Fourier transform, we know that $\hat{v}(k) = \widehat{\nabla^\perp \psi}(k) = \dfrac{\mathrm{i}(-k_2, k_1)}{|k|}\hat{\theta}(k)$, and by utilizing the Plancherel formula, we have

$$\frac{1}{2}\int_{\mathbf{R}^2} |v|^2 = \frac{1}{2}\int_{\mathbf{R}^2} |\theta|^2 \mathrm{d}x.$$

For the two-dimensional QG equation, it is evident that $\displaystyle\int_{\mathbf{R}^2} G(\theta)\mathrm{d}x$ is conserved. In particular, letting $G(\theta) = \dfrac{1}{2}\theta^2$ shows its kinetic energy is conserved. This is consistent with the three-dimensional Euler equation.

Recently, the following fractional Navier-Stokes equation is widely considered

$$\begin{cases} \partial_t u + u \cdot \nabla u + \nabla p = -\nu(-\Delta)^\alpha u, \\ \nabla \cdot u = 0, \end{cases} \qquad (1.3.8)$$

where $\nu > 0$, $\alpha > 0$ are real numbers. The existence and uniqueness of the fractional NS equation in the Besov space are established [223].

Recently, we also established the existence of solutions and their time decay of a class of high-order two-dimensional quasi-geostrophic equation [181]

$$\left(\frac{\partial}{\partial t} + \frac{\partial \psi}{\partial x}\frac{\partial}{\partial y} - \frac{\partial \psi}{\partial y}\frac{\partial}{\partial x}\right)q = \frac{1}{R_e}(-\Delta)^{1+\alpha}\psi, \qquad (1.3.9)$$

where $q = \Delta\psi - F\psi + \beta y$, $(x,y) \in \mathbf{R}^2$, $t \geqslant 0$.

1.4   Fractional nonlinear Schrödinger equation

In quantum mechanics, the Schrödinger equation of free particles plays an important role

$$i\hbar\frac{\partial}{\partial t}\psi(r,t) = -\frac{\hbar^2}{2m}\nabla^2\psi(r,t),$$

where $\psi(r,t)$ is the quantum ground state wave function describing the microscopic particles. After the wave function $\psi(r,t)$ is determined, the mean of any mechanical quantity of the particle and its probability distribution are completely determined. Hence determining the evolution of the wave function with time and identifying the possible wave functions under specific situations become the core issues in quantum mechanics. Taking into account the potential field $V(r,t)$, one gets

$$i\hbar\frac{\partial}{\partial t}\psi(r,t) = \left[-\frac{\hbar^2}{2m}\nabla^2 + V(r,t)\right]\psi(r,t).$$

This is the Schrödinger equation, which reveals the the basic law of the motion of matter in the microscopic world.

Consider the stable stochastic process. In the mid-1930s, P. Lévy and A.Y. Khintchine proposed that under which situation the probability distribution $p_N(X)$ of the summation $X = X_1 + \cdots + X_N$ of N independent and identically distributed random variables equals $p_i(X_i)$? The concept of stable roots here. Taking into consideration the central limit theorem, the traditional answer had been that each $p_i(X_i)$ satisfies the Gaussian distribution, i.e. the summation of Gaussian random variables is still a Gaussian random variable. Lévy and Khintchine showed the possibility of non-Gaussian distributions, i.e. the nowadays called Lévy α-stable probability distribution $(0 < \alpha \leqslant 2)$. When $\alpha = 2$, the distribution is the standard Gaussian distribution.

In quantum mechanics, Feynman path integrals are actually the Brownian-type quantum mechanics path based integrals. The Brownian motion is a Lévy α-stable stochastic process, and the Brownian-type path integral leads to the classical Schrödinger equation. Replacing the Brownian-type path with Levy-type quantum mechanics path, one gets the fractional Schrödinger equation [130]

$$i\hbar\frac{\partial}{\partial t}\psi(r,t) = D_\alpha(-\hbar^2\Delta)^{\alpha/2}\psi(r,t) + V(r,t)\psi(r,t), \qquad (1.4.1)$$

where α is the order of spatial derivative and D_α is a constant with dimensionless $[D_\alpha] = \mathrm{erg}^{1-\alpha}\cdot\mathrm{cm}^\alpha\cdot\mathrm{sec}^{-\alpha}$. This equation can also be written as the following operational form

$$i\hbar \frac{\partial \psi}{\partial t} = H_\alpha \psi,$$

where $H_\alpha = D_\alpha(-\hbar^2\Delta)^{\alpha/2} + V(r,t)$ is called the fractional Hamiltonian operator.

Consider the Fourier transform and its inverse transform of the three-dimensional case,

$$\varphi(p,t) = \int e^{-i\frac{px}{\hbar}} \psi(r,t)\mathrm{d}r, \quad \psi(r,t) = \frac{1}{(2\pi\hbar)^3} \int e^{i\frac{px}{\hbar}} \varphi(p,t)\mathrm{d}p.$$

The operation of the three-dimensional quantum Riesz fractional derivative $(-\hbar^2\Delta)^{\alpha/2}$ on a function $\psi(r,t)$ can be expressed as

$$(-\hbar^2\Delta)^{\alpha/2}\psi(r,t) = \frac{1}{(2\pi\hbar)^3} \int e^{i\frac{pr}{\hbar}} |p|^\alpha \varphi(p,t)\mathrm{d}p.$$

By using the integration by parts formula,

$$(\phi, (-\Delta)^{\alpha/2}\chi) = ((-\Delta)^{\alpha/2}, \chi)$$

we know that the fractional Hamilton operator H_α is hermitian operator under the dot product $(\phi, \chi) := \int_{-\infty}^{\infty} \phi^*(r,t)\chi(r,t)\mathrm{d}r$, where $*$ represents the complex conjugate. The average energy of the fractional quantum system with the Hamilton quantity H_α is

$$E_\alpha = \int_{\infty}^{\infty} \psi^*(r,t)H_\alpha\psi(r,t)\mathrm{d}r.$$

By using the integration by parts formula

$$E_\alpha = \int_{\infty}^{\infty} \psi^*(r,t)H_\alpha\psi(r,t)\mathrm{d}r = \int_{\infty}^{\infty} (H_\alpha^+\psi(r,t))\psi(r,t)\mathrm{d}r = E_\alpha^*,$$

which shows that the energy of the system is always real valued. Therefore the fractional Hamilton quantity in the above definition is Hermitian or self-adjoint under the dot product $(H_\alpha^+\phi, \chi) = (\phi, H_\alpha\chi)$.

Fractional nonlinear Schrödinger equation also has a certain parity structure. From the definition of the fractional Laplace operator,

$$(-\hbar^2\Delta)^{\alpha/2}e^{ipx/\hbar} = |p|^\alpha e^{ipx/\hbar},$$

hence $e^{ipx/\hbar}$ is the eigenfunction of the operator $(-\hbar^2\Delta)^{\alpha/2}$ whose eigenvalue is $|p|^\alpha$. On the other hand, the operator $(-\hbar^2\Delta)^{\alpha/2}$ is symmetric, i.e.

$$(-\hbar^2\Delta_r)^{\alpha/2} = \cdots = (-\hbar^2\Delta_{-r})^{\alpha/2} = \cdots.$$

From this, the Hamilton H_α is invariant under the space reflection transformation. Let $\hat{P}$ be the reflection operator, then the invariance can be
expressed as the commutativity of $\hat{P}$ and H_α, i.e., $\hat{P}H_\alpha = H_\alpha\hat{P}$. Under
these notations, the wave functions of quantum mechanical states with a
well-defined eigenvalue of the operator $\hat{P}$ can be divided into two classes.
Functions that invariant under the reflection transform $\hat{P}\psi_+(r) = \psi_+(r)$ are
called the even states, and functions that change signs under the reflection
transform $\hat{P}\psi_-(r) = -\psi_-(r)$ are called the odd states. If the state of a closed
fractional quantum mechanics system has a given parity, then this parity is
conserved.

In the study of the fractional Schrödinger equation, the case that H_α does
not depend on time is important in physics research. In this situation, the
equation (1.4.1) has a particular solution $\psi(r,t) = e^{-iEt/\hbar}\phi(t)$, where $\phi(r)$
satisfies $H_\alpha\phi(r) = E\phi(r)$, or

$$D_\alpha(-\hbar^2\Delta)^{\alpha/2}\phi(r) + V(r)\phi(r) = E\phi(r).$$

Usually, this equation is called the stationary fractional Schrödinger equation.

Consider the current density. From the equation (1.4.1),

$$\frac{\partial}{\partial t}\int \psi^*(r,t)\psi(r,t)\mathrm{d}r$$

$$=\frac{D_\alpha}{i\hbar}\int\left[\psi^*(r,t)(-\hbar^2\Delta)^{\alpha/2}\psi(r,t) - \psi(r,t)(-\hbar^2\Delta)^{\alpha/2}\psi^*(r,t)\right]\mathrm{d}r.$$

This equation can be simplified into

$$\frac{\partial\rho(r,t)}{\partial t} + \mathrm{div}j(r,t) = 0,$$

where $\rho(r,t) = \psi^*(r,t)\psi(r,t)$ is called the probability density and

$$j(r,t)=\frac{D_\alpha\hbar}{i}\left(\psi^*(r,t)(-\hbar^2\Delta)^{\alpha/2-1}\boldsymbol{\nabla}\psi(r,t) - \psi(r,t)(-\hbar^2\Delta)^{\alpha/2-1}\boldsymbol{\nabla}\psi^*(r,t)\right)$$

is called fractional probability current density with $\boldsymbol{\nabla} = \dfrac{\partial}{\partial r}$.

Introducing the momentum operator $\hat{p} = \dfrac{\hbar}{i}\boldsymbol{\nabla}$, the vector j can be written
as

$$j = D_\alpha\left(\psi(\hat{p}^2)^{\alpha/2-1}\hat{p}\psi^* + \psi^*(\hat{p}^{*2})^{\alpha/2-1}\hat{p}^*\psi\right), \quad 1 < \alpha \leqslant 2.$$

When $\alpha = 2$ and $D_\alpha = 1/2m$, the above derivation corresponds to the classical quantum mechanics and the classical Schrödinger equation. Thus, the
above discussion is the generalization of the classic system into the fractional

order system. Let $\hat{v} = \dfrac{\mathrm{d}}{\mathrm{d}t}\hat{r}$ be the coordinate operator, then

$$\hat{v} = \frac{\mathrm{d}}{\mathrm{d}t}\hat{r} = \frac{\mathrm{i}}{\hbar}[H_\alpha, r] = \frac{\mathrm{i}}{\hbar}(H_\alpha r - r H_\alpha),$$

and hence

$$\hat{v} = \alpha D_\alpha |\hat{p}^2|^{\alpha/2-1}\hat{p},$$

which yields

$$j = \frac{1}{\alpha}(\psi\hat{v}\psi^* + \psi^*\hat{v}\psi), \quad 1 < \alpha \leqslant 2.$$

To normalize the probability current density, one may let

$$\psi(r,t) = \sqrt{\frac{\alpha}{2v}}\,\mathrm{e}^{\frac{\mathrm{i}pr}{\hbar} - \frac{\mathrm{i}Et}{\hbar}}, \quad E = D_\alpha |p|^\alpha, \quad 1 < \alpha \leqslant 2.$$

The time fractional Schrödinger equation can also be considered. We only investigate the one-dimensional case and now the one dimensional classical Schrödinger equation is

$$\mathrm{i}\hbar\partial_t\psi = -\frac{\hbar^2}{2m}\partial_x^2\psi + V\psi.$$

Two types of generalizations can be made [172]

$$(\mathrm{i}T_p)^\nu D_t^\nu\psi = -\frac{L_p^2}{2N_m}\partial_x^2\psi + N_V\psi, \qquad (1.4.2)$$

and

$$\mathrm{i}(T_p)^\nu D_t^\nu\psi = -\frac{L_p^2}{2N_m}\partial_x^2\psi + N_V\psi, \qquad (1.4.3)$$

where, D_t^ν represents the ν-order Caputo fractional derivative, and its parameters are $T_p = \sqrt{G\hbar/c^5}$, $L_p = \sqrt{G\hbar/c^3}$, $N_\nu = V/E_p$, $E_p = M_p c^2$, $N_m = m/M_p$, $M_p = \sqrt{\hbar c/G}$.

1.5   Fractional Ginzburg-Landau equation

Here we derive the fractional Ginzburg-Landau equation (FGLE) from the Euler-Lagrange equation for fractal substance [211]. This equation can be used to describe the dynamics for substance having fractional dispersion. The classical Ginzburg-Landau equation (GLE) [132]

$$g\Delta Z = aZ - bZ^3,$$

can be derived as the variational Euler-Lagrange equation

$$\frac{\delta F(Z)}{\delta Z} = 0,$$

for the free energy functional

$$F(Z) = F_0 + \frac{1}{2}\int_\Omega [g(\boldsymbol{\nabla}Z)^2 + aZ^2 + \frac{b}{2}Z^4]\mathrm{d}V_3. \qquad (1.5.1)$$

Two fractional generalizations of the equation (1.5.1) are considered. One is the fractional generalization of the integral in $F(Z)$, and the other one is the fractional generalization of the derivatives in $F(Z)$.

The simplest generalization is to consider the following energy functional

$$F(Z) = F_0 + \frac{1}{2}\int_\Omega [g(\boldsymbol{\nabla}Z)^2 + aZ^2 + \frac{b}{2}Z^4]\mathrm{d}V_D, \qquad (1.5.2)$$

where $\mathrm{d}V_D$ is D-dimensional volume element $\mathrm{d}V_D = C_3(D,x)\mathrm{d}V_3$. Here, in the Riesz definition of fractional integral, we have $C_3(D,x) = (2^{3-D}\Gamma(3/2)$ $|x|^{D-3})/(\Gamma(D/2))$ and in the Riemann-Liouville definition of fractional integral, we have $C_3(D,x) = (|x_1 x_2 x_3|^{D/3-1})/(\Gamma^3(D/3))$.

Let

$$\mathcal{F}(Z(x),\boldsymbol{\nabla}Z(x)) = \frac{1}{2}\left[g(\boldsymbol{\nabla}Z)^2 + aZ^2 + \frac{b}{2}Z^4\right], \qquad (1.5.3)$$

then the Euler-Lagrange equation can be obtained

$$C_3(D,x)\frac{\partial\mathcal{F}}{\partial Z} - \sum_{k=1}^{3}\boldsymbol{\nabla}_k(C_3(D,x)\frac{\partial\mathcal{F}}{\partial\boldsymbol{\nabla}_k Z}) = 0.$$

From (1.5.2), the following generalized FGLE can be obtained

$$gC_3^{-1}(D,x)\boldsymbol{\nabla}_k(C_3(D,x)\boldsymbol{\nabla}_k Z) - aZ - bZ^3 = 0, \qquad (1.5.4)$$

or equivalently

$$g\Delta Z + E_k(D,x)\boldsymbol{\nabla}_k Z - aZ - bZ^3 = 0, \qquad (1.5.5)$$

where $E_k(D,x) = C_3^{-1}(D,x)\partial_k C_3(D,x)$.

Generalize the energy functional into the fractional form

$$F(Z) = F_0 + \int_\Omega \mathcal{F}(Z(x), D^\alpha Z(x))\mathrm{d}V_D, \qquad (1.5.6)$$

where D^α is the Riesz fractional derivative and $\mathcal{F}$ is given by

$$\mathcal{F}(Z(x), D^\alpha Z(x)) = \frac{1}{2}\left[g(D^\alpha Z)^2 + aZ^2 + \frac{b}{2}Z^4\right]. \qquad (1.5.7)$$

Its Euler-Lagrange equation is

$$C_3(D,x)\frac{\partial\mathcal{F}}{\partial Z} + \sum_{k=1}^{3} D^\alpha_{x_k}\left(C_3(D,x)\frac{\partial\mathcal{F}}{\partial D^\alpha_{x_k} Z}\right) = 0.$$

In general $D \neq 3\alpha$ and this equation is equivalent to

$$gC_3^{-1}(D,x)\sum_{k=1}^{3} D_{x_k}^{\alpha}(C_3(D,x)D_{x_k}^{\alpha}Z) + aZ + bZ^3 = 0. \qquad (1.5.8)$$

Such a generalized equation is called fractional Ginzburg-Landau equation.

Below we consider some special cases of the equation (1.5.8).

(1) In the one-dimensional case, $Z = Z(x)$. Using the formulas for fractional integration by parts

$$\int_{\infty}^{\infty} f(x)\frac{d^{\beta}g(x)}{dx^{\beta}}dx = \int_{\infty}^{\infty} g(x)\frac{d^{\beta}f(x)}{d(-x)^{\beta}}dx$$

$$\int_{\infty}^{\infty} f(x)D_x^{\alpha}g(x)dx = \int_{\infty}^{\infty} g(x)D_x^{\alpha}f(x)dx, \qquad (1.5.9)$$

we obtain the Euler-Lagrange equation

$$D_x^{\alpha}\left(C_1(D,x)\frac{\partial F}{\partial D_x^{\alpha}Z}\right) + C_1(D,x)\frac{\partial F}{\partial Z} = 0, \quad C_1(D,x) = \frac{|x|^{D-1}}{\Gamma(D)},$$

Using (1.5.7), we arrive at

$$C_1^{-1}(D,x)D_x^{\alpha}(C_1(D,x)D_x^{\alpha}Z) + aZ + bZ^3 = 0.$$

For the case $D = 1$, we have $C_1 = 1$ and hence

$$D_x^{2\alpha}Z + aZ + bZ^3 = 0,$$

where D_x^{α} is the Riesz fractional derivative operator.

$$(D_x^{\alpha}f)(x) = \frac{-1}{2\cos(\pi\alpha/2)\Gamma(n-\alpha)}\frac{\partial^n}{\partial x^n}$$

$$\left(\int_{-\infty}^{x} \frac{f(z)dz}{(x-z)^{\alpha-n+1}} + \int_{x}^{\infty} \frac{(-1)^n f(z)dz}{(z-x)^{\alpha-n+1}}\right).$$

(2) Consider

$$F = \frac{1}{2}g_1(D_x^{\alpha}Z)^2 + \frac{1}{2}g_2(D_x^{\beta}Z)^2 + \frac{a}{2}Z^2 + \frac{b}{4}Z^4.$$

Using (1.5.9), we get the following Euler-Lagrange equation

$$D_x^{\alpha}\left(C_1(D,x)\frac{\partial F}{\partial D_x^{\alpha}Z}\right) + D_x^{\beta}\left(C_1(D,x)\frac{\partial F}{\partial D_x^{\beta}Z}\right) + C_1(D,x)\frac{\partial F}{\partial Z} = 0,$$

and hence the fractional Ginzburg-Landau equation

$$g_1 C_1^{-1}D_x^{\alpha}(C_1(D,x)D_x^{\alpha}Z) + g_2 C_1^{-1}(D,x)D_x^{\beta}(C_1(D,x)D_x^{\beta}Z) + aZ + bZ^3 = 0.$$

In particular, when $D = 1$, $C_1 = 1$ and hence

$$g_1 D_x^{2\alpha} Z + g_2 D_x^{2\beta} Z + aZ + bZ^3 = 0, \qquad 1 \leqslant \alpha, \beta \leqslant 1.$$

(3) For a more general case, we consider

$$\mathcal{F} = \mathcal{F}(Z, D_{x_1}^{\alpha_1} Z, D_{x_2}^{\alpha_2} Z, D_{x_3}^{\alpha_3} Z).$$

In this case, the FGLE for fractal media is

$$g_1 C_3^{-1}(D, x) \sum_{k=1}^{3} D_{x_k}^{\alpha_k} (C_3(D, x) D_{x_k}^{\alpha_k} Z) + aZ + bZ^3 = 0.$$

Below we consider another generalization of the GLE. Consider the wave propagation in some media, whose wave vector $\vec{k}$ satisfies $\vec{k} = \vec{k}_0 + \vec{\kappa} = \vec{k}_0 + \vec{\kappa}_\parallel + \vec{\kappa}_\perp$, where $\vec{k}_0$ is the unperturbed wave vector and the subscripts $(\parallel, \perp)$ are taken respectively to the direction of $\vec{k}_0$. Considering a symmetric dispersion law $\omega = \omega(k)$ for wave propagation with $\kappa \ll k_0$, we have

$$\omega(k) = \omega(|\vec{k}_0 + \vec{\kappa}|) \approx \omega(k_0) + c(|\vec{k}_0 + \vec{\kappa}| - k_0) \approx \omega(k_0) + c\vec{\kappa}_\parallel + \frac{c}{2k_0} \vec{\kappa}_\perp^2 \quad (1.5.10)$$

where $c = \partial\omega/\partial k_0$. This equation is the momentum representation of the field Z in the dual space corresponding to the following equation in coordinate space

$$-i\frac{\partial Z}{\partial t} = ic\frac{\partial Z}{\partial x_1} + \frac{c}{2k_0}\Delta Z, \qquad (1.5.11)$$

where x_1 is the direction of $\vec{k}_0$. By comparing the two equations, one has the following correspondences between the dual space and space-time space

$$\omega(k) \leftrightarrow i\frac{\partial}{\partial t}, \quad \vec{\kappa}_\parallel \leftrightarrow -i\frac{\partial}{\partial x_1}, \quad (\vec{\kappa}_\perp)^2 \leftrightarrow -\Delta = -\frac{\partial^2}{\partial x_2^2} - \frac{\partial^2}{\partial x_3^2}.$$

Generalizing it into the nonlinear dispersion relation, one obtains

$$\omega(k, |Z|^2) \approx \omega(k, 0) + b|Z|^2 = \omega(|\vec{k}_0 + \vec{\kappa}|, 0) + b|Z|^2, \qquad (1.5.12)$$

with some constant $b = (\partial\omega(k, |Z|^2))/(\partial|Z|^2)$ at $|Z|^2 = 0$. Analogously, the following equation can be obtained

$$-i\frac{\partial Z}{\partial t} = ic\frac{\partial Z}{\partial x} + \frac{c}{2k_0}\Delta Z - \omega(k_0)Z - b|Z|^2 Z, \qquad (1.5.13)$$

which is also called the nonlinear Schrödinger equation, and all its coefficients can be complex numbers. Let $Z = Z(t, x_1 - t, x_2, x_3)$, then one has

$$-i\frac{\partial Z}{\partial t} = \frac{c}{2k_0}\Delta Z - \omega(k_0)Z - b|Z|^2 Z.$$

Generalizing the dispersion relation (1.5.12) into the fractional case, one obtains

$$\omega(k,|Z|^2) = \omega(\vec{k}_0, 0) + c\vec{k}_\| + c_\alpha(\vec{\kappa}_\perp^2)^{\alpha/2} + b|Z|^2, \quad 1 < \alpha < 2,$$

where c_α is a constant. By using the correspondence relation $(-\Delta)^{\alpha/2} \leftrightarrow (\vec{\kappa}_\perp^2)^{\alpha/2}$, we obtain

$$-\mathrm{i}\frac{\partial Z}{\partial t} = \mathrm{i}c\frac{\partial Z}{\partial x} - \frac{c}{2k_0}(-\Delta)^{\alpha/2}Z + \omega(k_0)Z + b|Z|^2 Z, \qquad (1.5.14)$$

which is called the fractional Ginzburg-Landau equation (FGLE) or the fractional nonlinear Schrödinger equation (FNLS). The first term of the equation on the right hand side describes the wave propagation in fractional media, and its fractional derivative can be caused by the super-diffusion wave propagation or other physical mechanisms. The remaining terms represent the interactions of the wave motions in the nonlinear media. Therefore, this equation can be used to depict the self focusing or related fractional processes.

In the one-dimensional case, the equation (1.5.14) can be simplified as

$$c\frac{\partial Z}{\partial t} = gD_x^\alpha Z + aZ + b|Z|^2 Z,$$

where g, b, c are constants. Let $x = x_3 - ct$, then the traveling wave solution $Z = Z(x)$ of the above equation satisfies

$$gD_x^\alpha Z + cD_x^1 Z + aZ + b|Z|^2 Z = 0,$$

or for the real value Z

$$gD_x^\alpha Z + cD_x^1 Z + aZ + bZ^3 = 0.$$

1.6   Fractional Landau-Lifshitz equation

The Landau-Lifshitz equation (LLE) plays an important role in the ferromagnetic theory, which describes the movement pattern of the magnetization vector. LLE was first proposed by Landau and Lifshitz when they studied the dispersive theory of the magnetization phenomena of the ferromagnetic body, which is also called the ferromagnetic chain equation [128]. Afterwards, the equation are often used in condensed matter physics. In the 1960s, Soviet physicists A.Z. Akhiezer *et al* studied the spin wave, travelling wave solutions of the ferromagnetic chain equation and so on in their monograph [7]. In 1974, Nakamura *et al* first obtained the soliton solution of the Landau-Lifshitz equation without Gilbert term in the one-dimensional case. Since the 1980s, many mathematicians turned to study this equation, and gained many important results. In China, at the frontier in this regard is the research team led by Y. Zhou and B. Guo, the academician, who obtained the global weak solutions of the initial and initial-boundary value problems of the Landau-Lifshitz equation, and the global smooth solution of the one-dimensional LLE [230]. Soon later, B. Guo and M. Hong studied the two-dimensional LLE, and got the global existence for small initial data and established the relationship between the LLE and heat flow of harmonic maps [104]. Recently, mathematical problems of the Landau-Lifshitz equation has increasingly attracted the attention of the mathematical community, and a large number of literature and monographs has been published. For further knowledge, the readers can refer to the recently published monograph by B. Guo and S. Ding [102] and the references therein.

The Landau-Lifshitz equation is of the following form

$$\frac{\partial M}{\partial t} = -\gamma M \times H_{\text{eff}} - \alpha^2 M \times (M \times H_{\text{eff}}),$$

where γ is called the gyromagnetic ratio, $\alpha > 0$ is a constant depending on the physical properties of the material, $M = (M_1, M_2, M_3)$ is the magnetization vector, and $H_{\text{eff}} - \dfrac{\delta E_{tot}}{\delta M}$ is the effective magnetic field acting on the magnetic moment. Here, E_{tot} represents the energy functional of the entire magnetic field, which consists of the following parts [47]

$$E_{tot} = E_{exc} + E_{ani} + E_{dem} + E_{app}.$$

Recently, DeSimone *et al* [63] gave the following two-dimensional model when studied the film micromagnetic theory, where

$$E_{tot} = \int_{\mathbf{R}^2} (|\xi \cdot \widehat{M\chi_\Omega}|^2 / |\xi|) \mathrm{d}\xi.$$

Now $\dfrac{\delta E_{tot}}{\delta M} = -\boldsymbol{\nabla}(-\Delta)^{-\frac{1}{2}} \mathrm{div} M$. If we only consider the Gilbert term (i.e., $\gamma = 0$), then the following equation is obtained

$$\frac{\partial M}{\partial t} - \boldsymbol{\nabla}(-\Delta)^{-\frac{1}{2}} \mathrm{div} M + \boldsymbol{\nabla}(-\Delta)^{-\frac{1}{2}} \mathrm{div} M \cdot M M = 0,$$

where $M = (M_1, M_2)$ represents the two-dimensional magnetization vector. It can be seen that this equation is a partial differential equation with fractional derivatives. We can also consider the following Landau-Lifshitz equation having exchange energy. In this case,

$$E_{tot} = \varepsilon \int_\Omega |\boldsymbol{\nabla} M|^2 \mathrm{d}x + \int_{\mathbf{R}^2} (|\xi \cdot \widehat{M\chi_\Omega}|^2 / |\xi|) \mathrm{d}\xi,$$

leading to

$$\frac{\partial M}{\partial t} = \varepsilon \Delta M + \boldsymbol{\nabla}(-\Delta)^{-\frac{1}{2}}\mathrm{div}M + \varepsilon|\boldsymbol{\nabla}M|^2 M - \boldsymbol{\nabla}(-\Delta)^{-\frac{1}{2}}\mathrm{div}M \cdot MM = 0.$$

In addition, recently B. Guo *et al* studied the initial value problem of the following fractional Landau-Lifshitz equation with periodic boundary values [108, 182]

$$\begin{cases} M_t = M \times (-\Delta)^\alpha M, & \mathbf{T}^d \times (0, T) \\ M(0, x) = M_0, & x \in \mathbf{T}^d. \end{cases} \tag{1.6.1}$$

Applying the vanishing viscosity method, the authors considered the following approximation problem and then proved the global existence of the weak solutions

$$M_t = \frac{M}{\max\{1, |M|\}} \times (-\Delta)^\alpha M - \beta \frac{M}{\max\{1, |M|\}} \times \Delta M + \varepsilon \Delta M. \tag{1.6.2}$$

Recently the authors obtained global existence of weak solutions by Galerkin approximation and local smooth solutions by vanishing viscosity method for the Landau-Lifshitz equation with or without Gilbert damping term. The following fractional Landau-Lifshitz-Gilbert equation can be also considered cf. [182, 183, 185],

$$M_t = \gamma M \times (-\Delta)^\alpha M + \beta M \times (M \times (-\Delta)^\alpha M). \tag{1.6.3}$$

1.7   Some applications of fractional differential equations

This section introduces some applications of fractional differential equations in applied disciplines, such as viscoelasticity mechanics, biology, cybernetics and statistics. In this section, we only introduce several applications of the fractional partial differential equations, from which we can catch a glimpse of how powerful the FPDEs are in applied scientific branches. The interested readers may refer to the literatures cited herein.

Viscoelasticity mechanics is one of the disciplines in which the fractional differential equations are extensively applied, and a lot of related research papers have been published [38,155,194]. Almost all deformed materials exhibit elastic and viscous properties through simultaneous storage and dissipation of mechanical energy. So any viscoelastic material may be treated as a linear system with the stress as excitation function and the strain as the response function. In mechanics of materials, the Hooke's low reads $\sigma(t) = E\epsilon(t)$ for a solid, and Newton's law reads $\sigma(t) = \eta d\epsilon(t)/dt$ for fluids, where σ is the stress and ϵ is the strain. Both of them are not universal laws, but merely mathematical models of the ideal solid and fluid. Neither of them can adequately describe the real situation in the real world. In fact, real materials are between the two limit cases. Two fundamental methods are employed to connect the above two models. One is the cascade connection and from this the Maxwell model in viscoelasticity mechanics is obtained; the other is the parallel connection from which the Voigt model is obtained. In the Maxwell model, when the stress is a constant, then strain will grow infinitely. However, in the Voigt model, the viscoelasticity does not reflect the experimentally observed stress relaxation. To remedy the disadvantages of these two models, Kelvin model and Zener model were proposed, both of which can give satisfactory qualitative descriptions of the viscoelasticity. But neither of them are satisfactory as far as quantitative descriptions are considered. Hence more complex rheological models are proposed for the viscoelasticity materials, leading to complicated differential equations of higher orders.

On the other hand, since the stress is proportional to the zeroth derivative of strain for solids and to the first derivative of strain for fluids, then G.W. Scott Blair [25, 26] proposed "intermediate" derivative models for such "intermediate" materials

$$\sigma(t) = E_0 D_t^\alpha \epsilon(t), \tag{1.7.1}$$

where $\alpha \in (0, 1)$ depends on the property of the material. Almost at the same time, Gerasimov [88] employed the Caputo fractional derivative to get the following model for $0 < \alpha < 1$,

$$\sigma(t) = \kappa_{-\infty} D_t^\alpha \epsilon(t). \tag{1.7.2}$$

By using the fractional derivative, we can get the generalized Maxwell model, Voigt model and Zener model. They are all special cases of the following general high order model

$$\sum_{k=0}^n a_k D^{\alpha_k} \sigma(t) = \sum_{k=0}^m b_k D^{\beta_k} \epsilon(t).$$

The fractional derivative is also successfully applied to statistics. Assume that we need to model the impact of the hereditary effects in steel wires to study the mechanical properties. To describe some basic disadvantages of the classical polynomial regression models, we consider the two main stages of the change of mechanical properties of such a steel wire. In the first stage, within a period of time after the wire installation, the performance enhancement can be observed, and in the second stage, then its performance gradually declines, getting worse and worse, until breaks down. The period of performance enhancement is shorter than the period of decline, and the process are generally asymmetric.

In the classical regression model, the linear regression can well describe the second stage, but can not well describe the stage of performance enhancement. The second order regression provides symmetrical regression curve, hence is not well consistent with its physical backgrounds of the process. The high order polynomial regression can give better interpolation within the time interval for which measurements are available, but cannot give a reasonable prediction of performance change of the wire properties. Of course, in practical problems, the exponential regression model, Logistic regression model and other models can be used. Here we would like to introduce the fractional derivative model.

Consider n experimental measured values $y_1, y_2, \cdots, y_n$, and assume that the interpolated function $y(t)$ satisfies the following fractional integral equation for $\alpha \in (0, m]$

$$y(t) = \sum_0^{m-1} a_k t^k - a_{m0} D_t^{-\alpha} y(t),$$

where α, a_k, $k = 0, \cdots, m$, are parameters to be determined and m is the smallest integer greater than or equal to α. Let $z(t) = y(t) - \sum_0^{m-1} a_k t^k$, then z satisfies the following initial value problem [179]

$$\begin{cases} {}_0 D_t^\alpha z(t) + a_m z(t) = -a_m \sum_{k=0}^{m-1} a_k t^k, \\ z^{(k)}(0) = 0, \quad k = 0, \cdots, m-1. \end{cases} \qquad (1.7.3)$$

Besides the fractional models introduced above, there are many other important fractional models in various fields, some of which are listed below without introducing their detailed physical background. We will list some equations below which are actively studied.

1. Space-time fractional diffusion equation [139].

$$\frac{\partial^\alpha u(x,t)}{\partial t^\alpha} = D_x^\beta u(x,t), \quad 0 \leqslant x \leqslant L, \quad 0 < t \leqslant T,$$

$$u(x,0) = f(x), \quad 0 \leqslant x \leqslant L,$$

$$u(0,t) = u(L,t) = 0.$$

where $D_x^\beta (1 < \beta \leqslant 2)$ is the Riemann-Liouville fractional derivative

$$D_x^\beta u(x,t) = \begin{cases} \dfrac{1}{\Gamma(2-\beta)} \dfrac{\partial^2}{\partial x^2} \displaystyle\int_0^x \dfrac{u(\xi,t)\mathrm{d}\xi}{(x-\xi)^{\beta-1}}, & 1 < \beta < 2, \\ \dfrac{\partial^2 u(x,t)}{\partial x^2}, & \beta = 2, \end{cases}$$

and $\partial^\alpha/\partial t^\alpha (0 < \alpha \leqslant 1)$ is the Caputo fractional derivative

$$\dfrac{\partial^\alpha u(x,t)}{\partial t^\alpha} = \begin{cases} \dfrac{1}{\Gamma(1-\beta)} \displaystyle\int_0^t \dfrac{\partial u(x,\eta)}{\partial \eta} \dfrac{\mathrm{d}\eta}{(t-\eta)^\alpha}, & 0 < \alpha < 1, \\ \dfrac{\partial u(x,t)}{\partial t}, & \alpha = 1. \end{cases}$$

When $\alpha = 1, \beta = 2$, this equation is the classical diffusion equation

$$\frac{\partial u(x,t)}{\partial t} = \frac{\partial^2 u(x,t)}{\partial x^2}.$$

When $\alpha < 1$, the solution of the equation is no longer a Markov process, whose behavior will depend on the behaviors of the solution at all the previous times.

2. Fractional Navier-Stokes equation [227].

$$\partial_t u + (-\Delta)^\beta u + (u \cdot \nabla)u - \nabla p = 0, \quad \text{in } \mathbf{R}_+^{1+d},$$
$$\nabla \cdot u = 0, \quad \text{in } \mathbf{R}_+^{1+d},$$
$$u|_{t=0} = u_0, \quad \text{in } \mathbf{R}^d,$$

where $\beta \in (1/2, 1)$. When the time fractional derivative is considered, the following fractional Navier-Stokes equation can be obtained [169]

$$\frac{\partial^\alpha}{\partial t^\alpha} u + (u \cdot \nabla)u = -\frac{1}{\rho}\nabla p + \nu \Delta u,$$
$$\nabla \cdot u = 0,$$

where $\partial^\alpha/\partial t^\alpha (0 < \alpha \leqslant 1)$ is the Caputo fractional derivative.

3. Fractional Burger's equation [24]

$$u_t + (-\Delta)^\alpha u = -a \cdot \nabla(u^r),$$

where $a \in \mathbf{R}^d, 0 < \alpha \leqslant 2, r \geqslant 1$.

4. The semi-linear fractional dissipative equation [164]

$$u_t + (-\Delta)^\alpha u = \pm \nu |u|^b u.$$

5. Fractional conduction-diffusion equation [164]

$$u_t + (-\Delta)^\alpha u = a \cdot \nabla(|u|^b u), \qquad a \in \mathbf{R}^d/\{0\}.$$

6. Fractional MHD equation [222, 229]

$$\begin{cases} \partial_t u + u \cdot \nabla u - b \cdot \nabla b + \nabla P = -(-\Delta)^\alpha u, \\ \partial_t b + u \cdot \nabla b - b \cdot \nabla u = -(-\Delta)^\beta b, \\ \nabla \cdot u = \nabla \cdot b = 0. \end{cases}$$

Chapter 2

Fractional Calculus and Fractional Differential Equations

This Chapter mainly introduces definitions and basic properties of fractional derivatives, including Riemann-Liouville fractional derivative, Caputo fractional derivative and fractional Laplace operator, etc. For the fractional Laplace operator, some basic tools of partial differential equations are introduced, such as pseudo-differential operators, fractional Sobolev spaces and commutators estimates, etc. Also, some existence results of fractional ordinary equations are obtained by iteration. For readers' convenience, some basics of Fourier transform, Laplace transform and Mittag-Leffler function are given at the end of the chapter.

2.1 Fractional integrals and derivatives

2.1.1 Riemann-Liouville fractional integrals

To introduce R-L fractional integral, consider first the following iteration integrals

$$D^{-1}[f](t) = \int_0^t f(\tau)\mathrm{d}\tau,$$

$$D^{-2}[f](t) = \int_0^t \mathrm{d}\tau_1 \int_0^{\tau_1} f(\tau)\mathrm{d}\tau,$$

$$\cdots$$

$$D^{-n}[f](t) = \int_0^t \mathrm{d}\tau_1 \int_0^{\tau_1} \mathrm{d}\tau_2 \cdots \int_0^{\tau_{n-1}} f(\tau)\mathrm{d}\tau.$$

$$\cdots$$

These multiple iteration integrals can all be expressed as

$$\int_0^t K_n(t,\tau)f(\tau)\mathrm{d}\tau,$$

for a certain kernel function $K_n(t,\tau)$. Obviously, $K_1(t,\tau) = 1$. When $n = 2$, then

$$\int_0^t d\tau \int_0^\tau f(\tau_1)d\tau_1 = \int_0^t f(\tau)d\tau \int_\tau^t d\tau_1$$

$$= \int_0^t (t-\tau)f(\tau)d\tau,$$

thus $K_1(t,\tau) = (t-\tau)$. When $n = 3$,

$$\int_0^t d\tau \int_0^\tau d\tau_1 \int_0^{\tau_1} f(\tau_2)d\tau_2 = \int_0^t d\tau \int_0^\tau (\tau - \tau_1)f(\tau_1)d\tau_1$$

$$= \int_0^t f(\tau)d\tau \int_\tau^t (\tau_1 - \tau)d\tau_1$$

$$= \int_0^t f(\tau)\frac{(t-\tau)^2}{2}d\tau,$$

hence $K_2(t,\tau) = (t-\tau)^2/2$. Generally, $K_n(t,\tau) = (t-\tau)^{n-1}/(n-1)!$ by induction, yielding

$$D^{-n}[f](t) = \frac{1}{\Gamma(n)}\int_0^t (t-\tau)^{n-1}f(\tau)d\tau, \qquad (2.1.1)$$

where $\Gamma(n) = (n-1)!$. Assume $f \in C[0,T]$, the space of continuous functions on $[0,T]$, then for arbitrary $t \in [0,T]$, the integral exists in the sense of Riemann integral for any $n \geqslant 1$. Certainly, this idea can be extended to the situation $0 < n < 1$, where the integral exists as a generalized integral. Extending n to a general complex number, one obtains the definition of the R-L integral.

Definition 2.1.1 *Suppose that f is piecewise continuous in $(0,\infty)$, and integrable in any finite subinterval of $[0,\infty)$. For any $t > 0$ and any complex number ν with $\mathrm{Re}\,\nu > 0$, the ν-th R-L fractional integral of f is defined by*

$$_0D_t^{-\nu}f(t) = \frac{1}{\Gamma(\nu)}\int_0^t (t-\tau)^{\nu-1}f(\tau)d\tau. \qquad (2.1.2)$$

Below, **C** will denote the class of functions of f such that (2.1.2) makes sense.

Example 2.1.1 *Let $f(t) = t^\mu$ and $\mu > -1$, then obviously $f \in$ **C**. By definition,*

$$_0D_t^{-\nu}t^\mu = \frac{1}{\Gamma(\nu)}\int_0^t (t-\tau)^{\nu-1}\tau^\mu d\tau$$

$$= \frac{B(\nu,\mu+1)}{\Gamma(\nu)}t^{\nu+\mu}$$

$$= \frac{\Gamma(\mu+1)}{\Gamma(\mu+\nu+1)}t^{\mu+\nu}, \quad \mathrm{Re}\,\nu > 0, t > 0,$$

where B and Γ are the Beta function and the Gamma function, respectively. When μ and ν are integers, it reduces to the classical situation, and is consistent to the multiple iteration integrals above.

Now, we give several discussions on this definition.

1. The class $\mathbf{C}$ includes functions which behave asymptotically like $\ln t$ or t^μ near $t = 0$ for $-1 < \mu < 0$, as well as functions like $f(\tau) = |\tau - a|^\mu$ for $\mu > -1$ and $0 < a < t$.

2. Rewrite the integral in (2.1.2) in the Stieltjes integral, we have

$$_0D_t^{-\nu}f(t) = \frac{1}{\Gamma(\nu+1)} \int_0^t f(\tau)dg(\xi),$$

where $g(\tau) = -(t-\tau)^\nu$ is a monotone increasing function on the closed interval $[0, t]$. If f is continuous in $[0, t]$, then $_0D_t^{-\nu}f(t) = \frac{1}{\Gamma(\nu+1)}f(\xi)t^\nu$ by the mean value theorem for some $\xi \in [0, t]$. Therefore, $\lim_{t\to 0} {_0D_t^{-\nu}}f(t) = 0$. If $f \in \mathbf{C}$, such limit does not necessarily hold. Indeed, from Example 2.1.1, when $\mu > -1$ and $\nu > 0$, there holds

$$\lim_{t\to 0} {_0D_t^{-\nu}}t^\mu = \begin{cases} 0, & \mu + \nu > 0 \\ \Gamma(\mu+1), & \mu + \nu = 0 \\ \infty, & \mu + \nu < 0. \end{cases}$$

3. In the symbol $_0D_t^{-\nu}$ of the definition of the R-L fractional integral, the left subscript 0 can be replaced by any constant c, leading to the following definition

$$_cD_t^{-\nu}f(t) = \frac{1}{\Gamma(\nu)} \int_c^t (t-\tau)^{\nu-1}f(\tau)d\tau.$$

We will use $D^{-\nu}$ to denote the operator $_0D_t^{-\nu}$ for simplicity in the rest unless otherwise stated.

4. Under certain assumptions

$$\lim_{\nu\to 0} D^{-\nu}f(t) = f(t), \tag{2.1.3}$$

and hence one can regard

$$D^0 f(t) = f(t). \tag{2.1.4}$$

When f is continuously differentiable, the conclusion obviously holds. Integrating by parts, one has

$$D^{-\nu}f(t) = \frac{1}{\Gamma(\nu+1)} \int_0^t (t-\tau)^\nu f'(\tau)d\tau + \frac{t^\nu f(0)}{\Gamma(\nu+1)},$$

and hence

$$\lim_{\nu \to 0} D^{-\nu} f(t) = \int_0^t f'(\tau) \mathrm{d}\tau + f(0) = f(t).$$

When $f(t)$ is only continuous for $t \geqslant 0$, the proof will be somewhat complicated. We should prove that for arbitrary $\varepsilon > 0$, there exists $\delta > 0$ such that when $0 < \nu < \delta$ there holds $|D^{-\nu} f(t) - f(t)| < \varepsilon$. For this purpose, rewritting $D^{-\nu} f(t)$ as

$$D^{-\nu} f(t) = \frac{1}{\Gamma(\nu)} \int_0^t (t - \tau)^{\nu-1} (f(\tau) - f(t)) \mathrm{d}\tau + \frac{f(t)}{\Gamma(\nu)} \int_0^t (t - \tau)^{\nu-1} \mathrm{d}\tau$$

$$= \frac{1}{\Gamma(\nu)} \int_0^{t-\eta} (t - \tau)^{\nu-1} (f(\tau) - f(t)) \mathrm{d}\tau$$

$$+ \frac{1}{\Gamma(\nu)} \int_{t-\eta}^t (t - \tau)^{\nu-1} (f(\tau) - f(t)) \mathrm{d}\tau + \frac{f(t)t^\nu}{\Gamma(\nu+1)}. \quad (2.1.5)$$

Since f is continuous, for any $\tilde{\varepsilon} > 0$, there exists $\tilde{\delta} > 0$, such that for $|t - \tau| < \tilde{\delta}$, there holds $|f(\tau) - f(t)| < \tilde{\varepsilon}$. Hence the second term of the right hand side of (2.1.5) can be estimated as

$$|I_2| < \frac{\tilde{\varepsilon}}{\Gamma(\nu)} \int_{t-\tilde{\delta}}^t (t - \tau)^{\nu-1} \mathrm{d}\tau < \frac{\tilde{\varepsilon}\tilde{\delta}^\nu}{\Gamma(\nu+1)},$$

where we have used $\Gamma(\nu+1) = \nu\Gamma(\nu)$. Therefore, $|I_2| \to 0$ when $\tilde{\varepsilon} \to 0$.

Let $\varepsilon > 0$ be arbitrarily given. There always exists $0 < \eta < t$ such that $|I_2| < \varepsilon/3$ holds, for every $\nu > 0$. Fixed η, then the first term of the right hand side of (2.1.5) can be estimated as

$$|I_1| \leqslant \frac{M}{\Gamma(\nu)} \int_0^{t-\eta} (t - \tau)^{\nu-1} \mathrm{d}\tau \leqslant \frac{M}{\Gamma(\nu+1)} (\eta^\nu - t^\nu).$$

For such a fixed η, when $\nu \to 0$, the right hand side tends to zero, i.e., there exists $\delta_1 > 0$ such that for $0 < \nu < \delta_1$, one has $|I_1| < \varepsilon/3$. For the third term of the right hand side, $|I_3| \leqslant Mt^\nu/\Gamma(\nu+1)$, and hence there exists $\delta_2 > 0$, such that for $0 < \nu < \delta_2$ we have $|I_3| < \varepsilon/3$. Summarizing we have $\limsup_{\nu \to 0} |D^{-\nu} f(t) - f(t)| = 0$, completing the proof.

Theorem 2.1.1 Let $f \in C([0, \infty))$ be a continuous function and $\mu, \nu > 0$. Then, for any $t > 0$,

$$D^{-\nu}\{D^{-\mu} f(t)\} = D^{-\mu-\nu} f(t) = D^{-\mu}[D^{-\nu} f(t)].$$

Proof By definition,

$$D^{-\nu}\{D^{-\mu}f(t)\} = \frac{1}{\Gamma(\nu)} \int_0^t (t-x)^{\nu-1} \left[\frac{1}{\Gamma(\mu)} \int_0^x (x-y)^{\mu-1} f(y) dy \right] dx$$

$$= \frac{1}{\Gamma(\nu)\Gamma(\mu)} \int_0^t \int_y^t (t-x)^{\nu-1}(x-y)^{\mu-1} dx f(y) dy.$$

Substituting $x = (t-y)\xi + y$, we have

$$D^{-\nu}[D^{-\mu}f(t)] = \frac{1}{\Gamma(\nu)\Gamma(\mu)} \int_0^t \int_0^1 \xi^{\mu-1}(1-\xi)^{\nu-1} d\xi (t-y)^{\nu+\mu-1} f(y) dy$$

$$= \frac{B(\mu,\nu)}{\Gamma(\nu)\Gamma(\mu)} \int_0^t (t-y)^{\nu+\mu-1} f(y) dy$$

$$= D^{-\mu-\nu} f(t),$$

where we have used $B(\nu,\mu) = \Gamma(\mu)\Gamma(\nu)/\Gamma(\nu+\mu)$. The second equality is proved similarly.

From above, one can see that when $\nu = n \geqslant 0$ is an integer, $D^{-n}f(t)$ represents the n-fold integral of f. For any real number $\mu = n + \nu$ with $n \geqslant 0$, one has from this theorem

$$D^{-\mu}f(t) = D^{-n}[D^{-\nu}f(t)] = D^{-\nu}[D^{-n}f(t)].$$

It shows that the μ-th ($\mu = n+\nu$) R-L integral of f is equal to firstly taking n-fold integral and then the ν-th R-L integral of f or firstly taking the ν-th R-L integral and then the n-fold integral of f. Now consider the derivatives of R-L fractional integrals and the R-L fractional integrals of derivatives.

Theorem 2.1.2 *Let n be a positive integer, $\nu > 0$ and $D^n f \in C([0,\infty))$.*

(1) *When $D^n f \in \mathbf{C}$, then*

$$D^{-\nu-n}[D^n f(t)] = D^{-\nu} f(t) - R_n(t,\nu);$$

(2) *When $D^n f \in C([0,\infty))$, then for $t > 0$*

$$D^n[D^{-\nu} f(t)] = D^{-\nu}[D^n f(t)] + R_n(t, \nu - n),$$

where $R_n(t,\nu) = \sum_{k=0}^{n-1} t^{\nu+k} D^k f(0)/\Gamma(\nu+k+1)$.

Proof First we prove (1). When $n = 1$, let $\eta > 0, \delta > 0$, then $(t-\tau)^{\nu-1}$ and $f(\tau)$ are both continuous and differentiable in $[\delta, t - \eta]$. Integrating by parts then yields

$$\int_{\delta}^{t-\eta} (t-\tau)^{\nu}[Df(\tau)]d\tau$$

$$= \nu \int_{\delta}^{t-\eta} (t-\tau)^{\nu-1}f(\tau)d\tau + \eta^{\nu}f(t-\eta) - (t-\delta)^{\nu}f(\delta).$$

Letting δ, η tend to zero respectively, dividing both sides by $\Gamma(\nu+1)$, and using the property of Gamma function (2.7.1), one can prove the conclusion (1) for $n = 1$. For a general $n \geqslant 2$, applying repeatedly the conclusion of the case $n = 1$, we have

$$D^{-(\nu+n-1)-1}[D^1 D^{n-1}f(t)] = D^{-(\nu+n-1)}[D^{n-1}f(t)] - \frac{D^{n-1}f(0)}{\Gamma(\nu+n)}t^{\nu+n-1}$$

$$= D^{-\nu+n-2}[D^{n-2}f(t)]$$

$$- \frac{D^{n-2}f(0)}{\Gamma(\nu+n-1)}t^{\nu+n-2} - \frac{D^{n-1}f(0)}{\Gamma(\nu+n)}t^{\nu+n-1}$$

$$= \cdots$$

$$= D^{-\nu}f(t) - R_n(t,\nu).$$

Hence the conclusion (1) holds for a general positive integer.

We prove the (2). Let $\tau = t - \xi^{1/\nu}$ to obtain

$$D^{-\nu}f(t) = \frac{1}{\Gamma(\nu+1)} \int_0^{t^{\nu}} f(t - \xi^{1/\nu})d\xi.$$

Hence for $t > 0$,

$$D[D^{-\nu}f(t)] = \frac{1}{\Gamma(\nu+1)} \left[\nu t^{\nu-1}f(0) + \int_0^{t^{\nu}} f'(t - \xi^{1/\nu})d\xi \right].$$

Letting $t - \xi^{1/\nu} = \tau$ in the equation then completes the proof for (2) when $n = 1$. For a general positive integer $n \geqslant 2$, one can complete proof by induction similarly.

It shows that in general, D^n and $D^{-\nu}$ do not commute. But we have the following.

Corollary 2.1.1   *Under the assumptions of Theorem 2.1.2, if moreover* $D^k f(0) = 0$ *for all* $k = 0, 1, \cdots, n-1$, *then*

$$D^{-\nu-n}[D^n f(t)] = D^{-\nu}f(t)$$

and

$$D^n[D^{-\nu}f(t)] = D^{-\nu}[D^n f(t)].$$

Theorem 2.1.3 *Let n be a positive integer, $\nu > n$ and $D^n f$ be continuously differentiable in $[0, \infty)$, then for arbitrary $t \in [0, \infty)$, we have*

$$D^n[D^{-\nu}f(t)] = D^{-(\nu-n)}f(t).$$

Proof Let $\nu > n$. First, when $n = 1$ we have by definition D^{n-1} $[D^{-\nu}f(t)] = D^{-(\nu-1)-1}f(t)$. Indeed, similar equality holds for a general $n \geqslant 2$ with $n < \nu$, i.e.,

$$D^{n-1}[D^{-\nu}f(t)] = D^{-(\nu-n)-1}f(t).$$

We will show that this also holds when $n-1$ is replaced with n. Differentiating the expression by D, we have

$$\begin{aligned} D^n[D^{-\nu}f(t)] &= D[D^{-(\nu-n)-1}f(t)] \\ &= D^{-(\nu-n)-1}[Df(t)] + \frac{f(0)}{\Gamma(\nu-n+1)}t^{\nu-n} \\ &= D^{-(\nu-n)}f(t), \end{aligned}$$

where we have used (2) and (1) of Theorem 2.1.2 in the second and third step, respectively.

Theorem 2.1.4 *Let n and m be positive integers, $\nu, \mu > 0$ and $\nu - \mu = m - n$. Assume f is r-th continuously differentiable in $[0, \infty)$, then for arbitrary $t \in [0, \infty)$*

$$D^{-\nu}[D^m f(t)] = D^{-\mu}[D^n f(t)] + sgn(n-m)\sum_{k=s}^{r-1}\frac{t^{\nu-m+k}D^k f(0)}{\Gamma(\nu-m+k+1)}, \quad (2.1.6)$$

where $r = \max\{m, n\}$ and $s = \min\{m, n\}$, and for arbitrary $t \in (0, \infty)$, there holds

$$D^n[D^{-\mu}f(t)] = D^m[D^{-\nu}f(t)].$$

Proof If $m = n$, the theorem holds obviously. Now assume $n > m$ and denote $\sigma = n - m$, then using Theorem 2.1.2, we can see that

$$D^{-\nu}[D^m f(t)] = D^{-\nu-\sigma}[D^{\sigma+m}f(t)] + \sum_{k=0}^{\sigma-1}\frac{t^{\nu+k}D^{k+m}f(0)}{\Gamma(\nu+k+1)}.$$

(2.1.6) then follows since $\nu + \sigma = \mu$ and $\sigma + m = n$. On the other hand, from Theorem 2.1.3, we have

$$D^\sigma[D^{-\nu-\sigma}f(t)] = D^{-\nu}f(t),$$

since $\nu > 0$ here. Differentiating this formula m times then yields

$$D^{m+\sigma}[D^{-\nu-\sigma}f(t)] = D^m[D^{-\nu}f(t)],$$

This completes the proof.

Let $n \geqslant 1$ be a positive integer, when f and g are n-th continuously differentiable, the classical Leibniz rule holds

$$D^n(f(t)g(t)) = \sum_{k=0}^{n} C_n^k D^k f(t) D^{n-k} g(t).$$

To extend the Leibniz rule to the R-L fractional integrals, we first consider the following example.

Example 2.1.2 *Assume $\nu > 0$ and $f \in$ C, then*

$$D^{-\nu}[t^n f(t)] = \sum_{k=0}^{n} C_{-\nu}^k D^k t^n D^{-\nu-k} f(t).$$

Indeed, by definition,

$$D^{-\nu}[t^n f(t)] = \frac{1}{\Gamma(\nu)} \int_0^t (t-\tau)^{\nu-1} [\tau^n f(\tau)] d\tau.$$

Writting

$$\tau^n = [t - (t-\tau)]^n = \sum_{k=0}^{n} (-1)^k C_n^k t^{n-k} (t-\tau)^k,$$

and applying the generalized binomial coefficients (2.7.2), we have

$$\begin{aligned}
D^{-\nu}[t^n f(t)] &= \frac{1}{\Gamma(\nu)} \sum_{k=0}^{n} (-1)^k C_n^k t^{n-k} \int_0^t (t-\tau)^{\nu+k-1} f(\tau) d\tau \\
&= \frac{1}{\Gamma(\nu)} \sum_{k=0}^{n} (-1)^k C_n^k \Gamma(\nu+k) t^{n-k} D^{-\nu-k} f(t) \\
&= \sum_{k=0}^{n} C_{-\nu}^k [D^k t^n][D^{-\nu-k} f(t)],
\end{aligned}$$

where $C_{-\nu}^k = (-1)^k \Gamma(k+\nu)/k!\Gamma(\nu)$.

Theorem 2.1.5 *Suppose that f is continuous on $[0, T]$ and g is analytical at t for arbitrary $t \in [0, T]$. Then for any $\nu > 0$ and $0 < t \leqslant T$, there holds*

$$D^{-\nu}[f(t)g(t)] = \sum_{k=0}^{\infty} C_{-\nu}^k [D^k g(t)][D^{-\nu-k} f(t)].$$

Proof The idea of proof is illustrated in the above example. From the assumptions of f and g, we have $fg \in$ C. Therefore, the fractional integral

$D^{-\nu}\{f(t)g(t)\}$ exists for arbitrary $\nu > 0$. Since g is analytic, it can be expanded in Taylor series

$$g(\tau) = g(t) + \sum_{k=1}^{\infty} \frac{D^k g(t)}{k!}(\tau - t)^k,$$

which converges uniformly on $\tau \in [0, t]$. Substituting this into the expression of $D^{-\nu}[f(t)g(t)]$, we have

$$D^{-\nu}[f(t)g(t)] = \frac{1}{\Gamma(\nu)} \int_0^t (t - \tau)^{\nu-1}[f(\tau)g(\tau)]d\tau$$

$$= g(t)D^{-\nu}f(t) + \frac{1}{\Gamma(\nu)} \int_0^t (t-\tau)^\nu f(\tau)$$

$$\left[\sum_{k=1}^{\infty}(-1)^k \frac{D^k g(t)}{k!}(\tau - t)^{k-1}\right] d\tau.$$

Since f is continuous on $[0, T]$ and $\nu > 0$, $(t - \tau)^\nu f(\tau)$ is bounded on $[0, t]$, and hence interchanging the order of integration and summation yields

$$D^{-\nu}[f(t)g(t)] = g(t)D^{-\nu}f(t) + \sum_{k=1}^{\infty}(-1)^k \frac{\Gamma(\nu+k)}{k!\Gamma(\nu)}[D^k g(t)][D^{-\nu-k}f(t)]$$

$$= \sum_{k=0}^{\infty} C_{-\nu}^k [D^k g(t)][D^{-\nu-k}f(t)].$$

This completes the proof.

2.1.2 R-L fractional derivatives

Based on the R-L fractional integral, the R-L derivative can be defined naturally.

Definition 2.1.2 *Let $f \in \mathbf{C}$ and $\mu > 0$. Suppose that m is the smallest integer greater than μ and $m = \mu + \nu$ for $\nu \in (0, 1]$. Then the μ-th fractional derivative of f is defined by*

$$D^\mu f(t) = D^m[D^{-\nu}f(t)], \quad \mu > 0, t > 0,$$

where D^m represents the traditional m-th derivative.

Consider the special case $\mu = n$. In this case, $m = n + 1, \nu = 1$ and by definition

$$D^\mu f(t) = D^{n+1}[D^{-1}f(t)] = D^n f(t),$$

where the right hand side is understood in the classical sense of derivatives. Namely, when $\mu = 0, 1, 2, \cdots$ are integers, the fractional derivative reduces to the conventional derivative. This is why it does not make confusions to use the conventional derivative symbol D in fractional derivative. When $\mu = n$ is an integer, the condition $f \in \mathbf{C}$ is not necessary for the existence of $D^n f(t)$. For example, $f(t) = t^{-1}$ does not belong $\mathbf{C}$, however $Df(t)$ obviously exists. Indeed, in this situation, $f(t)$ has arbitrary integer-order derivatives.

To give some insight for the differences between the R-L fractional integrals and the R-L fractional derivatives, we consider the following two examples.

Example 2.1.3 (Continuation of Example 2.1.1) *Consider $f(t) = t^\lambda, \lambda > -1$, obviously, $f \in \mathbf{C}$. Assume $\mu > 0$ and m is the smallest integer greater than μ, then by definition $D^\mu[t^\lambda] = D^m[D^{-\nu}t^\lambda]$ for $\nu = m - \mu > 0$. But from Example 2.1.1, we have*

$$D^{-\nu}t^\lambda = \frac{\Gamma(\lambda + 1)}{\Gamma(\lambda + \nu + 1)}t^{\lambda + \nu}, \quad t > 0.$$

From this, we can see that

$$D^\mu t^\lambda = \frac{\Gamma(\lambda + 1)}{\Gamma(\lambda + \nu + 1)}D^m t^{\lambda + \nu} = \frac{\Gamma(\lambda + 1)}{\Gamma(\lambda - \mu + 1)}t^{\lambda - \mu}, \quad t > 0.$$

Comparing this example with Example 2.1.1, we see that the fractional derivative of t^λ of the order μ can be written in the form of a fractional integral in Example 2.1.1 by replacing the integrating order ν with $-\mu$. Namely, if $D^{-\nu}t^\lambda$ represents the ν-th R-L integral of t^λ, then the μ-th R-L derivative $D^\mu t^\lambda$ of t^λ can be expressed as $D^\mu t^\lambda = [D^{-\nu}t^\lambda]|_{\nu=-\mu}$ and vice versa. However, it does not always hold for general functions in $\mathbf{C}$. This is illustrated in the following example.

Example 2.1.4 *Consider $f(t) = \mathrm{e}^t$, then by definition*

$$D^\nu \mathrm{e}^t = D^\nu \sum_{k=0}^\infty \frac{t^k}{k!} = \sum_{k=0}^\infty \frac{t^{k-\nu}}{\Gamma(k - \nu + 1)}. \tag{2.1.7}$$

When $\nu = n$ is a positive integer, then $D^n \mathrm{e}^t = \mathrm{e}^t$, which is the correct answer. But by definition of the R-L fractional integral,

$$D^{-1}\mathrm{e}^t = \int_0^t \mathrm{e}^\tau \mathrm{d}\tau = \mathrm{e}^t - 1.$$

Therefore, we cannot find $D\mathrm{e}^t$ from $D^{-1}\mathrm{e}^t$ and vice versa.

The following is a generalization of Theorem 2.1.2.

Theorem 2.1.6   (1) *Suppose that* $f \in C([0, \infty))$ *and if* $p \geqslant q \geqslant 0$, *that* $D^{p-q} f(t)$ *exists, then*

$$D^p[D^{-q} f(t)] = D^{p-q} f(t). \tag{2.1.8}$$

(2) *When* $D^n f \in C([0, \infty))$ *and* $0 \leqslant k - 1 \leqslant q < k$, *then for all* $t > 0$

$$D^{-p}[D^q f(t)] = D^{q-p} f(t) - \sum_{j=1}^{k} \frac{t^{p-j}}{\Gamma(1 + p - j)} D^{q-j} f(0). \tag{2.1.9}$$

Proof First, we prove (2.1.8). When $p = q = n \geqslant 1$ are integers, the equality holds obviously. Taking now $k - 1 \leqslant p < k$ and using Theorem 2.1.1, we have

$$D^{-k} f(t) = D^{-(k-p)}[D^{-p} f(t)],$$

and therefore

$$D^p[D^{-p} f(t)] = D^k\{D^{-(k-p)}[D^{-p} f(t)]\} = D^k[D^{-k} f(t)] = f(t).$$

This proves the theorem for $p = q$. For general p, q, two cases must be considered: $q \geqslant p \geqslant 0$ and $p > q \geqslant 0$. When $q \geqslant p \geqslant 0$, by Theorem 2.1.1, we have

$$D^p[D^{-q} f(t)] = D^p\{D^{-p}[D^{-(q-p)} f(t)]\} = D^{-(q-p)} f(t) = D^{p-q} f(t).$$

When $p > q \geqslant 0$, let m, n be integers such that $0 \leqslant m - 1 \leqslant p < m$ and $0 \leqslant n - 1 \leqslant p - q < n$. Obviously, $n \leqslant m$. By definition of the R-L fractional derivative and Theorem 2.1.1,

$$
\begin{aligned}
D^p[D^{-q} f(t)] &= D^m\{D^{-(m-p)}[D^{-q} f(t)]\} \\
&= D^m[D^{p-q-m} f(t)] = D^n D^{m-n}[D^{p-q-m} f(t)] \\
&= D^n[D^{p-q-n} f(t)] = D^{p-q} f(t).
\end{aligned}
$$

Here, in the first and the second equalities, we have used Theorem 2.1.2 and the fact that $m - p > 0$, $m > 0$ and $q > 0$, respectively. The third step is obvious since m and n are both integers. In the last two steps, we have used Definition 2.1.2 for $p - q - m < 0$, $m - n \geqslant 0$ and $p - q - n < 0$, $n \geqslant 1$, respectively.

To prove (2.1.9), we first consider the case when $p = q$. By assumption $0 \leqslant k - 1 \leqslant p < k$. From the definition of R-L fractional integral,

$$
\begin{aligned}
D^{-p}[D^p f(t)] &= \frac{1}{\Gamma(p)} \int_0^t (t - \tau)^{p-1} D^p f(\tau) d\tau \\
&= D\left[\frac{1}{\Gamma(p+1)} \int_0^t (t - \tau)^p D^p f(\tau) d\tau \right].
\end{aligned}
\tag{2.1.10}
$$

The integral of $\{\cdots\}$ is given by

$$\frac{1}{\Gamma(p+1)} \int_0^t (t-\tau)^p D^p f(\tau)\mathrm{d}\tau$$

$$= \frac{1}{\Gamma(p+1)} \int_0^t (t-\tau)^p D^k[D^{-(k-p)}f(\tau)]\mathrm{d}\tau$$

$$= \frac{1}{\Gamma(p-k+1)} \int_0^t (t-\tau)^{p-k} D^{-(k-p)}f(\tau)\mathrm{d}\tau$$

$$\quad - \sum_{j=1}^k \frac{D^{k-j}[D^{-(k-p)}f(\tau)]|_{\tau=0}}{\Gamma(p+2-j)} t^{p-j+1}$$

$$= D^{-(p-k+1)}[D^{-(k-p)}f(t)] - \sum_{j=1}^k \frac{D^{p-j}f(0)}{\Gamma(p+2-j)} t^{p-j+1}$$

$$= D^{-1}f(t) - \sum_{j=1}^k \frac{D^{p-j}f(0)}{\Gamma(p+2-j)} t^{p-j+1},$$

where we have used Definition 2.1.2 and $k - p > 0$ in the first equality, integration by parts k times in the second inequality, Definition 2.1.2 in the third inequality and finally Theorem 2.1.2 in the last inequality. Since $D^p f(t)$ is integrable, $D^{p-j}f(t)$ is bounded at the endpoint $t = 0$ for each $j = 1, 2, \cdots, k$, and hence all the terms in the above formula exist. Using (2.1.10), if the fractional derivative $D^p f(t)$ of $f(t)$ is integrable, then

$$D^{-p}[D^p f(t)] = f(t) - \sum_{j=1}^k \frac{t^{p-j}[D^{p-j}f(0)]}{\Gamma(p-j+1)}, \quad (k-1 \leqslant p < k).$$

When $p \neq q$, two cases must be considered: $q < p$ and $q > p$. When $q < p$, Theorem 2.1.1 can be applied. When $q > p$, (2.1.8) can then be applied. Recalling Example 2.1.3, we have in both cases

$$D^{-p}[D^q f(t)] = D^{q-p}\{D^{-q}[D^q f(t)]\}$$

$$= D^{q-p}\left\{ f(t) - \sum_{j=1}^k \frac{[D^{q-j}f(0)]}{\Gamma(q-j+1)} t^{q-j} \right\}$$

$$= D^{q-p}\left\{ f(t) - \sum_{j=1}^k \frac{[D^{q-j}f(0)]}{\Gamma(p-j+1)} t^{p-j} \right\}.$$

The proof is complete.

Comparing Theorem 2.1.6 with Theorem 2.1.2, we can see that (2.1.9) reduces to (1) of Theorem 2.1.2 when $q = n$ and $p = \nu + n$ and (2.1.8) reduces to (2) of Theorem 2.1.2 when $q = \nu$ and $p = n$. Comparing (2.1.8) with (2.1.9), we can see that unless $D^{p-j}f(0) = 0$ for $0 \leqslant k-1 \leqslant p < k$, the R-L fractional derivative D^p and integral D^{-q} do not commute in general. Similar to Corollary 2.1.1, the commutativity of fractional derivatives can be considered. By definition of the fractional derivative, we have

$$
D^n[D^{k-\alpha}f(t)] = \frac{D^{n+k}}{\Gamma(\alpha)} \int_0^t (t-\tau)^{\alpha-1}f(\tau)d\tau
$$
$$
= D^{n+k-\alpha}f(t), \quad 0 < \alpha \leqslant 1. \tag{2.1.11}
$$

Denoting $p = k - \alpha$ leads to $D^n[D^p f(t)] = D^{n+p}f(t)$. On the other hand, by definition of the R-L fractional integral and integration by parts, we have

$$
D^{-n}[f^{(n)}(t)] = \frac{1}{(n-1)!} \int_0^t (t-\tau)^{n-1}f^{(n)}(\tau)d\tau
$$
$$
= f(t) - \sum_{j=0}^{n-1} \frac{f^{(j)}(a)t^j}{\Gamma(j+1)}.
$$

Using the conclusion (1) of Theorem 2.1.6, one gets

$$
\begin{aligned}
D^p[f^{(n)}(t)] &= D^{p+n}\{D^{-n}[f^{(n)}(t)]\} \\
&= D^{p+n}\left[f(t) - \sum_{j=0}^{n-1} \frac{f^{(j)}(0)t^j}{\Gamma(j+1)} \right] \\
&= D^{p+n}\left[f(t) - \sum_{j=0}^{n-1} \frac{f^{(j)}(0)t^{j-p-n}}{\Gamma(j+1-n-p)} \right].
\end{aligned} \tag{2.1.12}
$$

This shows that D^n and D^p do not commute in general, except $f^{(k)}(0) = 0$, $k = 0, 1, \cdots, n-1$.

Furthermore, we can consider the commutativity of the R-L fractional derivatives D^p and D^q. Assume $m - 1 \leqslant p < m$ and $n - 1 \leqslant q < n$, then by definition of fractional derivative and (2.1.9), we have

$$
\begin{aligned}
D^p[D^q f(t)] &= D^m\{D^{-(m-p)}[D^q f(t)]\} \\
&= D^m\left[D^{p+q-m} - \sum_{j=0}^{n} \frac{D^{q-j}f(0)t^{m-p-j}}{\Gamma(1+m-p-j)} \right] \\
&= D^{p+q}f(t) - \sum_{j=0}^{n} \frac{D^{q-j}f(0)t^{-p-j}}{\Gamma(1-p-j)}.
\end{aligned} \tag{2.1.13}
$$

Similarly, there holds

$$D^q[D^p f(t)] = D^{p+q} f(t) - \sum_{j=0}^{m} \frac{D^{p-j} f(0) t^{-q-j}}{\Gamma(1-q-j)}. \qquad (2.1.14)$$

Comparing these two formulas, it can be seen that the R-L fractional derivative operators D^q and D^p do not commute except $p = q$ or the summations of (2.1.13) and (2.1.14) are zero, i.e.,

$$D^{p-j} f(0) = 0, \quad j = 1, 2, \cdots, m,$$
$$D^{q-j} f(0) = 0, \quad j = 1, 2, \cdots, n.$$

We continue to introduce the Leibniz rule for the fractional derivatives. For this purpose, we introduce the subclass $\mathscr{C}$ of $\mathbf{C}$. We say that $f \in \mathscr{C}$ if $f \in \mathbf{C}$ has both a fractional integral and a fractional derivative of any order. Let $\eta(t)$ be analytical in a neighborhood of the origin, the family of functions $\mathscr{C}$ can be defined as the space of all functions of the form $t^\lambda \eta(t)$ and $t^\lambda (\ln t) \eta(t)$ with $\lambda > -1$. For example, polynomials, exponentials, sine and cosine functions all belong to $\mathscr{C}$.

Consider a simple case below. Assume $\mu > 0$ and n be a positive integer, then the R-L fractional integral of $t^n f(t)$ exists if $f \in \mathbf{C}$. Let m be the smallest integer greater than μ, then by definition of fractional derivative

$$D^\mu[t^p f(t)] = D^m[D^{-m+\mu} t^n f(t)].$$

From Example 2.1.2, we have

$$D^{-(m-\mu)}[t^n f(t)] = \sum_{k=0}^{n} C_{\nu-m}^k [D^k t^n][D^{\mu-m-k} f(t)]. \qquad (2.1.15)$$

We can show that if $f \in \mathscr{C}$, then for arbitrary $l = 0, 1, 2, \cdots$, there holds

$$D^l[D^{\mu-m-k} f(t)] = D^{l+\mu-m-k} f(t).$$

Hence, if $f \in \mathscr{C}$,

$$D^\mu[t^n f(t)] = \sum_{k=0}^{n} C_{\mu-m}^k D^m \left\{ [D^k t^n][D^{-m+\mu-k} f(t)] \right\}$$

$$= \sum_{k=0}^{n} C_{\mu-m}^k \sum_{j=0}^{m} C_m^j [D^{j+k} t^n][D^{\mu-j-k} f(t)].$$

Denote $r = j + k, s = k$, then

$$D^\mu[t^n f(t)] = \sum_{r=0}^{n} \left(\sum_{s=0}^{r} C_{\mu-m}^s C_m^{r-s} \right) [D^r t^n][D^{\mu-r} f(t)]$$

$$= \sum_{r=0}^{n} C_\mu^r [D^r t^n][D^{\mu-r} f(t)], \quad \mu > 0. \tag{2.1.16}$$

Theorem 2.1.7 *Suppose that $f \in \mathscr{C}$ and g is analytical at t for arbitrary $t \in [0, T]$. Then for any $\nu > 0$ and $0 < t \leqslant T$, there holds*

$$D^\mu[f(t)g(t)] = \sum_{k=0}^{\infty} C_\nu^k [D^k g(t)][D^{-\nu-k} f(t)].$$

Proof   The proof is similar to Theorem 2.1.5, and hence omitted.

2.1.3   Laplace transforms of R-L fractional derivatives

The Laplace transform $\mathscr{L}$ is an important tool in fractional calculus. Readers may refer to Appendix B for the definitions and properties. The purpose of this section is to apply Laplace transform to fractional integrals and derivatives, and compare them with the conventional integrals and derivatives. Let $F(s)$ and $G(s)$ are the Laplace transforms of f and g, respectively, then there holds

$$\mathscr{L} \left\{ \int_0^t f(t - \tau)g(\tau)d\tau \right\} = F(s)G(s). \tag{2.1.17}$$

Let $\mu > 0$. If $f \in \mathbf{C}$, then the μ-th R-L fractional integral $D^{-\mu} f(t)$ is, by definition, a convolution of the kernel functions $t^{\mu-1}$ and f. Hence, if f is at most exponentially increasing, then

$$\mathscr{L}[D^{-\mu} f(t)] = \frac{1}{\Gamma(\mu)} \mathscr{L}[t^{\mu-1}]\mathscr{L}[f(t)] = s^{-\mu} F(s), \tag{2.1.18}$$

where $F(s)$ is the Laplace transform of $f(t)$.

Example 2.1.5 *The following Laplace transforms hold*

$$\mathscr{L}[D^{-\mu} t^\nu] = \frac{\Gamma(\nu + 1)}{s^{\mu+\nu+1}}, \quad \mu > 0, \nu > -1,$$

$$\mathscr{L}[D^{-\mu} e^{at}] = \frac{1}{s^\mu(s - a)}, \quad \mu > 0,$$

$$\mathscr{L}[D^{-\mu} \cos at] = \frac{1}{s^{\mu-1}(s^2 + a^2)}, \quad \mu > 0.$$

Compare the Laplace transforms of the following two case. In the first case, we first take the R-L fractional order derivative and then the conventional derivative, while in the second case, we first take the conventional derivative and then the R-L fractional derivative. Let $f \in C([0, \infty))$ and $Df \in \mathbf{C}$ grow at most exponentially, then using (2.1.18) and the properties of Laplace transform, we have for $\mu > 0$

$$\mathscr{L}\{D^{-\mu}[Df(t)]\} = s^{-\mu}\mathscr{L}[Df(t)] = s^{-\mu}[sF(s) - f(0)], \qquad (2.1.19)$$

while using Theorem 2.1.2, we have

$$\begin{aligned} \mathscr{L}\{D[D^{-\mu}f(t)]\} &= \mathscr{L}\{D^{-\mu}[Df(t)]\} + \frac{s(0)}{\Gamma(\mu)}\mathscr{L}\left[t^{\mu-1}\right] \\ &= s^{-\mu}[sF(s) - f(0)] + s^{-\mu}f(0) \\ &= s^{1-\mu}F(s). \end{aligned} \qquad (2.1.20)$$

This shows that the Laplace transforms in the two cases are different. Furthermore, when $\mu \to 0$, the right term of (2.1.19) tends to $sF(s) - f(0)$, while the right term of (2.1.20) tends to $sF(s)$. The underlying reason may be that $\lim_{\mu \to 0} \dfrac{t^{\mu-1}}{\Gamma(\mu)} = 0$, but $\lim_{\mu \to 0} \mathscr{L}\left[t^{\mu-1}/\Gamma(\mu)\right] = 1$. This shows that the Laplace transform $\mathscr{L}$ and the limit operator $\lim$ do not commute. Further distinguishes can be seen in forthcoming chapters and sections.

Now, consider the Laplace transform of the R-L fractional derivative. Let $f \in \mathscr{C}$ be of the form

$$f(t) = t^\lambda \sum_{n=0}^{\infty} a_n t^n \quad \text{or} \quad t^\lambda(\ln t) \sum_{n=0}^{\infty} a_n t^n, \qquad \lambda > -1.$$

For simplicity, we only consider $f(t) = t^\lambda \eta(t)$. By definition, we have

$$D^\mu f(t) = t^{\lambda-\mu} \sum_{n=0}^{\infty} a_n \frac{\Gamma(n+\lambda+1)}{\lambda(n+\lambda+1-\mu)} t^n.$$

If f grows at most exponentially, the Laplace transform $F(s)$ exists and can be written as

$$F(s) = \frac{1}{s^{\lambda+1}} \sum_{n=0}^{\infty} a_n \Gamma(n+\lambda+1) s^{-n}.$$

Moreover, if $\lambda - \mu > -1$, then the Laplace transform of $D^\mu f(t)$ exists and

$$\mathscr{L}[D^\mu f(t)] = \sum_{n=0}^{\infty} a_n \frac{\Gamma(n+\lambda+1)}{s^{n+\lambda-\mu+1}}.$$

Comparing these two formulas, one can see that $\mathscr{L}[D^\mu f(t)] = s^\mu F(s)$ for $\mu < \lambda + 1$. When $\mu \leqslant 0$, it reduces to the R-L fractional integral (refer to (2.1.18)). When $\mu > 0$, let m be the smallest integer greater than μ, then $\mu - m \leqslant 0$. If $f \in \mathscr{C}$, by definition of the R-L fractional derivative, $D^\mu f(t) = D^m[D^{-(m-\mu)}f(t)]$ exists. Using the properties of Laplace transform, we have

$$\mathscr{L}[D^\mu f(t)] = \mathscr{L}\{D^m[D^{-(m-\mu)}f(t)]\}$$

$$= s^m \mathscr{L}[D^{-(m-\mu)}f(t)] - \sum_{k=0}^{m-1} s^{m-k-1} D^k[D^{-(m-\mu)}f(t)]|_{t=0}$$

$$= s^m[s^{-(m-\mu)}F(s)] - \sum_{k=0}^{m-1} s^{m-k-1} D^{k-(m-\mu)}f(0)$$

$$= s^\mu F(s) - \sum_{k=0}^{m-1} s^{m-k-1} D^{k-(m-\mu)}f(0),$$

where $m - 1 < \mu \leqslant m$. This is the Laplace transform of the R-L fractional derivative. By comparing this formula to (2.1.18), we can see the differences and similarities between the R-L fractional integrals and R-L fractional derivatives. In particular, we can compare the fractional order case with the integer order case. When μ is an integer, this reduces to the situation of integer order case.

2.1.4   Caputo's definitions of fractional derivatives

Caputo's fractional derivative is another method for computing fractional derivatives. It was introduced by M. Caputo in his paper [35]. See also [35, 36, 78, 79]. The μ-th Caputo' fractional derivative of f is defined by

$$_a^C D_t^\mu f(t) = \frac{1}{\Gamma(n-\mu)} \int_a^t \frac{f^{(n)}(\tau)}{(t-\tau)^{\mu+1-n}} d\tau, \qquad (n-1 < \mu < n). \qquad (2.1.21)$$

Here, we denote the Caputo's fractional derivative by $_a^C D_t^\mu$ to distinguish it from the R-L fractional derivative. Without confusions, the R-L fractional derivative is still denoted by D. When $a = 0$, the μ-th Caputo's fractional derivative is simplified as $^C D^\mu$. The obvious difference between the R-L fractional derivative and the Caputo's fractional derivative is the order of differentiation. In the R-L fractional derivative, it first takes the fractional order and then the integer order conventional derivative, while in the Caputo's fractional derivative, it first takes the integer order conventional derivative and then the fractional order derivative.

First, we observe that when $\mu \to n$, the Caputo's fractional derivative reduces to the n-the derivative in the classical sense. Indeed, assume that

$0 \leqslant n-1 < \mu < n$ and f is a $n+1$ times continuously differentiable function in $[0, T]$, then by definition and integration by parts, we have

$$^{C}D^{\mu}f(t) = \frac{f^{(n)}(0)t^{n-\mu}}{\Gamma(n-\mu+1)} + \int_{0}^{t} \frac{(t-\tau)^{n-\mu}f^{(n+1)}(\tau)}{\Gamma(n-\mu+1)}d\tau.$$

By dominated convergence, taking $\mu \to n$ then yields

$$\lim_{\mu \to n} {}^{C}D^{\mu}f(t) = f^{(n)}(0) + \int_{0}^{t} f^{(n+1)}(\tau)d\tau = f^{(n)}(t), \quad n = 1, 2, \cdots.$$

This shows that, similar to the R-L approaches, the Caputo approach also provides an interpolation between the integer order derivatives.

Now, we make a simple comparison between the Riemann-Liouville fractional derivative and the Caputo's fractional derivative. The R-L fractional derivative and Caputo's fractional derivative can both be expressed by the R-L fractional integrals. The ν-th R-L fractional integral can be written as

$$D^{-\nu}f(t) = \frac{1}{\Gamma(\nu)} \int_{0}^{t} \frac{f(\tau)d\tau}{(t-\tau)^{1-\nu}}, \quad \nu > 0.$$

By R-L fractional integral $D^{-\nu}$, the R-L fractional derivative can be written as

$$^{RL}_{0}D^{\mu}_{t}f(t) = \frac{1}{\Gamma(\nu)} \frac{d^{n}}{dt^{n}} \int_{0}^{t} \frac{f(\tau)d\tau}{(t-\tau)^{1-\nu}} = \frac{d^{n}}{dt^{n}}[D^{-\nu}f(t)],$$

for $\nu = n - \mu > 0$. Similarly, for $\nu = n - \mu > 0$, by the R-L integral $D^{-\nu}$, the Caputo's fractional derivative can be written as

$$^{C}_{0}D^{\mu}_{t}f(t) = \frac{1}{\Gamma(\nu)} \int_{0}^{t} \frac{f^{(n)}d\tau}{(t-\tau)^{1-\nu}} = D^{-\nu}\left[\frac{d^{n}}{dt^{n}}f(t)\right].$$

Therefore, the R-L fractional derivative takes a fractional integral first and then integer order derivatives, while the Caputo fractional derivative takes an integer order derivative first and then fractional integral. They are related by

$$^{RL}_{0}D^{\mu}_{t}f(t) = {}^{C}_{0}D^{\mu}_{t}f(t) + \sum_{k=0}^{n-1} \frac{t^{k-\mu}}{\Gamma(k-\mu+1)}f^{(k)}(0),$$

for $t > 0$ and $\mu \in (n-1, n]$. The right hand side is equivalent to the Grunwald-Letnikov definition of fractional derivative, which requires that the function $f(t)$ be n times continuously differentiable. But the Riemann-Liouville definition provides an excellent opportunity to weaken the conditions on the function $f(t)$. It is enough to require that $f(t)$ is integrable, then (2.1.2) exists for all $t > 0$ and can be differentiated k times.

Although the R-L fractional derivative weakens the conditions on the function $f(t)$, Caputo's fractional derivatives are more widely used in initial values problems of differential equations and have stronger physical interpretations. This can be illustrated via the Laplace transform. The Laplace transform of the R-L fractional derivative is from the last section

$$\mathscr{L}\left[{}_0^{RL}D_t^\mu f(t)\right](s) = s^\mu F(s) - \sum_{k=0}^{n-1} ({}_0^{RL}D_t^{\mu-k-1} f(t))|_{t=0} \cdot s^k. \qquad (2.1.22)$$

Hence in general, to solve a initial value problem of a fractional differential equation, we have to know the fractional initial conditions $({}_0^{RL}D_t^{\mu-k-1} f(t))|_{t=0}$, $k = 0, \cdots, n-1$ of $f(t)$. Although the initial value problems with such fractional initial conditions can be solved mathematically, their solutions are practically useless, since there is no known physical interpretation for such types of initial conditions. For a specific physical system, the initial conditions are the measurable conditions of a system but not the fractional derivative conditions. On the other hand, the Laplace transform of the Caputo's fractional derivative is (cf. [35])

$$\mathscr{L}[{}_0^C D_t^\mu f(t)](s) = s^\mu F(s) - \sum_{k=0}^{n-1} s^{\mu-k-1}(D_t^k f(t))|_{t=0}. \qquad (2.1.23)$$

To solve an initial value problem of the Caputo type, like the integer order differential equations, only initial values of integer order derivatives of unknown functions at the initial time required. The Caputo fractional derivative can better reconcile the well-established and polished mathematical theory with practical needs.

2.1.5 Weyl's definition for fractional derivatives

The Weyl fractional calculus was first introduced by Weyl in [219]. Let f be of the Schwartz class, the μ-th Weyl fractional integral of f is defined as

$$_tW_\infty^{-\mu} f(t) = \frac{1}{\Gamma(\mu)} \int_t^\infty (\tau - t)^{\mu-1} f(\tau) d\tau, \quad \text{Re}\,\mu > 0, t > 0. \qquad (2.1.24)$$

Usually, we use to $W^{-\mu}$ to simplify $_tW_\infty^{-\mu}$. Let $\tau = t + \xi$, then

$$W^{-\nu} f(t) = \frac{1}{\Gamma(\nu)} \int_0^\infty \xi^{\nu-1} f(t + \xi) d\xi,$$

and hence

$$
\begin{aligned}
D[W^{-\nu}f(t)] &= D\left[\frac{1}{\Gamma(\nu)}\int_0^\infty \xi^{\nu-1}f(t+\xi)\mathrm{d}\xi\right] \\
&= \frac{1}{\Gamma(\nu)}\int_0^\infty \xi^{\nu-1}\frac{\partial}{\partial t}f(t+\xi)\mathrm{d}\xi \\
&= \frac{1}{\Gamma(\nu)}\int_0^\infty \xi^{\nu-1}Df(t+\xi)\mathrm{d}\xi \\
&= W^{-\nu}[Df(t)].
\end{aligned}
$$

Similarly, for a general positive integer n, one obtains

$$
D^n[W^{-\nu}f(t)] = W^{-\nu}[D^n f(t)]. \tag{2.1.25}
$$

Now we consider the composition of two Weyl integral operators. If f is rapidly decreasing, then $W^{-\mu}f(t)$ is also rapidly decreasing, hence for any arbitrary $\nu > 0$

$$
\begin{aligned}
W^{-\nu}\left[W^{-\mu}f(t)\right] &= \frac{1}{\Gamma(\mu)}W^{-\nu}\left[\int_t^\infty (\tau-t)^{\mu-1}f(\tau)\mathrm{d}\tau\right] \\
&= \frac{1}{\Gamma(\mu)\Gamma(\nu)}\int_t^\infty (\xi-t)^{\nu-1}\mathrm{d}\xi\left[\int_\xi^\infty (\tau-\xi)^{\mu-1}f(\tau)\mathrm{d}\tau\right].
\end{aligned}
$$

By definition of the Beta function, we have

$$
W^{-\nu}\left[W^{-\mu}f(t)\right] = \frac{B(\mu,\nu)}{\Gamma(\mu)\Gamma(\nu)}\int_t^\infty (\tau-t)^{\mu+\nu-1}f(\tau)\mathrm{d}\tau,
$$

yielding that

$$
W^{-\nu}W^{-\mu} = W^{-(\mu+\nu)}. \tag{2.1.26}
$$

The Weyl fractional derivative is defined by the Weyl fractional integral, just like the R-L fractional derivative is defined from the R-L fractional integral. Let $L = -D$, then (2.1.25) can be written as

$$
L^n W^{-\nu} = W^{-\nu}L^n. \tag{2.1.27}
$$

For a rapidly decreasing function f, by integration by parts and (2.1.27), we have

$$
\begin{aligned}
W^{-\mu}f(t) &= \frac{1}{\Gamma(\mu)}\int_t^\infty (\tau-t)^{\mu-1}f(\tau)\mathrm{d}\tau \\
&= W^{-(\mu+n)}[L^n f(t)] \\
&= L^n[W^{-(\mu+n)}f(t)].
\end{aligned}
$$

Applying the operator L^m at both sides then yields

$$
L^m[W^{-\mu}f(t)] = L^{m+n}[W^{-(\mu+n)}f(t)]. \tag{2.1.28}
$$

Definition 2.1.3 *Let $\mu > 0$ and $n = [\mu] + 1$ be the smallest integer greater than μ. Denote $\nu = n - \mu$, assume the $-\nu$-th Weyl integral of the function f exists and is n-times continuously differentiable, then the μ-th Weyl derivative of f is defined as*

$$W^\mu f(t) = L^n[W^{-(n-\mu)}f(t)]. \tag{2.1.29}$$

Example 2.1.6 *We give two examples here.*

1. *Let $\mu > 0$ and $a > 0$, then by definition, we have $W^{-\mu}e^{-at} = a^{-\mu}e^{-at}$. Let $n = [\mu] + 1$ be the smallest integer greater than μ and $\nu = n - \mu$. First, by definition of the Gamma function*

$$\begin{aligned} W^{-\nu}e^{-at} &= \frac{1}{\Gamma(\nu)} \int_t^\infty (\tau - t)^{\nu-1}e^{-a\tau}d\tau \\ &= a^{-\nu}e^{-at}\Gamma(\nu) \int_0^\infty t^{\nu-1}e^{-t}dt \\ &= a^{-\nu}e^{-at}. \end{aligned}$$

The result then follows from (2.1.29).

2. *Let $\lambda > \nu > 0$, then by definition of the Weyl integral, definition of the Beta function, change of variable and $B(\nu, \lambda - \nu) = \Gamma(\nu)\Gamma(\lambda - \nu)/\Gamma(\lambda)$, we then have $W^{-\nu}t^{-\lambda} = \Gamma(\lambda - \nu)t^{\nu-\lambda}/\Gamma(\lambda)$. Let n be such that $0 < n - \mu < \lambda$, then by (2.1.29), the μ-th Weyl derivative of $t^{-\lambda}$ is given by*

$$W^\mu t^{-\lambda} = L^n[W^{-(n-\mu)}t^{-\lambda}] = \Gamma(\lambda + \mu)t^{-\mu-\lambda}/\Gamma(\lambda).$$

Proposition 2.1.1 *For any arbitrary μ, there holds $W^{-\mu}W^\mu = I = W^\mu W^{-\mu}$.*

Proof First, when $\mu = n$ is a positive integer, by integration by parts n-times, we have

$$W^{-n}[L^n f(t)] = \frac{1}{\Gamma(n)} \int_t^\infty (\tau - t)^{n-1}L^n f(\tau)d\tau = f(t),$$

yielding the result. More generally, let $n = [\mu] + 1$ be the smallest positive integer greater than μ. Using (2.1.26), we know

$$\begin{aligned} W^\mu \left[W^{-\mu}f(t)\right] &= L^n\{W^{-(n-\mu)}[W^{-\mu}f(t)]\} \\ &= L^n[W^{-n}f(t)] \\ &= f(t). \end{aligned}$$

Similarly, by definition, (2.1.29), (2.1.27) and (2.1.26), we have

$$
\begin{aligned}
W^{-\mu}\left[W^{\mu}f(t)\right] &= W^{-\mu}\left[L^{n}W^{-(n-\mu)}f(t)\right] \\
&= L^{n}\left[W^{-\mu}W^{-(n-\mu)}f(t)\right] \\
&= L^{n}\left[W^{-n}f(t)\right] \\
&= f(t).
\end{aligned}
$$

This completes the proof.

Similarly to (2.1.26), we can prove the law of exponents of the Weyl fractional derivative. We shall define $W^{0} = I$, the identity operator.

Proposition 2.1.2 *Let μ and ν be real numbers, then the Weyl fractional derivative satisfies the following exponential relation*

$$
W^{\mu}W^{\nu} = W^{\mu+\nu}.
$$

Proof   The proof is omitted.

Finally, we consider the Leibniz rule of the Weyl fractional integral. To illustrate, we first consider $W^{-\mu}\{t^{n}f(t)\}$ and note

$$
\tau^{n} = [(\tau - t) + t]^{n} = \sum_{k=1}^{n} C_{n}^{k}(\tau - t)^{k}t^{n-k}.
$$

Therefore,

$$
\begin{aligned}
W^{-\mu}[t^{n}f(t)] &= \frac{1}{\Gamma(\mu)}\sum_{k=0}^{n} C_{n}^{k}t^{n-k}\int_{t}^{\infty}(\tau - t)^{\mu-1}(\tau - t)^{k}f(\tau)d\tau \\
&= \sum_{k=0}^{n}\frac{\Gamma(\mu + k)}{\Gamma(\mu)}C_{n}^{k}t^{n-k}W^{-\mu-k}f(t) \\
&= \sum_{k=0}^{n}\frac{\Gamma(\mu + k)}{\Gamma(\mu)k!}[D^{k}t^{n}][W^{-\mu-k}f(t)].
\end{aligned}
$$

By using the generalized binomial formula, we then arrive at the more familiar form

$$
W^{-\mu}[t^{n}f(t)] = \sum_{k=0}^{n}C_{-\mu}^{k}[L^{k}t^{n}][W^{-\mu-k}f(t)]. \tag{2.1.30}
$$

More generally, similarly to Theorem 2.1.5, we have

Theorem 2.1.8 *Let f and g are two rapidly decreasing functions, and g is an entire function, then for arbitrary $\mu > 0$,*

$$W^{-\mu}[f(t)g(t)] = \sum_{k=0}^{\infty} C_{-\nu}^{k}[L^{k}g(t)][W^{-\nu-k}f(t)].$$

Proof   The proof is omitted for simplicity.

2.2   Fractional Laplacian

As is well known, the standard Laplace operator (Laplacian) $\Delta = \partial_{x_1}^2 + \cdots + \partial_{x_d}^2$ in a d-dimensional domain possess an explanation in terms of the diffusion and Brownian motion. This explanation has enormous success both in Mathematics and Physics. In recent years, there has been a plenty of work on anomalous diffusion, with standard Laplace operator replaced by the so-called fractional Laplace operator, with the aim of extending the diffusion theory by taking into account the long range interactions. As we will see, such a Laplacian is non-local and do not act by pointwise differentiation but by a global integration with respect to a singular kernel. This section consists of the definition and basic properties of the fractional Laplacian, pseudodifferential operator, Riesz and Bessel potentials and fractional Sobolev spaces, and finally commutator estimates for the fractional Laplacian. These are very fundamental topics in analysis and partial differential equations with fractional Laplcian.

2.2.1   Definition and properties

Let $f \in \mathcal{S}(\mathbf{R}^d)$ be a function in the Schwartz class, then $-\Delta f = \mathcal{F}^{-1}(|\xi|^2 \mathcal{F}u)$. The $(-\Delta)^{\alpha/2}f$ can be defined naturally via Fourier transform

$$\widehat{(-\Delta)^{\alpha/2}f}(\xi) = |\xi|^\alpha \hat{f}(\xi).$$

The fractional Laplacian on a torus can be similarly defined. The interest in these fractional Laplacian operators has a long history in probability since the fractional Laplacian $(-\Delta)^\alpha$ for $\alpha \in (0,2)$ are infinitesimal generators of stable Lévy processes. Indeed, let $X = \{X_t : t \geqslant 0, \mathbb{P}_x, x \in \mathbf{R}^d\}$ be a rotational invariant α-stable process in $\mathbf{R}^d$, then X is a Lévy process, and for arbitrary $x \in \mathbf{R}^d$ and $\xi \in \mathbf{R}^d$, $\mathbb{E}_x\left[e^{i\xi \cdot (X_t - X_0)}\right] = e^{-t|\xi|^\alpha}$. For such a process, the generator is $-(-\Delta)^\alpha$, and can be represented by

$$- (-\Delta)^{\alpha/2}f(x) = c_{d,\alpha} \lim_{\varepsilon \downarrow 0} \int_{|y-x|>\varepsilon} \frac{f(y) - f(x)}{|x - y|^{d+\alpha}} dy, \qquad (2.2.1)$$

where $c_{d,\alpha} = \dfrac{2^{\alpha-1}\alpha\Gamma((d+\alpha)/2)}{\pi^{d/2}\Gamma(1-\alpha/2)}$ is a normalization constant. By a change of variable, this is equivalent to

$$-(-\Delta)^{\alpha/2}f(x) = \frac{1}{2}c_{d,\alpha}\lim_{\varepsilon\downarrow 0}\int_{|y-x|>\varepsilon}\frac{f(x+y)+f(x-y)-2f(x)}{|y|^{d+\alpha}}\mathrm{d}y.$$

This formula is very useful in studying local properties of equations involving the fractional Laplacian and regularity for critical semilinear problems. When $\alpha \in (0,2)$, the operators $(-\Delta)^{-\alpha/2}$ is defined to be the inverse of $(-\Delta)^{\alpha/2}$ and are given by the standard convolution

$$(-\Delta)^{-\alpha/2}f(x) = c_{d,-\alpha}\int_{\mathbf{R}^d}|x-y|^{-d+\alpha}f(y)\mathrm{d}y, \qquad (2.2.2)$$

in terms of the Riesz potential, where $c_{d,-\alpha} = \dfrac{\Gamma((d-\alpha)/2)}{\pi^{d/2}2^\alpha\Gamma(\alpha/2)}$.

In what follows, we consider some properties of the fractional Laplacian, which proves very useful in partial differential equations. For notational simplicity, we denote $\Lambda = (-\Delta)^{\frac{1}{2}}$ and hence $\Lambda^\alpha = (-\Delta)^{\alpha/2}$. The following discussion is based on $\mathbf{R}^2$ or $\mathbf{T}^2$, but can be extended to $\mathbf{R}^d$ or $\mathbf{T}^d$ without essential difficulties. First we prove (2.2.1) from the Riesz potential (2.2.2). The following several theorems are modified from [58].

Proposition 2.2.1 *Let $0 < \alpha < 2, x \in \mathbf{R}^2$, and $f \in \mathcal{S}$ is a function in the Schwartz class, then*

$$\Lambda^\alpha f(x) = c_\alpha P.V.\int_{\mathbf{R}^2}\frac{f(x)-f(y)}{|x-y|^{2+\alpha}}\mathrm{d}y, \qquad (2.2.3)$$

where $c_\alpha > 0$ is a constant.

Proof By definition of the Riesz's potential, Λ^α can be expressed

$$\begin{aligned}
\Lambda^\alpha f(x) =& \Lambda^{\alpha-2}(-\Delta f) = c_\alpha\int_{\mathbf{R}^2}\frac{-\Delta f(y)}{|x-y|^\alpha}\mathrm{d}y\\
=& c_\alpha\int_{\mathbf{R}^2}\frac{\Delta_y[f(x)-f(y)]}{|x-y|^\alpha}\mathrm{d}y\\
=& \lim_{\varepsilon\to 0}c_\alpha\int_{|x-y|\geqslant\varepsilon}\frac{\Delta_y[f(x)-f(y)]}{|x-y|^\alpha}\mathrm{d}y\\
=&: \lim_{\varepsilon\to 0}c_\alpha\Lambda_\varepsilon^\alpha\theta,
\end{aligned}$$

where $c_\alpha = \dfrac{\Gamma\left(\dfrac{\alpha}{2}\right)}{\pi 2^{2-\alpha}\Gamma\left(1-\dfrac{\alpha}{2}\right)}$. By Green's formula, we have

$$\Lambda_\varepsilon^\alpha f(x) = \tilde{c}_\alpha \int_{|x-y|\geqslant\varepsilon} \frac{f(x)-f(y)}{|x-y|^{2+\alpha}}\mathrm{d}y$$

$$+ \int_{|x-y|=\varepsilon} [f(x)-f(y)]\frac{\partial\dfrac{1}{|x-y|^\alpha}}{\partial n}\mathrm{d}S_y$$

$$- \int_{|x-y|=\varepsilon} \frac{1}{|x-y|^\alpha}\frac{\partial(f(x)-f(y))}{\partial n}\mathrm{d}S_y$$

$$= I_1 + I_2 + I_3,$$

where $\tilde{c}_\alpha > 0$ is a constant and n is the unit external normal vector. When $\varepsilon \to 0$,

$$I_2 = \frac{1}{\varepsilon^{\alpha+1}}\int_{|x-y|=\varepsilon}[f(x)-f(y)]\mathrm{d}S_y = O(\varepsilon^{2-\alpha}) \to 0,$$

$$I_3 = \frac{1}{\varepsilon^\alpha}\int_{|x-y|=\varepsilon}\frac{\partial[f(x)-f(y)]}{\partial n}\mathrm{d}S_y = O(\varepsilon^{2-\alpha}) \to 0,$$

and I_1 is what we want, yielding the result.

Proposition 2.2.2 Let $0 < \alpha < 2$, $x \in \mathbf{T}^2$ and $f \in \mathcal{S}$ be a Schwartz function, then

$$\Lambda^\alpha f(x) = c_\alpha \sum_{k\in\mathbf{Z}^2} P.V. \int_{\mathbf{T}^2} \frac{f(x)-f(y)}{|x-y-k|^{2+\alpha}}\mathrm{d}y, \qquad (2.2.4)$$

where $c_\alpha > 0$ is a constant.

Proof From the definition,

$$\Lambda^\alpha f(x) = \sum_{|k|>0} |k|^\alpha \hat{f}(k)\mathrm{e}^{\mathrm{i}k\cdot x} = -\sum_{|k|>0} |k|^\alpha \widehat{\Delta f}(k)\mathrm{e}^{\mathrm{i}k\cdot x}.$$

Let $\chi \in C^\infty$ be a truncated function

$$\chi(x) = \begin{cases} 0, & \text{when } |x| \leqslant 1 \\ 1, & \text{when } |x| \geqslant 2, \end{cases}$$

and $\varphi_\varepsilon(x) = \varepsilon^{-2}\varphi(\frac{x}{\varepsilon})$ be a standard approximation of the identity with

$$0 \leqslant \varphi \leqslant C^\infty, \quad \text{supp}\varphi \subset B_1 \text{ and } \int \varphi = 1.$$

Let $\Phi_\varepsilon(x) = (|x|^{\alpha-2})_\varepsilon * \varphi_\varepsilon(x)$, where $(|x|^{\alpha-2})_\varepsilon = |x|^{\alpha-2} * \chi\left(\dfrac{|x|}{\varepsilon}\right)$, then

$$
\begin{aligned}
\Lambda^\alpha f(x) &= -\lim_{\varepsilon \to 0} \sum \Phi_\varepsilon(k)\widehat{\Delta f}(k)e^{ik\cdot x} \\
&= -\lim_{\varepsilon \to 0}\left(\sum \Phi_\varepsilon(k)e^{ik\cdot x}\right) * \left(\sum \widehat{\Delta f}(k)e^{ik\cdot x}\right).
\end{aligned}
$$

Taking Poisson summation then yields

$$
\begin{aligned}
\Lambda^\alpha f(x) &= -\lim_{\varepsilon \to 0}\left(\sum \widehat{\Phi}_\varepsilon(x-k)\right) * \Delta f(x) \\
&= \lim_{\varepsilon \to 0} \sum \int_{\mathbf{T}^2} \widehat{\Phi}_\varepsilon(x-y-k)\Delta(f(x)-f(y))\mathrm{d}y \\
&= \lim_{\varepsilon \to 0} \sum \int_{\mathbf{T}^2} \Delta(\widehat{\Phi}_\varepsilon)(x-y-k)(f(x)-f(y))\mathrm{d}y. \qquad (2.2.5)
\end{aligned}
$$

Noting

$$
\begin{aligned}
\widehat{\Phi}_\varepsilon(\eta) &= \widehat{(|x|^{\alpha-2})}_\varepsilon(\eta)\cdot\widehat{\varphi}_\varepsilon(\eta) = \widehat{(|x|^{\alpha-2})}_\varepsilon(\eta)\cdot\widehat{\varphi}(\varepsilon\eta), \\
\Delta\widehat{\Phi}_\varepsilon(\eta) &= \Delta((\widehat{|x|^{\alpha-2}})_\varepsilon)(\eta)\cdot\widehat{\varphi}(\varepsilon\eta) + O(\varepsilon), \\
\widehat{(|x|^{\alpha-2})}_\varepsilon(\eta) &= \frac{c_\alpha}{|\eta|^\alpha} - \int e^{-i\eta\cdot x}|x|^{\alpha-2}\left(1-\chi\left(\frac{|x|}{\varepsilon}\right)\right)\mathrm{d}x, \\
\Delta((\widehat{|x|^{\alpha-2}})_\varepsilon)(\eta) &= \frac{\tilde{c}_\alpha}{|\eta|^{\alpha+2}} - \int e^{-i\eta\cdot x}|x|^\alpha\left(1-\chi\left(\frac{|x|}{\varepsilon}\right)\right)\mathrm{d}x,
\end{aligned}
$$

there exists $\delta > 0$ such that

$$
\sum_k \Delta(\widehat{\Phi}_\varepsilon)(y-k) = \tilde{c}_\alpha \sum_k \frac{1}{|y-k|^{\alpha+2}} + O\left(\sum_k \frac{1}{|y-k|^{2+\delta}}O(\varepsilon^\delta)\right).
$$

Substituting this formula into (2.2.5) completes the proof.

The positive property is often useful in PDEs, which was firstly presented by A. Cordoba and D. Cordoba [58], and then extended to the general situation by Ju [120]. See also [57]. First consider the Laplacian $\Delta = (\partial^2_{x_1} + \partial^2_{x_2})$ in $\mathbf{R}^2$, using the chain rule, we obtain

$$
\Delta(f^2) - 2f\Delta f = 2|\nabla f|^2 \geqslant 0, \qquad (2.2.6)
$$

which can also be rewritten in the following form

$$
2f(-\Delta)f \geqslant (-\Delta)(f^2(x)).
$$

This pointwise positivity plays an important role in *a priori* estimates of PDEs, and is often essential. The results derived from Cordoba-Cordoba and Ju extend (2.2.6) to fractional Laplacian.

Lemma 2.2.1 *Let $0 < \alpha < 2$, $x \in \mathbf{R}^2$ or $\mathbf{T}^2$ and $f \in \mathcal{S}$ be a Schwartz function, then there holds the following pointwise estimate*

$$2f\Lambda^\alpha f(x) \geqslant \Lambda^\alpha(f^2)(x). \tag{2.2.7}$$

Proof From Proposition 2.2.1, we have

$$
\begin{aligned}
2f\Lambda^\alpha f(x) &= 2c_\alpha P.V. \int \frac{[f^2(x) - f(y)f(x)]}{|x - y|^{\alpha+2}} \mathrm{d}y \\
&= c_\alpha P.V. \int \frac{[f(x) - f(y)]^2}{|x - y|^{\alpha+2}} \mathrm{d}y + c_\alpha P.V. \int \frac{[f^2(x) - f^2(y)]}{|x - y|^{\alpha+2}} \mathrm{d}y \\
&\geqslant \Lambda^\alpha(f^2)(x).
\end{aligned}
$$

This completes the proof.

Proposition 2.2.3 *Let $0 < \alpha < 2$, $x \in \mathbf{R}^2$ or $\mathbf{T}^2$ and $f, \Lambda^\alpha f \in L^p$ for $p = 2^n$, then there holds*

$$\int |f|^{p-2} f\Lambda^\alpha f \mathrm{d}x \geqslant \frac{1}{p} \int |\Lambda^{\frac{\alpha}{2}}(f^{\frac{p}{2}})|^2 \mathrm{d}x. \tag{2.2.8}$$

Proof The situations for $\alpha = 0$ and $\alpha = 2$ obviously hold. When $0 < \alpha < 2$, repeatedly using (2.2.7), we have

$$
\begin{aligned}
\int |f|^{p-2} f\Lambda^\alpha f \mathrm{d}x &\geqslant \frac{1}{2} \int |f|^{p-2} \Lambda^\alpha f^2 \mathrm{d}x = \frac{1}{2} \int |f|^{p-4} f^2 \Lambda^\alpha f^2 \mathrm{d}x \\
&\geqslant \frac{1}{4} \int |f|^{p-4} \Lambda^\alpha f^4 \mathrm{d}x \geqslant \cdots \geqslant \frac{1}{2^{n-1}} \int |f|^{2^{n-1}} \Lambda^\alpha f^{2^{n-1}} \mathrm{d}x.
\end{aligned}
$$

Using the Parseval's identity, one completes the proof.

Because of the restriction $p = 2^n$, this theorem can not be well applied to many situations. To generalize this result to arbitrary $p \geqslant 2$, we first prove the following lemma, which can be regarded as a generalization of Lemma 2.2.1. See [120].

Lemma 2.2.2 *Assume $\alpha \in [0, 2]$, $\beta + 1 \geqslant 0$ and $f \in \mathcal{S}$, then there holds*

$$|f(x)|^\beta f(x)\Lambda^\alpha f(x) \geqslant \frac{1}{\beta + 2} \Lambda^\alpha |f(x)|^{\beta+2}. \tag{2.2.9}$$

Proof We consider the case $\alpha \in (0, 2)$. Similar to the proof of Lemma 2.2.1, by the Riesz potential, we have

$$\Lambda^\alpha f(x) = c_\alpha P.V. \int \frac{f(x) - f(y)}{|x - y|^{2+\alpha}} \mathrm{d}y,$$

which yields

$$|f(x)|^\beta f(x)\Lambda^\alpha f(x) = c_\alpha P.V. \int \frac{|f(x)|^{\beta+2} - |f(y)|^\beta f(x)f(y)}{|x-y|^{2+\alpha}}dy. \quad (2.2.10)$$

When $\beta + 1 > 0$, by using the Young's inequality,

$$|f(y)|^\beta f(x)f(y) \leqslant |f(x)|^{\beta+1}|f(y)| \leqslant \frac{\beta+1}{\beta+2}|f(x)|^{\beta+2} + \frac{1}{\beta}|f(y)|^{\beta+2},$$

hence

$$|f(x)|^\beta f(x)\Lambda^\alpha f(x) \geqslant c_\alpha \frac{1}{\beta+2}P.V. \int \frac{|f(x)|^{\beta+2} - |f(y)|^{\beta+2}}{|x-y|^{2+\alpha}}dy$$

$$=\frac{1}{\beta+2}\Lambda^\alpha |f(x)|^{\beta+2}.$$

When $\beta + 1 = 0$, directly estimating (2.2.10) yields the conclusion.

Remark 2.2.1   *When $\alpha \in [0,2]$, $\beta, \gamma > 0$, if $f \in \mathcal{S}$ and $f \geqslant 0$, then there holds the pointwise estimate*

$$f^\beta(x)\Lambda^\alpha f^\gamma(x) \geqslant \frac{\gamma}{\beta+\gamma}\Lambda^\alpha f^{\beta+\gamma}(x). \quad (2.2.11)$$

Theorem 2.2.1   *Let $\alpha \in [0,2]$ and $f, \Lambda^\alpha f \in L^p$, then for any arbitrary $p \geqslant 2$, there holds*

$$\int |f|^{p-2}f\Lambda^\alpha f dx \geqslant \frac{2}{p}\int \left(\Lambda^{\frac{\alpha}{2}}|f|^{\frac{p}{2}}\right)^2 dx.$$

Proof   When $\alpha = 0$ or $\alpha = 2$, and $p = 2$, the theorem obviously holds. Let $p > 2$ and $\alpha \in (0,2)$, and assume $f \in \mathcal{S}$. Let $\beta = \frac{p}{2} - 1$, then $\beta + 1 > 0$, using the lemma above, we obtain

$$\int |f(x)|^{p-2}f(x)\Lambda^\alpha f(x)dx = \int |f(x)|^{\frac{p}{2}}|f(x)|^\beta f(x)\Lambda^\alpha f(x)dx$$

$$\geqslant \int \frac{2}{p}|f(x)|^{\frac{p}{2}}\Lambda^\alpha |f(x)|^{\frac{p}{2}}dx$$

$$=\frac{2}{p}\int \left(\Lambda^{\frac{\alpha}{2}}|f|^{\frac{p}{2}}\right)^2 dx.$$

The proof is complete.

These estimates can be developed for complex functions, which are very useful for complex partial differential equations, such as Ginzburg-landau equation and nonlinear Schrödinger equation.

Proposition 2.2.4 *Let $\alpha \in [0,2]$ and $f \in \mathcal{S}$ be complex, then there holds the following pointwise estimate*

$$f^*(x)\Lambda^\alpha f(x) + f(x)\Lambda^\alpha f^*(x) \geqslant \Lambda^\alpha |f|^2(x).$$

Proof   We only consider the case $\alpha \in (0,2)$. From the definition, we have

$$\Lambda^\alpha f(x) = c_\alpha \ P.V. \int_{\mathbf{R}^d} \frac{f(x) - f(y)}{|x-y|^{d+\alpha}} dy, \qquad (2.2.12)$$

hence

$$
\begin{aligned}
&f^*(x)\Lambda^\alpha f(x) + f(x)\Lambda^\alpha f^*(x) \\
&= c_\alpha \ P.V. \int_{\mathbf{R}^d} \frac{(f(x)-f(y))f^*(x) + (f^*(x)-f^*(y))f(x)}{|x-y|^{d+\alpha}} dy \\
&\geqslant c_\alpha \ P.V. \int_{\mathbf{R}^d} \frac{|f(x)|^2 - |f(y)|^2}{|x-y|^{d+\alpha}} dy \\
&= \Lambda^\alpha |f|^2(x).
\end{aligned}
$$

Moreover, this theorem obviously holds for $\alpha = 0, 2$.

Proposition 2.2.5 *Let $\alpha \in [0,2]$, $\beta + 1 \geqslant 0$ and $f \in \mathcal{S}$, then there holds the pointwise estimate*

$$|f(x)|^\beta (f^*(x)\Lambda^\alpha f(x) + f(x)\Lambda^\alpha f^*(x)) \geqslant \frac{2}{\beta+2} \Lambda^\alpha |f(x)|^{\beta+2}.$$

Proof   Using (2.2.12) and the Young's inequality

$$|f(x)|^\beta f^*(x)f(y) \leqslant \frac{\beta+1}{\beta+2}|f(x)|^{\beta+2} + \frac{1}{\beta+2}|f(y)|^{\beta+2},$$

one obtains

$$
\begin{aligned}
&|f(x)|^\beta (f^*(x)\Lambda^\alpha f(x) + f(x)\Lambda^\alpha f^*(x)) \\
&= c_\alpha \ P.V. \int_{\mathbf{R}^d} \frac{2|f(x)|^{\beta+2} - |f(x)|^\beta (f^*(x)f(y) + f(x)f^*(y))}{|x-y|^{d+\alpha}} dy \\
&\geqslant \frac{2}{\beta+2} c_\alpha \ P.V. \int_{\mathbf{R}^d} \frac{|f(x)|^{\beta+2} - |f(y)|^{\beta+2}}{|x-y|^{d+\alpha}} dy \\
&= \frac{2}{\beta+2} \Lambda^\alpha |f|^{\beta+2}(x).
\end{aligned}
$$

When $\alpha = 0, 2$ or $\beta = 0$, the conclusion obviously holds.

Lemma 2.2.3 *Let $\alpha \in [0,2]$, $p \geqslant 2$, $f, \Lambda^\alpha f \in L^p$, then*

$$\int |f|^{p-2}(f^* \Lambda^\alpha f + f \Lambda^\alpha f^*) dx \geqslant \frac{4}{p} \int (\Lambda^{\frac{\alpha}{2}} |f|^{\frac{p}{2}})^2 dx.$$

Proof   When $p > 2$ and $\alpha \in (0, 2)$,

$$\int |f(x)|^{p-2}(f^*\Lambda^\alpha f + f\Lambda^\alpha f^*)\mathrm{d}x = \int |f(x)|^{\frac{p}{2}}|f(x)|^{\frac{p}{2}-2}(f^*\Lambda^\alpha f + f\Lambda^\alpha f^*)\mathrm{d}x$$

$$\geqslant \frac{4}{p}\int |f(x)|^{\frac{p}{2}}\Lambda^\alpha |f|^{\frac{p}{2}}(x)\mathrm{d}x$$

$$= \frac{4}{p}\int (|\Lambda^{\frac{\alpha}{2}}|f|^{\frac{p}{2}})^2\mathrm{d}x.$$

For $\alpha = 0, 2$ or $p = 2$, the theorem obviously holds.

2.2.2   Pseudo-differential operator

The research of pseudo-differential operators (PsDO) started with the work of Kohn and Nirenberg in 1960s, cf. [127]. Before this, the work on PsDO focused on singular integral and Fourier analysis; after this, the PsDO are widely popularized, among which, Hörmander's work is striking. At present, the theory of PsDO becomes a powerful tool in partial differential equations with variable coefficients and distributions of singularity set, especially in the field of PDEs. This section simply introduces some concepts and properties of PsDO. For more details, readers are referred to monographs [9, 85, 113, 114, 186, 204, 212].

The function $a \in C^\infty$ is called a slowly increasing function, if there holds

$$\forall \alpha \in (\mathbf{N})^d, \exists M_\alpha \in \mathbf{N}, \exists C_\alpha > 0, s.t. |\partial^\alpha a(x)| \leqslant C_\alpha(1 + |x|)^M_\alpha, \forall x \in \mathbf{R}^d.$$

For a slowly increasing function $a(\xi)$, define the operator $a(D)$ in $\mathcal{S}'(\mathbf{R}^d)$ by $\widehat{(a(D)u)}(\xi) = a(\xi)\hat{u}(\xi)$, where $a(\xi)$ is called the symbol of the operator $a(D)$. Using the Fourier transform, for $u \in \mathcal{S}$, one gets

$$(a(D)u)(x) = \frac{1}{(2\pi)^d}\int e^{ix\cdot\xi}a(\xi)\hat{u}(\xi)\mathrm{d}\xi.$$

Taking into account the inverse Fourier transform, one can see that, at the frequency ξ, the effect of $a(D)$ is multiplying the complex amplitude $\hat{u}(\xi)$ by the coefficient $a(\xi)$ in the phase space. More generally, one can generalize $a(\xi)$ to a function $a(x, \xi)$ depending on x, and leads to the following definition.

Definition 2.2.1   *The PsDO is defined as a mapping $u \mapsto T_a u$ by the following*

$$(T_a u)(x) = a(x, D)u(x) = \frac{1}{(2\pi)^d}\int_{\mathbf{R}^d} e^{ix\cdot\xi}a(x, \xi)\hat{u}(\xi)\mathrm{d}\xi, \qquad (2.2.13)$$

where

$$\hat{u}(\xi) = \int_{\mathbf{R}^d} u(x)e^{-ix\cdot\xi}\mathrm{d}x$$

is the Fourier transform of u and $a(x, \xi)$ is called the symbol of the operator $a(x, D)$.

For the definition above, generally, some additional conditions should be included for $a(x, \xi)$, which leads to the definition of the symbol class, denoted by S^m.

Definition 2.2.2 *Let $m \in \mathbf{R}$, a function a belongs to S^m and is said to be of order m, if $a(x, \xi) \in C^\infty(\mathbf{R}^d \times \mathbf{R}^d)$ and satisfies the differential inequalities*

$$|\partial_x^\alpha \partial_\xi^\beta a(x, \xi)| \leqslant C_{\alpha,\beta}(1 + |\xi|)^{m-|\beta|}, \tag{2.2.14}$$

for all multi-indices α and β. Define $S^{-\infty} = \bigcap_m S^m$.

Remark 2.2.2 *We can also define a more general symbol class $S^m_{\rho,\delta}$. Let $\rho, \delta \in [0, 1], m \in \mathbf{R}, S^m_{\rho,\delta}$ is defined as the set of functions C^∞ satisfying*

$$|\partial_x^\alpha \partial_\xi^\beta a(x, \xi)| \leqslant C_{\alpha,\beta}\langle\xi\rangle^{m-\rho|\beta|+\delta\alpha},$$

for all multi-indices α and β, where $\langle\xi\rangle = (1 + |\xi|^2)^{1/2}$.

Example 2.2.1 1. *The symbol of the Laplacian operator $\Delta = \partial_1^2 + \cdots + \partial_d^2$ is $a(\xi) = -|\xi|^2$;*

2. *The symbol of the fractional Laplacian operator $(-\Delta)^{\alpha/2}$ is $a(\xi) = |\xi|^\alpha$;*

3. *The symbol of partial differential operator $L = \sum\limits_{|\alpha|\leqslant m} a_\alpha(x)\partial_x^\alpha$ for $a_\alpha \in$*

$C^\infty(\mathbf{R}^d)$ *is $a(x, \xi) = \sum\limits_{|\alpha|\leqslant m} a_\alpha(x)(i\xi)^\alpha$, and a is called a differential symbol.*

4. *If $\varphi \in \mathcal{S}$, then $\varphi(\xi) \in S^{-\infty}$;*

5. *The function $a(x, \xi) = e^{ix\cdot\xi}$ is not a symbol.*

It is easy to see that the symbol of the differential operator L is the characteristic polynomials of L. In particular, if $a(x, \xi) = a_1(\xi)$ does not depend on x, then $a(x, D) = a(D)$ is a multiplier operator $\widehat{a(D)u}(\xi) = a_1(\xi)\hat{u}(\xi)$. If $a(x, \xi) = a_2(x)$ does not depend on ξ, then $a(x, D)$ is reduced to a multiplication operator $(a_2(x, D)u)(x) = a_2(x)u(x)$.

For a given symbol $a \in S^m$, it is not difficult to show that the operator T_a maps $\mathcal{S}$ to itself. Firstly, if $u \in \mathcal{S}$, then the integral (2.2.13) is absolutely convergent, and $T_a u$ is infinitely differentiable. In fact, $T_a u$ is also rapidly decreasing. Noting that $(I - \Delta_\xi)e^{ix\cdot\xi} = (1 + |x|^2)e^{ix\cdot\xi}$, one can define an invariant derivative operator $L_\xi = (1+|x|^2)^{-1}(I-\Delta_\xi)$ such that $(L_\xi)^N e^{ix\cdot\xi} = e^{ix\cdot\xi}$. Substituting this formula into (2.2.13), and integrating by parts, we have

$$(T_a u)(x) = \frac{1}{(2\pi)^d} \int (L_\xi)^N [a(x, \xi)\hat{u}(\xi)]e^{ix\cdot\xi}.$$

Hence $T_a u$ is rapidly decreasing. From this, one can show that T_a maps $\mathcal{S}$ to itself, and the mapping is continuous. Indeed, if $\{a_k\}$ satisfies the inequality (2.2.14) uniformly and is pointwise convergent to the symbol $a \in S^m$ in S^m, then $T_{a_k}(u) \to T_a(u)$ is convergent in $\mathcal{S}$, where $u \in \mathcal{S}$.

We also hope that T_a can be extended to more wide function class $\mathcal{S}'$. The definition (2.2.13) can be rewritten in the following form

$$(T_a u)(x) = \frac{1}{(2\pi)^d} \iint a(x,\xi) e^{i\xi \cdot (x-y)} u(y) dy d\xi. \qquad (2.2.15)$$

However, even if $f \in \mathcal{S}$, this integral is not necessarily absolutely convergent. To avoid such a situation, we can use symbols with compact support to approximate a general symbol. Fix $\gamma \in C_c^\infty(\mathbf{R}^d \times \mathbf{R}^d)$, and $\gamma(0,0) = 1$. Let $a_\varepsilon(x,\xi) = a(x,\xi) \gamma(\varepsilon x, \varepsilon \xi)$, then if $a \in S^m$, we have $a_\varepsilon \in S^m$ and satisfies the inequality (2.2.14) uniformly for $0 < \varepsilon \leqslant 1$. On the other hand, from the definition of $T_a u$, $T_{a_\varepsilon}(u) \to T_a(u)$ in $\mathcal{S}$ as $\varepsilon \to 0$ for arbitrary $u \in \mathcal{S}$, denoted by $T_{a_\varepsilon} \to T_a$. In this case, for the symbol a with compact support, the integral (2.2.15) is absolutely convergent, and

$$(T_a u)(x) = \lim_{\varepsilon \to 0} \frac{1}{(2\pi)^d} \iint a_\varepsilon(x,\xi) e^{i\xi \cdot (x-y)} u(y) dy d\xi.$$

Consider the integral expression of a PsDO. The purpose is to derive the kernel function of a PsDO. First assume $a \in S^{-\infty}$, one obtains for $u \in \mathcal{S}$

$$\begin{aligned}(T_a u)(x) &= \frac{1}{(2\pi)^d} \iint a(x,\xi) e^{i\xi \cdot (x-y)} u(y) dy d\xi \\ &= \frac{1}{(2\pi)^d} \int u(y) dy \int e^{i\xi \cdot (x-y)} a(x,\xi) d\xi.\end{aligned}$$

The kernel function K of the operator T_a can then be given by the following oscillatory integral

$$K(x,y) = \frac{1}{(2\pi)^d} \int e^{i\xi \cdot (x-y)} a(x,\xi) d\xi = (\mathcal{F}_\xi^{-1} a)(x-y),$$

where $\mathcal{F}_\xi^{-1}$ represents the inverse Fourier transform with respect to ξ. $K(x,y)$ is called the Schwartz kernel of the operator $T_a = a(x,D)$.

Proposition 2.2.6 $K(x,y)$ is smooth away from the diagonal $\Delta = \{(x,y) \in \mathbf{R}^d \times \mathbf{R}^d : x = y\}$ and

$$|K(x,y)| \leqslant A_N |x-y|^{-N}, \qquad \forall |x-y| \geqslant 1, \forall N > 0. \qquad (2.2.16)$$

Proof For any arbitrary $\alpha \geqslant 0$,

$$(x-y)^\alpha K(x,y) = \frac{1}{(2\pi)^d} \int e^{i\xi \cdot (x-y)} D_\xi^\alpha a(x,\xi) d\xi,$$

where $D^\alpha = D_1^{\alpha_1} \cdots D_d^{\alpha_d}$, $D_j = -i\partial_{x_j}$. From the definition of the symbol class S^m, we can see that when $|\alpha| > m+d$, this integral is absolutely convergent, hence $(x-y)^\alpha K$ is continuous. Similarly, evaluate j-th derivative of the formula above, as long as $|\alpha| \geqslant m+j+d$, and then the integral is absolutely convergent, and $(x-y)^\alpha K \in C^j(\mathbf{R}^d \times \mathbf{R}^d)$. Simultaneously, there exists a constant $A_\alpha > 0$ such that $|x-y|^\alpha |K(x-y)| \leqslant A_\alpha$, where $|\alpha| > m+d$. In particular, (2.2.16) holds.

For a operator A mapping $\mathcal{S}$ to itself, we can define an operator A^* mapping $\mathcal{S}$ to itself, such that $\langle Au, v \rangle = \langle u, A^*v \rangle$ for all $u, v \in \mathcal{S}$. By density argument, one can see that if A^* exists, then it is unique. Such an operator A^* is called the adjoint operator of A. In the situation of the PsDO defined by (2.2.13), the dual operator of T_a can be defined as the operator T_a^* such that

$$\langle T_a u, v \rangle = \langle u, T_a^* v \rangle, \quad \forall u, v \in \mathcal{S}. \tag{2.2.17}$$

Noting that $\langle u, v \rangle = \int u(x)\overline{v(x)}dx$, one immediately has

$$(T_a^* v)(y) = \lim_{\varepsilon \to 0} \frac{1}{(2\pi)^d} \int \int \overline{a_\varepsilon(x,\xi)} e^{i(y-x)\cdot\xi} v(x) dx d\xi.$$

Using the invariant derivative, it is not difficult to verify T_a^* maps $\mathcal{S}$ to itself. Hence, using the duality (2.2.17), one can extend T_a to a continuous mapping which maps $\mathcal{S}'$ to itself $\mathcal{S}'$.

The boundness estimate of operators is a key problem in the theory of PDEs, many important results of which ultimately attribute to the boundness of an operator in a certain norm.

Theorem 2.2.2 *Let $a \in S^0$, then the PsDO $T_a = a(x, D)$ satisfies*

$$\|T_a(u)\|_{L^2} \leqslant A\|u\|_{L^2}, \quad \forall u \in \mathcal{S}. \tag{2.2.18}$$

Hence T_a can be extended to a bound operator mapping L^2 to itself.

Proof The proof is divided into three steps. First, assume $a(x,\xi)$ is compactly supported in x. Integrating by parts yields

$$(i\lambda)^\alpha \hat{a}(\lambda, \xi) = \int_{\mathbf{R}^\mu} \partial_x^\alpha a(x,\xi) e^{-ix\cdot\lambda} dx$$

and $|(i\lambda)^\alpha \hat{a}(\lambda, \xi)| \leqslant C_\alpha$ uniformly in ξ. Therefore for arbitrary $N \geqslant 0$,

$$\sup_\xi |\hat{a}(\lambda, \xi)| \leqslant A_N (1 + |\lambda|)^{-N}. \tag{2.2.19}$$

On the other hand,

$$\begin{aligned}
(T_a u)(x) &= (2\pi)^{-d} \int a(x, \xi) e^{ix \cdot \xi} \hat{u}(\xi) d\xi \\
&= (2\pi)^{-2d} \int \int \hat{a}(\lambda, \xi) e^{i\lambda \cdot x} e^{ix \cdot \xi} \hat{u}(\xi) d\lambda d\xi \\
&= \int (T^\lambda u)(x) d\lambda,
\end{aligned}$$

where $(T^\lambda u)(x) = (2\pi)^{-d} e^{i\lambda \cdot x} (T_{\hat{a}(\lambda, \xi)} u)(x)$. For a fixed λ, $T_{\hat{a}(\lambda, \xi)}$ is a multiplier operator, yielding from the Plancherel's theorem

$$\|T_{\hat{a}(\lambda, \xi)} u\|_{L^2} \leqslant \sup_\xi |\hat{a}(\lambda, \xi)| \cdot \|\hat{u}\|_{L^2} = (2\pi)^d \sup_\xi |\hat{a}(\lambda, \xi)| \cdot \|u\|_{L^2}.$$

Using (2.2.19), we have $\|T^\lambda\| \leqslant (2\pi)^d A_N (1 + |\lambda|)^{-N}$. From $T_a = \int T^\lambda d\lambda$, letting $N > d$ yields

$$\|T_a\| \leqslant A_N \int (1 + |\lambda|)^{-N} d\lambda < \infty.$$

Secondly, we show the following auxiliary conclusion. For arbitrary $x_0 \in \mathbf{R}^d$,

$$\int_{|x - x_0| \leqslant 1} |(T_a u)(x)|^2 dx \leqslant A_N \int_{\mathbf{R}^d} \frac{|u(x)|^2}{(1 + |x - x_0|)^N} dx, \quad \forall N \geqslant 0. \tag{2.2.20}$$

Let $x_0 = 0$ and $B(r) = B(0, r)$ be the ball of radius r, centered at the origin in $\mathbf{R}^d$. Decompose $u = u_1 + u_2$ such that $supp(u_1) \subset B(3)$ and $supp(u_2) \subset B(2)^c$, for smooth functions u_1 and u_2 with $|u_1|, |u_2| \leqslant |u|$. Fix $\eta \in C_c^\infty$ such that $\eta \equiv 1$ in $B(1)$, then $\eta T_a(u_1) = T_{\eta a}(u_1)$ in $B(1)$ and $\eta(x) a(x, \xi)$ has compact support in x. Using the results of the first step, one has

$$\int_{B(1)} |T_a u_1|^2 \leqslant \int_{\mathbf{R}^d} |T_{\eta a} u_1|^2 \leqslant A \int_{\mathbf{R}^d} |u_1|^2 \leqslant A \int_{B(3)} |u|^2. \tag{2.2.21}$$

For u_2, using the Schwartz kernel to obtain

$$(T_a u_2)(x) = \int_{B(2)^c} K(x, y) u_2(y) dy.$$

When $x \in B(1)$, we have $|x - y| \geqslant 1$ for $y \in B(2)^c$ and hence there exists a constant such that $|x - y| \geqslant c(1 + |y|)$. Using Proposition 2.2.6, we obtain

$$|(T_a u_2)(x)| \leqslant A \int_{B(2)^c} |u(y)||x - y|^{-N} dy$$

$$\leqslant A_N \int |u(y)|(1 + |y|)^{-N} dy.$$

Letting $N > n$ and using the Schwartz's inequality to obtain

$$\int_{B(1)} |(T_a u_2)(x)|^2 dx \leqslant A \int \frac{|u(x)|^2}{(1 + |x|)^N} dx. \qquad (2.2.22)$$

Combining (2.2.21) and (2.2.22), (2.2.20) holds when $x_0 = 0$.

When $x_0 \neq 0$, let τ_h be the translation operator such that $(\tau_h u)(x) = u(x - h)$ for $h \in \mathbf{R}^d$. Then it is easy to verify $\tau_h T_a \tau_{-h} = T_{a_h}$, where $a_h(x, \xi) = a(x - h, \xi)$. Since a_h and a satisfy the same estimate in (2.2.14) independent of h, (2.2.20) also holds for a_h independent of h. Setting $h = x_0$, we see that (2.2.20) holds and the coefficient A_N is independent of x_0.

Finally, we prove (2.2.18) without assuming that $a(x, \xi)$ is compactly supported in x. Integrating (2.2.20) in $\mathbf{R}^d$ with respective to x_0 and exchanging the orders of integration, one obtains

$$|B(1)| \int |(T_a u)(x)|^2 dx \leqslant A_N \iint \frac{|u(x)|^2}{(1 + |x - x_0|)^N} dx dx_0 \leqslant A\|u\|_{L^2}^2,$$

i.e.,

$$\|T_a u\|_{L^2} \leqslant A\|u\|_{L^2},$$

completing the proof of the theorem.

Theorem 2.2.3 *If $a_1 \in S^{m_1}, a_2 \in S^{m_2}$, then there exists a symbol $b \in S^{m_1 + m_2}$ such that $T_b = T_{a_1} \circ T_{a_2}$, and $b \sim \sum_{\alpha} \frac{1}{\alpha!} \partial_\xi^\alpha a_1 \partial_x^\alpha a_2$.*

The proof is omitted here. After a simple calculation, b can be given by ([212, Vol.II])

$$b(x, \xi) = (2\pi)^{-d} \int e^{-i(x-y)(\xi-\eta)} a_1(x, \eta) a_2(y, \xi) dy d\eta.$$

From the boundness of S^0 in L^2, it is easy to show

Corollary 2.2.1 *Let $a \subset S^m$, then $T_a = u(x, D) : H^s(\mathbf{R}^d) \to H^{s-m}(\mathbf{R}^d)$ defined by a is a bounded linear operator.*

Proof By definition, the symbol of $\mathcal{J}^m = (I - \Delta)^{-m/2}$ is $\langle\xi\rangle e^{-m} \in S^{-m}$ and $\mathcal{J}^{-m} : H^s(\mathbf{R}^d) \to H^{s-m}(\mathbf{R}^d)$ is a bounded linear operator. By symbolic calculus, there exists $b \in S^0$ such that $T_b = T_a \circ \mathcal{J}^m : H^{s-m} \to H^{s-m}$ is a bounded linear operator. Hence $T_a = T_b \circ \mathcal{J}^{-m} : H^s \to H^{s-m}$ is a bounded linear operator.

From the theory of singular integral, the following conclusions can be drawn, cf. [204, 212].

Theorem 2.2.4   *Let $a \in S^0$, then T_a can be extended to a bounded linear operator from $L^p(1 < p < \infty)$ to itself. Similarly, if $a \in S^m$, then $T_a : W^{s,p} \to W^{s-m,p}$ is a bounded linear operator.*

Theorem 2.2.5   *Let $\sigma \in C^\infty(\mathbf{R}^d \times \mathbf{R}^d - (0,0))$ satisfy*

$$|\partial_\xi^\alpha \partial_\eta^\beta \sigma(\xi, \eta)| \leqslant C_{\alpha,\beta}(|\xi| + |\eta|)^{-(|\alpha|-|\beta|)}, \quad \forall (\xi, \eta) \neq (0,0), \alpha, \beta \in (\mathbf{Z}^+)^d. \tag{2.2.23}$$

Let $\sigma(D)$ be the following bilinear operator

$$\sigma(D)(a, h)(x) = \iint e^{i\langle x, \xi + \eta\rangle} \sigma(\xi, \eta) \hat{a}(\xi) \hat{h}(\eta) d\xi d\eta,$$

then

$$\|\sigma(D)(a, h)\|_2 \leqslant C\|a\|_\infty \|h\|_2.$$

Remark 2.2.3   *For the proof of this theorem, readers can refer to [48, p.154]. The result can also be generalized to $L^p(1 < p < \infty)$, see the literature [62, p.382]. Indeed, they verified that when $a(\cdot)$ is fixed, the linear operator $T(\cdot) = \sigma(D)(a, \cdot)$ is a Calderon-Zygmund operator, and the norm can be bounded by $C\|a\|_\infty$. In this theorem, we only need to assume (2.2.23) for $|\alpha|, |\beta| \leqslant k$, where k only depends on m, q. Without loss of generality, we can assume $k \geqslant m$.*

2.2.3   Riesz potential and Bessel potential

Both Riesz potential and Bessel potential are often used in PDEs. For the sake of completeness, we simply introduce Riesz potential and Bessel potential in the case of $\mathbf{R}^d$, for the further discussion, One can refer to Stein's monograph [203] and Miao's monograph [59]. Denote $\mathcal{I}_d = (-\Delta)^{-\frac{1}{2}}$ and $\mathcal{J}_d = (I - \Delta)^{-\frac{1}{2}}$.

Definition 2.2.3   *The Riesz potential of f can be defined as*

$$\mathcal{I}_d^\alpha f = (-\Delta)^{-\frac{\alpha}{2}} f(x) = \frac{1}{\gamma(\alpha)} \int_{\mathbf{R}^d} |x - y|^{-d+\alpha} f(y) dy, \quad n > \alpha > 0, \tag{2.2.24}$$

where

$$\gamma(\alpha) = \pi^{d/2} 2^\alpha \Gamma\left(\frac{\alpha}{2}\right) / \Gamma\left(\frac{d}{2} - \frac{\alpha}{2}\right).$$

The Bessel potential of f can be defined as

$$\mathcal{J}_d^\alpha f = (I - \Delta)^{-\frac{\alpha}{2}} f(x) = G_\alpha * f = \int_{\mathbf{R}^d} G_\alpha(x - y)f(y)\mathrm{d}y, \quad \alpha > 0,$$

where

$$G_\alpha(x) = \frac{1}{(4\pi)^{\alpha/2}} \frac{1}{\Gamma(\alpha/2)} \int_0^\infty e^{-\pi|x|^2/\delta} e^{-\delta/(4\pi)} \delta^{\frac{-d+\alpha}{2}} \frac{\mathrm{d}\delta}{\delta}.$$

Theorem 2.2.6 *Let $0 < \alpha < d$, then*

1. for any arbitrary $\varphi \in \mathcal{S}(\mathbf{R}^d)$, we have

$$\int_{\mathbf{R}^d} |x|^{-d+\alpha} \overline{\varphi(x)} \mathrm{d}x = \int_{\mathbf{R}^d} \gamma(\alpha)(2\pi|x|)^{-\alpha} \overline{\hat{\varphi}}(x)\mathrm{d}x.$$

Namely, in the sense of $\mathcal{S}'$, $\mathcal{F}(|x|^{-d+\alpha}) = \gamma(\alpha)(2\pi)^{-\alpha}|x|^{-\alpha}$

2. for any arbitrary $f, g \in \mathcal{S}(\mathbf{R}^d)$, then

$$\int_{\mathbf{R}^d} \mathcal{I}_d^\alpha(f)\bar{g}(x)\mathrm{d}x = \int_{\mathbf{R}^d} (2\pi|x|)^{-\alpha} \hat{f}(x)\bar{\hat{g}}(x)\mathrm{d}x.$$

Namely in the sense of $\mathcal{S}'$, $\widehat{\mathcal{I}_d^\alpha f}(x) = (2\pi)^{-\alpha}|x|^{-\alpha}\hat{f}(x)$.

From the theorem, the following two further identities can be obtained, which reflect essential properties of the Riesz operator I_d^α,

$$\mathcal{I}_d^\alpha(\mathcal{I}_d^\beta f) = \mathcal{I}_d^{\alpha+\beta} f, \quad \forall f \in \mathcal{S}, \alpha > 0, \beta > 0, \alpha + \beta < d,$$

$$\Delta(\mathcal{I}_d^\alpha f) = \mathcal{I}_d^\alpha(\Delta f) = -\mathcal{I}_d^{\alpha-2}(f), \quad \forall f \in \mathcal{S}, d > 3, 2 \leqslant \alpha \leqslant d.$$

Theorem 2.2.7 *Let $0 < \alpha < d$, $1 \leqslant p \leqslant q < \infty$, $1/q = 1/p - \alpha/d$, then*

1. if $f \in L^p(\mathbf{R}^d)$, then the integral defined by (2.2.24) is absolutely convergent for a.e. $x \in \mathbf{R}^d$,

2. if $1 < p$, then

$$\|\mathcal{I}_d^\alpha(f)\|_q \leqslant C_{p,q}\|f\|_p, \tag{2.2.25}$$

3. if $f \in L^1(\mathbf{R}^d)$, then $m\{x : |\mathcal{I}_d^\alpha| > \lambda\} \leqslant \left(\frac{C\|f\|_1}{\lambda}\right)^q$ holds for any arbitrary $\lambda > 0$. Namely $\mathcal{I}_d^\alpha$ is of weak type $(1, q)$.

The condition $1/q = 1/p - \alpha/d$ can be obtained by scaling, the treatment can be referred to Appendix A. In fact, if (2.2.25) holds for f, then this formula also holds for $g(x) = f(x/\delta)$ and

$$\|\mathcal{I}_d^\alpha(g)\|_q \leqslant C_{p,q}\|g\|_p. \tag{2.2.26}$$

However, in this case

$$\|g\|_p = \delta^{\frac{d}{p}}\|f\|_p, \quad \|\mathcal{I}_d^\alpha(g)\|_q = \delta^{\alpha+\frac{d}{q}}\|\mathcal{I}_d^\alpha(f)\|_q,$$

hence the necessary condition for (2.2.26) is $1/q = 1/p - \alpha/d$. The inequality (2.2.25) is also called the Hardy-Littlewood-Sobolev (HLS) inequality [204].

2.2.4   Fractional Sobolev space

Let $\Omega \subset \mathbf{R}^d$ be a smooth domain of $\mathbf{R}^d$, define the Sobolev norm $\| \cdot \|_{m,p}$ as follows. When $1 \leqslant p < \infty$, we define

$$\|u\|_{m,p} := \left(\sum_{0 \leqslant |\alpha| \leqslant m} \|D^\alpha u\|_p^p \right)^{1/p}$$

and when $p = \infty$, we define

$$\|u\|_{m,\infty} := \max_{0 \leqslant |\alpha| \leqslant m} \|D^\alpha u\|_\infty,$$

where m is a positive integer and $\|u\|_p$ is the L^p norm of u. For arbitrary positive integer m and $1 \leqslant p \leqslant \infty$, $W^{m,p}$ is defined by

$$W^{m,p}(\Omega) = \{ u \in L^p(\Omega) : \ D^\alpha u \in L^p(\Omega), \forall \, 0 \leqslant |\alpha| \leqslant m \}, \qquad (2.2.27)$$

where D represents the weak derivative. The space $W^{m,p}$ is a Banach space under the norm $\| \cdot \|_{m,p}$. When $p = 2$, $W^{m,p}$ is a separable Hilbert space under the inner product

$$(u,v)_m = \sum_{0 \leqslant |\alpha| \leqslant m} (D^\alpha u, D^\alpha v),$$

where $(u,v) := \int_\Omega u(x)\overline{v(x)}\mathrm{d}x$ is the inner product on $L^2(\Omega)$.

Another approach to introduce Sobolev space is to consider the completeness of the class of smooth functions under certain norm. For arbitrary positive integer m and $1 \leqslant p \leqslant \infty$, define $H^{m,p}$ to be the completeness of the space $C^\infty(\Omega)$ under the norm $\| \cdot \|_{m,p}$. However, we can prove that when $1 \leqslant p < \infty$, $H^{m,p} = W^{m,p}$, cf. Meyers and Serrin [163]. It also shows that the space $C^\infty(\Omega)$ is dense in $W^{m,p}(\Omega)$. In particular, when $\Omega = \mathbf{R}^d$, $C_c^\infty(\mathbf{R}^d)$ is still dense in $W^{m,p}(\mathbf{R}^d)$.

It is worth noting that, this conclusion does not hold when $p = \infty$. A simple example is given in [3]. Let $\Omega = \{x \in \mathbf{R}, -1 < x < 1\}$ and $u(x) = |x|$. In this case, when $x \neq 0$, $u'(x) = x/|x|$, hence $u \in W^{1,\infty}$. However, such u does not belong to $H^{1,\infty}$. In fact, for arbitrary $0 < \varepsilon < \dfrac{1}{2}$, there does not exist $\phi \in C^\infty$ such that $\|\phi' - u'\|_\infty < \varepsilon$, since L^∞ is not separable.

The Sobolev space satisfies the following embedding theorem, which plays an important role in PDEs, whose proof can be found in [3, 203]. For more details on embedding theorems and embedding inequalities, one may refer to [3, 89, 214, 232].

Theorem 2.2.8 *Let m be a positive integer, and $1/q = 1/p - m/d$, then*

 1. *if $q < \infty$, then $W^{m,p}(\mathbf{R}^d) \hookrightarrow L^q(\mathbf{R}^d)$ continuously.*

 2. *if $q = \infty$, then the restrictions on any arbitrary compact set $\mathbf{R}^d$ of the functions in $W^{m,p}$ all belong to $L^r(\mathbf{R}^d)$ for all $r < \infty$.*

 3. *if $p > \dfrac{d}{k}$, then after possibly modifying the function on a null set, $f \in W^{m,p}(\mathbf{R}^d)$ is a continuous function.*

We consider the Fourier characterization of functions in $H^1(\mathbf{R}^d) = W^{1,2}(\mathbf{R}^d)$. Let $\hat{f}$ be the Fourier transform of $f \in L^2(\mathbf{R}^d)$, then $f \in H^1(\mathbf{R}^d)$ if and only if $|\xi|\hat{f}(\xi) \in L^2(\mathbf{R}^d)$. In this case, $\widehat{\nabla f}(\xi) = i\xi\hat{f}(\xi)$ holds and

$$\|f\|_{H^1(\mathbf{R}^d)}^2 \sim \int_{\mathbf{R}^d} (1 + |\xi|^2)|\hat{f}(\xi)|^2 \mathrm{d}\xi. \qquad (2.2.28)$$

Indeed, if $f \in H^1$, there exists a sequence of functions $\{f_k\}_{k=1}^\infty$ in C_c^∞ such that f_k converges to f in H^1. For f_k, integrating by parts, we have $\widehat{\nabla f_k}(\xi) = i\xi\hat{f}_k(\xi)$. From the Plancherel's theorem, we can see that $\hat{f}_k$ and $\widehat{\nabla f_k}$ converge to $\hat{f}$ and $\widehat{\nabla f}$ in L^2, respectively. On the other hand, up to a subsequence, $\xi\hat{f}_k(\xi)$ and $i\xi\hat{f}_k(\xi)$ converge to $\xi\hat{f}(\xi)$ and $\widehat{\nabla f}(\xi)$ a.e.. Therefore, $\widehat{\nabla f}(\xi) = i\xi\hat{f}(\xi)$. By Plancherel theorem, (2.2.28) is obviously established.

The Fourier transform can also well depict the integer order Sobolev space. It is easy to show that the following two norms are equivalent

$$\left[\sum_{|\alpha| \leqslant m} \|\partial^\alpha f\|_{L^2}^2 \right]^{1/2} \sim \left[\int (1 + |\xi|^2)^m |\hat{f}(\xi)|^2 \mathrm{d}\xi \right]^{1/2}.$$

Hence $f \in H^m(\mathbf{R}^d)$ if and only if $(1 + |\cdot|^2)^{\frac{m}{2}}\hat{f}(\cdot) \in L^2(\mathbf{R}^d)$. In another perspective, $H^m(\mathbf{R}^d)$ is nothing but $L^2(\mathbf{R}^d)$ with the *usual* Lebesgue measure replaced by $(1 + |\xi|^2)^m \mathrm{d}\xi$. Using Fourier transform, it is easy to define fractional Sobolev space. When $p = 2$, the Sobolev space H^s of order s can be defined as

$$H^s = H^s(\mathbf{R}^d) = \{f \in \mathcal{S}'(\mathbf{R}^d) : \hat{f} \text{ is a function and } \|f\|_{H^s}^2 < \infty\}, \quad (2.2.29)$$

where

$$\|f\|_{H^s}^2 := \int_{\mathbf{R}^d} (1 + |\xi|^2)^s |\hat{f}(\xi)|^2 \mathrm{d}\xi < \infty.$$

Obviously $H^0 = L^2$. It is easy to verify that H^s is a Banach space as well as a Hilbert space under the inner product

$$\langle f, g \rangle = \int \hat{f}(\xi)\overline{\hat{g}(\xi)}(1 + |\xi|^2)^s \mathrm{d}\xi.$$

When $p \neq 2$, the definition of $W^{s,p}$ space is more complex. When Ω is a smooth domain of $\mathbf{R}^d$, $W^{s,p}$ is defined by complex interpolation. Let $s > 0$, $m = [s] + 1$ be the smallest integer greater than s, define

$$W^{s,p}(\Omega) = [L^p(\Omega), W^{m,p}(\Omega)]_{s/m}.$$

We have the following description. Let $s = [s] + \lambda$ with $0 < \lambda < 1$, the fractional Sobolev space can be defined as the completion of the set

$$\left\{ u \in C^{\infty}(\Omega) : \frac{|\partial^{\alpha}(u(x) - u(y))|}{|x - y|^{\frac{d}{p} + \lambda}} \in L^p(\Omega \times \Omega), \ \ \forall \alpha \in (\mathbf{Z} \cup \{0\})^n, |\alpha| = [s] \right\}$$

under the norm

$$\|u\|_{W^{s,p}(\Omega)} = \|u\|_{W^{[s,p]}(\Omega)} + \left(\sum_{|\alpha|=[s]} \int_{\Omega \times \Omega} \frac{|\partial^{\alpha}(u(x) - u(y))|^p}{|x - y|^{d + p\lambda}} \mathrm{d}x \mathrm{d}y \right)^{1/p}.$$

When $s = m$ is a positive integer, such $W^{s,p}(\Omega)$ and the integer order Sobolev space $W^{m,p}$ defined by (2.2.27) is equivalent.

When $\Omega = \mathbf{R}^d$, the fractional Sobolev space $W^{s,p}$ is defined as

$$W^{s,p} := \{f \in \mathcal{S}' : \text{ there exists } g \in L^p(\mathbf{R}^d) \text{ such that } (1 + |\cdot|^2)^{s/2} \hat{f}(\cdot) = \hat{g}(\cdot)\},$$

with norm $\|f\|_{W^{s,p}} = \|(I - \Delta)^{s/2} f\|_{L^p}$. Such a space is also called Bessel potential space. When $s = m$ is a positive integer, this definition reduces to the ordinary Sobolev space. When $p = 2$, it reduces to the fractional Sobolev space defined by (2.2.29).

The norm $\| \cdot \|_{W^{s,p}}$ is well-defined. For this purpose, we only need to show that if $\mathcal{J}_d^s(g_1) = \mathcal{J}_d^s(g_2)$, then $g_1 = g_2$. In fact, for arbitrary $\varphi \in \mathcal{S}$, by Fubini's theorem

$$\int \mathcal{J}_d^s(g)\varphi(x)\mathrm{d}x = \iint G_s(x - y)g(y)\varphi(x)\mathrm{d}x\mathrm{d}y = \int g\mathcal{J}_d^s(\varphi)\mathrm{d}x.$$

On the other hand, the map $\mathcal{J}_d^s : \mathcal{S} \to \mathcal{S}$ is surjective. For a given $\psi \in \mathcal{S}$, let $\hat{\varphi}(\xi) = \hat{\psi}(\xi)(1 + |\xi|^2)^{-s/2}$, then $\hat{\varphi} \in \mathcal{S}$, hence $\varphi \in \mathcal{S}$. Noting that $\hat{\psi}(\xi) = (1 + |\xi|^2)^{s/2}\hat{\varphi}(\xi)$, we immediately have $\psi = \mathcal{J}_d^s(\varphi)$. Finally, since $\mathcal{J}_d^s(g_1) = \mathcal{J}_d^s(g_2)$, we then have $\int (g_1 - g_2)\mathcal{J}_d^s(\varphi) = 0$. Therefore $g_1 = g_2$ from the surjectivity.

Such defined space $W^{s,p}$ is a Banach space. Assume f_n is a Cauchy sequence of $W^{s,p}$, then there exists $g_n \in L^p$ such that $f_n = \mathcal{J}_d^s g_n$. By definition, g_n is a Cauchy sequence in L^p and hence there exists $g \in L^p$ such that $g_n \to g$ and $\|f_n - \mathcal{J}^s g\|_{s,p} = \|\mathcal{J}^{-s} f_n - g\|_p \to 0$ as $n \to \infty$. Let $f = \mathcal{J}^s g$, then obviously $f \in W^{s,p}$, completing the proof.

It follows from the theory of Fourier multiplier, when $0 \leqslant \beta \leqslant \alpha$,

$$W^{\alpha,p} \hookrightarrow W^{\beta,p}, \quad \text{and} \|f\|_{W^{\beta,p}} \leqslant \|f\|_{W^{\alpha,p}}.$$

When $\beta \geqslant \alpha \geqslant 0$, $\mathcal{J}_d^{\beta-\alpha}$ is an isomorphism from $W^{\alpha,p}$ to $W^{\beta,p}$.

Similarly, when we consider Riesz potential, it leads to the definition of homogeneous fractional Sobolev space $\dot{W}^{s,p}$. When $p = 2$ and $s \in \mathbf{R}$, then for a tempered distribution f on $\mathbf{R}^d$, we define the norm $\|\cdot\|_{s,2} := \|\cdot\|_{\dot{W}^{s,2}}$ as

$$\|f\|_{\dot{W}^{s,2}} = \|\Lambda^s f\|_{L^2} = \left(\int_{\mathbf{R}^d} |\xi|^{2s} |\hat{f}(\xi)|^2 \mathrm{d}\xi \right)^{1/2}.$$

The homogeneous fractional Sobolev space can then be defined by

$$\dot{W}^{s,2} = \{f \in \mathcal{S}' : \|f\|_{\dot{W}^{s,2}} < \infty\}.$$

When $1 \leqslant p \leqslant \infty$ and $s \in \mathbf{R}$, the space $\dot{W}^{s,p}$ can also be defined as

$$\dot{W}^{s,p} := \{f \in \mathcal{S}' : \text{ there exists } g \in L^p(\mathbf{R}^d) \text{ such that } |\cdot|^s \hat{f}(\cdot) = \hat{g}(\cdot)\}.$$

The $\|\cdot\|_{s,p} := \|\cdot\|_{\dot{W}^{s,p}}$ norm of f is defined by $\|f\|_{s,p} = \|\Lambda^s f\|_p$. For $p = 2$, we denote $\dot{H}^s = \dot{W}^{s,2}$.

In summary, for $s \in \mathbf{R}$, the nonhomogeneous and homogeneous fractional Sobolev space can be defined by $W^{s,p} = \mathcal{J}_d^s(L^p(\mathbf{R}^d))$ and $\dot{W}^{s,p} = \mathcal{I}_d^s(L^p(\mathbf{R}^d))$ by Bessel potential $\mathcal{J}_d^s = (I - \Delta)^{-s/2}$ and Riesz potential $\mathcal{I}_d^s = (-\Delta)^{-s/2}$, respectively. When $s = m$ is an integer, they reduce to the integer order Sobolev spaces.

Lemma 2.2.4 *Let $1 < p < \infty, s \geqslant 0$. Then $f \in W^{s,p}(\mathbf{R}^d)$ if and only if $f \in L^p(\mathbf{R}^d)$ and $\mathcal{I}_d^{-s} f \in L^p(\mathbf{R}^d)$. The norm $\|\cdot\|_{s,p}$ and $\|f\|_p + \|f\|_{s,p}$ are equivalent.*

Proof   The inequality $\|f\|_p + \|f\|_{s,p} \leqslant c\|f\|_{s,p}$ obviously holds from $1 + |\xi|^{2s} \leqslant (1 + |\xi|^2)^s$. On the other hand, for $(1 + |\xi|^2)^{s/2}/(1 + |\xi|^s)$, using the Mihlin's multiplier theorem, the reverse inequality holds.

In particular, when $s \geqslant 0$ and $1 < p < \infty$, $W^{s,p} = L^p \cap \dot{W}^{s,p}$, cf. [21]. We can extend the interpolation theory and embedding theorem of Sobolev space to the fractional Sobolev space. For the proof and further discussions, readers are referred to [3, 21, 212, 214].

Lemma 2.2.5 *Let $s \in \mathbf{R}$, $\theta \in (0,1)$ and $p \in (1,\infty)$, then*

$$[L^p(\mathbf{R}^d), W^{s,p}(\mathbf{R}^d)]_\theta = W^{\theta s,p}(\mathbf{R}^d).$$

More generally, for $s_1, s_2 \in \mathbf{R}$, $\theta \in (0,1)$ and $p \in (1,\infty)$, there holds

$$[W^{s_1,p}(\mathbf{R}^d), W^{s_2,p}(\mathbf{R}^d)]_\theta = W^{(1-\theta)s_1+\theta s_2,p}(\mathbf{R}^d).$$

Theorem 2.2.9 *Let $1 < p < \infty$, $-\infty < s < \infty$, then*
1. *$W^{s,p}$ is a Banach space;*
2. *$\mathcal{S} \subset W^{s,p} \subset \mathcal{S}'$;*
3. *$W^{s+\varepsilon,p} \hookrightarrow W^{s,p}(\varepsilon > 0)$;*
4. *$W^{s,p}(\mathbf{R}^d) \hookrightarrow L^{\frac{dp}{d-sp}}(\mathbf{R}^d)$, $s < d/p$;*
5. *$W^{s,p}(\mathbf{R}^d) \hookrightarrow C(\mathbf{R}^d) \hookrightarrow L^\infty(\mathbf{R}^d)$, $s > d/p$.*

At the end of this section, we simply discuss the relationship between the space $H^{1/2}$ and the operator $\Lambda = (-\Delta)^{1/2}$. For this purpose, we consider the definition domain of the operator $\Lambda = (-\Delta)^{1/2}$. If T is a distribution, it is infinitely differentiable in the sense of weak derivatives. Hence, for a distribution T, it makes sense to discuss ΔT. However, to ensure ΛT makes sense, only requiring T to be a distribution is not adequate. To illustrate this, we first recall come concepts about distribution. Let $C_c^\infty(\mathbf{R}^d)$ be the space of infinitely differential complex valued functions compactly supported in $\mathbf{R}^d$. The space of $\mathcal{D}(\mathbf{R}^d)$ of test functions is defined to be $C_c^\infty(\mathbf{R}^d)$ with the topology induced by the limit of a sequence of elements in $\mathcal{D}(\mathbf{R}^d)$. A sequence $\phi_k \in C_c^\infty(\mathbf{R}^d)$ is said to be convergent to $\phi \in C_c^\infty(\mathbf{R}^d)$ in $\mathcal{D}$ if and only if there exists a given compact set $K \subset \mathbf{R}^d$, such that for arbitrary k, $\cup \mathrm{supp}\phi_k \subset K$ and for arbitrary multi-index α, $D^\alpha \phi_k \to D^\alpha \phi$ uniformly as $k \to \infty$. Under this topology, $\mathcal{D}(\mathbf{R}^d)$ becomes a complete locally convex topological vector space satisfying the Heine-Borel property. A distribution T is a continuous linear functional in $\mathcal{D}$. Here, the continuity means if $\phi_k \to \phi$ in $\mathcal{D}$, then $T(\phi_k) \to T(\phi)$. The space of all distributions on $\mathbf{R}^d$ is denoted by $\mathcal{D}'(\mathbf{R}^d)$. Equivalently, the vector space $\mathcal{D}'$ is the continuous dual space of the topological vector space $\mathcal{D}$. If $T_j \in \mathcal{D}'$, we call $T_j \to T \in \mathcal{D}'$ if $T_j(\phi) \to T(\phi)$ for any $\phi \in \mathcal{D}$.

The meaning of the product of two distributions is not clear. However, distributions can be multiplied by and taken convolution with C^∞ functions. Consider $T \in \mathcal{D}'$ and $\psi \in \mathcal{D}$, then the product of them ψT is defined by $\psi T(\phi) := T(\psi\phi)$. Such a ψT is a distribution. In fact, if $\phi \in C_c^\infty$, then $\psi\phi \in C^\infty$. In addition, if $\phi_k \to \phi$ in $\mathcal{D}$, then $\psi\phi_k \to \psi\phi$ in $\mathcal{D}$. The convolution of the distribution T with a C_c^∞ function j is defined by $(j*T)(\phi) := T(j_{\mathbf{R}}*$

$\phi) = T\left(\int_{\mathbf{R}^d} j(y)\phi(\cdot + y)dy\right)$ for all $\phi \in \mathcal{D}$, where $j_{\mathbf{R}}(x) = j(-x)$. Here, j is required to be compactly supported. Otherwise $j_{\mathbf{R}} * \phi$ is not compactly supported and fails to define a distribution.

Now we turn to discuss ΛT. By definition, ΛT is defined by

$$(\Lambda T)(\phi) := T(\Lambda\phi), \quad \forall \phi \in \mathcal{D}(\mathbf{R}^d). \qquad (2.2.30)$$

Since Λ is a nonlocal operator, $\Lambda\phi$ is not compactly supported in general and hence ΛT is not a distribution. However, when T is a distribution defined by a function, then (2.2.30) defines a distribution. Indeed, as long as $f \in H^{1/2}(\mathbf{R}^d)$, then Λf is a distribution, i.e., the mapping

$$\phi \mapsto \Lambda f(\phi) := \int_{\mathbf{R}^d} |\xi|\hat{f}(\xi)\hat{\phi}(-\xi)d\xi$$

makes sense. In this case, we have $|\xi|^{1/2}\hat{f} \in L^2(\mathbf{R}^d)$ and the mapping is continuous in $\mathcal{D}$. Let $\phi_k \to \phi$ in $\mathcal{D}$, then from the Schwartz's inequality and the Plancherel's theorem,

$$|\Lambda f(\phi_k - \phi)| \leqslant c\|\hat{f}\|_2 \left(\int_{\mathbf{R}^d} |\xi|^2 |\hat{\phi}_k(\xi) - \hat{\phi}(\xi)|^2 d\xi\right)^{1/2}$$
$$= c\|f\|_2 \|\boldsymbol{\nabla}(\phi_k - \phi)\|_2.$$

It follows when $k \to \infty$, $\|\boldsymbol{\nabla}(\phi_k - \phi)\|_2 \to 0$ and $\Lambda f(\phi_k - \phi) \to 0$, showing $\Lambda f \in \mathcal{D}'(\mathbf{R}^d)$ is a distribution.

2.2.5   Commutator estimates

In this section, we consider the commutator estimates of the fractional Laplacian. For this purpose, we first consider the following proposition.

Lemma 2.2.6 *Let k be an integer, and β, γ be multi-indices. If $|\beta| + |\gamma| = k$, then for arbitrary $f, g \in C_0(\mathbf{R}^d) \cap H^p(\mathbf{R}^d)$, there holds*

$$\|(D^\beta f)(D^\gamma g)\|_{L^2} \leqslant C\|f\|_{L^\infty}\|g\|_{H^k} + C\|f\|_{H^k}\|g\|_{L^\infty}.$$

Proof Let $|\beta| = l, |\gamma| = m$, then $l + m = k$. Using the interpolation estimates yields

$$\|D^l u\|_{L^{2k/l}} \leqslant C\|u\|_{L^\infty}^{1-l/k}\|D^k u\|_{L^2}^{l/k}. \qquad (2.2.31)$$

It follows from Hölder inequality that

$$\|(D^\beta f)(D^\gamma g)\|_{L^2} \leqslant \|D^\beta f\|_{L^{2k/l}}\|D^\gamma g\|_{L^{2k/m}}$$
$$\leqslant C\|f\|_{L^\infty}^{1-l/k}\|f\|_{H^k}^{l/k}\|g\|_{L^\infty}^{1-m/k}\|y\|_{H^k}^{m/k}.$$

Noting that $1 - l/k = m/k$, the result follows from Young's inequality.

Theorem 2.2.10 *Let k be an integer and $f, g \in L^\infty \cap H^k$, then there holds*

$$\|fg\|_{H^k} \leqslant C\|f\|_{L^\infty}\|g\|_{H^k} + C\|f\|_{H^k}\|g\|_{L^\infty},$$

and for arbitrary multi-index α with $|\alpha| \leqslant k$, then

$$\|D^\alpha(fg) - fD^\alpha g\|_{L^2} \leqslant C\|\nabla f\|_{H^{k-1}}\|g\|_{L^\infty} + C\|\nabla f\|_{L^\infty}\|g\|_{H^{k-1}}.$$

Proof   From Lemma 2.2.6, it follows the first inequality. For the second one, from Leibniz formula

$$D^\alpha(fg) = \sum_{\beta+\gamma=\alpha} C_\alpha^\beta (D^\beta f)(D^\gamma g).$$

Hence, if $\alpha = k$,

$$
\begin{aligned}
D^\alpha(fg) - fD^\alpha g &= \sum_{\beta+\gamma=\alpha, \beta>0} C_\alpha^\beta (D^\beta f)(D^\gamma g) \\
&= \sum_{|\beta|+|\gamma|=k-1} C_{j\beta\gamma}(D^\beta D_j f)(D^\gamma g),
\end{aligned}
$$

where $C_{j\beta\gamma}$ is a constant only depending on j, β, γ. Let $u = D_j f$, then the second inequality follows from Proposition 2.2.6.

Lemma 2.2.7 *Let $f = (f_1, \cdots, f_\mu) \in L^\infty \cap H^k$, then if $|\beta_1| + \cdots + |\beta_\mu| = k$, there holds*

$$\|D^{\beta_1}f_1 \cdots D^{\beta_\mu}f_\mu\|_{L^2} \leqslant C\sum_\nu \left(\|f_1\|_{L^\infty} \cdots \widehat{\|f_\nu\|}_{L^\infty} \cdots \|f_\mu\|_{L^\infty} \right) \|f\|_{H^k},$$

where, $\widehat{\cdot}$ represents that the term is deleted from the expression.

Proof   From the generalized Hölder inequality, it follows

$$\|D^{\beta_1}f_1 \cdots D^{\beta_\mu}f_\mu\|_{L^2} \leqslant \|D^{\beta_1}f_1\|_{L^{2k/|\beta_1|}} \cdots \|D^{\beta_\mu}f_\mu\|_{L^{2k/|\beta_\mu|}}.$$

Then from the interpolation inequality (2.2.31),

$$\|D^{\beta_1}f_1 \cdots D^{\beta_\mu}f_\mu\|_{L^2} \leqslant C\|f_1\|_{L^\infty}^{1-|\beta_1|/k}\|f_1\|_{H^k}^{|\beta_1|/k} \cdots \|f_\mu\|_{L^\infty}^{1-|\beta_\mu|/k}\|f_\mu\|_{H^k}^{|\beta_\mu|/k}.$$

Noting that $|\beta_1| + \cdots + |\beta_\mu| = k$, we have from the Young's inequality that

$$\|f_1\|_{H^k}^{|\beta_1|/k} \cdots \|f_\mu\|_{H^k}^{|\beta_\mu|/k} \leqslant \|f_1\|_{H^k} + \cdots + \|f_\mu\|_{H^k},$$

and by repeatedly using the Young's inequality

$$\|f_1\|_{L^\infty}^{1-|\beta_1|/k} \cdots \|f_\mu\|_{L^\infty}^{1-|\beta_\mu|/k} \leqslant C\sum_\nu \left(\|f_1\|_{L^\infty} \cdots \widehat{\|f_\nu\|}_{L^\infty} \cdots \|f_\mu\|_{L^\infty} \right).$$

This completes the proof.

Proposition 2.2.7 *Let F be a smooth function, and $F(0) = 0$. Then for arbitrary $u \in H^k \cap L^\infty$, there holds*

$$\|F(u)\|_{H^k} \leqslant C_k(\|u\|_{L^\infty})(1 + \|u\|_{H^k}).$$

Proof It follows from the chain rule

$$D^\alpha F(u) = \sum_{\beta_1 + \cdots + \beta_\mu = \alpha} C_\beta D^{\beta_1} u \cdots D^{\beta_\mu} u F^{(\mu)}(u).$$

By Hölder inequality, there holds

$$\|D^k F(u)\|_{L^2} \leqslant C_k(\|u\|_{L^\infty}) \sum \|D^{\beta_1} u \cdots D^{\beta_\mu} u\|.$$

The result follows from Proposition 2.2.7.

In what follows, we will generalize the inequalities to a more general fractional operator.

Theorem 2.2.11 *Let $s > 0$ and $1 < p < \infty$, then*

$$\|\mathcal{J}^s(fg) - f(\mathcal{J}^s g)\|_{L^p} \leqslant c(\|\nabla f\|_{L^\infty}\|\mathcal{J}^{s-1}g\|_{L^p} + \|\mathcal{J}^s f\|_{L^p}\|g\|_{L^\infty}). \quad (2.2.32)$$

Proof Define the real valued C^∞ functions Φ_j in $\mathbf{R}$ such that

$$0 \leqslant \Phi_j \leqslant 1, \quad j = 1, 2, 3, \quad \Phi_1 + \Phi_2 + \Phi_3 = 1,$$

and

$$supp\Phi_1 \subset \left[-\frac{1}{3}, \frac{1}{3}\right], \quad supp\Phi_2 \subset \left[\frac{1}{4}, 4\right], \quad supp\Phi_3 \subset [3, \infty).$$

Then by definition of the operator $\mathcal{J}^s$, one obtains

$$[\mathcal{J}^s(fg) - f(\mathcal{J}^s g)](x)$$
$$= c \int \int e^{i\langle x, \xi + \eta \rangle} \{(1 + |\xi + \eta|^2)^{\frac{s}{2}} - (1 + |\eta|^2)^{\frac{s}{2}}\}\hat{f}(\xi)\hat{g}(\eta)d\xi d\eta$$
$$= c \sum_{j=1}^{3} \sigma_j(D)(f, g)(x),$$

where

$$\sigma_j(\xi, \eta) = [(1 + |\xi + \eta|^2)^{\frac{s}{2}} - (1 + |\eta|^2)^{\frac{s}{2}}]\Phi_j(|\xi|/|\eta|).$$

First, we consider $\sigma_1(D)(f, g)$. Rewrite the formula as

$$\sigma_1(\xi, \eta) = (1 + |\eta|^2)^{\frac{s}{2}} \left\{[1 + (1 + |\eta|^2)^{-1}\langle \xi, \xi + 2\eta \rangle]^{\frac{s}{2}} - 1\right\}\Phi_1(|\xi|/|\eta|)$$
$$= c_1(1 + |\eta|^2)^{\frac{s}{2} - 1}\langle \xi, \xi + 2\eta \rangle \Phi_1$$
$$\quad + c_2(1 + |\eta|^2)^{\frac{s}{2} - 2}\langle \xi, \xi + 2\eta \rangle^2 \Phi_1 +$$
$$\quad + c_r(1 + |\eta|^2)^{\frac{s}{2} - r}\langle \xi, \xi + 2\eta \rangle^r \Phi_1 + \cdots. \quad (2.2.33)$$

After multiplied by $\hat{f}(\xi)\hat{g}(\xi)$, the r-th term can be written as

$$\langle \sigma_{1,r}(\xi,\eta), ((\widehat{\boldsymbol{\nabla}/i})f)(\xi)\rangle(\widehat{\mathcal{J}^{s-1}g})(\eta),$$

where

$$\sigma_{1,r}(\xi,\eta) = c_r(1+|\eta|^2)^{-r+\frac{1}{2}}\langle \xi, \xi+2\eta\rangle^{r-1}(\xi+2\eta)\Phi_1 \in \mathbf{R}^d.$$

It is easy to see that $\sigma_{1,r}$ satisfies the condition (2.2.23) of Theorem 2.2.5. Also if $\Phi \neq 0$, then we require $|\xi| \leqslant |\eta|/3$ to ensure convergence of the series (2.2.33). Then Theorem 2.2.5 yields

$$\|\sigma_1(D)(f,g)\|_p \leqslant c\|\boldsymbol{\nabla} f\|_\infty \|\mathcal{J}^{s-1}g\|_p. \tag{2.2.34}$$

Secondly, we consider $\sigma_3(D)(f,g)$. Let $\sigma_3 = \sigma_{3,1} - \sigma_{3,2}$ with

$$\sigma_{3,1}(\xi,\eta) = [(1+|\xi+\eta|^2)^{s/2} - 1]\Phi_3,$$

and

$$\sigma_{3,1}(\xi,\eta) = [(1+|\eta|^2)^{s/2} - 1]\Phi_3,$$

then

$$\sigma_{3,1}(\xi,\eta)\hat{f}(\xi)\hat{g}(\eta) = (1+|\xi|^2)^{-s/2}[(1+|\xi+\eta|^2)^{s/2} - 1]\widehat{(\mathcal{J}^s f)}(\xi)\hat{g}(\eta)\Phi_3.$$

Since $\Phi_3 \neq 0$ only if $|\xi| \geqslant 3|\eta|$ and $\sigma_{3,1}$ satisfies the condition (2.2.23) of Theorem 2.2.5, one has

$$\|\sigma_{3,1}(D)(f,g)\|_p \leqslant c\|\mathcal{J}^s f\|_p \|g\|_\infty. \tag{2.2.35}$$

Define the operator G, such that

$$\widehat{(Gh)}(\eta) = \eta|\eta|^{-2}(1+|\eta|^2)^{\frac{1}{2}-\frac{s}{2}}[(1+|\eta|^2)^{\frac{s}{2}} - 1]\hat{h}(\eta).$$

Then using the Mihlin's multiplier theorem, we can see that G is a bounded operator in L^p, cf. [21]. In this case, $\sigma_{3,2}$ can be expressed as

$$\sigma_{3,2}(\xi,\eta)\hat{f}(\xi)\hat{g}(\eta) = |\xi|^2\langle \xi, (\widehat{\boldsymbol{\nabla} f})(\xi)\rangle\langle \eta, (\widehat{G\mathcal{J}^{s-1}g})(\eta)\rangle\Phi_3,$$

which, since $|\xi|^2\xi_j\eta_k\Phi_3$ satisfies the condition (2.2.23), has the bound

$$\|\sigma_{3,2}(D)(f,g)\|_p \leqslant c\|\boldsymbol{\nabla} f\|_\infty \|\mathcal{J}^{s-1}g\|_p. \tag{2.2.36}$$

Finally, we estimate $\sigma_2(D)(f,g)$. Since $\xi+\eta$ may be zero, any negative powers of $1+|\xi+\eta|$ fails to satisfy the condition (2.2.23). Divide σ_2 into two parts $\sigma_2 = \sigma_{2,1} - \sigma_{2,2}$, where $\sigma_{2,1}(\xi,\eta) = (1+|\xi+\eta|^2)^{\frac{s}{2}}\Phi_2$ and $\sigma_{2,2}(\xi,\eta) = (1+|\eta|^2)^{\frac{s}{2}}\Phi_2$. Since

$$\sigma_{2,2}(\xi,\eta)\hat{f}(\xi)\hat{g}(\eta) = (1+|\eta|^2)^{s/2}(1+|\xi|^2)^{-s/2}\widehat{(\mathcal{J}^s f)}(\xi)\hat{g}(\eta)\Phi_2,$$

the assumption of Theorem 2.2.5 is satisfied and hence

$$\|\sigma_{2,2}(D)(f,g)\|_p \leqslant c\|\mathcal{J}^s f\|_p \|g\|_\infty. \tag{2.2.37}$$

For $\sigma_{2,1}$, it can be rewritten as

$$\sigma_{2,1}(\xi,\eta)\hat{f}(\xi)\hat{g}(\eta) = (1+|\xi+\eta|^2)^{s/2}(1+|\xi|^2)^{-s/2}\widehat{(\mathcal{J}^s f)}(\xi)\hat{g}(\eta)\Phi_2.$$

Denote $\tilde{\sigma}_{2,1}(\xi,\eta) = (1+|\xi+\eta|^2)^{s/2}(1+|\xi|^2)^{-s/2}\Phi_2$, hence $|\tilde{\sigma}_{2,1}| \leqslant C$. Furthermore, when $s > 2$, from the definition of Φ_2, one has

$$|\partial_\eta \tilde{\sigma}_{2,1}(\xi,\eta)| \leqslant \frac{(1+|\xi+\eta|^2)^{s/2-1}|\eta|}{(1+|\xi|^2)^{s/2-1}(1+|\xi|^2)} \leqslant \frac{c|\eta|}{1+|\xi|^2}.$$

Also from the definition of Φ_2, $|\eta|^2 \leqslant c|\xi|^2$ holds, hence

$$|\partial_\eta \tilde{\sigma}_{2,1}(\xi,\eta)| \leqslant \frac{c|\eta|}{1+|\xi|^2} \leqslant \frac{c}{|\xi|+|\eta|}, \tag{2.2.38}$$

satisfying the condition of Theorem 2.2.5. In fact, as long as s is big enough such that and negative powers of $1+|\xi+\eta|^2$ do not appear in (2.2.38), then the discussion above is applicable. From Remark 2.2.3, when s is big enough such that $s \geqslant k(m,p)$, the estimate (2.2.37) still holds.

When s is not big enough, the discussion above is not applicable. To overcome this problem, we need to extend $\sigma_{2,1}(\xi,\eta)$ to the complex valued case $\sigma_{2,1}^s(\xi,\eta)$ for a complex s with $0 \leqslant \operatorname{Re} s \leqslant k$, and then apply the complex interpolation theory. When $s = k + it, t \in \mathbf{R}$, Remark 2.2.3 still applies to yield

$$\|\sigma_{2,1}^{k+it}(D)(f,g)\|_p \leqslant C(t)\|\mathcal{J}^s f\|_p \|g\|_\infty, \tag{2.2.39}$$

where $C(t)$ depends on t. Since $|\alpha|, |\beta| \leqslant k$ in (2.2.23), we know $C(t) = O(t^k)$. To apply the complex interpolation theory, we need to handle the case when $s = it$. For this purpose, first note

$$\mathcal{J}^s(fg) = \sum_{j=1}^{3} \sigma_{j,1}^s(D)(f,g), \tag{2.2.40}$$

for $\sigma_{j,1}^s = (1+|\xi+\eta|^2)^{s/2}\Phi_j$. When $j = 1, 3$, by definition of Φ_1 and Φ_3, the condition (2.2.23) is easily verified, yielding

$$\|\sigma_{j,1}^{it}(D)(f,g)\|_p \leqslant C(t)\|f\|_p\|g\|_\infty, \quad j = 1, 3. \tag{2.2.41}$$

Moreover, from the Mihlin's theorem [21], one has

$$\|\mathcal{J}^{it}(fg)\|_p \leqslant C(t)\|fg\|_p \leqslant C(t)\|f\|_p\|g\|_\infty, \tag{2.2.42}$$

where $C(t) = O(t^k)$. Using (2.2.40)-(2.2.42), we have

$$\|\sigma_{2,1}^{it}(D)(f,g)\|_p \leqslant C(t)\|f\|_p\|g\|_\infty. \tag{2.2.43}$$

Using the complex interpolation theory between (2.2.39) and (2.2.43), we can see that the similar estimate holds for arbitrary $s(0 \leqslant s \leqslant k)$. Since we have proved that the conclusion holds when $s \geqslant k$, therefore

$$\|\sigma_{2,1}(D)(f,g)\|_p \leqslant c\|\mathcal{J}^s f\|_p\|g\|_\infty. \tag{2.2.44}$$

Combining the estimates (2.2.34), (2.2.35), (2.2.36), (2.2.37) and (2.2.44), we complete the proof.

Theorem 2.2.12 *When $s > 0$, $1 < p < \infty$, $L_s^p \cap L^\infty$ is an algebra. In particular, we have*

$$\|fg\|_{s,p} \leqslant c(\|f\|_\infty\|g\|_{s,p} + \|f\|_{s,p}\|g\|_\infty). \tag{2.2.45}$$

Proof   The proof is similar to the proof of the theorem above, hence we omit it here.

Remark 2.2.4 *When s is a positive integer, the results (2.2.32) and (2.2.45) are well known, and can be proved by applying the Leibniz rule and the Gagliardo-Nirenberg inequality. When $\dfrac{d}{p} < s < 1$, the proof can be referred to Strichartz [207].*

Theorem 2.2.13 *Let $s > 0$, $p \in (1,\infty)$. If $f,g \in \mathcal{S}$, then*

$$\|\mathcal{J}^s(fg) - f(\mathcal{J}^s g)\|_p \leqslant C\{\|\nabla f\|_{p_1}\|g\|_{s-1,p_2} + \|f\|_{s,p_3}\|g\|_{p_4}\}, \tag{2.2.46}$$

and

$$\|\mathcal{J}^s(fg)\|_p \leqslant C\{\|f\|_{p_1}\|g\|_{s,p_2} + \|f\|_{s,p_3}\|g\|_{p_4}\}, \tag{2.2.47}$$

where $p_2, p_3 \in (1,+\infty)$ satisfies

$$\frac{1}{p} = \frac{1}{p_1} + \frac{1}{p_2} = \frac{1}{p_3} + \frac{1}{p_4}.$$

This theorem follows by using the approach of the theorem above. When $p_1 = p_4 = \infty$, this theorem reduces to Theorem 2.2.11 and 2.2.12. For the homogeneous operator Λ, we have

Theorem 2.2.14 *Let $s > 0$, $p \in (1,\infty)$. If $f,g \in \mathcal{S}$, then that*

$$\|\Lambda^s(fg) - f(\Lambda^s g)\|_{L^p} \leqslant C\{\|\nabla f\|_{L^{p_1}}\|y\|_{W^{s-1,p_2}} + \|f\|_{\dot{W}^{s,r_0}}\|g\|_{L^{p_4}}\}, \tag{2.2.48}$$

and the following product estimate holds

$$\|\Lambda^s(fg)\|_{L^p} \leqslant C\{\|f\|_{L^{p_1}}\|g\|_{\dot{W}^{s,p_2}} + \|f\|_{\dot{W}^{s,p_3}}\|g\|_{L^{p_4}}\},$$

where $p_2, p_3 \in (1, +\infty)$ satisfies

$$\frac{1}{p} = \frac{1}{p_1} + \frac{1}{p_2} = \frac{1}{p_3} + \frac{1}{p_4}.$$

Proof We only need to apply Theorem 2.2.13 to $f_\varepsilon(x) = f(x/\varepsilon)$ and $g_\varepsilon = g(x/\varepsilon)$ and then let $\varepsilon \to 0$.

Theorem 2.2.15 *Let $s_j < d/p(j = 1, 2)$, $s_1 + s_2 = s + \dfrac{d}{p}$, $0 < s \leqslant$ $\min\{s_1, s_2\}$, then*

$$\|fg\|_{W^{s,p}} \leqslant c\|f\|_{W^{s_1,p}}\|g\|_{W^{s_2,p}}.$$

Similarly, in the homogeneous case, there holds

$$\|fg\|_{\dot{W}^{s,p}} \leqslant c\|f\|_{\dot{W}^{s_1,p}}\|g\|_{\dot{W}^{s_2,p}}.$$

Proof Let $\dfrac{d}{p_2} = s_1 = \dfrac{d}{p} - s_2 + s$ and $\dfrac{d}{p_2} = \dfrac{d}{p} - s_1$, then $\dfrac{1}{p_1} + \dfrac{1}{p_2} = \dfrac{1}{p}$. Using the Sobolev's embedding theorem, we have

$$\|g\|_{W^{s,p_2}} \leqslant c\|g\|_{W^{s_2,p}} \quad \text{and} \quad \|f\|_{L^{p_2}} \leqslant c\|f\|_{W^{s_1,p}}.$$

The proposition then follows by interchanging f and g and using the result of (2.2.47). The homogeneous case also holds by Theorem 2.2.14.

Theorem 2.2.16 *Let $q > 1, p \in [q, +\infty)$, and $\dfrac{1}{p} + \dfrac{\sigma}{d} = \dfrac{1}{q}$, then there exists a constant $C > 0$ such that for any $f \in \mathcal{S}'$ and $\hat{f}$ is a function, then*

$$\|f\|_{L^p} \leqslant C\|\Lambda^\sigma f\|_{L^q}.$$

Proof When $q = 2$, the proof is given by [197]. Since $\hat{f}$ is a function, $\hat{f}(\xi) = |\xi|^{-\sigma}|\xi|^\sigma \hat{f}(\xi)$ holds. Using the inverse Fourier transform, we can see that $f = \mathcal{I}_d^\sigma(\Lambda^\sigma f)$, where $\mathcal{I}_d^\sigma$ is a Riesz potential operator. The proposition then follows from the boundedness of the Riesz operator $\mathcal{I}_d^\sigma$ in Theorem 2.2.7.

2.3   An existence theorem

We consider the following fractional ordinary differential equation

$$D^\mu y(t) = f(t, y), \quad \mu \in (n-1, n], t \in (0, t], \tag{2.3.1}$$

where $n \geqslant 1$ is an integer and D^μ represents either the Riemann-Liouville derivative ${}^{RL}{}_0D_t^\mu$ or the Caputo derivative ${}^C{}_0D_t^\mu$. To solve (2.3.1), we need to prescribe n initial data for the unknown function $y(t)$ at initial time $t = 0$. With different choice of the derivative, the initial data is different. From (2.1.22) and (2.1.23), we see that, when $D^\mu = {}^{RL}{}_0D_t^\mu$, the fractional derivatives of $y(t)$ at time $t = 0$ should be prescribed, while when $D^\mu = {}^C{}_0D_t^\mu$, the integer order derivatives of $y(t)$ at time $t = 0$ should be prescribed.

Before we state and prove the existence theory for (2.3.1), we consider the following two examples when $D^\mu = {}^{RL}{}_0D_t^\mu$. In the first example, $f(t, y)$ depends only on y linearly

$$
\begin{cases}
D^\mu y(t) = \lambda y(t) \\
D^{\mu-n+k} y(t) = b_{k+1}, \quad k = 0, \cdots, n-1,
\end{cases}
\tag{2.3.2}
$$

where λ is a complex number and $b_1, \cdots, b_n$ are known initial data. By the Laplace transform (2.1.22), we have

$$
s^\mu Y(s) - \sum_{k=0}^{n-1} s^k b_{n-k} = \lambda Y(s),
$$

where $Y(s)$ is the Laplace transform of $y(t)$. Therefore,

$$
\begin{aligned}
Y(s) &= \sum_{j=1}^{n} b_j s^{n-j} \frac{s^{n-\mu-j}}{1 - \lambda/s^\mu} \\
&= \sum_{j=1}^{n} b_j \sum_{k=0}^{\infty} \lambda^k s^{-k\mu} s^{n-\mu-j}.
\end{aligned}
$$

Taking inverse Laplace transform, we have

$$
\begin{aligned}
y(t) &= \sum_{j=1}^{n} b_j \sum_{k=0}^{\infty} \frac{\lambda^k t^{k\mu+(\mu-n+j)-1}}{\Gamma(k\mu+(\mu-n+j))} \\
&= \sum_{j=1}^{n} b_j t^{(\mu-n+j)-1} E_{\mu,\mu-n+j}(\lambda t^\mu),
\end{aligned}
$$

where $E_{\mu,\mu-n+j}(\cdot)$ is the two-parameter Mittag-Leffler function. In particular, when $\mu = n = 1$, we obtain $y(t) = b_1 E_{1,1}(\lambda t) = b_1 e^{\lambda t}$, which is well-known. Similar results can be obtained when $D^\mu = {}^{RL}{}_0D_t^\mu$ and initial data in (2.3.2) should be replaced with $y^{(k)} = b_{k+1}$ for $k = 0, \cdots, n-1$.

In the second example, we consider

$$
\begin{cases}
D^\mu y(t) = f(t) \\
D^{\mu-n+k} y(t) = b_{k+1}, \quad k = 0, \cdots, n-1,
\end{cases}
\tag{2.3.3}
$$

where $b_1, \cdots, b_n$ are known initial data. By Laplace transform, we obtain

$$Y(s) = s^{-\mu}F(s) + \sum_{k=0}^{n-1} b_{n-k}s^{k-\mu}.$$

By taking inverse Laplace transform, we obtain

$$y(t) = \frac{1}{\Gamma(\mu)} \int_0^t (t-\tau)^{\mu-1} f(\tau)d\tau + \sum_{k=0}^{n-1} \frac{b_{n-k}}{\Gamma(\mu-k)} t^{\mu-k-1}. \qquad (2.3.4)$$

It is important to note that from Example 2.1.3

$$D^{\mu-k}\left(\frac{t^{\mu-l-1}}{\Gamma(\mu-l)}\right) = \begin{cases} t^{k-l-1}/\Gamma(k-l), & k > l \\ 0 & k \leqslant l, \end{cases}$$

and

$$D^{\mu-k-1}\left(\frac{t^{\mu-l-1}}{\Gamma(\mu-l)}\right) = \begin{cases} t^{k-l}/\Gamma(1+k-l), & k > l \\ 1 & k = l \\ 0 & k < l. \end{cases}$$

We see (2.3.4) is indeed a solution satisfying the initial data. Furthermore, by a simple argument, the solution can be shown unique in $L^1(0,T)$ for a given $T > 0$.

We now state an existence and uniqueness theorem for the fractional ordinary differential equation with Rieman-Liouville derivative. Similar theorem for the case of Caputo derivative can be obtained, which is omitted for simplicity. In this R-L derivative case, the initial value problem is proposed as finding $y(t)$ on $[0, T]$ such that

$$\begin{cases} D^\mu y(t) = f(t, y), & \mu \in (n-1, n], \ 0 < t < T < \infty, \\ D^{\mu-(n-k)-1}y(t)|_{t=0} = b_k, & k = 1, \cdots, n. \end{cases} \qquad (2.3.5)$$

Assume that $f(t, y)$ is defined in a domain G of a plane (t, y), and let $R(h, K) \subset G$ be a region such that

$$0 < t < h, \quad \left|t^{1-\sigma_1}y(t) - \sum_{i=1}^n b_i \frac{t^{\sigma_i-\sigma_1}}{\Gamma(\sigma_i)}\right| \leqslant K,$$

for some constants h and K.

Theorem 2.3.1 *Let $f(t, y)$ be a real-valued continuous function, defined in the domain G, satisfying the Lipshitz condition with respect to y, i.e.,*

$$|f(t, y_1) - f(t, y_2)| \leqslant L|y_1 - y_2|,$$

such that

$$|f(t,y)| \leqslant M < \infty, \qquad \forall (t,y) \in G.$$

Let $K \geqslant \dfrac{Mh^n}{\Gamma(1+\mu)}$, *then there exists a unique and continuous solution* $y(t)$ *in the region* $R(h,K)$ *of the problem* (2.3.5). *Furthermore, if* $\tilde{y}$ *is a solution of* (2.3.5) *satisfying the initial conditions*

$$D^{\mu-(n-k)-1}\tilde{y}(t)|_{t=0} = \tilde{b}_k = b_k + \delta_k, \qquad k = 1,\cdots,n, \qquad (2.3.6)$$

where $\delta_k (k = 1,2,\cdots,n)$ *are small constants, then for* $0 < t \leqslant h$ *there holds*

$$|y(t) - \tilde{y}(t)| \leqslant \sum_{i=1}^{n} |\delta_i| t^{\mu-(n-i)-1} E_{\mu,\mu-(n-i)}(Lt^\mu),$$

where $E_{\alpha,\beta}(z)$ *is the Mittag-Leffler function.*

Proof   By applying (2.3.4), we reduce (2.3.5) to the following equivalent integral equation

$$y(t) = \frac{1}{\Gamma(\mu)} \int_0^t (t-\tau)^{\mu-1} f(\tau, y(\tau)) d\tau + \sum_{i=1}^{n} \frac{b_i t^{\mu-(n-i)-1}}{\Gamma(\mu-(n-i))}. \qquad (2.3.7)$$

Consider the iterative sequence

$$y_0(t) = \sum_{i=1}^{n} \frac{b_i t^{\mu-(n-i)-1}}{\Gamma(\mu-(n-i))},$$

$$y_m(t) = \sum_{i=1}^{n} \frac{b_i t^{\mu-(n-i)-1}}{\Gamma(\mu-(n-i))} + \frac{1}{\Gamma(\mu)} \int_0^t (t-\tau)^{\mu-1} f(\tau, y_{m-1}(\tau)) d\tau, \quad m=1,2,\cdots.$$

$$(2.3.8)$$

We need to show that $\lim_{m\to\infty} y_m(t)$ exists and it is the solution of equation (2.3.7). First, for $0 < t \leqslant h$, we obviously have $(t, y_m(t)) \in R(h,K)$ for all m. Indeed,

$$\left| t^{n-\mu} y_m(t) - \sum_{i=1}^{n} \frac{b_i t^{i-1}}{\Gamma(\mu-(n-i))} \right| \leqslant \left| \frac{t^{n-\mu}}{\Gamma(\mu)} \int_0^t (t-\tau)^{\mu-1} f(\tau, y_{m-1}(\tau)) d\tau \right|$$

$$\leqslant \frac{Mt^n}{\Gamma(1+\mu)} \leqslant \frac{Mh^n}{\Gamma(1+\mu)} \leqslant K, \qquad (2.3.9)$$

and similarly

$$\left| t^{n-\mu} y_1(t) - \sum_{i=1}^{n} \frac{b_i t^{i-1}}{\Gamma(\mu-(n-i))} \right| \leqslant \frac{Mh^n}{\Gamma(1+\mu)} \leqslant K.$$

Next, we show by induction that for all m

$$|y_m(t) - y_{m-1}(t)| \leqslant \frac{ML^{m-1}t^{m\mu}}{\Gamma(1 + m\mu)}. \tag{2.3.10}$$

When $m = 1$, this holds since from (2.3.9), we have

$$|y_1(t) - y_0(t)| \leqslant \frac{Mt^\mu}{\Gamma(1 + \mu)}, \quad 0 < t \leqslant h.$$

Assume that (2.3.10) holds for $m - 1$, we will show that it also holds for m. Using (2.3.8) and Example 2.1.3, we obtain

$$
\begin{aligned}
|y_m(t) - y_{m-1}(t)| &\leqslant \frac{L}{\Gamma(\mu)} \int_0^t (t - \tau)^\mu |y_{m-1}(\tau) - y_{m-2}(\tau)| \, d\tau \\
&\leqslant \frac{ML^{m-1}}{\Gamma(1 + (m-1)\mu)} \frac{1}{\Gamma(\mu)} \int_0^t (t - \tau)^{\mu-1} \tau^{(m-1)\mu} d\tau \\
&= \frac{ML^{m-1}}{\Gamma(1 + (m-1)\mu)} D_t^{-\mu} t^{(m-1)\mu} \\
&= \frac{ML^{m-1}}{\Gamma(1 + m\mu)}.
\end{aligned}
$$

Consider the series

$$y^*(t) = \lim_{m \to \infty} (y_m(t) - y_0(t)) = \sum_{j=1}^\infty (y_j(t) - y_{j-1}(t)). \tag{2.3.11}$$

By definition of the Mittag-Leffler function,

$$M \sum_{j=1}^\infty \frac{L^{j-1}h^{j\mu}}{\Gamma(1 + j\mu)} = \frac{M}{L} \left(E_{\mu,1}(Lh^{mu}) - 1 \right).$$

Applying the estimate (2.3.10), it is obvious that (2.3.11) converges uniformly for $0 < t \leqslant h$. Since each term of the series is continuous for $[0, h]$, $y^*(t)$ is a continuous function for $t \in [0, h]$. Let $y(t) = y_0(t) + y^*(t)$, then $y(t)$ is continuous. Letting $m \to \infty$ in (2.3.8) then yilds (2.3.7).

Uniqueness follows from the Lipschitz condition of f in y. Let $y(t)$ and $\tilde{y}(t)$ be two continuous solutions, then $z(t) = y(t) - \tilde{y}(t)$ satisfies

$$z(t) = \frac{1}{\Gamma(\mu)} \int_0^t (t - \tau)^{\mu-1} [f(\tau, y(\tau)) - f(\tau, \tilde{y}(\tau))] d\tau.$$

Since $z(t)$ is continuous for $[0, h]$, then $|z(t)| \leqslant B$ for some positive constant B and hence

$$|z(t)| \leqslant \frac{BLt^\mu}{\Gamma(1 + \mu)}, \quad 0 \leqslant t \leqslant h.$$

By iteration, one has

$$|z(t)| \leqslant \frac{BL^j t^{j\mu}}{\Gamma(\mu)}, \qquad j = 1, 2, \cdots,$$

where the bound is the nothing but j-th term of the Mittag-Leffler function $E_{\mu,1}(Lt^\mu)$, and hence for all $t \in [0, h]$, there holds $\lim_{j\to\infty} L^j t^{j\mu}/\Gamma(1 + j\mu) = 0$. Uniqueness then follows.

Next we prove the continuous dependence part of the theorem. This is proved by induction since

$$y(t) = \lim_{m\to\infty} y_m(t), \qquad \tilde{y}(t) = \lim_{m\to\infty} \tilde{y}_m(t),$$

where y_m and $\tilde{y}_m$ are the iterative processes of y and $\tilde{y}$, respectively. When $m = 0$, it is easy to know

$$|y_0(t) - \tilde{y}_0(t)| \leqslant \sum_{i=1}^{n} |\delta_i| \frac{t^{\mu-(n-i)-1}}{\Gamma(\mu - (n - i))}.$$

When $m = 1$, by definition of the Riemann-Liouville derivative and the Lipshitz condition of $f(t, y)$, there holds

$$|y_1(t) - \tilde{y}_1(t)|$$

$$= \left| \sum_{i=1}^{n} \delta_i \frac{t^{\mu-(n-i)-1}}{\Gamma(\mu - (n - i))} + \frac{1}{\Gamma(\mu)} \int_0^t (t - \tau)^{\mu-1} \{f(\tau, y_0(\tau)) - f(\tau, \tilde{y}_0(\tau))\} \right|$$

$$\leqslant \sum_{i=1}^{n} |\delta_i| \frac{t^{\mu-(n-i)-1}}{\Gamma(\mu - (n - i))} + \frac{L}{\Gamma(\mu)} \int_0^t (t - \tau)^{\mu-1} |y_0(\tau) - \tilde{y}_0(\tau)| \mathrm{d}\tau$$

$$\leqslant \sum_{i=1}^{n} |\delta_i| \frac{t^{\mu-(n-i)-1}}{\Gamma(\mu - (n - i))} + \frac{L}{\Gamma(\mu)} \int_0^t (t - \tau)^{\mu-1} \left\{ \sum_{i=1}^{n} |\delta_i| \frac{\tau^{\mu-(n-i)-1}}{\Gamma(\mu - (n - i))} \right\} \mathrm{d}\tau$$

$$\leqslant \sum_{i=1}^{n} |\delta_i| \frac{t^{\mu-(n-i)-1}}{\Gamma(\mu - (n - i))} + LD^{-\mu} \left\{ \sum_{i=1}^{n} |\delta_i| \frac{t^{\mu-(n-i)-1}}{\Gamma(\mu - (n - i))} \right\}$$

$$= \sum_{i=1}^{n} |\delta_i| \frac{t^{\mu-(n-i)-1}}{\Gamma(\mu - (n - i))} + L \sum_{i=1}^{n} |\delta_i| \frac{t^{2\mu-(n-i)-1}}{\Gamma(2\mu - (n - i))}$$

$$= \sum_{i=1}^{n} |\delta_i| t^{\mu-(n-i)-1} \left\{ \sum_{k=0}^{1} \frac{L^k t^{k\mu}}{\Gamma(k\mu + \mu - (n - i))} \right\}.$$

Similarly, according to the induction, we obtain

$$|y_m(t) - \tilde{y}_m(t)| \leqslant \sum_{i=1}^{n} |\delta_i| t^{\mu-(n-i)-1} \left\{ \sum_{k=0}^{m} \frac{L^k t^{k\mu}}{\Gamma((k+1)\mu - (n - i))} \right\}.$$

Letting $m \to \infty$ then yields

$$|y(t) - \tilde{y}(t)| \leqslant \sum_{i=1}^{n} |\delta_i| t^{\mu-(n-i)-1} \left\{ \sum_{k=0}^{\infty} \frac{L^k t^{k\mu}}{\Gamma((k+1)\mu - (n-i))} \right\}$$
$$= \sum_{i=1}^{n} |\delta_i| t^{\mu-(n-i)-1} E_{\mu,\mu-(n-i)} (Lt^{\mu}),$$

completing the proof.

Furthermore, Let us consider example from partial differential equation

Example 2.3.1 *Let $\alpha \in (0,1)$. Consider the following fractional diffusion equation*

$$\begin{cases} {}_0D_t^{\alpha} u(x,t) = \lambda^2 \partial_x^2 u(x,t), & t > 0, -\infty < x < \infty, \\ {}_0D_t^{\alpha-1} y(t)|_{t=0} = \varphi(x), & \lim_{x \to \pm\infty} u(x,t) = 0. \end{cases} \quad (2.3.12)$$

Taking into account the boundary conditions at infinity and applying the Fourier transform with respect to variable x, one obtains

$$\begin{cases} {}_0D_t^{\alpha} \hat{u}(\xi,t) + \lambda^2 \xi^2 \hat{u}(\xi,t) = 0 \\ {}_0D_t^{\alpha-1} \hat{u}(\xi,t)|_{t=0} = \hat{\varphi}(\xi). \end{cases}$$

Upon the Laplace transform, one has

$$U(\xi,s) = \frac{\hat{\varphi}(\xi)}{s^{\alpha} + \lambda^2 \xi^2},$$

where $U(\xi,s)$ is the Laplace transform for $\hat{u}(\xi,t)$. Applying the inverse Laplace transform to obtain

$$\hat{u}(\xi,t) = \hat{\varphi}(\xi) t^{\alpha-1} E_{\alpha,\alpha}(-\lambda^2 \xi^2 t^{\alpha}),$$

and then the inverse Fourier transform to obtain a solution of the problem (2.3.12)

$$u(x,t) = \int_{-\infty}^{\infty} G(x - x', t)\varphi(x')dx',$$

where

$$G(x,t) = \frac{1}{\pi} \int_0^{\infty} t^{\alpha-1} E_{\alpha,\alpha}(-\lambda^2 \xi^2 t^{\alpha}) \cos \xi x d\xi.$$

After careful calculations, we have [179]

$$G(x,t) = \frac{1}{2\lambda} t^{\frac{\alpha}{2}-1} W(-z; -\rho, \rho), \quad z = \frac{|x|}{\lambda t^{\alpha/2}},$$

where $W(z; \lambda, \mu)$ is the Wright function

$$W(z; \alpha, \beta) = \sum_{k=0}^{\infty} \frac{z^k}{k! \Gamma(\alpha k + \beta)}.$$

In particular, when $\alpha = 1$, the fractional Green function reduces to

$$G(x, t) = \frac{1}{2\lambda\sqrt{\pi t}} e^{-\frac{x^2}{4\lambda^2 t}}.$$

2.4   Distributed order differential equations

In this section, we introduce some fractional differential equations of distributed order. The study of properties of distributed order differential equations and their applications has been developed extensively in recent years, although the idea of fractional differential equations with distributed order was first introduced by Caputo [36] and solved by him later in 1995 [37]. The distributed order differential equations have been used to model the input-output relationship of linear time-invariant system, to study the rheological properties of composite materials, to model the dissipation in seismology and in metallurgy and to model ultraslow and lateral processes. See, for example, [14, 15, 42, 142, 171].

We still use $_0^C D_t^\mu$ to denote the Riemann-Liouville. Integrating $_0^C D_t^\mu f(t)$ w.r.t. μ, the order of differentiation, we obtain the distributed-order differential operator

$$_0 D_t^\phi f(t) = \int_\lambda^\eta \phi(\mu) {_0^C} D_t^\mu f(t) \mathrm{d}\mu,$$

where ϕ is a continuous function in $[\lambda, \eta] \subset [0, k]$ for $k \in \mathbf{N}$ and $\phi = 0$ outside of $[\lambda, \eta]$. In the definition, we can replace the Caputo derivative with the Riemann-Liouville derivative in applications. This definition can also be generalized to the case when ϕ is a $\mathcal{E}'(\mathbf{R})$ distribution, where $\mathcal{E}'(\mathbf{R})$ is the space of compactly supported distributions on the space $\mathcal{E}(\mathbf{R})$ of smooth functions. When $\phi \in \mathcal{E}'(\mathbf{R})$, $\mathrm{supp}\phi \subset [0, k]$, $_0 D_t^\phi f$ is defined as an element of $\mathcal{S}'_+(\mathbf{R})$ by

$$\left\langle \int_{\mathrm{supp}\phi} \phi(\mu) {_0^C} D_t^\mu f(t) \mathrm{d}\mu, \psi(t) \right\rangle = \left\langle \phi(\mu), \left\langle {_0^C} D_t^\mu f(t), \psi(t) \right\rangle \right\rangle, \quad \psi \in \mathcal{S}(\mathbf{R}).$$

Here, $h \in \mathcal{S}'_+(\mathbf{R})$ if $h \in \mathcal{S}'(\mathbf{R})$ and $\mathrm{supp} h \subset [0, \infty)$, where $\mathcal{S}'(\mathbf{R})$ is the space of tempered distributions, i.e., the dual of the space $\mathcal{S}(\mathbf{R})$ of rapidly decreasing smooth functions. For a detailed analysis, we refer the authors to [12]. By using the Laplace transformation for the Caputo derivative, we have for Re $s > 0$,

$$\mathscr{L}[_0D_t^\phi f(t)] = \mathscr{L}[f](s) \int_0^k \phi(\mu)s^\mu d\mu - \sum_{l=0}^{k-1} \frac{u^{(l)}(t)|_{t=0}}{s^{l+1}} \int_l^k \phi(\mu)s^\mu d\mu. \quad (2.4.1)$$

Example 2.4.1   *Let* $f \in AC_{loc}^2([0, \infty))$, *i.e.,* f *is continuous and* f' *is absolutely continuous on* $[0, T]$ *for any* $T > 0$. *When* $k = 2$ *and* $\phi(\mu) = \sum_{j=1}^k a_j\delta(\mu - \mu_j)$ *for* $a_j \in \mathbf{R}$ *and* $\mu_j \in [0, 2]$, *we have*

$$_0D_t^\phi f(t) = \sum_{j=1}^k a_j {_0^C}D_t^{\mu_j} f(t), \quad t > 0,$$

which reduces to a linear combination of Caputo derivatives of different orders. In this case, we have for Re $s > 0$,

$$\mathscr{L}[_0D_t^\phi f(t)] = \mathscr{L}[f](s) \sum_{j=1}^k a_j s^{\mu_j} - \frac{f(t)|_{t=0}}{s} \sum_{j=1}^k a_j s^{\mu_j} - \frac{f'(t)|_{t=0}}{s^2} \sum_{1 < \mu_j \leqslant 2} a_j s^{\mu_j}.$$

Besides the time-fractional differential equation of distributed order, one may also consider the space-fractional differential derivatives. For example, we may consider

$$\int_0^2 a(\alpha)(-\Delta)^{\alpha/2}u(t, x)d\alpha,$$

for some positively integrable function $a(\cdot)$, and accordingly consider the distributed-order space-fractional differential equations

$$\frac{\partial}{\partial t}u(t, x) = \int_0^2 a(\alpha)(-\Delta)^{\alpha/2}u(t, x)d\alpha, \quad t > 0, x \in \mathbf{R}^d. \quad (2.4.2)$$

A distribution $G(t, x)$, which satisfies the equation (2.4.2) in the weak sense with initial data

$$G(0, x) = \delta(x), \quad x \in \mathbf{R}^d \quad (2.4.3)$$

is called a fundamental solution of the Cauchy problem (2.4.2) and (2.4.3). Let $\mathcal{B}(\xi) = -\int_0^2 a(\alpha)|\xi|^\alpha d\alpha$, then $G(t, x)$ is given by

$$G(t, x) = \mathcal{F}^{-1}(e^{t\mathcal{B}(\xi)}). \quad (2.4.4)$$

In the particular case of $a(\alpha) = \delta(\alpha - 2)$, we have the classical heat equation, whose fundamental solution is given by

$$G_2(t, x) = \frac{1}{(4\pi t)^{d/2}} e^{-\frac{|x|^2}{4t}}, \qquad (2.4.5)$$

while in the case of $a(\alpha) = \delta(\alpha - 1)$, the fundamental solution corresponds to the Cauchy-Poisson probability density

$$G_1(t, x) = \frac{\Gamma(d + 1/2)}{\pi^{(d+1)/2}} \frac{1}{(|x|^2 + t^2)^{(d+1)/2}}. \qquad (2.4.6)$$

For a general case of $a(\alpha) = \delta(\alpha - \alpha_0)$, $0 < \alpha_0 < 2$, the fundamental solution is the Lévy α_0-stable probability density

$$G_{\alpha_0}(t, x) = \frac{1}{(2\pi)^d} \int_{\mathbf{R}^d} e^{-t|\xi|^{\alpha_0}} e^{ix \cdot \xi} d\xi. \qquad (2.4.7)$$

Remark 2.4.1 *One may also consider the distributed order fractional derivatives when the Lebesgue measure is replaced by a general finite Borel measure. Let ν be a finite Borel measure with $\nu(0, k) > 0$, one may define*

$$_0 D_t^\nu f(t) = \int_0^k {}_0^C D_t^\mu f(t) \nu(d\mu).$$

2.4.1   Distributed order diffusion-wave equation

We now consider the following time-fractional differential equations of distributed order

$$_0 D_t^\phi u(x, t) - \Delta u(x, t) = f(t, x), \quad x \in \mathbf{R}^d, t \in [0, \infty). \qquad (2.4.8)$$

Here, we assume that $\phi(\mu) \geqslant 0$ and is not zero everywhere. The equation (2.4.8) is a generalization of the fractional differential equations and is important from the viewpoint of applications. When $\phi(\mu) = \delta(\mu - \mu_1)$, we obtain the fractional differential equation

$$_0^C D_t^{\mu_1} u = \Delta u + f.$$

In particular, when $\mu_1 = 1$ (resp. $\mu_1 = 2$), we obtain the heat equation (resp. wave equation) in $\mathbf{R}^d$. When $\phi(\mu) = a_1 \delta(\mu - \mu_1) + a_2 \delta(\mu - \mu_2)$ with $0 < \mu_1 < \mu_2 \leqslant 1$, $a_1 > 0$, $a_2 > 0$, $a_1 + a_2 = 1$, we obtain

$$a_1 {}_0^C D_t^{\mu_1} u + a_2 {}_0^C D_t^{\mu_2} u = \Delta u + f,$$

which describes a sub-diffusion process with retardation [42].

We consider the more general equation (2.4.8) with initial data

$$u^{(l)}(0, x) = u_l(x), \quad x \in \mathbf{R}^d, l = 0, 1, \cdots, k - 1. \tag{2.4.9}$$

By linear superposition principle, we may divide the equation (2.4.8) and (2.4.9) into the following several equations satisfied by u_j, $j = 0, 1, \cdots, k$. Let u_j, $j = 0, \cdots, k - 1$ satisfy

$$\begin{cases} {}_0D_t^\phi u(x,t) - \Delta u(x,t) = 0, & x \in \mathbf{R}^d, t \in [0, \infty), \\ u^{(j)}(0, x) = u_j(x); \\ u^{(l)}(0, x) = 0, & l \in \{0, 1, \cdots, k - 1\}, l \neq j, x \in \mathbf{R}^d. \end{cases} \tag{2.4.10}$$

and u_k satisfy

$$\begin{cases} {}_0D_t^\phi u - \Delta u = f, \\ u^{(l)}(0, x) = 0, & l \in \{0, 1, \cdots, k - 1\}, x \in \mathbf{R}^d. \end{cases} \tag{2.4.11}$$

Then the solution of (2.4.8) and (2.4.9) is given by $u = \displaystyle\sum_{j=0}^{k} u_j(t, x)$. We will use the Laplace transform method to find a solution of the equations (2.4.8) and (2.4.9). Applying the Laplace transformation, we obtain

$$\tilde{u} \int_0^k \phi(\mu)s^\mu d\mu - \sum_{l=0}^{k-1} \frac{u_l(x)}{s^{l+1}} \int_l^k \phi(\mu)s^\mu d\mu = \Delta\tilde{u} + \tilde{f}, \tag{2.4.12}$$

where $\tilde{u}(s, x) = \mathscr{L}[u](s, x)$ is the Laplace transform of the u and $\tilde{f}$ is the Laplace transform of f in the variable t. Let $\tilde{B}_l(s) := \displaystyle\int_l^k \phi(\mu)s^\mu d\mu \neq 0$, we can write the equation in the new form

$$\tilde{u} = \frac{u_0(x)}{s} + \sum_{l=1}^{k-1} \frac{u_l(x)}{s^{l+1}} \frac{\tilde{B}_l(s)}{\tilde{B}_0(s)} + \frac{\tilde{f}}{\tilde{B}_0(s)} + \frac{1}{\tilde{B}_0(s)}\Delta\tilde{u}. \tag{2.4.13}$$

In particular,

$$\tilde{u}_j = \frac{u_j(x)}{s^{j+1}} \frac{\tilde{B}_j(s)}{\tilde{B}_0(s)} + \frac{1}{\tilde{B}_0(s)}\Delta\tilde{u}_j, \quad k = 0, 1, \cdots, k - 1, \tag{2.4.14}$$

and

$$\tilde{u}_k = \frac{\tilde{f}}{\tilde{B}_0(s)} + \frac{1}{\tilde{B}_0(s)}\Delta\tilde{u}_k. \tag{2.4.15}$$

By taking inverse Laplace transform, we obtain from (2.4.13)

$$u(t, x) = u_0(x) + \sum_{l=1}^{k-1} \frac{u_l(x)}{\Gamma(l + 1)} t^l *_t b_l(t) *_t b_0(t) + f(t, x) *_t h_0(t) + b_0(t) *_t \Delta u, \tag{2.4.16}$$

where $b_0(t) = \mathscr{L}^{-1}[1/\tilde{B}_0(s)](t)$, $b_l(t) = \mathscr{L}^{-1}[\tilde{B}_l(s)](t)$ and $*_t$ denotes the convolution in the t variable. Equation (2.4.13) and (2.4.14) are of the form of an abstract Volterra equation

$$x(t) = w(t) + b(t) * \mathcal{A}x(t), \quad t \in J,$$

where $b(t)$ is the scalar kernel, $w \in C(J, X)$ is a continuous function from interval J to space X and $\mathcal{A}$ is an unbounded operator on a dense set of $X_{\mathcal{A}}$ equipped with the graph norm $\|x\|_{\mathcal{A}} = \|x\| + \|\mathcal{A}x\|$. Resolvents for such problems and their applications to well-posedness are introduced in [181], for example.

One remark should be placed here. From (2.4.13) and (2.4.14), we know that between two consecutive integers, it does not matter how many different fractional orders are taken in the given equation. For the initial value problem, we need only to give the initial values of the unknown solution just for the integer order derivatives less than k. As for the number of initial conditions, it depends on the support of the weight function ϕ. If for some least $k > 0$, such that supp$\phi \subset [0, k]$, then k initial conditions should be placed on the unknown function. Indeed, if supp$\phi \subset [0, k-1]$, then the Cauchy problem with k given initial data becomes ill-posed. In such a case, $\tilde{B}_{k-1}(s) = 0$, and then u_{k-1} satisfies

$$\tilde{u}_j = \frac{1}{\tilde{B}_0(s)} \Delta \tilde{u}_j,$$

whose solution vanishes in $\mathbf{R}^d$.

Special attention is focused on the cases when $k = 1$ and $k = 2$. The equation (2.4.8) is then completed by the following initial conditions

$$\begin{cases} u(x,0) = u_0(x), & \textit{if } \text{supp}\phi \subset [0,1] \\ u(x,0) = u_0(x), \ \partial_t u(x,0) = u_1(x), & \textit{if } \text{supp}\phi \subset [0,2]. \end{cases} \tag{2.4.17}$$

Then u is a solution to the initial value problem (2.4.8) and (2.4.17) with suitable assumptions on u_0 and u_1, if $u \in AC^1_{loc}([0,\infty); H^2(\mathbf{R}^d))$, i.e., u is locally absolutely continuous in time on $[0,\infty)$ with values in $H^2(\mathbf{R}^d)$ when supp$\phi \subset [0,1]$ and $u \in AC^2_{loc}([0,\infty); H^2(\mathbf{R}^d))$ when supp$\phi \subset [0,2]$, respectively, and satisfies (2.4.8) and (2.4.17). According to the support of ϕ, we can divide (2.4.8) into the following three cases:

1. distributed-order diffusion-wave equation, if $0 \leqslant a \leqslant 1 < b \leqslant 2$;
2. distributed-order diffusion equation, if $b \leqslant 1$; and
3. distributed-order wave equation, if $a > 1$.

For more detailed studies for the cases $k = 1$ or $k = 2$ can be found in recent papers by Atanackovic $et\ al$ [11,12].

2.4.2   Initial boundary value problem of distributed order

In the following, we consider a initial boundary value problem of the distributed order time-fractional differential equation of the form

$$\int_0^1 \phi(\mu)_0^C D_t^\mu u(t) d\mu = \frac{\partial^2}{\partial x^2} u \qquad (2.4.18)$$

with the following initial and boundary conditions

$$\begin{cases} u(0,x) = u_0(x), \\ u(t,0) = u(t,1) = 0. \end{cases} \qquad (2.4.19)$$

Here, $f(x)$ denotes the initial distribution of the temperature. We employ the separation of variables method to find out a solutions (2.4.18) and (2.4.24). Let $u(t,x) = X(x)T(t)$, then we obtain from (2.4.18)

$$\frac{1}{T(t)} \int_0^1 \phi(\mu)^C{}_0 D_t^\mu T(t) d\mu = \frac{1}{X} X'' = -\lambda.$$

Consider the eigenvalue problem for X:

$$\begin{cases} X'' + \lambda X = 0, \\ X(0) = X(1) = 0. \end{cases}$$

By standard ODE theory, we know λ has discrete eigenvalues $\lambda_n = n^2 \pi^2$ with eigenfunction $X_n(x) = \sin(n\pi x)$. Fixing λ_n, we solve the equation of T by Laplace transform. Let $\tilde{T}_n(s)$ denote the Laplace transform of $T_n(t)$ and $T_n(0) = 1$, then we have

$$\tilde{T}_n(s)\tilde{B}_0(s) - \frac{1}{s}\tilde{B}_0(s) + (n\pi)^2 \tilde{T}_n(s) = 0,$$

which yields

$$\tilde{T}_n(s) = \frac{\displaystyle\int_0^1 \phi(\mu)s^{\mu-1}d\mu}{\tilde{B}_0(s) + (n\pi)^2}.$$

To find the inverse Laplace transform for $\tilde{T}_n(s)$, we consider

$$\tilde{T}_n(\mu, s) = \frac{s^{\mu-1}}{\tilde{B}_0(s) + (n\pi)^2},$$

whose inverse transform is given by $T_n(\mu,t)$ and the solution $T_n(t)$ can thus be expressed as

$$T_n(t) = \int_0^1 \phi(\mu)T_n(\mu,t)d\mu.$$

By the complex inversion formula

$$T_n(\mu, t) = \frac{1}{2\pi i} \int_{\gamma-i\infty}^{\gamma+i\infty} \tilde{T}_n(\mu, s) e^{st} ds, \quad (t > 0),$$

where s is taken to be complex and γ is an arbitrary real number which lies to the right of all poles and branch points in the integral. Due to $s^{\mu-1}$ in the numerator and s^μ in the denominator, the integral has a branch point. Any poles will be simple and are zeros of $g(s) := \tilde{B}_0(s) + (n\pi)^2$. The function $g(s)$ is analytic in any region not containing the origin, so any poles will be isolated. Since $\phi(\mu) \geqslant 0$ for all $\mu \in [0, 1]$ and not zero everywhere, it can be proved that all the zeros of $g(s)$ lie on the negative real axis. The inverse transform $T_n(\mu, t)$ can be computed using residues. Due to the branch point at the origin, the usual Bromwich contour cannot be used. In essence, a path of integration is then chosen that excludes the branch points. This is referred to as a Hankel contour. We let $\Gamma_{R,\varepsilon} = \sum_{k=0}^{5} \Gamma_{R,r}^{(k)}$, with $\Gamma_{R,\varepsilon}^{(0)} = \{\gamma + i\eta : -R \leqslant \eta \leqslant R\}$, $\Gamma_{R,\varepsilon}^{(1)} = \{\gamma + Re^{i\theta} : \pi/2 \leqslant \theta \leqslant \pi\}$, $\Gamma_{R,\varepsilon}^{(2)} = \{re^{\pi i} : -R + \gamma \leqslant r \leqslant -\varepsilon\}$, $\Gamma_{R,\varepsilon}^{(3)} = \{\varepsilon e^{-i\theta} : -\pi \leqslant \theta \leqslant \pi\}$, $\Gamma_{R,\varepsilon}^{(4)} = \{-re^{-\pi i} : \varepsilon \leqslant r \leqslant R - \gamma\}$ and $\Gamma_{R,\varepsilon}^{(5)} = \{\gamma + Re^{i\theta} : \pi/2 \leqslant \theta \leqslant 3\pi/2\}$, where $\varepsilon > 0$ is sufficient small and $R > 0$ is sufficiently large. By using the residue formula, we have

$$\frac{1}{2\pi i} \int_{\Gamma_{R,\varepsilon}} \frac{e^{st} s^{\mu-1}}{\tilde{B}_0(s) + (n\pi)^2} ds = Res\left(\frac{e^{st} s^{\mu-1}}{\tilde{B}_0(s) + (n\pi)^2}\right). \tag{2.4.20}$$

By letting $R \to \infty$ and $\varepsilon \to 0$, we obtain

$$T_n(\mu, t) = Res\left(\frac{e^{st} s^{\mu-1}}{\tilde{B}_0(s) + (n\pi)^2}\right)$$
$$- \frac{1}{2\pi i} \lim_{R\to\infty,\varepsilon\to 0}\left[\int_{\Gamma_{R,\varepsilon}^{(2)}} + \int_{\Gamma_{R,\varepsilon}^{(4)}} \frac{e^{st} s^{\mu-1}}{\tilde{B}_0(s) + (n\pi)^2} ds\right], \tag{2.4.21}$$

since the contributions along $\Gamma_{R,\varepsilon}^{(1)}$, $\Gamma_{R,\varepsilon}^{(3)}$, $\Gamma_{R,\varepsilon}^{(5)}$ vanish as $R \to \infty$ and $\varepsilon \to 0$. Since the only possible poles are on the negative real axis, the contribution from the residue part also vanishes. This leaves only the contributions along the integral path $-\infty \to 0$ and $0 \to -\infty$, which yield

$$T_n(\mu, t) = -\frac{1}{2\pi i} \lim_{R\to\infty,\varepsilon\to 0}\left[\int_{\Gamma_{R,\varepsilon}^{(2)}} + \int_{\Gamma_{R,\varepsilon}^{(4)}} \frac{e^{st} s^{\mu-1}}{\tilde{B}_0(s) + (n\pi)^2} ds\right]$$

$$= \frac{1}{2\pi i} \left\{ \int_0^\infty \frac{e^{-rt} r^{\mu-1} e^{i\pi\mu} dr}{\int_0^1 \phi(\mu) r^\mu e^{i\pi\mu} d\mu + (n\pi)^2} - \int_0^\infty \frac{e^{-rt} r^{\mu-1} e^{-i\pi\mu} dr}{\int_0^1 \phi(\mu) r^\mu e^{-i\pi\mu} d\mu + (n\pi)^2} \right\}$$

$$= \frac{1}{\pi} \Im \left\{ \int_0^\infty \frac{e^{-rt} r^{\mu-1} e^{i\pi\mu} dr}{\int_0^1 \phi(\mu) r^\mu e^{i\pi\mu} d\mu + (n\pi)^2} \right\}.$$

$$(2.4.22)$$

The solution to (2.4.18) and (2.4.24) is given by

$$u(t,x) = \sum_{n=1}^\infty a_n T_n(t) \sin(n\pi x), \tag{2.4.23}$$

where $\{a_n\}_{n=1}^\infty$ are the Fourier coefficients of $u_0(x)$, i.e., $a_n = 2 \int_0^1 u_0(x) \sin(n\pi x) dx$.

Remark 2.4.2 *The equation (2.4.18) with the following initial and Neumann boundary condition*

$$\begin{cases} u(0,x) = u_0(x), \\ \partial_x u(t,0) = \partial_x u(t,1) = 0 \end{cases} \tag{2.4.24}$$

can be similarly handled and we finally obtain

$$u(t,x) = \sum_{n=0}^\infty a_n B_n(t) \cos(n\pi x),$$

where $\{a_n\}_{n=0}^\infty$ are the coefficients of $u_0(x)$.

2.5   Appendix A: the Fourier transform

The Fourier transform is a powerful tool in the analysis, one of whose advantages is to transform the differentiation operation and convolution into the product operation in phase space. In this section, the Fourier transform and its basic properties are introduced. For further knowledge about the Fourier transform, readers may refer to, for example, [204-206].

What follows is divided into several parts. First, we introduce the definition of the Fourier transform in L^1, and by applying the continuity method we extend the Fourier transform to L^2. The key point is the Plancherel identity. Next introduce the Fourier transform in L^p for $1 < p < 2$ by applying the interpolation and Hausdorff-Young inequality. To extend the Fourier transform to distributions, we first consider the Fourier transform in the Schwartz space and then use the duality method. For the sake of convenience, some properties of the Fourier transform are listed in Table 2.5.1 below.

Definition 2.5.1 *If $f(x) \in L^1(\mathbf{R}^d)$, the Fourier transform of f is defined by*

$$\mathcal{F}f(\xi) = \hat{f}(\xi) = \int_{\mathbf{R}^d} f(x)e^{-ix\cdot\xi}dx, \tag{2.5.1}$$

where $x \cdot \xi$ is the inner product in $\mathbf{R}^d$.

By this definition, as long as $f \in L^1(\mathbf{R}^d)$, $\mathcal{F}f(\xi)$ makes sense. First of all, we give two useful theorems, cf. [228].

Theorem 2.5.1 *Let $\hat{f} \in L^1(\mathbf{R}^d)$, then $\hat{f}(\xi)$ is a uniformly continuous function on $\mathbf{R}^d$.*

Theorem 2.5.2 (Riemann-Lebesgue Lemma) *Suppose $\hat{f} \in L^1(\mathbf{R}^d)$, then $\lim_{|\xi|\to\infty} \hat{f}(\xi) = 0$.*

In addition, there holds the following properties for $f \in L^1(\mathbf{R}^d)$.
(1) $\mathcal{F}$ is a linear operator on $L^1(\mathbf{R}^d)$, and

$$\|\mathcal{F}f\|_{L^\infty} \leqslant \|f\|_{L^1}.$$

Furthermore, if $f(x) \geqslant 0$, then $\|\mathcal{F}f\|_{L^\infty} = \|f\|_{L^1} = \hat{f}(0)$.
(2) Let τ_a as translation operator such that $\tau_a f(\cdot) = f(\cdot - a)$, then

$$\mathcal{F}(\tau_a f)(\xi) = e^{-ia\xi}\mathcal{F}f(\xi).$$

(3) Let δ_λ be the scaling operator such that $(\delta_\lambda f)(x) = f(x/\lambda)$, then

$$\widehat{\delta_\lambda f}(\xi) = \lambda^d \hat{f}(\lambda\xi), \quad \lambda > 0.$$

(4) Let x_k be the k-th coordinate of x and $x_k f \in L^1(\mathbf{R}^d)$, then

$$\frac{\partial \hat{f}(\xi)}{\partial \xi} = (\widehat{-ix_k f})(\xi).$$

If $f, \dfrac{\partial f}{\partial x} \in L^1(\mathbf{R}^d)$, then

$$\mathcal{F}\left(\frac{\partial f}{\partial x_k}\right)(\xi) = i\xi_k \hat{f}(\xi).$$

(5) If $f, g \in L^1$, then $\widehat{f * g} = \hat{f}\hat{g}$. Indeed, when $f, g \in L^1$, then $f * g \in L^1(\mathbf{R}^d)$ by Fubini Theorem and

$$\widehat{f * g}(\xi) = \iint e^{-ix\cdot\xi} f(x - y)g(y)\mathrm{d}y\mathrm{d}x$$
$$= \iint e^{-i(x-y)\cdot\xi} f(x - y)e^{-iy\cdot\xi}g(y)\mathrm{d}x\mathrm{d}y$$
$$= \hat{f}(\xi)\hat{g}(\xi).$$

(6) Multiplication formula. Let $f, g \in L^1$, then

$$\int_{\mathbf{R}^d} \hat{f}(x)g(x)\mathrm{d}x = \int_{\mathbf{R}^d} f(x)\hat{g}(x)\mathrm{d}x.$$

In this case, since $f, g \in L^1$, then $\hat{f}, \hat{g} \in L^\infty$, and the integrals of both sides makes sense.

The above properties show that the Fourier transform maps L^1 into L^∞. However, not every L^∞ function is a Fourier transform of an L^1 function, such as the constant function. However, we have the following (cf. [118])

Theorem 2.5.3 Let $f \in L^1(\mathbf{R}^d) \cap C^1(\mathbf{R}^d)$, then

$$\lim_{N\to\infty} \frac{1}{(2\pi)^d} \int_{-N}^{N} \hat{f}(\xi)e^{ix\cdot\xi}\mathrm{d}\xi = f(x),$$

where the left hand side denotes the Cauchy principal value integral.

This theorem states that any function $f \in L^1(\mathbf{R}^d) \cap C^1(\mathbf{R}^d)$ can be decomposed into a superposition of simple harmonic waves $e^{ix\cdot\xi}$ while $\hat{f}(\xi)$ denotes the complex amplitude of harmonic waves at frequency ξ. Therefore, $\hat{f}$ is also called the spectrum of f in applied sciences. This leads to the definition of the inverse Fourier transform. Let $g(\xi) \in L^1(\mathbf{R}^d)$, then

$$\mathcal{F}^{-1}f(x) = \frac{1}{(2\pi)^d} \int_{\mathbf{R}^d} g(\xi)e^{ix\cdot\xi}\mathrm{d}\xi \tag{2.5.2}$$

is called the inverse Fourier transform of g.

Theorem 2.5.4 Let $f \in L^1(\mathbf{R}^d)$, $\hat{f} \in L^1(\mathbf{R}^d)$, then

$$\mathcal{F}^{-1}(\mathcal{F}f)(x) = \mathcal{F}(\mathcal{F}^{-1}f)(x) = f(x), \quad a.e. \ x \in \mathbf{R}^d.$$

Under the conditions of this theorem, we know that $\mathcal{F}^{-1}(\mathcal{F}f)(x)$ is uniformly continuous and tends to zero as $|x| \to \infty$ according to Theorem 2.5.1 and Theorem2.5.2. Furthermore, there is always a continuous function $\tilde{f}(x)$

in the equivalence class of f such that $\mathcal{F}^{-1}(\mathcal{F}\tilde{f})(x) = \tilde{f}(x)$ for any $x \in \mathbf{R}^d$. The Fourier transform and the inverse Fourier transform are generalizations of the Fourier series. Here we draw an analogy between the Fourier transform and the Fourier series of periodic function $f(x)$ in $(-l, l)$. In the one dimensional case, f is expanded in Fourier series,

$$f(x) = \sum_{n=-\infty}^{\infty} c_n e^{in\pi x/l},$$

with

$$c_n = \frac{1}{2l} \int_{-l}^{l} f(x) e^{-in\pi x/l} dx$$

being the Fourier coefficient. The coefficients c_n can be viewed as a discrete Fourier transform, and the Fourier series expansion of f can be thought of as discrete inverse Fourier transform. In fact, letting $l \to \infty$ we can get formally the Fourier transform in Definition 2.5.1.

We now introduce the Fourier transform of periodic functions. Let a_1, $a_2, \cdots, a_n$ are positive integers, and $\mathbf{T}_a^d$ be the d-dimensional periodic box with the period in the ith direction being $2\pi a_i$. Let also $\mathbf{Z}_a^d = \mathbf{Z}/a_1 \times \cdots \times \mathbf{Z}/a_N$ be the dual lattice of $\mathbf{T}_a^d$. For function u in $\mathbf{T}_a^d$, it can be expressed in terms of Fourier series

$$u(x) = \sum_{\xi \in \mathbf{Z}_a^d} \hat{u}_\xi e^{i\xi \cdot x},$$

where

$$\hat{u}_\xi := \frac{1}{|\mathbf{T}_a^d|} \int_{\mathbf{T}_a^d} e^{-i\xi \cdot y} u(y) dy, \quad \xi \in \mathbf{Z}_a^d.$$

The property (5) of the Fourier transform of functions in L^1 can be generalized as follows

(5') Let $\hat{f}(n)$ and $\hat{g}(n)$ denote the Fourier transform of f and g, respectively, then

$$\widehat{fg}(n) = \sum_{n_1+n_2=n} \hat{f}(n_1)\hat{g}(n_2).$$

After discussing the Fourier transform in $L^1(\mathbf{R}^d)$, we consider the Fourier transform on L^2. For $f \in L^2(\mathbf{R}^d)$, the integral in the definition 2.5.1 is not necessarily convergent. However, its Fourier transform can be defined by the continuity method, which requires the following Plancherel identity. Since $L^1 \cap L^2$ is a dense linear subspace in L^2, the Fourier transform is firstly defined in $L^1 \cap L^2$, and then extend to L^2 by the Hahn-Banach extension theorem.

Lemma 2.5.1 *Let $f \in L^1 \cap L^2$, then $\hat{f} \in L^2(\mathbf{R}^d)$ and $\|\hat{f}\|_{L^2} = (2\pi)^d \|f\|_{L^2}$.*

Proof The proof can be found in [60, 228].

The coefficient $(2\pi)^d$ is caused by the definition of the Fourier transform and appropriately modifying the definition can eliminate this coefficient. According to Lemma 2.5.1, $\mathcal{F}$ is a bounded linear operator on $L^1 \cap L^2 \to L^2$, hence by Hahn-Banach theorem there exists a unique bounded extension $\tilde{\mathcal{F}}$ on $L^2(\mathbf{R}^d)$ such that $\tilde{\mathcal{F}}|_{L^1 \cap L^2} f = \mathcal{F}f$ for all $f \in L^1 \cap L^2$ and $\|\tilde{\mathcal{F}}\| \leqslant \|\mathcal{F}\|$. The extension is called the Fourier transform on $L^2(\mathbf{R}^d)$, still denoted by $\mathcal{F}$. Indeed, for $f \in L^2(\mathbf{R}^d)$, there exists a sequence f_k in $L^2 \cap L^1$ such that $f_k \to f$ in L^2 as $k \to \infty$. It follows from Lemma 2.5.1 that $\|\hat{f}_k - \hat{f}_l\|_{L^2} \to 0$ and hence there exists a function $\hat{f}$ in L^2 such that $\hat{f}_k \to \hat{f}$ as $k \to \infty$. Then the Fourier transform on $f \in L^2$ is defined to be $\hat{f}$. As will be shown in the following theorem, mapping $f \mapsto \hat{f}$ is not only isometric in the sense of Lemma 2.5.1, but also unitary. I.e., $\mathcal{F}$ is a invertible isometric transformation in L^2.

Theorem 2.5.5 $\mathcal{F}$ *is a unitary transformation on $L^2(\mathbf{R}^d)$.*

Proof Since $\mathcal{F}$ is a isometric linear operator on L^2 by Lemma 2.5.1, we only need to show $\mathcal{F}$ is surjective. Since $\mathcal{F}$ is isometric and L^2 is closed, the range $\mathcal{R}(\mathcal{F})$ is a closed subspace of L^2. Let $\varphi \in L^2$ such that $\int_{\mathbf{R}^d} \hat{f}(\xi)\overline{\varphi(\xi)}d\xi = 0$ for all $f \in L^2$. Applying product formula then yields $\int_{\mathbf{R}^d} f(x)\hat{\overline{\varphi}}(x)dx = 0$ for all $f \in L^2$. Direct calculation shows that $\hat{\overline{\varphi}}(x) = \overline{\hat{\varphi}(-x)}$, and $\int_{\mathbf{R}^d} f(x)\overline{\hat{\varphi}(-x)}dx = 0$ for all $f \in L^2$. In particular, taking $f(x) = \hat{\varphi}(-x)$ yields $\|\varphi\|_{L^2} = 0$. This shows that $\mathcal{F}$ is surjective, and hence $\mathcal{F}$ is unitary on L^2.

This theorem is called Plancherel theorem and the product formula is called Plancherel identity. Analogously, applying polarization identity

$$(f, g) = \frac{1}{2}\left\{\|f + g\|_{L^2}^2 - i\|f + ig\|_{L^2}^2 - (1 - i)\|f\|_{L^2}^2 - (1 - i)\|g\|_{L^2}^2\right\},$$

one obtains the following Parseval identity

$$(f, g) = \int_{\mathbf{R}^d} f(x)\overline{g(x)}dx = \frac{1}{(2\pi)^d}\int_{\mathbf{R}^d} \hat{f}(\xi)\overline{\hat{g}(\xi)}d\xi = \frac{1}{(2\pi)^d}(\hat{f}, \hat{g}).$$

After discussing the Fourier transform in L^1 and L^2, we generalize the Fourier transform to $L^p(\mathbf{R}^d)$.

Theorem 2.5.6 *Let* $1 < p < 2$, $f \in L^p(\mathbf{R}^d) \cap L^1(\mathbf{R}^d)$ *and* $\frac{1}{p} + \frac{1}{q} = 1$, *then* $\mathcal{F}f \in L^q(\mathbf{R}^d)$ *and*

$$\|\mathcal{F}f\|_{L^q} \leqslant C_p^n \|f\|_{L^p}. \tag{2.5.3}$$

The inequality (2.5.3) is called the Hausdorff-Young inequality. The proof of this theorem can be found in [131], and the proof of inequality (2.5.3) employs the Riesz-Thorin interpolation theorem. Details can be found in [59, 189]. As a remarkable note, the necessary and sufficient condition for the equality in (2.5.3) holds is that f is a Gauss function of the form $f(x) = Ae^{-\langle x, Mx \rangle} + Bx$, where $A \in \mathbb{C}$, M is an arbitrary real symmetric positive definite matrix and B is an arbitrary vector in $\mathbb{C}^n$.

Since $L^p \cap L^1$ is a sublinear space of L^1, we can also define the Fourier transform on $L^p(\mathbf{R}^d)$ for $1 < p < 2$ by continuity method. Let $f \in L^p(\mathbf{R}^d)$ for $1 < p < 2$, the Fourier transform of f is denoted by $\hat{f}$ belonging to $\in L^q(\mathbf{R}^d)$ and satisfies (2.5.3). However, unlike the case of $p = 2$, $\mathcal{F} : L^p \to L^q$ is not surjective, so the Fourier transform $\mathcal{F} : L^p \to L^q$ is not invertible.

On the other hand, Theorem 2.5.6 shows that the index q in (2.5.3) is not arbitrary, which should be the conjugate index of p, i.e., $\frac{1}{p} + \frac{1}{q} = 1$. This can be also seen from scaling. Let $f \in L^p$, then $g(x) = (\delta_\lambda f)(x) = f(x/\lambda) \in L^p$ for all $\lambda > 0$ and hence $\|\hat{g}\|_{L^q} \leqslant C_p^n \|g\|_{L^p}$ with the same constant C_p^n as in (2.5.3). This is equivalent to the inequality in terms of f

$$\lambda^{d - \frac{d}{q}} \|\hat{f}\|_{L^q} \leqslant C_p^n \lambda^{\frac{d}{p}} \|f\|_{L^p}, \quad \forall \lambda > 0.$$

Since $\lambda > 0$ is arbitrary, it follows that $d - \frac{d}{q} = \frac{d}{p}$. Otherwise, the ratio $\|\hat{f}\|_{L^q} / \|f\|_{L^p}$ can be taken arbitrarily large. Similar to the property (5), we can show

(5") Let $f \in L^p(\mathbf{R}^d)$, $g \in L^q(\mathbf{R}^d)$ and $1 + \frac{1}{r} = \frac{1}{p} + \frac{1}{q}$ with $1 \leqslant p, q, r \leqslant 2$, then

$$\widehat{f * g}(\xi) = \hat{f}(\xi)\hat{g}(\xi).$$

Proof   It follows $f * g \in L^r(\mathbf{R}^d)$ from Young's inequality of convolution. Hence the Fourier transform of $f*g$ makes sense and $\widehat{f * g} \in L^{r'}(\mathbf{R}^d)$. On the other hand, from (2.5.3), it follows that $\hat{f} \in L^{p'}(\mathbf{R}^d)$ and $\hat{g} \in L^{q'}(\mathbf{R}^d)$ and hence $\hat{f}\hat{g} \in L^{r'}(\mathbf{R}^d)$ by Hölder inequality. Therefore, both sides in (2.5) make sense. When both f and g belong to $L^1(\mathbf{R}^d)$, (5") reduces to (5). Otherwise by an approximation process, one can also establishes (5").

However, there are counter examples showing that inequalities similar to (2.5.3) when $p > 2$ does not exist. Therefore, the continuity method fails when $p > 2$. When regarded as a distribution, functions in $L^p(\mathbf{R}^d)$ can also be defined, as discussed in what follows. We first consider the Fourier transform of functions in Schwartz class.

Definition 2.5.2 *The Schwartz class of functions $\mathcal{S}$ is defined as*

$$\mathcal{S} = \{\phi: \ \phi \in C^\infty(\mathbf{R}^d), \ \sup_{x \in \mathbf{R}^d} |x^\alpha \partial^\beta \phi| < \infty, \ \ \forall \alpha, \beta \in \mathbb{N}^d\}.$$

Note that under the usual scalar multiplication and addition operations, $\mathcal{S}$ is a vector space, which is also called the *calss of rapidly decreasing functions*. It turns out to be a Hausdorff locally convex topological space under the family of seminorms

$$\rho_{\alpha,\beta} = \sup_{x \in \mathbf{R}^d} |x^\alpha \partial^\beta \phi|, \quad \forall \alpha, \beta \in \mathbb{N}^d.$$

A sequence $\{\phi_\nu(x)\} \in \mathcal{S}$ is said to converge to zero if $x^\alpha \partial^\beta \phi_\nu(x) \to 0$ uniformly as $\nu \to \infty$ for any multi-indices α, β.

Definition 2.5.3 *The space $\mathcal{S}'$ of tempered distributions is defined as the space consisting of continuous linear functionals on the Schwartz space $\mathcal{S}$.*

In other words, F is a tempered distribution if and only if $\lim_{\nu \to \infty} F(\phi_\nu) = 0$ whenever $\lim_{\nu \to \infty} \rho_{\alpha,\beta}(\phi_\nu) = 0$ for all multi-indices α and β. $F(\phi)$ is also denoted as $\langle \phi, F \rangle$.

Example 2.5.1 1. $\phi(x) = e^{-x^2} \in \mathcal{S}$ is a rapidly decreasing function.
2. $\delta(x) \in \mathcal{S}'$ is a tempered distribution.
3. $\langle \phi, \delta \rangle = \int_{\mathbf{R}^d} \delta(x)\phi(x)dx = \phi(0)$.

Theorem 2.5.7 *If $\phi \in \mathcal{S}$, then $\mathcal{F}\phi \in \mathcal{S}$.*

Proof Indeed, $\mathcal{F}$ is a continuous linear operator from $\mathcal{S}$ to itself. Linearity is obvious. Next we show that $\mathcal{F}\phi \in \mathcal{S}$. For any multi-indices α, β, it follows by definition of Fourier transform that

$$(i\xi)^\alpha \partial^\beta \hat{\phi}(\xi) = \mathcal{F}(\partial^\alpha((-ix)^\beta \phi))(\xi),$$

and hence

$$\sup_{y \in \mathbf{R}^d} |(i\xi)^\alpha \partial^\beta \hat{\phi}(\xi)| \leqslant \int_{\mathbf{R}^d} |\partial^\alpha (ix)^\beta \phi(x)| \, dx < \infty.$$

Finally, for any multi-indices α, β, there holds

$$\sup_{y \in \mathbf{R}^d} |(i\xi)^\alpha \partial^\beta \hat{\phi}(\xi)| \leqslant \int_{\mathbf{R}^d} \frac{(1+|x|^2)^d |\partial^\alpha (ix)^\beta \phi(x)|}{(1+|x|^2)^d} dx$$

$$\leqslant C \sup_{x \in \mathbf{R}^d} |(1+|x|^2)^d |\partial^\alpha (ix)^\beta \phi(x)||$$

$$\leqslant C \sum_{|\tilde{\beta}| \leqslant |\beta|+2d, |\tilde{\alpha}| \leqslant |\alpha|} \sup_{x \in \mathbf{R}^d} |x^{\tilde{\beta}} \partial^{\tilde{\alpha}} \phi(x)|.$$

This shows that for any multi-indices α, β, $\xi^\alpha \partial^\beta \hat{\phi}(\xi)$ is bounded and hence belongs to $\mathcal{S}$. It is also shown that $\hat{\phi}_\nu \to 0$ in $\mathcal{S}$ as $\phi_\nu \to 0$ in $\mathcal{S}$. Similarly, we can show that $\mathcal{F}^{-1}$ is a continuous linear mapping from $\mathcal{S}$ to itself. Therefore, $\mathcal{S}$ is indeed an isomorphism on $\mathcal{S}(\mathbf{R}^d)$.

Now we turn to consider the Fourier transform on $\mathcal{S}'$.

Definition 2.5.4 *Let T be a tempered distribution, $\hat{T}$ is called the Fourier transform of T and denoted as $\mathcal{F}T = \hat{T}$, if there exists a tempered distribution $\hat{T}$ on $\mathcal{S}'$ such that $\langle \hat{T}, \phi \rangle = \langle T, \mathcal{F}\phi \rangle$ for all $\phi \in \mathcal{S}$.*

Theorem 2.5.7 shows that if $\phi \in \mathcal{S}$, then $\mathcal{F}\phi \in \mathcal{S}$, thus $\langle \hat{T}, \phi \rangle = \langle T, \mathcal{F}\phi \rangle$ in the definition makes sense. Furthermore, if $\phi_\nu \to 0$ in $\mathcal{S}$, then $\mathcal{F}\phi_\nu \to 0$. Similarly, the inverse Fourier transform $\mathcal{F}^{-1}T$ of a tempered distribution T is defined as $\mathcal{F}^{-1}T$ such that $\langle \mathcal{F}^{-1}T, \phi \rangle = \langle T, \mathcal{F}^{-1}\phi \rangle$ for all $\phi \in \mathcal{S}$. Let T be an absolutely integrable function on $\mathbf{R}^d$. Regarding T as a tempered distribution, then for any $\phi \in \mathcal{S}$ there holds

$$\langle \mathcal{F}T, \phi \rangle = \langle T, \mathcal{F}\phi \rangle$$

$$= \int_{\mathbf{R}^d} \left(\int_{\mathbf{R}^d} \phi(\xi) e^{-ix\cdot\xi} d\xi \right) T(x) dx$$

$$= \int_{\mathbf{R}^d} \left(\int_{\mathbf{R}^d} T(x) e^{-ix\cdot\xi} dx \right) \phi(\xi) d\xi.$$

Thus $\mathcal{F}T = \int_{\mathbf{R}^d} T(x) e^{-ix\cdot\xi} dx$ reduces to the classical Fourier transform in $L^1(\mathbf{R}^d)$.

If T is a tempered distribution and ψ is a slowly increasing infinitely differentiable function on $\mathbf{R}^d$, i.e., all derivatives of ψ grow at most as fast as polynomials, then $T\psi$ is a tempered distribution and $\mathcal{F}(T\psi) = \mathcal{F}T * \mathcal{F}\psi$. In particular, the Fourier transform of the constant function $\psi(x) = (2\pi)^{-d}$ is the delta distribution. Indeed, by Cauchy integral theorem, we have

$$\mathcal{F}(e^{-x^2/2m}) = m^{d/2}(2\pi)^{d/2} e^{-m\xi^2/2}.$$

When $m \to \infty$, we have $e^{-x^2/2m} \to 1$ and $m^{d/2} e^{-m\xi^2/2} \to (2\pi)^{d/2}\delta(x)$ in $\mathcal{S}'$, hence by continuity, we have $\mathcal{F}(1)(\xi) = (2\pi)^d \delta(\xi)$.

Example 2.5.2 1. $\mathcal{F}[e^{-x^2}] = \sqrt{\pi}e^{-\xi^2/4}$.

    2. $\mathcal{F}[\delta(x)] = 1(\xi)$.

    3. $\mathcal{F}(1)(\xi) = (2\pi)^d \delta(\xi)$.

We end the introduction of Fourier transform by the following table, which lists some Fourier transforms of some functions or distributions and the related properties.

Table 2.5.1   Fourier transform and the related properties

Primitive function $f(x)$	The Fourier transform $\hat{f}(\xi)$	Function(f, g)	Transform$(\hat{f}(\xi), \hat{g}(\xi))$
$\delta(x)$	1	$af(x) + bg(x)$	$a\hat{f}(\xi) + b\hat{g}(\xi)$
$e^{-a\lvert x \rvert}$	$\dfrac{2a}{a^2 + \xi^2}$	$\dfrac{df}{dx}$	$i\xi\hat{f}(\xi)$
$H(x)$	$\pi\delta(\xi) + \dfrac{1}{i\xi}$	$xf(x)$	$i\dfrac{d\hat{f}}{d\xi}$
$H(a - \lvert x \rvert)$	$\dfrac{2}{\xi}\sin a\xi$	$f(x - a)$	$e^{-ia\xi}\hat{f}(\xi)$
1	$2\pi\delta(\xi)$	$e^{iax}f(x)$	$\hat{f}(\xi - a)$
$e^{-x^2/2}$	$\sqrt{2\pi}e^{-\xi^2/2}$	$f(ax)$	$\dfrac{1}{a}\hat{f}\left(\dfrac{\xi}{a}\right)$

2.6   Appendix B: Laplace transform

As we have already shown, when $f \in L^1(\mathbf{R})$ then its Fourier transform exists in the classical sense. However, many simple functions cannot satisfy this strict requirement to be integrable. To remedy this drawback, the Laplace transform is proposed, which can be regarded as a generalization of the Fourier transform.

Definition 2.6.1 *Let f be a function defined in $\mathbf{R}^+$. If*

$$F(s) = \mathscr{L}[f](s) = \int_0^{+\infty} f(t)e^{-st}dt, \quad s \in \mathbb{C} \tag{2.6.1}$$

is convergent in a certain region of $\mathbb{C}$, then F is called the Laplace transform of f and f is called the inverse Laplace transform of F, denoted as $f = \mathscr{L}^{-1}[F]$.

It can be seen from the definition that the requirement of the function in the Laplace transform is much weaker that in the Fourier transform. For example, the Fourier transform of a Heaviside function H does not exist in the classic sense, but its Laplace transform does and

$$\mathscr{L}[H](s) = \int_0^{+\infty} e^{-st}dt = \frac{1}{s}, \quad \text{Re } s > 0,$$

where $H(t) = 0$ for $t \geqslant 0$ and $H(t) = 0$ otherwise. A function f grows at most exponentially, if there exist $M > 0$ and $\sigma > 0$ such that $|f(t)| \leqslant Me^{\sigma t}$ for any $t \geqslant 0$. It can be shown that if f grows at most exponentially and is piecewise continuous on any finite interval of $[0, \infty)$, then the Laplace transform of f exists for $\Sigma = \{\text{Re } s \geqslant \sigma_1 > \sigma\}$, the integral (2.6.1) converges absolutely and uniformly in Σ and $F(s)$ is an analytical function.

In practical problems, we usually need to find f when its Laplace transform F is given. In general, this is not an easy task. It generally reduces to the following complex integral, This integral is also called the Bromwich integral, Fourier-Mellin integral or the Mellin's inverse integral,

$$f(t) = \frac{1}{2\pi i} \int_{\gamma - i\infty}^{\gamma + i\infty} F(s)e^{st}ds, \quad t > 0,$$

where the path of integration is a vertical line parallel to the imaginary axis such that γ is greater than the real part of all singularities of $F(s)$. This ensures that the path is in the region of convergence. In particular, when all the singularities are in the left half plane, we can take $\gamma = 0$, and the integral reduces to the inverse Fourier transform of F. In practice, when $F(s)$ satisfies certain conditions, the inverse Laplace function can be obtained by Cauchy residue theorem in the complex integration theory and the residue formula is useful in computing such a integral.

The table lists some Laplace transforms of some frequently used functions and some properties of the Laplace transform.

Table 2.6.1   Laplace transform and the related properties

Function $f(t)$	Transform F(s)	Function	Transform
$t^{m-1}e^{at}$	$\dfrac{\Gamma(m)}{(s-a)^m}\,(m>0)$	$af(t) + bg(t)$	$aF(s) + bG(s)$
$\cos \omega t$	$\dfrac{s}{s^2 + \omega^2}$	$\underbrace{\int_0^t d\tau \cdots \int_0^\tau f(\tau')d\tau'}_{n\,\text{orders}}$	$s^{-n}F(s)$
$\sin \omega t$	$\dfrac{\omega}{s^2 + \omega^2}$	$f^n(t)$	$s^n F(s) - \displaystyle\sum_{j=0}^{n-1} s^{n-1-j}f^j(0)$
$t^m\,(m > -1)$	$\dfrac{\Gamma(m+1)}{s^{m+1}},\ \text{Re } s > 0$	$f(ct)$	$\dfrac{1}{c}F(s/c)$
$\delta(t - a)$	e^{-as}	$tf(t)$	$-\dfrac{dF}{df}$
$H(t - a)$	$\dfrac{1}{s}e^{-as}$	$\dfrac{f(t)}{t}$	$\displaystyle\int_s^\infty F(s')ds'$
$(\pi t)^{\frac{1}{2}}e^{-a^2/4t}$	$\dfrac{1}{\sqrt{s}}e^{-a\sqrt{s}}$	$\displaystyle\int_0^t g(t - \tau)f(\tau)d\tau$	$F(s)G(s)$

2.7 Appendix C: Mittag-Leffler function

2.7.1 Gamma function and Beta function

Let z be a complex number with Re $z > 0$, then the integral

$$\Gamma(z) = \int_0^\infty e^{-t} t^{z-1} dt$$

converges absolutely and is known as the Gamma function or the Euler integral of the second kind. It is obvious that $\Gamma(1) = 1$. Using integration by parts, one obtains

$$\int_0^\infty e^{-t} t^z dt = -e^{-t} t^z |_{t=0}^{t=\infty} + z \int_0^\infty e^{-t} t^{z-1} dt,$$

yielding a fundamental property of the Gamma function

$$\Gamma(z+1) = z\Gamma(z). \tag{2.7.1}$$

By employing this property, the Γ function can be generalize into the case Re $z < 0$. When $-m < \text{Re } z \leqslant -m+1$, we define

$$\Gamma(z) = \frac{\Gamma(z+m)}{z(z+1)\cdots(z+m-1)}.$$

Proposition 2.7.1 $\Gamma(n+1) = n!, \quad \forall n \in \mathbb{N}.$

Proposition 2.7.2 *There holds the Euler's reflection formula* $\Gamma(z)\Gamma(1-z) = \dfrac{\pi}{\sin(\pi z)}$ *for any z with Re $z \notin \mathbb{Z}$.*

Corollary 2.7.1 $\Gamma(1/2) = \sqrt{\pi}.$

Definition 2.7.1 *The Beta function, also called the Euler integral of the first kind, is a special function defined by*

$$B(z, w) = \frac{\Gamma(z)\Gamma(w)}{\Gamma(z+w)}, \quad \text{Re } z > 0, \text{Re } w > 0.$$

By definition, it is obvious that $B(z, w) = B(w, z)$. The Beta function is related to the Gamma function by the formula $B(z, w) = \Gamma(z)\Gamma(w)/\Gamma(z+w)$ for Re z, Re $w > 0$.

Proposition 2.7.3 *The Gamma function satisfies the Legendre formula*

$$\Gamma(z)\Gamma\left(z + \frac{1}{2}\right) = \sqrt{\pi} 2^{1-2z} \Gamma(2z).$$

Proof   Indeed, consider the identity for Re $z > 0$ that

$$B(z, z) = \int_0^1 (t(1 - t))^{z-1} dt.$$

Let $s = 4t(1 - t)$, then

$$B(z, z) = 2 \int_0^{1/2} (t(1 - t))^{z-1} dt$$

$$= \frac{1}{2^{2z-1}} \int_0^1 s^{z-1}(1 - s)^{-1/2} ds = 2^{1-2z} B(z, 1/2),$$

yielding the Legendre formula.

Employing the Beta function, the classical binomial coefficient

$$C_n^k = \frac{n!}{k!(n - k)!} = \frac{\Gamma(1 + n)}{\Gamma(1 + k)\Gamma(1 + n - k)}$$

can be generalized into

$$C_{-\nu}^\mu = \frac{\Gamma(1 - \nu)}{\Gamma(1 + \mu)\Gamma(1 - \nu - \mu)}, \tag{2.7.2}$$

where the μ, ν are complex numbers. Specially, when $\mu = k$ is a positive integer, Proposition 2.7.2 then implies

$$C_{-\nu}^k = \frac{\Gamma(1 - \nu)}{k!\Gamma(1 - \nu - k)} = (-1)^k \frac{\Gamma(k + \nu)}{k!\Gamma(\nu)} = (-1)^k C_{\nu+k-1}^k.$$

2.7.2   Mittag-Leffler function

The Mittag-Leffler function with two parameters is a special function defined by

$$E_{\alpha,\beta}(z) = \sum_{k=1}^\infty \frac{z^k}{\Gamma(k\alpha + \beta)}, \quad \text{Re } \alpha > 0.$$

This function is named after Magnus Gustaf (Gösta) Mittag-Leffler, a Swedish mathematician. When $\beta = 1$, we usually denote $E_\alpha(z) = E_{\alpha,1}(z)$ to be the Mittag-Leffler function with one parameter α. When α, β are real and positive, this series converges for all z and hence the Mittag-Leffler function is an entire function.

Consider the fractional linear differential equation ${}_0^C D_t^\nu y = \sigma y$, $\nu \in (0, 1]$, with initial data $y|_{t=0} = y_0$. Denote $Y(s) = \mathscr{L}[y](s)$. Taking the Laplace transform of the above equation yields

$$s^\nu Y(s) - s^{\nu-1} y_0 = \sigma Y(s).$$

Therefore, one obtains

$$Y(s) = \frac{s^{\nu-1}y_0}{s^\nu - \sigma} = y_0 \sum_{k=0}^{\infty} \frac{\sigma^k}{s^{1+\nu k}}.$$

Then taking the inverse Laplace transform, we get the solution in terms of the Mittag-Leffler function

$$y(t) = y_0 \sum_{k=0}^{\infty} \frac{\sigma^k t^{\nu k}}{\Gamma(\nu k + 1)} = y_0 E_\nu(\sigma t^\nu).$$

When σ in this example is replaced with σi^ν, then the solution is given by $y(t) = y_0 E_\nu(\sigma i^\nu t^\nu)$. When $\sigma = 1$, $_0^C D_t^1$ reduces to the classical derivative d/dt, and the solution is given by $y(t) = y_0 e^{\sigma t} = y_0 E_1(\sigma t)$.

Chapter 3

Fractional Partial Differential Equations

3.1 Fractional diffusion equation

This section mainly discusses the estimates of fractional dissipative equation with Fractional Laplacian. Consider the following fractional diffusion equation

$$\begin{cases} u_t + (-\Delta)^\alpha u = 0, & (t, x) \in (0, \infty) \times \mathbf{R}^d, \\ u(0) = \varphi(x), & x \in \mathbf{R}^d. \end{cases} \tag{3.1.1}$$

The solution of this equation can be obtained by the semigroup method

$$u(t) = \mathcal{S}^\alpha(t)\varphi = e^{-t(-\Delta)^\alpha}\varphi.$$

Next, we will prove the kernel function derived from $\mathcal{S}^\alpha(t)$ is a bounded linear operator on $L^p(\mathbf{R}^d)$ for $1 \leqslant p \leqslant \infty$.

Applying the Fourier transform, the solution of equation (3.1.1) can be written as

$$u(t, x) = \mathcal{F}^{-1}(e^{-t|\xi|^{2\alpha}}\widehat{\varphi}(\xi)) = \mathcal{F}^{-1}(e^{-t|\xi|^{2\alpha}}) * \varphi(x) = K_t * \varphi, \tag{3.1.2}$$

where

$$K_t(x) = \frac{1}{(2\pi)^d} \int_{\mathbf{R}^d} e^{ix \cdot \xi} e^{-t|\xi|^{2\alpha}} d\xi.$$

Clearly, when $\alpha = 1$, $K_t(x)$ is the Gaussian kernel function and when $\alpha = \frac{1}{2}$, $K_t(x)$ is the Poisson kernel function.

According to (3.1.2) and using Young's convolution inequality, one obtains

$$\|f * g\|_{L^p} \leqslant \|f\|_{L^1}\|g\|_{L^p}, \quad \forall f \in L^1(\mathbf{R}^d), g \in L^p(\mathbf{R}^d), \forall p \in [1, \infty].$$

It is clear that to obtain the (p, p) type estimate, we need only the L^1 estimate of the kernel function $K_t(x)$. To this end, we first notice that by scaling

$$K_t(x) = \frac{1}{(2\pi)^d} t^{-\frac{d}{2\alpha}} \int_{\mathbf{R}^d} e^{i\frac{x}{t^{1/2\alpha}}\cdot\eta} e^{-|\eta|^{2\alpha}} d\eta$$

$$= : t^{-\frac{d}{2\alpha}} K\left(\frac{x}{t^{1/2\alpha}}\right).$$

Hence, we need only to consider the property of the kernel function $K(x)$

$$K(x) = (2\pi)^{-d} \int_{\mathbf{R}^d} e^{ix\cdot\xi} e^{-|\xi|^{2\alpha}} d\xi.$$

Noticing $e^{-|\xi|^{2\alpha}} \in L^1(\mathbf{R}^d)$ and taking advantage of the properties of Fourier transform, we know $K \in L^\infty(\mathbf{R}^d) \cap C(\mathbf{R}^d)$. According to the Riemann-Lebesgue lemma, $\lim_{|x|\to\infty} K(x) = 0$, i.e., $K \in L^\infty(\mathbf{R}^d) \cap C_0(\mathbf{R}^d)$. Here $C_0(\mathbf{R}^d)$ is the continuous function that tends to zero at infinity. Similarly, since $|\xi|^\nu e^{-|\xi|^{2\alpha}} \in L^1(\mathbf{R}^d)$, then for any $\nu > 0$ one has $(-\Delta)^{\nu/2} K \in L^\infty(\mathbf{R}^d) \cap C_0(\mathbf{R}^d)$. Since $i\xi e^{-|\xi|^{2\alpha}} \in (L^1(\mathbf{R}^d))^d$, one has $\nabla K \in L^\infty(\mathbf{R}^d) \cap C_0(\mathbf{R}^d)$. Indeed, the function $e^{-|\xi|^{2\alpha}} \in \mathcal{S}(\mathbf{R}^d)$, the Schwartz space of rapidly decreasing functions, hence $K \in \mathcal{S}(\mathbf{R}^d)$ from the properties of Fourier transform.

Lemma 3.1.1   *The kernel function $K(x)$ satisfies the pointwise estimate*

$$|K(x)| \leqslant C(1 + |x|)^{-d-2\alpha}, \quad x \in \mathbf{R}^d, \ \alpha > 0,$$

thus $K \in L^p(\mathbf{R}^d)$ for all $p \in [1, \infty]$.

Proof Introducing the invariant derivative $L(x, D) = x \cdot D/|x|^2 = x \cdot \nabla_\xi/i|x|^2$, then $L(x, D)e^{ix\cdot\xi} = e^{ix\cdot\xi}$. Its conjugate operator is defined by $L^*(x, D) = -x \cdot \nabla_\xi/i|x|^2$. Introducing the truncation of $C^\infty(\mathbf{R}^d)$ function $\chi(\xi)$ such that $\chi(\xi) = 1$ when $|\xi| \leqslant 1$ and $\chi(\xi) = 0$ when $|\xi| > 2$, the kernel function can be written as

$$K(x) = (2\pi)^{-d} \int_{\mathbf{R}^d} e^{ix\cdot\xi} L^*(e^{-|\xi|^{2\alpha}}) d\xi$$

$$= (2\pi)^{-d} \int_{\mathbf{R}^d} e^{ix\cdot\xi} \chi(\xi/\delta) L^*(e^{-|\xi|^{2\alpha}}) d\xi$$

$$+ (2\pi)^{-d} \int_{\mathbf{R}^d} e^{ix\cdot\xi} (1 - \chi(\xi/\delta)) L^*(e^{-|\xi|^{2\alpha}}) d\xi =: I + II,$$

where $\delta > 0$ is to be determined. Obviously

$$|I| \leqslant \frac{C}{|x|} \int_{|\xi|\leqslant 2\delta} |\xi|^{2\alpha-1} d\xi \leqslant \tilde{C}|x|^{-1}\delta^{2\alpha+d-1}.$$

For sufficiently large N (e.g., $N > [2\alpha] + d$), by integration by parts, we know

$$|II| \leqslant (2\pi)^{-d} \int_{\mathbf{R}^d} \left| e^{ix \cdot \xi} (L^*)^{N-1} (1 - \chi(\xi/\delta)) L^*(e^{-|\xi|^{2\alpha}}) \right| d\xi$$

$$\leqslant C|x|^{-N} \int_{|\xi| \geqslant \delta} \sum_{j=1}^{N} |\xi|^{2j\alpha - N} e^{-|\xi|^{2\alpha}} d\xi$$

$$+ C|x|^{-N} \sum_{k=1}^{N-1} C_k \delta^{-k} \int_{\delta \leqslant |\xi| \leqslant 2\delta} \sum_{l=1}^{N-k} C_l |\xi|^{2j\alpha - N + k} e^{-|\xi|^{2\alpha}} d\xi$$

$$\leqslant C|x|^{-N} \int_{|\xi| \geqslant \delta} |\xi|^{2\alpha - N} e^{-|\xi|^{2\alpha}} d\xi + C|x|^{-N} \int_{|\xi| \geqslant \delta} |\xi|^{2\alpha - N} |\xi|^{2\alpha(N-1)} e^{-|\xi|^{2\alpha}} d\xi$$

$$+ C|x|^{-N} \sum_{k=1}^{N-1} \int_{\delta \leqslant |\xi| \leqslant 2\delta} (|\xi|^{2\alpha - N} e^{-|\xi|^{2\alpha}} + |\xi|^{2\alpha(N-k) - N} e^{-|\xi|^{2\alpha}}) d\xi.$$

For arbitrary $k = 1, 2, \cdots, N - 1$, one has $|\xi|^{2\alpha(N-1)} e^{-|\xi|^{2\alpha}} \leqslant C$ and $|\xi|^{2\alpha(N-k-1)} e^{-|\xi|^{2\alpha}} \leqslant C$, yielding

$$|II| \leqslant C|\xi|^{-N} \left(\int_{|\xi| \geqslant \delta} |\xi|^{2\alpha - N} d\xi + \int_{\delta \leqslant |\xi| \leqslant \delta} |\xi|^{2\alpha - N} d\xi \right) \leqslant C|x|^{-N} \delta^{2\alpha - N + d}.$$

Therefore, we obtain the estimate

$$|K(x)| \leqslant C|x|^{-1} \delta^{2\alpha + d - 1} + C|x|^{-N} \delta^{2\alpha - N + d}.$$

Choosing $\delta = |x|^{-1}$ then completes the proof.

The above technique is frequently used in the analysis of the theory of harmonic analysis and partial differential equations. Similarly, this technique can also be used to prove the following.

Lemma 3.1.2 *For arbitrary $\nu > 0$, there exists $C > 0$ such that for any $x \in \mathbf{R}^d$,*

$$|(-\Delta)^{\nu/2} K(x)| \leqslant C(1 + |x|)^{-d-\nu}.$$

Thus $(-\Delta)^{\nu} K \in L^p(\mathbf{R}^d)$ for arbitrary $1 \leqslant p \leqslant \infty$.

Remark 3.1.1 *1. Similarly, there holds the estimate $|\nabla K(x)| \leqslant C(1 + |x|)^{-d-1}$, and hence $\nabla K \in L^p(\mathbf{R}^d)(1 \leqslant p \leqslant \infty)$.*

2. According to the above lemma, for arbitrary $p \in [1, \infty]$, $0 < t < \infty$, the kernel function $K_t(x)$ satisfies

$$K_t \in L^p(\mathbf{R}^d), \quad (-\Delta)^{\nu/2} K_t \in L^p(\mathbf{R}^d), \quad \nabla K_t \in L^p(\mathbf{R}^d).$$

Proposition 3.1.1 *Let $\alpha > 0$, and the initial value $\varphi \in L^1(\mathbf{R}^d)$, then there hold the following estimates*

$$\lim_{t \to \infty} t^{\frac{d}{2\alpha}} \|u(\cdot, t)\|_{L^2}^2 \leqslant A(d, \alpha) \|\varphi\|_{L^1}^2; \tag{3.1.3}$$

$$\lim_{t \to \infty} t^{\frac{d+2}{2\alpha}} \|\boldsymbol{\nabla} u(\cdot, t)\|_{L^2}^2 \leqslant B(d, \alpha) \|\varphi\|_{L^1}^2, \tag{3.1.4}$$

where constants $A(d, \alpha) = \displaystyle\int_{\mathbf{R}}^d e^{-2|\eta|^{2\alpha}} \mathrm{d}\eta$ and $B(d, \alpha) = \displaystyle\int_{\mathbf{R}}^d |\eta|^2 e^{-2|\eta|^{2\alpha}} \mathrm{d}\eta$.

Proof Using the Plancherel theorem and changing of variables, we have

$$\lim_{t \to \infty} t^{\frac{d}{2\alpha}} \|u(\cdot, t)\|_{L^2}^2 = \lim_{t \to \infty} t^{\frac{d}{2\alpha}} \|\hat{u}(\cdot, t)\|_{L^2}^2$$

$$= \lim_{t \to \infty} t^{\frac{d}{2\alpha}} \int_{\mathbf{R}}^d e^{-2|\xi|^{2\alpha} t} |\hat{\varphi}(\xi)|^2 \mathrm{d}\xi = \lim_{t \to \infty} \int_{\mathbf{R}}^d e^{-2|\eta|^{2\alpha}} |\hat{\varphi}(\eta t^{-\frac{1}{2\alpha}})|^2 \mathrm{d}\eta.$$

For arbitrary $t \in [0, \infty)$, since

$$\int_{\mathbf{R}}^d e^{-2|\eta|^{2\alpha}} |\hat{\varphi}(\eta t^{-\frac{1}{2\alpha}})|^2 \mathrm{d}\eta \leqslant \|\hat{\varphi}\|_{L^\infty}^2 \int_{\mathbf{R}}^d e^{-2|\eta|^{2\alpha}} \mathrm{d}\eta \leqslant A(d, \alpha) \|\varphi\|_{L^1}^2,$$

using the dominated convergence theorem then leads to (3.1.3).

Similarly, using the Plancherel theorem, we have

$$\lim_{t \to \infty} t^{\frac{d+2}{2\alpha}} \|\boldsymbol{\nabla} u(\cdot, t)\|_{L^2}^2 = \lim_{t \to \infty} t^{\frac{d+2}{2\alpha}} \int_{\mathbf{R}}^d |\xi|^2 e^{-2|\xi|^{2\alpha} t} |\hat{\varphi}(\xi)|^2 \mathrm{d}\xi$$

$$= \lim_{t \to \infty} \int_{\mathbf{R}}^d |\eta|^2 e^{-2|\eta|^{2\alpha}} |\hat{\varphi}(\eta t^{-\frac{1}{2\alpha}})|^2 \mathrm{d}\eta \leqslant B(d, \alpha) \|\varphi\|_{L^1}^2.$$

We complete the proof.

Proposition 3.1.2 *Let $\alpha \in (0, 1]$ and the initial data $\varphi \in L^2(\mathbf{R}^d)$, then the solution u of (3.1.1) satisfies the estimate*

$$\|\boldsymbol{\nabla} u(t)\|_{L^\infty}^2 \leqslant C t^{-\frac{d+2}{4\alpha}}.$$

Proof From (3.1.2) and $\widehat{K_t}(\xi) = e^{-|\xi|^{2\alpha} t}$, one has

$$\|\boldsymbol{\nabla} u\|_{L^\infty} \leqslant \int_{\mathbf{R}^d} |\xi| |\hat{u}(\xi)| \mathrm{d}\xi = \int_{\mathbf{R}^d} |\xi| e^{-|\xi|^{2\alpha} t} |\hat{\varphi}(\xi)| \mathrm{d}\xi$$

$$\leqslant \|\varphi\|_{L^2} \left(\int_{\mathbf{R}^d} |\xi|^2 e^{-2|\xi|^{2\alpha} t} \mathrm{d}\xi \right)^{1/2} \leqslant C \left(\int_0^\infty r^{d+1} e^{-2r^{2\alpha} t} \mathrm{d}r \right) \leqslant C t^{-\frac{d+2}{4\alpha}}.$$

3.2 Fractional nonlinear Schrödinger equation

This section mainly considers fractional nonlinear Schrödinger equation, which is divided into two parts. The first one considers the space fractional nonlinear schrödinger equation, while the second one concerns the time fractional nonlinear Schrödinger equation.

3.2.1 Space fractional nonlinear Schrödinger equation

Now we consider the following fractional nonlinear Schrödinger equation with periodic boundary conditions

$$\begin{cases} iu_t + (-\Delta)^\alpha u + \beta |u|^\rho u = 0, & x \in \mathbf{R}^n, t > 0, \\ u(x,0) = u_0(x), & x \in \mathbf{R}^n, \\ u(x + 2\pi e_i, t) = u(x,t), & x \in \mathbf{R}^n, t > 0, \end{cases} \tag{3.2.1}$$

where, $e_i = (0, \cdots, 0, 1, 0, \cdots, 0)$, $i = 1, \cdots, n$ is an orthonormal basis in $\mathbf{R}^n$, $i = \sqrt{-1}$ is the imaginary unit, $\alpha \in (0,1), \beta \in \mathbf{R}, \beta \neq 0$ and $\rho > 0$ is a real number. For convenience, we denote $\Omega = (0, 2\pi) \times \cdots (0, 2\pi) \subset \mathbf{R}^n$.

When $\alpha = 1$, equation (3.2.1) is the classical nonlinear Schrödinger equation, and has been extensively studied in recent decades. The existence and uniqueness of weak solutions for the initial-boundary value problems can be referred to [135]. The global existence of smooth solutions can be found in [101]. In this section, we mainly take advantage of the energy method to study the existence and uniqueness of smooth solution of the fractional nonlinear Schrödinger equation. Specifically, we will prove the following thereom [103]

Theorem 3.2.1 *Let* $\alpha > \dfrac{n}{2}$. *If* ρ *is an even number, suppose that* $\rho > 0$ *when* $\beta > 0$ *and* $0 < \rho < \dfrac{4\alpha}{n}$ *when* $\beta < 0$. *If* ρ *is not a even number, suppose that* $\rho > 2[\alpha] + 1$ *when* $\beta > 0$ *and* $2[\alpha] + 1 < \rho < \dfrac{4\alpha}{n}$ *when* $\beta < 0$. *Then, for arbitrary* $u_0 \in H^{4\alpha}$, *there exists a unique global smooth solution* u *of* (3.2.1) *such that*

$$u \in L^\infty(0, T; H^{4\alpha}(\Omega)), \quad u_t \in L^\infty(0, T; H^{2\alpha}(\Omega)).$$

Theorem 3.2.2 *Let* $\alpha > 0$ *and* $u_0 \in H^\alpha(\Omega)$. *When* $\beta > 0$, *suppose that* $\rho > 0$ *if* $\alpha \geqslant \dfrac{n}{2}$ *and* $0 < \rho < \dfrac{4\alpha}{n - 2\alpha}$ *if* $\alpha < \dfrac{n}{2}$. *When* $\beta < 0$, *suppose that* $0 < \rho < \dfrac{4\alpha}{n}$. *Then, there exists a global solution* u *of the equation* (3.2.1) *such that*

$$u \in L^\infty(0,T; H^{4\alpha}(\Omega) \cap L^{\rho+2}(\Omega)), \quad u_t \in L^\infty(0,T; H^{-\alpha}(\Omega)). \qquad (3.2.2)$$

Below we introduce some notations. Since u is spatial periodic, it can be expanded by using Fourier series $u = \sum\limits_{k \in \mathbf{Z}^n} a_k e^{i<k,x>}$, where a_k is the Fourier coefficient of u. Thus $\partial_{x_j} u = \sum\limits_{k \in \mathbf{Z}^n} ik_j a_k e^{i<k,x>}$. Here the fractional Laplacian $(-\Delta)^\alpha$ can be expressed as $(-\Delta)^\alpha u = \sum\limits_{k \in \mathbf{Z}^n} |k|^{2\alpha} a_k e^{i<k,x>}$. Let

$$A = \left\{ u \,\Big|\, u = \sum_{k \in \mathbf{Z}^n} a_k e^{i<k,x>} : \sum_{k \in \mathbf{Z}^n} |k|^{2\alpha} |a_k|^2, \sum_{k \in \mathbf{Z}^n} |a_k|^2 < \infty \right\}$$

and H^α denote the completion of A under the norm

$$\|u\|_{H^\alpha} = \left(\sum_{k \in \mathbf{Z}^n} |k|^{2\alpha} |a_k| \right)^{1/2} + \left(\sum_{k \in \mathbf{Z}^n} |a_k|^2 \right)^{1/2}.$$

Clearly, H^α is a Banach space. It is easy to prove H^α is a Hilbert space under the inner product

$$(u,v)_{H^\alpha} = ((-\Delta)^{\alpha/2} u, (-\Delta)^{\alpha/2} v) = \sum_{k \in \mathbf{Z}^n} |k|^{2\alpha} a_k \bar{b}_k.$$

Hereinafter, the norm of function space $H = L^2(\Omega)$ is usually denoted as $\|\cdot\|$, its inner product is expressed as $(\cdot,\cdot)$; the norm of $L^p(\Omega)$ is denoted by $\|\cdot\|_{L^p(\Omega)}$. Obviously $\|\cdot\|_{L^2(\Omega)} = \|\cdot\|$. $H^{-\alpha}$ denotes the dual space of H^α. In order to study the problem (3.2.1), we introduce the following Banach space $V = H^\alpha(\Omega) \cap L^{\rho+2}(\Omega)$, whose norm is given by

$$\|v\|_V = \|v\|_{H^\alpha(\Omega)} + \|v\|_{L^{\rho+2}(\Omega)}.$$

Definition 3.2.1 *The space $L^p(0,T;X)$ consists of all the measurable functions $f : [0,T] \to X$ with*

$$\|f\|_{L^p(0,T;X)} = \left(\int_0^T \|f\|_X^p dt \right)^{\frac{1}{p}} < \infty$$

for $1 \leqslant p < \infty$, and when $p = \infty$,

$$\|f\|_{L^\infty(0,T;X)} = \sup_{0 \leqslant t \leqslant T} \|f\|_X < \infty.$$

Let $C([0,T];X)$ denote the space of all the continuous functions $f : [0,T] \to X$ whose norm is given by $\|f\|_{C([0,T];X)} = \max_{0 \leqslant t \leqslant T} \|f\|_X$.

What follow are some *a priori* estimates and the proofs of the theorems.

Lemma 3.2.1 *Suppose $\alpha > 0$, $\rho > 0$, if u is a solution of equation (3.2.1), then*

$$\sup_{0 \leqslant t < \infty} \|u(t)\| = \|u_0\|. \tag{3.2.3}$$

This lemma is obvious. Multiplying the equation by $\bar{u}$, integrating with respect to the space variable x over Ω, and taking the imaginary part, one has $\dfrac{\mathrm{d}}{\mathrm{d}t} \|u(t)\|^2 = 0$.

In what follows, T denotes an arbitrary positive constant, and C a constant depending only on initial value and T.

Lemma 3.2.2 *Let $\alpha > 0$. Suppose $\rho > 0$ when $\beta > 0$ and $0 < \rho < \dfrac{4\alpha}{n}$ when $\beta < 0$. Then the solution u satisfies the estimate*

$$\sup_{0 \leqslant t < \infty} (\|(-\Delta)^{\alpha/2} u\| + \|u\|_{L^{\rho+2}}) \leqslant C(\|u_0\|_{H^\alpha}, \|u_0\|_{L^{\rho+2}}).$$

Proof Multiplying the equation by $\bar{u}_t$ and integrating over Ω, then

$$(iu_t, u_t) + ((-\Delta)^\alpha u, u_t) + (\beta |u|^\rho u, u_t) = 0.$$

Taking the real part yields

$$\frac{\mathrm{d}}{\mathrm{d}t} \int_\Omega |(-\Delta)^{\alpha/2} u|^2 + \frac{2\beta}{\rho+2} |u|^{\rho+2} \mathrm{d}x = 0,$$

and hence

$$\|(-\Delta)^{\alpha/2} u\|^2 + \frac{2\beta}{\rho+2} \|u\|^{\rho+2}_{L^{\rho+2}(\Omega)} = \|(-\Delta)^{\alpha/2} u_0\|^2 + \frac{2\beta}{\rho+2} \|u_0\|^{\rho+2}_{L^{\rho+2}(\Omega)} = E(u_0). \tag{3.2.4}$$

If $\beta > 0$, then

$$\|(-\Delta)^{\alpha/2} u\|^2 \leqslant E(u_0) \leqslant C(\|u_0\|_{H^\alpha(\Omega)}, \|u_0\|_{L^{\rho+2}(\Omega)}),$$

and

$$\|u\|_{L^{\rho+2}(\Omega)} \leqslant C(\|u_0\|_{H^\alpha(\Omega)}, \|u_0\|_{L^{\rho+2}(\Omega)}).$$

When $\beta < 0$, let $\theta = \dfrac{n\rho}{2\alpha(\rho+2)} < 1$, then by the Gagliardo-Nirenberg inequality

$$\|u\|^{\rho+2}_{L^{\rho+2}(\Omega)} \leqslant C\|(-\Delta)^{\alpha/2} u\|^{\theta(\rho+2)} \|u\|^{(1-\theta)(\rho+2)} \leqslant C\|(-\Delta)^{\alpha/2} u\|^{\frac{n\rho}{2\alpha}},$$

where $\dfrac{1}{\rho+2} = \theta\left(\dfrac{1}{2} - \dfrac{\alpha}{n}\right) + (1-\theta)\dfrac{1}{2}$. Since $\rho < \dfrac{4\alpha}{n}$, i.e., $\dfrac{n\rho}{2\alpha} < 2$, then

$$\frac{2|\beta|}{\rho+2}\|u\|_{L^{\rho+2}(\Omega)}^{\rho+2} \leqslant \frac{1}{2}\|(-\Delta)^{\alpha/2}u\|^2 + C. \tag{3.2.5}$$

Therefore, using equation (3.2.4) and inequality (3.2.5), we know

$$\|(-\Delta)^{\alpha/2}u\|^2 \leqslant C(\|u_0\|_{H^\alpha(\Omega)}, \|u_0\|_{L^{\rho+2}(\Omega)}),$$
$$\|u\|_{L^{\rho+2}(\Omega)} \leqslant C(\|u_0\|_{H^\alpha(\Omega)}, \|u_0\|_{L^{\rho+2}(\Omega)}),$$

completing the proof.

Lemma 3.2.3 *Let $\alpha > \dfrac{n}{2}$ and ρ satisfies the conditions of lemma 3.2.2, then u satisfies*

$$\sup_{0\leqslant t<\infty} (\|u_t\| + \|(-\Delta)^\alpha u\|) \leqslant C(\|u_0\|_{H^{2\alpha}(\Omega)}). \tag{3.2.6}$$

Proof Differentiate the equation with respect to time t, multiply the resulting equation by u_t, and then integrate with respect to x over Ω to obtain

$$(iu_{tt}, u_t) + ((-\Delta)^\alpha u_t, u_t) + \left(\frac{\mathrm{d}}{\mathrm{d}t}(\beta|u|^\rho u), u_t\right) = 0.$$

Taking the imaginary part yields

$$\frac{1}{2}\frac{\mathrm{d}}{\mathrm{d}t}\|u_t\|^2 + \Im\left(\frac{\mathrm{d}}{\mathrm{d}t}(\beta|u|^\rho u), u_t\right) = 0. \tag{3.2.7}$$

Moreover, since

$$\begin{aligned}
\Im\left(\frac{\mathrm{d}}{\mathrm{d}t}(\beta|u|^\rho u), u_t\right) &= \Im\int_\Omega \frac{\mathrm{d}}{\mathrm{d}t}(\beta|u|^\rho u)\bar{u}_t \mathrm{d}x \\
&= \Im\int_\Omega \beta|u|^\rho|u_t|^2 \mathrm{d}x + \Im\int_\Omega \frac{\rho\beta}{2}|u|^{\rho-2}(|u_t|^2|u|^2 + u^2\bar{u}_t^2)\mathrm{d}x \\
&= \Im\int_\Omega \frac{\rho\beta}{2}|u|^{\rho-2}(u^2\bar{u}_t^2)\mathrm{d}x,
\end{aligned}$$

$$\tag{3.2.8}$$

by (3.2.7) and (3.2.8), we have

$$\frac{1}{2}\frac{\mathrm{d}}{\mathrm{d}t}\|u_t\|^2 + \Im\int_\Omega \frac{\rho\beta}{2}|u|^{\rho-2}(u^2\bar{u}_t^2)\mathrm{d}x = 0.$$

Integrating the above equation with respect to time from 0 to t yields

$$\begin{aligned}
\|u_t\|^2 &= -\int_0^t \Im\int_\Omega \rho\beta|u|^{\rho-2}(u^2\bar{u}_t^2)\mathrm{d}x\mathrm{d}s + \|u_t(x,0)\|^2 \\
&\leqslant C\int_0^t \int_\Omega |u|^2|u_t|^2\mathrm{d}x\mathrm{d}s + \|u_t(x,0)\|^2.
\end{aligned}$$

$$\tag{3.2.9}$$

Using (3.2.1) as well as Sobolev embedding inequality $\|u\|_{L^\infty} \leqslant C\|u\|_{H^\alpha(\Omega)} \leqslant C$ for $\alpha > \dfrac{n}{2}$ gives

$$\|u_t(x,0)\| \leqslant C\|(-\Delta)^\alpha u_0\| + C\|\beta|u_0|^\rho u_0\| \leqslant C(\|u_0\|_{H^{2\alpha}(\Omega)}).$$

Thus it follows from (3.2.9) that

$$\|u_t\|^2 \leqslant C \int_0^t \|u\|_{L^\infty(\Omega)}^\rho \|u_t\|^2 \mathrm{d}s + \|u_t(x,0)\|^2 \leqslant C \int_0^t \|u_t\|^2 \mathrm{d}s + C(\|u_0\|_{H^{2\alpha}(\Omega)}).$$

Taking advantage of Gronwall inequality, we have

$$\|u_t\|^2 \leqslant C(\|u_0\|_{H^{2\alpha}}),$$

and hence

$$\|(-\Delta)^\alpha u\| \leqslant \|u_t\| + \|\beta|u|^\rho u\|$$
$$\leqslant C(\|u_0\|_{H^{2\alpha}(\Omega)}) + C\|u\|_{L^\infty(\Omega)}^\rho \|u\| \leqslant C(\|u_0\|_{H^{2\alpha}(\Omega)}),$$

completing the proof.

Lemma 3.2.4 *Let $\alpha > \dfrac{n}{2}$. Suppose that ρ satisfies the conditions of Lemma 3.2.2 if ρ is an even number. If ρ is not an even number, suppose that $\rho > [\alpha]$ when $\beta > 0$ and $[\alpha] < \rho < \dfrac{4\alpha}{n}$ when $\beta < 0$. Then the solution u satisfies the estimate*

$$\sup_{0 \leqslant t < \infty} \|(-\Delta)^{\alpha/2} u_t\| \leqslant C(\|u_0\|_{H^{3\alpha}(\Omega)}).$$

Proof Differentiating the equation (3.2.1) with respect to time variable, multiplying by $\bar{u}_{tt}$, and then integrating with respect to the spatial variable x over Ω, we have

$$(iu_{tt}, u_{tt}) + ((-\Delta)^\alpha u_t, u_{tt}) + \left(\frac{\mathrm{d}}{\mathrm{d}t}(\beta|u|^\rho u), u_{tt}\right).$$

By integration by parts, we have

$$\frac{\mathrm{d}}{\mathrm{d}t}\|(-\Delta)^{\alpha/2} u_t\|^2 + 2\mathrm{Re}\left(\frac{\mathrm{d}}{\mathrm{d}t}(\beta|u|^\rho u), u_{tt}\right) = 0.$$

Since

$$2\mathrm{Re}\left(\frac{\mathrm{d}}{\mathrm{d}t}(\beta|u|^\rho u), u_{tt}\right) = \int_\Omega \left(\frac{\rho}{2}+1\right)\beta|u|^\rho \frac{\mathrm{d}}{\mathrm{d}t}|u_t|^2 \mathrm{d}x$$
$$+ \int_\Omega \frac{\rho\beta}{4}|u|^{\rho-2}\left(u^2 \frac{\mathrm{d}}{\mathrm{d}t}\bar{u}_t^2 + \bar{u}^2 \frac{\mathrm{d}}{\mathrm{d}t}u_t^2\right)\mathrm{d}x,$$

one has

$$\frac{\mathrm{d}}{\mathrm{d}t}\|(-\Delta)^{\alpha/2}u_t\|^2 + \int_\Omega \left(\frac{\rho}{2}+1\right)\beta|u|^\rho \frac{\mathrm{d}}{\mathrm{d}t}|u_t|^2\mathrm{d}x$$

$$+\int_\Omega \frac{\rho\beta}{4}|u|^{\rho-2}\left(u^2\frac{\mathrm{d}}{\mathrm{d}t}\bar{u}_t^2 + \bar{u}^2\frac{\mathrm{d}}{\mathrm{d}t}u_t^2\right)\mathrm{d}x.$$

This implies that

$$\frac{\mathrm{d}}{\mathrm{d}t}\|(-\Delta)^{\alpha/2}u_t\|^2 + \frac{\mathrm{d}}{\mathrm{d}t}\int_\Omega \left(\frac{\rho}{2}+1\right)\beta|u|^\rho|u_t|^2\mathrm{d}x$$

$$+\frac{\mathrm{d}}{\mathrm{d}t}\int_\Omega \frac{\rho\beta}{4}|u|^{\rho-2}(u^2\bar{u}_t^2 + \bar{u}^2u_t^2)\mathrm{d}x$$

$$=-\left(\frac{\rho}{2}+1\right)\beta\int_\Omega \frac{\mathrm{d}}{\mathrm{d}t}(|u|^\rho)|u_t|^2\mathrm{d}x - \frac{\rho\beta}{4}\int_\Omega \frac{\mathrm{d}}{\mathrm{d}t}(|u|^{\rho-2}u^2)\bar{u}_t^2\mathrm{d}x \qquad (3.2.10)$$

$$-\frac{\rho\beta}{4}\int_\Omega \frac{\mathrm{d}}{\mathrm{d}t}(|u|^{\rho-2}\bar{u}^2)u_t^2\mathrm{d}x$$

$$\leqslant C\int_\Omega |u|^{\rho-1}|u_t|^3\mathrm{d}x \leqslant C\|u\|_{L^\infty(\Omega)}^{\rho-1}\|u_t\|_{L^3(\Omega)}^3 \leqslant C\|u_t\|_{L^3(\Omega)}^3.$$

Let $\theta = \dfrac{n}{6\alpha} < \dfrac{1}{3}$, then $\dfrac{1}{3} = \theta\left(\dfrac{1}{2}-\dfrac{\alpha}{n}\right)+(1-\theta)\dfrac{1}{2}$. By the Gagliardo-Nirenberg inequality and (3.2.6), we have

$$\begin{aligned}\|u_t\|_{L^3(\Omega)}^3 &\leqslant C\|u_t\|^{3(1-\theta)}\|(-\Delta)^{\alpha/2}u_t\|^{3\theta}\\ &\leqslant C\|(-\Delta)^{\alpha/2}u_t\|^{3\theta} \leqslant C\|(-\Delta)^{\alpha/2}u_t\|^2 + C.\end{aligned} \qquad (3.2.11)$$

Then from (3.2.10) and (3.2.11), we have

$$\|(-\Delta)^{\alpha/2}u_t\|^2 + \int_\Omega (\frac{\rho}{2}+1)\beta|u|^\rho|u_t|^2\mathrm{d}x + \int_\Omega \frac{\rho\beta}{4}|u|^{\rho-2}(u^2\bar{u}_t^2 + \bar{u}^2u_t^2)\mathrm{d}x$$

$$\leqslant\|(-\Delta)^{\alpha/2}u_t(x,0)\|^2 + \int_\Omega \left(\frac{\rho}{2}+1\right)|\beta||u_0|^\rho|u_t(x,0)|^2\mathrm{d}x$$

$$+\int_\Omega \frac{\rho\beta}{4}|u_0|^{\rho-2}(u_0^2\bar{u}_t(x,0)^2 + \bar{u}_0^2u_t^2(x,0))\mathrm{d}x + C\int_0^t \|(-\Delta)^{\alpha/2}u_t\|^2\mathrm{d}s + C$$

$$\leqslant C + C\int_0^t \|(-\Delta)^{\alpha/2}u_t\|^2\mathrm{d}s.$$

$$(3.2.12)$$

Indeed, from (3.2.1), we have

$$\begin{aligned}\|(-\Delta)^{\alpha/2}u_t(x,0)\| &\leqslant\|(-\Delta)^{3\alpha/2}u(x,0)\| + \|(-\Delta)^{\alpha/2}(\beta|u_0|^\rho u_0)\|\\ &\leqslant C\|u_0\|_{H^{3\alpha}(\Omega)} + C\||u_0|^\rho u_0\|_{H^{[\alpha]+1}(\Omega)} \lesssim C\|u_0\|_{H^{q_m}(\Omega)},\end{aligned}$$

where $\rho > [\alpha]$ if ρ is not an even number. On the other hand,

$$\int_\Omega |\beta| \left(\frac{\rho}{2} + 1\right) |u_0|^\rho |u_t(x,0)|^2 \mathrm{d}x + \int_\Omega \frac{\rho\beta}{4} |u_0|^{\rho-2} (u_0^2 \bar{u}_t(x,0)^2 + \bar{u}_0^2 u_t(x,0)^2) \mathrm{d}x$$

$$\leqslant C \|u_0\|_{L^\infty(\Omega)}^\rho \|u_t(x,0)\|^2 \leqslant C(\|u_0\|_{H^{2\alpha}(\Omega)}),$$

and from (3.2.11), we have

$$\int_\Omega |\beta| \left(\frac{\rho}{2} + 1\right) |u|^\rho |u_t|^2 \mathrm{d}x + \int_\Omega \frac{\rho\beta}{4} |u|^{\rho-2} (u^2 \bar{u}_t^2 + \bar{u}^2 u_t^2) \mathrm{d}x$$

$$\leqslant C |u|^\rho |u_t|^2 \mathrm{d}x \leqslant C \left(\int_\Omega |u|^{3\rho} \mathrm{d}x\right)^{1/3} \left(\int_\Omega |u_t|^3 \mathrm{d}x\right)^{2/3}$$

$$\leqslant C \|(-\Delta)^{\alpha/2} u_t\|^{2\theta} \leqslant \frac{1}{2} \|(-\Delta)^{\alpha/2} u_t\|^2 + C.$$

Then using (3.2.12) and Gronwall inequality, we have

$$\|(-\Delta)^{\alpha/2} u_t\|^2 \leqslant C + C \int_0^t \|(-\Delta)^{\alpha/2} u_t\|^2 \mathrm{d}s \leqslant C(\|u_0\|_{H^{3\alpha}(\Omega)}),$$

completing the proof.

Lemma 3.2.5 *Let* $\alpha > \dfrac{n}{2}$. *If* ρ *is an even number, suppose* ρ *obeys the hypothesis of lemma 3.2.2; if* ρ *is not an even number, suppose* $\rho > 2[\alpha] + 1$ *when* $\beta > 0$, *and suppose* $2[\alpha] + 1 < \rho < \dfrac{4\alpha}{n}$ *when* $\beta < 0$. *Then, there holds for the solution* u *of equation* (3.2.1) *that*

$$\sup_{0 \leqslant t < \infty} (\|u_{tt}\| + \|(-\Delta)^\alpha u_t\|) \leqslant C(\|u_0\|_{H^{4\alpha}(\Omega)}).$$

Proof Differentiating twice the equation with respect to time variable, multiplying the resulting equation by $\bar{u}_{tt}$, and integrating with respect to x over Ω, we have

$$(iu_{tt}, u_{tt}) + ((-\Delta)^\alpha u_{tt}, u_{tt}) + \left(\frac{\mathrm{d}^2}{\mathrm{d}t^2}(\beta |u|^\rho u), u_{tt}\right) = 0.$$

Taking the imaginary part, we have

$$\frac{1}{2} \frac{\mathrm{d}}{\mathrm{d}t} \|u_{tt}\|^2 + \Im\left(\frac{\mathrm{d}^2}{\mathrm{d}t^2}(\beta |u|^\rho u), u_{tt}\right) = 0. \tag{3.2.13}$$

By direct computation

$$\Im\left(\frac{\mathrm{d}^2}{\mathrm{d}t^2}(\beta |u|^\rho u), u_{tt}\right) = \Im\left(\frac{\rho^2}{2} + \rho\right) \beta(|u|^{\rho-2} |u_t|^2 u, u_{tt})$$

$$+ \Im\left(\frac{\rho^2}{4} + \frac{\rho}{2}\right) \beta(|u|^{\rho-2} u_t^2 \bar{u}, u_{tt})$$

$$+ \Im\left(\frac{\rho^2}{4} - \frac{\rho}{2}\right) \beta(|u|^{\rho-4} u_t^2 u^3, u_{tt})$$

$$+\Im\frac{\beta\rho}{2}(|u|^{\rho-2}u^2\bar{u}_{tt}, u_{tt}).$$

The first term of the right hand side can be estimated as

$$\Im\left(\frac{\rho^2}{2}+\rho\right)\beta(|u|^{\rho-2}|u_t|^2u, u_{tt}) \leqslant C\int_\Omega |u|^{\rho-1}|u_t|^2|u_{tt}|\mathrm{d}x$$

$$\leqslant C\|u\|_{L^\infty(\Omega)}^{\rho-1}\|u_t\|_{L^4(\Omega)}^2\|u_{tt}\|$$

$$\leqslant C\|u_t\|_{L^4(\Omega)}^4 + C\|u_{tt}\|^2.$$

In a similar way, the second and the third term of the right side can be estimated as

$$\Im\left(\frac{\rho^2}{4}+\frac{\rho}{2}\right)\beta(|u|^{\rho-2}u_t^2\bar{u}, u_{tt}) + \Im\left(\frac{\rho^2}{4}-\frac{\rho}{2}\right)\beta(|u|^{\rho-4}u_t^2u^3, u_{tt})$$

$$\leqslant C\|u_t\|_{L^4(\Omega)}^4 + C\|u_{tt}\|^2.$$

The last term can be estimated as

$$\Im\frac{\beta\rho}{2}(|u|^{\rho-2}u^2\bar{u}_{tt}, u_{tt}) \leqslant C\|u_{tt}\|^2. \tag{3.2.14}$$

According to (3.2.13) and (3.2.14), we have

$$\|u_{tt}\|^2 \leqslant C\int_0^t \|u_t\|_{L^4(\Omega)}^4\mathrm{d}s + C\int_0^t \|u_{tt}\|^2\mathrm{d}s + \|u_{tt}(x,0)\|^2. \tag{3.2.15}$$

Let $\theta = \dfrac{n}{8\alpha} < \dfrac{1}{4}$, then $\dfrac{1}{4} = \theta\left(\dfrac{1}{2}-\dfrac{\alpha}{n}\right) + (1-\theta)\dfrac{1}{2}$. Taking advantage of Gagliardo-Nirenberg inequality, Lemma 3.2.3 and 3.2.4, we have

$$\|u_t\|_{L^4(\Omega)} \leqslant C\|u_t\|^{1-\theta}\|(-\Delta)^{\alpha/2}u_t\|^\theta \leqslant C(\|u_0\|_{H^{3\alpha}(\Omega)}).$$

From the equation (3.2.1) and Lemma 3.2.3, we know

$$\|u_{tt}(x,0)\| \leqslant \|(-\Delta)^\alpha((-\Delta)^\alpha u_0 + \beta|u_0|^\rho u_0)\| + \|\frac{\mathrm{d}}{\mathrm{d}t}(\beta|u|^\rho u)\|$$

$$\leqslant C\|(-\Delta)^{2\alpha}u_0\| + C\|(-\Delta)^\alpha(\beta|u_0|^\rho u_0)\| + C\||u_0|^\rho|u_t(x,0)|\|$$

$$\leqslant C(\|u_0\|_{H^{4\alpha}(\Omega)}) + C\|(-\Delta)^\alpha(|u_0|^\rho u_0)\| + C\|u_t(x,0)\|$$

$$\leqslant C(\|u_0\|_{H^{4\alpha}(\Omega)}) + C\|(-\Delta)^\alpha(|u_0|^\rho u_0)\|. \tag{3.2.16}$$

If $\alpha \geqslant \max\left\{\dfrac{n}{2}, 1\right\}$, then

$$\|(-\Delta)^\alpha(|u_0|^\rho u_0)\| \leqslant C\|(-\Delta)^{[\alpha]+1}(|u_0|^\rho u_0)\| \leqslant C(\|u_0\|_{H^{4\alpha}(\Omega)}),$$

where $\rho > 2[\alpha] + 1$ when ρ is not an even number.

When $n=1$ and $\frac{1}{2} < \alpha < 1$,

$$\|(-\Delta)^\alpha(|u_0|^\rho u_0)\| \leqslant C\|\Delta(|u_0|^\rho u_0)\| \leqslant C(\|u_0\|_{H^{4\alpha}}).$$

Thus, from (3.2.16), we know

$$\|u_{tt}(x,0)\| \leqslant C(\|u_0\|_{H^{4\alpha}}).$$

Furthermore, according to (3.2.15), we know

$$\|u_{tt}\|^2 \leqslant C\int_0^t \|u_{tt}\|^2 \mathrm{d}s + C(\|u_0\|_{H^{4\alpha}(\Omega)}),$$

which implies by Gronwall inequality that

$$\|u_{tt}\|^2 \leqslant C(\|u_0\|_{H^{4\alpha}(\Omega)}).$$

But

$$\left\|\frac{\mathrm{d}}{\mathrm{d}t}(|u|^\rho u)\right\| = \left\|\frac{\rho}{2}|u|^{\rho-2}(u\bar{u}_t + \bar{u}u_t)u + |u|^\rho u_t\right\|$$
$$\leqslant C\|u\|_{L^\infty(\Omega)}^\rho \|u_t\| \leqslant C(\|u_0\|_{H^{2\alpha}(\Omega)}),$$

then there holds

$$\|(-\Delta)^\alpha u_t\| \leqslant C\|u_{tt}\| + C\left\|\frac{\mathrm{d}}{\mathrm{d}t}(|u|^\rho u)\right\| \leqslant C(\|u_0\|_{H^{4\alpha}(\Omega)}).$$

Therefore

$$\sup_{0\leqslant t<\infty} \|(-\Delta)^\alpha u_t\| \leqslant C(\|u_0\|_{H^{4\alpha}(\Omega)}),$$

completing the proof.

Lemma 3.2.6 *Supposing α and ρ satisfy the conditions of Lemma 3.2.5, the solution u of equation (3.2.1) satisfies the following a priori estimate*

$$\sup_{0\leqslant t<\infty} \|(-\Delta)^{2\alpha}u\| \leqslant C(\|u_0\|_{H^{4\alpha}}).$$

Proof Let $\alpha \geqslant \max\left\{\frac{n}{2}, 1\right\}$. Applying equation (3.2.1), Lemma 3.2.3 and Lemma 3.2.5, we obtain

$$\|(-\Delta)^{2\alpha}u\| \leqslant C\|(-\Delta)^\alpha u_t\| + C\|(-\Delta)^\alpha(|u|^\rho u)\|$$
$$\leqslant C\|(-\Delta)^\alpha u_t\| + C\|(-\Delta)^{[\alpha]+1}(|u|^\rho u)\| \leqslant C(\|u_0\|_{H^{4\alpha}}). \tag{3.2.17}$$

When $n = 1$ and $\dfrac{1}{2} < \alpha < 1$, by (3.2.1) and Lemma 3.2.5, we know

$$
\begin{aligned}
\|(-\Delta)^{2\alpha}u\| &\leqslant C\|(-\Delta)^{\alpha}u_t\| + C\|(-\Delta)^{\alpha}(|u|^p u)\| \\
&\leqslant C(\|u_0\|_{H^{4\alpha}(\Omega)}) + C(\|\Delta(|u|^p u)\|) \\
&\leqslant C(\|u_0\|_{H^{4\alpha}(\Omega)}) + C\|u\|_{L^{\infty}(\Omega)}^p \|\Delta u\| + C\|u\|_{L^{\infty}(\Omega)}^{p-1}\||\nabla u|^2\| \\
&\leqslant C(\|u_0\|_{H^{4\alpha}(\Omega)}) + C\|\Delta u\| + C\|\nabla u\|_{L^4(\Omega)}^2.
\end{aligned}
$$

$$(3.2.18)$$

Let $\theta = \dfrac{2}{4\alpha} < 1$, then by Gagliardo-Nirenberg inequality and Lemma 3.2.1, we know

$$
C\|\Delta u\| \leqslant C\|(-\Delta)^{2\alpha}u\|^{\theta}\|u\|^{1-\theta} \leqslant \frac{1}{4}\|(-\Delta)^{2\alpha}u\| + C.
$$

Let $\delta = \dfrac{1}{16\alpha - 4} < \dfrac{1}{4}$, then by the Gagliardo-Nirenberg inequality, we can similarly obtain

$$
\begin{aligned}
C\|\nabla u\|_{L^4(\Omega)}^2 &\leqslant C\|(-\Delta)^{2\alpha}u\|^{2\delta}\|\nabla u\|^{2(1-\delta)} \\
&\leqslant C\|(-\Delta)^{2\alpha}u\|^{2\delta}\|(-\Delta)^{\alpha}u\|^{2(1-\delta)} \leqslant \frac{1}{4}\|(-\Delta)^{2\alpha}u\| + C.
\end{aligned}
$$

$$(3.2.19)$$

Thus we conclude that when $n = 1$ and $\dfrac{1}{2} < \alpha < 1$,

$$
\|(-\Delta)^{2\alpha}u\| \leqslant C(\|u_0\|_{H^{4\alpha}(\Omega)}).
$$

Therefore, taking advantage of (3.2.17) and the above inequality, we complete the proof.

Before the theorem 3.2.1 is proved, we take advantage of the Faedo-Galerkin method to prove the existence of the weak solution of equation (3.2.1). In doing this, three lemmas are given below.

Lemma 3.2.7 *Let B_0, B and B_1 be three Banach spaces. Assume that $B_0 \subset B \subset B_1$ and B_0 and B_1 are reflexive. Suppose also that B_0 is compactly embedded in B. Denote*

$$
W = \{v | v \in L^{p_0}(0, T; B_0), v' = \frac{dv}{dt} \in L^{p_1}(0, T; B_1)\},
$$

where $T < \infty$ and $1 < p_i < \infty$, $i = 0, 1$, then W is a Banach space when equipped with the norm

$$
\|v\|_{L^{p_0}(0,T;B_0)} + \|v'\|_{L^{p_1}(0,T;B_1)},
$$

and W is compactly embedded to $L^{p_0}(0, T; B)$.

Lemma 3.2.8 *Suppose Q is a bounded domain in $\mathbf{T}_x^n \times \mathbf{R}_t$, $g_\mu, g \in L^q(Q)(1 < q < \infty)$ and $\|g_\mu\|_{L^q(Q)} \leqslant C$. Furthermore, suppose that $g_\mu \to g$ a.e. in Q, then $g_\mu \rightharpoonup g$ weakly in $L^q(Q)$.*

Lemma 3.2.9 *Supposing X is a Banach space, if $g \in L^p(0,T;X)$ and $\dfrac{\partial g}{\partial t} \in L^p(0,T;X)(1 \leqslant p \leqslant \infty)$, then $g \in C([0,T];X)$ after possibly being redefined on a set of measure zero.*

In what follows, we prove Theorem 3.2.2.

Proof of theorem 3.2.2. We prove this theorem in steps.

In the first step, we fix a positive integer m and seek a function $u_m = u_m(t)$ of the form

$$u_m(t) = \sum_{|j|=1}^{m} g_{jm}(t)w_j, \quad w_j = e^{i<j,x>}, j \in \mathbf{Z}^n,$$

where $g_{jm}(t)(|j| = 0, 1, \cdots, m)$ satisfy the following approximating equations

$$(iu_{m,t}, w_j) + ((-\Delta)^\alpha u_m, w_j) + (\beta|u_m|^\rho u_m, w_j) = 0, \quad 0 \leqslant |j| \leqslant m, \quad (3.2.20)$$

with the initial conditions

$$u_m(0) = u_{0m} \in Span\{w_j, 0 \leqslant |j| \leqslant m\}, \quad u_{0m} \to u_0(m \to \infty) \text{ in } H^\alpha(\Omega). \quad (3.2.21)$$

Then (3.2.20) and (3.2.21) are a system of nonlinear ordinary differential equations. According to standard existence theory for nonlinear ODEs, there exists a unique solution $u_m(t)$ for $0 \leqslant t \leqslant t_m$. By the *a priori* estimates given above, we obtain that $t_m = T$.

In the second step, we make several *a priori* estimates. Taking into consideration Lemma 3.2.2 and Lemma 3.2.1, we obtain

$$u_m \in L^\infty(0,T; H^\alpha(\Omega) \cap L^{\rho+2}(\Omega)). \quad (3.2.22)$$

For arbitrary $\varphi \in H^\alpha(\Omega)$, one has

$$(iu_{m,t}, \varphi) + ((-\Delta)^\alpha u_m, \varphi) + (\beta|u_m|^\rho u_m, \varphi) = 0. \quad (3.2.23)$$

Thus

$$\begin{aligned}
|(u_{m,t}, \varphi)| &\leqslant |((-\Delta)^\alpha u_m, \varphi)| + |(\beta|u_m|^\rho u_m, \varphi)| \\
&\leqslant C\|(-\Delta)^{\alpha/2}u_m\|\|(-\Delta)^{\alpha/2}\varphi\| + C\|u_m\|_{L^{\rho+2}(\Omega)}^{\rho+1}\|\varphi\|_{L^{\rho+2}(\Omega)} \\
&\leqslant C\|(-\Delta)^{\alpha/2}\varphi\| + C\|\varphi\|_{L^{\rho+2}(\Omega)}.
\end{aligned}$$

$$(3.2.24)$$

By Sobolev embedding theorem, one has $\|\varphi\|_{L^{\rho+2}(\Omega)} \leqslant C\|(-\Delta)^{\alpha/2}\varphi\|$, and from (3.2.23) and (3.2.24), we know $|(u_{m,t}, \varphi)| \leqslant C\|(-\Delta)^{\alpha/2}\varphi\|$ for all $\varphi \in H^\alpha(\Omega)$. Therefore,

$$u_{m,t} \in L^\infty(0, T; H^{-\alpha}(\Omega)). \qquad (3.2.25)$$

In the third step, we pass to the limit $m \to \infty$. By (3.2.22) and (3.2.25), there exists a subsequence $\{u_\mu\}$ of $\{u_m\}$ such that

$$
\begin{aligned}
u_\mu &\rightharpoonup u, \quad \text{weakly in} \quad L^\infty(0, T; H^\alpha(\Omega)); \\
u_{\mu,t} &\rightharpoonup u_t, \quad \text{weakly in} \quad L^\infty(0, T; H^{-\alpha}(\Omega)).
\end{aligned}
\qquad (3.2.26)
$$

Using (3.2.22), we know $\{u_m\}$ is bounded in $L^2(0, T; H^\alpha(\Omega))$, and from (3.2.25), we obtain $\{u_{m,t}\}$ is bounded in $L^2(0, T; H^{-\alpha}(\Omega))$. Let

$$W = \{v|v \in L^2(0, T; H^\alpha(\Omega)), v_t \in L^2(0, T; H^{-\alpha}(\Omega))\},$$

equipped with the norm

$$\|v\|_W = \|v\|_{L^2(0,T;H^\alpha(\Omega))} + \|v_t\|_{L^2(0,T;H^{-\alpha}(\Omega))}.$$

Since $H^\alpha(\Omega)$ is compactly embedded to $L^2(\Omega)$, Lemma 3.2.7 shows that W is compactly embedded into $L^2(0, T; L^2(\Omega))$. But $u_m \in W$, then there exists a subsequence u_μ such that $u_\mu \to u$ strongly and a.e. in $L^2(0, T; L^2(\Omega))$. By (3.2.22) and Lemma 3.2.6, we know

$$|u_\mu|^\rho u_\mu \rightharpoonup |u|^\rho u \quad \text{weakly in } L^\infty(0, T; L^{\frac{\rho+2}{\rho+1}}(\Omega)). \qquad (3.2.27)$$

Fixing j and using (3.2.20), we obtain

$$(iu_{\mu,t}, w_j) + ((-\Delta)^\alpha u_\mu, w_j) + (\beta|u_\mu|^\rho u_\mu, w_j) = 0. \qquad (3.2.28)$$

By (3.2.26) and (3.2.27), there exists a subsequence u_μ such that

$$
\begin{aligned}
((-\Delta)^\alpha u_\mu, w_j) &\rightharpoonup ((-\Delta)^\alpha u, w_j) \quad && \text{weakly in } L^\infty(0, T); \\
(u_{\mu,t}, w_j) &\rightharpoonup (u_t, w_j) && \text{weakly in } L^\infty(0, T); \\
(\beta|u_\mu|^\rho u_\mu, w_j) &\rightharpoonup (\beta|u|^\rho u, w_j) && \text{weakly in } L^\infty(0, T).
\end{aligned}
$$

From (3.2.28), we know that for any fixed j

$$(iu_t, w_j) + ((-\Delta)^\alpha u, w_j) + (\beta|u|^\rho u, w_j) = 0,$$

then

$$(iu_t, v) + ((-\Delta)^\alpha u, v) + (\beta|u|^\rho u, v) = 0, \quad \forall v \in H^\alpha(\Omega).$$

Therefore, u satisfies equations (3.2.1) and (3.2.2). By (3.2.22), (3.2.25) and Lemma 3.2.9, we have $u_\mu \in C([0, T]; H^{-\alpha}(\Omega))$ and hence $u_\mu(0) \rightharpoonup u(0)$ weakly in $H^{-\alpha}(\Omega)$. Finally, by (3.2.21), we know $u_\mu(0) \rightharpoonup u_0$ weakly in $H^\alpha(\Omega)$. Therefore, $u(0) = u_0$.

Proof of Theorem 3.2.1 By the *a priori* estimates from Lemma 3.2.1 to Lemma 3.2.6 and theorem 3.2.2, there exists a global smooth solution u for (3.2.1) such that

$$u \in L^\infty(0, T; H^{4\alpha}(\Omega)), \quad u_t \in L^\infty(0, T; H^{2\alpha}(\Omega)).$$

We now prove the uniqueness. Let u and v be two solutions of equation (3.2.1) with the same initial data. Let $w = u - v$, then

$$iw_t + (-\Delta)^\alpha w + \beta(|u|^\rho u - |v|^\rho v) = 0.$$

Taking inner product of this equation with w, we obtain

$$i(w_t, w) + ((-\Delta)^\alpha w, w) + \beta((|u|^\rho u - |v|^\rho v), w) = 0.$$

Taking the imaginary part, we have

$$\frac{1}{2}\frac{d}{dr}\|w\|^2 + \Im\beta((|u|^\rho u - |v|^\rho v), u - v) = 0.$$

Since

$$\Im\beta((|u|^\rho u - |v|^\rho v), u - v) \leqslant C|(|u|^\rho(u - v) + (|u|^\rho u - |v|^\rho v)v, u - v)|$$
$$\leqslant C\|u\|_{L^\infty(\Omega)}\|u - v\|^2 + C\|v\|_{L^\infty(\Omega)}\||u|^\rho$$
$$- |v|^\rho|\|u - v\| \leqslant C\|w\|^2,$$

the Gronwall inequality implies $\|w\|^2 = 0$, yielding $w = 0$. The proof of the Theorem 3.2.1 is then complete.

3.2.2   Time fractional nonlinear Schrödinger equation

The main purpose of this section is to consider the time fractional nonlinear Schrödinger equations (1.4.2) and (1.4.3) with time fractional derivative,

$$(iT_p)^\nu D_t^\nu \psi = -\frac{L_p^2}{2N_m}\partial_x^2\psi + N_V\psi, \tag{3.2.29}$$

and

$$i(T_p)^\nu D_t^\nu \psi = -\frac{L_p^2}{2N_m}\partial_x^2\psi + N_V\psi, \tag{3.2.30}$$

where D_t^ν denotes the ν-order Caputo fractional derivative. Setting $\alpha = N_V/T_p^\nu$ and $\beta = L_p^2/2N_m(T_p)^\nu$, then the equation (3.2.29) can be rewritten as

$$D_t^\nu \psi = -\frac{\beta}{i^\nu}\partial_x^2\psi + \frac{\alpha}{i^\nu}\psi.$$

On the other hand for $0 < \nu < 1$,

$$D_t^{1-\nu} D_t^\nu y(t) = \frac{d}{dt}y(t) - \frac{[D_t^\nu y(t)]_{t=0}}{t^{1-\nu}\Gamma(\nu)}. \tag{3.2.31}$$

We obtain

$$\partial_t \psi = -\frac{\beta}{i^\nu}\partial_x^2(D_t^{1-\nu}\psi) + \frac{\alpha}{i^\nu}(D_t^{1-\nu}\psi) + \frac{[D^\nu\psi(t)]_{t=0}}{t^{1-\nu}\Gamma(\nu)}. \tag{3.2.32}$$

In this equation, since the Hamiltonian depends on time, we cannot expect the probability conservation. Meanwhile, the Hamiltonian is nonlocal in time, we cannot expect the inversion invariance with time. Finally, since $0 < \nu < 1$, the last term on the RHS will tend to infinity as time tends to zero. Consider the nonlocal term in (3.2.32)

$$D_t^{1-\nu}\psi(t,x) = \frac{1}{\Gamma(1-\nu)}\int_0^t \frac{d}{d\tau}\psi(\tau,x)\frac{d\tau}{(t-\tau)^\nu}.$$

To give a possible physical interpretation for this term, we first recall the interpretation of the first-order time derivative in the classical quantum mechanics $\dfrac{\partial}{\partial t} = \dfrac{E}{i\hbar}$, where E is the energy operator (Hamiltonian). So the inner product $\displaystyle\int_{-\infty}^\infty \psi(t,x) * D_t^{1-\nu}\psi(t,x)dx$ can be interpreted as the weighted time average of the energy of the wave function, the weighting function being $(t-\tau)^{-\nu}$.

Denote $\widetilde{\psi} = D_t^{1-\nu}\psi$. For the classical free particle Schrödinger equation, the probability current density and corresponding equation are respectively

$$P = \psi\psi^*, \qquad \partial_t P = \partial_t\psi\psi^* + \psi\partial_t\psi^*.$$

Similarly, we can obtain the probability current density equation of fractional Schrödinger equation as

$$\partial_t P = \left(-\frac{\beta}{i^\nu}\partial_x^2\widetilde{\psi} + \frac{[D_t^\nu\psi(t,x)]_{t=0}}{t^{1-\nu}\Gamma(\nu)}\right)\psi^* + \psi\left(-\frac{\beta}{(-i)^\nu}\partial_x^2\widetilde{\psi}^* + \frac{[D_t^\nu\psi^*(t,x)]_{t=0}}{t^{1-\nu}\Gamma(\nu)}\right).$$

Rearrange above equation to obtain

$$\partial_t P + \beta\partial_x\left(\frac{\partial_x\widetilde{\psi}\psi^*}{i^\nu} + \frac{\partial_x\widetilde{\psi}^*\psi}{(-i)^\nu}\right) = \beta\left(\frac{\partial_x\widetilde{\psi}\partial_x\psi^*}{i^\nu} + \frac{\partial_x\widetilde{\psi}^*\partial_x\psi}{(-i)^\nu}\right)$$
$$+ \frac{\psi^*[D_t^\nu\psi(t,x)]_{t=0} + \psi[D_t^\nu\psi^*(t,x)]_{t=0}}{t^{1-\nu}\Gamma(\nu)}. \tag{3.2.33}$$

In this equation, the right hand side term can be regarded as a source of the probability current equation. If the Hamiltonian does not depend on time, i.e., if $\nu \to 1$, the right hand side of (3.2.33) would be zero. The probability current of fractional equation (the left second term) is

$$J = \frac{\beta}{i^\nu}(\partial_x \widetilde{\psi})\psi^* + \frac{\beta}{(-i)^\nu}\psi(\partial_x \widetilde{\psi}^*).$$

Since the right hand of (3.2.33) does not equal to zero, the probability is not conserved for the time fractional Schrödinger equation. Denoting the right hand items (3.2.33) as $S(x,t)$, we then obtain

$$\partial_t P + \partial_x J = S.$$

Integrating this equation with respect to space variable, and letting the wave function and its first derivative equal to zero at infinity, we obtain

$$\partial_t \int_{-\infty}^{\infty} P\,\mathrm{d}x = \int_{-\infty}^{\infty} S\,\mathrm{d}x.$$

1. *Free particle fractional Schrödinger equation.*

The time fractional Schrödinger equation for a free particle is given by

$$(iT_p)^\nu D_t^\nu \psi = -\frac{L_p^2}{2N_m}\partial_x^2 \psi.$$

Performing the Fourier transform and letting $\Psi(\xi,t) = \mathcal{F}(\psi(x,t))$, we can obtain

$$D_t^\nu \Psi = \frac{(L_p\xi)^2}{2N_m(iT_p)^\nu}\Psi.$$

Letting $\omega = (L_p\xi)^2/2N_m T_p^\nu$ and using the Mittag-Leffler function, the solution can be expressed as

$$\Psi = \Psi_0 E_\nu(\omega(-it)^\nu), \quad \text{or} \quad \Psi = \frac{\Psi_0}{\nu}\{e^{-i\omega^{1/\nu}t} - \nu F_\nu(\omega(-i)^\nu,t)\},$$

where the function F_ν is defined as

$$F_\nu(\rho,t) = \frac{\rho\sin(\nu\pi)}{\pi}\int_0^\infty \frac{e^{-rt}r^{\nu-1}\mathrm{d}r}{r^{2\nu} - 2\rho\cos(\nu\pi)r^\nu + \rho^2}.$$

Taking advantage of the inverse Fourier transform, we obtain

$$\psi(x,t) = \mathcal{F}^{-1}\Psi(\xi,t) = \frac{1}{2\pi}\int_{\mathbf{R}} e^{ix\xi}\frac{\Psi_0}{\nu}\left\{e^{-i\omega^{1/\nu}t} - \nu F_\nu(\omega(-i)^\nu,t)\right\}\mathrm{d}\xi$$
$$=: \psi_S(x,t) + \psi_D(d,t),$$

where the first term

$$\psi_S(x,t) = \frac{1}{2\pi\nu}\int_{\mathbf{R}} e^{ix\xi}\Psi_0 e^{-i\omega^{1/\nu}t}\mathrm{d}\xi$$

is oscillating and the second term

$$\psi_D(x,t) = \frac{-1}{2\pi} \int_{\mathbf{R}} e^{ix\xi} \Psi_0 F_\nu(\omega(-i)^\nu, t) d\xi$$

decays to zero as time goes to infinity. When $\nu \to 1$, we have $\psi_D \to 0$, and then the solution reduces to the classic integer order Schrödinger equation. ψ_0 can be normalized such that

$$\int_{\mathbf{R}} \psi(x,0)\psi^*(x,0)dx = 1.$$

As to the total probability as time evolves, we have the probability limit when time goes to infinity

$$\lim_{t\to\infty} \int_{\mathbf{R}} \psi(x,t)\psi^*(x,t)dx$$

$$= \lim_{t\to\infty} \int_{\mathbf{R}} \mathcal{F}^{-1}\left(\frac{\Psi_0}{\nu}\left\{ e^{-i\omega^{1/\nu}t} - \nu F_\nu(\omega(-i)^\nu, t) \right\} \right)$$

$$\mathcal{F}^{-1}\left(\frac{\Psi_0}{\nu}\left\{ e^{-i\omega^{1/\nu}t} - \nu F_\nu(\omega(-i)^\nu, t) \right\} \right)^*$$

$$= \frac{2\pi}{\nu^2} \lim_{t\to\infty} \int_{\mathbf{R}} \Psi_0\left\{ e^{-i\omega^{1/\nu}t} - \nu F_\nu(\omega(-i)^\nu, t) \right\}$$

$$\left(\Psi_0\left\{ e^{-i\omega^{1/\nu}t} - \nu F_\nu(\omega(-i)^\nu, t) \right\} \right)^* d\xi$$

$$= \frac{2\pi}{\nu^2} \lim_{t\to\infty} \int_{\mathbf{R}} \Psi_0 e^{-i\omega^{1/\nu}t} \Psi_0^* e^{i\omega^{1/\nu}t} d\xi$$

$$= \frac{2\pi}{\nu^2} \lim_{t\to\infty} \int_{\mathbf{R}} \Psi_0 \Psi_0^* d\xi$$

$$= \frac{1}{\nu^2} \lim_{t\to\infty} \int_{\mathbf{R}} \psi_0 \psi_0^* dx.$$

Therefore, using the normalization condition, we obtain

$$\lim_{t\to\infty} \int_{\mathbf{R}} \psi(x,t)\psi^*(x,t)dx = \frac{1}{\nu^2} > 1.$$

2. *Potential well situation.*

Finally, let us consider the following ideal situation in which the particles are in a infinitely deep potential well $V(x) = 1$ for $1 < x < a$ and $V(x) = \infty$ otherwise. In this case, the equation can be written as

$$\begin{cases} (iT_p)^\nu D_t^\nu \psi = -\frac{L_p^2}{2N_m} \partial_x^2 \psi, \\ \psi(0,t) = 0, \quad \psi(a,t) = 0. \end{cases}$$

This can be solved by separation of variables. Let $\psi(x,t) = X(x)T(t)$, then we obtain

$$(iT_p)^\nu \frac{D_t^\nu T}{T} = -\frac{L_p^2}{2N_m}\frac{\partial_x^2 X}{X} = \lambda.$$

Under the boundary conditions $X(0) = X(a) = 0$, we have

$$X_n = c_n \sin\left(\frac{n\pi x}{a}\right), \quad \lambda_n = \left(\frac{n\pi L_p^2}{a}\right)^2 \frac{1}{2N_m}.$$

Normalizing to obtain the eigenfunction

$$\psi_n(x) = \sqrt{2/a}\sin(n\pi x/a), \quad \int_0^a |\psi_n|^2 \mathrm{d}x = 1.$$

Now, the equation of T can be written as $D_t^\nu T = \dfrac{\lambda_n}{(iT_p)^\nu}T$. Letting $T(0) = 1$ and using the Mittag-Leffler function, the solution can be written as

$$T_n(t) = E_\nu(\omega_n(-\mathrm{i}t)^\nu),$$

or

$$T_n(t) = \frac{1}{\nu}\left\{e^{-\mathrm{i}\omega^{1/\nu}t} - \nu F_\nu((-\mathrm{i}\omega)^\nu, t)\right\}, \quad \omega_n = \lambda_n/T_p^\nu.$$

It is easy to know $\lim_{t\to\infty}|T(t)| = \dfrac{1}{\nu}$. Then the solution is given by

$$\psi_n(x,t) = \sqrt{\frac{2}{a}}\sin(n\pi x/a)\frac{1}{\nu}\left\{e^{-\mathrm{i}\omega^{1/\nu}t} - \nu F_\nu((-\mathrm{i}\omega)^\nu, t)\right\}.$$

Similar to the free particle situation,

$$\lim_{t\to\infty}\int_0^a \psi_n(x,t)\psi_n(x,t)^* \mathrm{d}x = \frac{1}{\nu^2}.$$

3.2.3   Global well-posedness of the one-dimensional fractional nonlinear Schrödinger equation

This section considers the following one-dimensional fractional nonlinear Schrödinger equation [107],

$$\begin{cases} iu_t + (-\triangle)^\alpha u + |u|^2 u = 0, & (t,x) \in \mathbf{R}\times\mathbf{R}, \dfrac{1}{2} < \alpha < 1 \\ u(x,0) = u_0(x) \in H^s(\mathbf{R}). \end{cases} \tag{3.2.34}$$

We will obtain the global well-posedness of equation in L^2. We have shown the global well-posedness of periodic problem (3.2.34) in $H^{4\alpha}$. In what follows, we will prove the posedness of Cauchy problem (3.2.34) in L^2 with

$\frac{1}{2} < \alpha < 1$. Unlike the nonlinear Schrödinger equation, Strichartz estimates are not enough for solving the fractional nonlinear Schröinger equation in L^2, we also need the local smoothing effect and maximal function estimates. Thus, we will use the Bourgain's space to consider the well-posedness for (3.2.34).

To this end, we introduce some notations. We usually use its integral equivalent formulation to study the problem

$$u(t) = S(t)u_0 - \mathrm{i} \int_0^t S(t-t')|u|^2 u(t')\mathrm{d}t',$$

where $S(t) = \mathcal{F}_x^{-1} \mathrm{e}^{\mathrm{i}t|\xi|^{2\alpha}} \mathcal{F}_x$ is the semigroup of equation (3.2.34). First we define

$$\|f\|_{L_x^p L_t^q} = \left(\int_{-\infty}^{\infty} \left(\int_{-\infty}^{\infty} |f(x,t)|^q \mathrm{d}t \right)^{\frac{p}{q}} \mathrm{d}x \right)^{\frac{1}{p}},$$

$$\|f\|_{L_t^q L_x^p} = \left(\int_{-\infty}^{\infty} \left(\int_{-\infty}^{\infty} |f(x,t)|^p \mathrm{d}x \right)^{\frac{q}{p}} \mathrm{d}t \right)^{\frac{1}{q}}.$$

For $s, b \in \mathbf{R}$, spaces $X_{s,b}$ and $\bar{X}_{s,b}$ are defined to be the complete of the Schwartz function space in $\mathbf{R}^2$ under the norms [29, 122, 124]

$$\|u\|_{X_{s,b}} = \|S(-t)u\|_{H_x^s H_t^b} = \|\langle \xi \rangle^s \langle \tau - \phi(\xi) \rangle^b \hat{u}(\xi, \tau)\|_{L_\xi^2 L_\tau^2},$$

$$\|u\|_{\bar{X}_{s,b}} = \|S(t)u\|_{H_x^s H_t^b} = \|\langle \xi \rangle^s \langle \tau + \phi(\xi) \rangle^b \hat{u}(\xi, \tau)\|_{L_\xi^2 L_\tau^2},$$

where $\phi(\xi) = |\xi|^{2\alpha}$. We easily obtain $\|u\|_{X_{s,b}} = \|\bar{u}\|_{\bar{X}_{s,b}}$.

Denote by $\hat{u}(\tau, \xi) = \mathcal{F}u$ the Fourier transform of u with respect to variables t and x, and by $\mathcal{F}_{(\cdot)}u$ the Fourier transform only in the variable $(\cdot)$, respectively. Denote by $\int_\star \mathrm{d}\delta$ the convolution integral

$$\int_{\xi = \xi_1 + \xi_2 + \xi_3; \tau = \tau_1 + \tau_2 + \tau_3} \mathrm{d}\tau_1 \mathrm{d}\tau_2 \mathrm{d}\tau_3 \mathrm{d}\xi_1 \mathrm{d}\xi_2 \xi_3.$$

Let

$$\sigma = \tau - |\xi|^{2\alpha}, \sigma_1 = \tau_1 - |\xi_1|^{2\alpha}, \bar{\sigma}_2 = \tau_2 + |\xi_2|^{2\alpha}, \sigma_3 = \tau_3 - |\xi_3|^{2\alpha}, \bar{\sigma}_4 = \tau_4 + |\xi_4|^{2\alpha},$$

$$-\xi_4 = \xi = \xi_1 + \xi_2 + \xi_3, \quad -\tau_4 = \tau = \tau_1 + \tau_2 + \tau_3,$$

then

$$\sigma - \sigma_1 - \bar{\sigma}_2 - \sigma_3 = -|\xi|^{2\alpha} + |\xi_1|^{2\alpha} - |\xi_2|^{2\alpha} + |\xi_3|^{2\alpha},$$

or

$$\sigma_1 + \bar{\sigma}_2 + \sigma_3 + \bar{\sigma}_4 = -|\xi_1|^{2\alpha} + |\xi_2|^{2\alpha} - |\xi_3|^{2\alpha} + |\xi_4|^{2\alpha},$$

Let $\psi \in C_0^\infty(\mathbf{R})$ with $\psi = 1$ on $[-1/2, 1/2]$ and supp$\psi \subset [-1, 1]$. Denote $\psi_\delta(\cdot) = \psi(\delta^{-1}(\cdot))$ for some $\delta \in \mathbf{R} \setminus \{0\}$. Below we often use $A \sim B$ to denote the statement that there exists a constant $C_1 > 0$ such that $A \leqslant C_1 B$ and $B \leqslant C_1 A$, $A \ll B$ the statement that there exists a large enough constant $C_2 > 0$ such that $A \leqslant \dfrac{1}{C_2} B$ and $A \lesssim B$ the statement that there exists $C_3 > 0$ such that $A \leqslant C_3 B$. In what follows, $a+$ and $a-$ denote $a + \varepsilon$ and $a - \varepsilon$, respectively, for some $0 < \varepsilon \ll 1$.

We will prove the following

Theorem 3.2.3 *For $1/2 < \alpha < 1$, the Cauchy problem (3.2.34) is globally wee-posed in L^2.*

In order to establish the local well-posedness of the equation, we need to establish some linear and trilinear estimates. For this, we need to employ the $[k; Z]$ multiplier method (refer to [216]). Let Z be arbitrary Abelian additivity group with an invariant measure $d\xi$. For arbitrary integer $k \geqslant 2$, $\Gamma_k(Z)$ denotes the following "hyperplane"

$$\Gamma_k(Z) = \{(\xi_1, ..., \xi_k) \in Z^k : \xi_1 + ... + \xi_k = 0\},$$

which is endowed with

$$\int_{\Gamma_k(Z)} f = \int_{Z^{k-1}} f(\xi_1, ..., \xi_{k-1}, -\xi_1 - ... - \xi_{k-1}) d\xi_1 ... d\xi_{k-1}.$$

Define $[k; Z]$ multiplier as the function $m: \Gamma_k(Z) \to \mathbb{C}$. If m is a $[k; Z]$ multiplier, we define $\|m\|_{[k;Z]}$ to be the best constant such that

$$\left| \int_{\Gamma_k(Z)} m(\xi) \prod_{j=1}^k f_j(\xi_j) \right| \leqslant \| m \|_{[k;Z]} \prod_{j=1}^k \| f_j \|_{L_2(Z)},$$

for all test function f_j defined on Z. In this way, $\|m\|_{[k;Z]}$ determines a norm of m. When m is defined on all of Z^k, by restricting to $\Gamma_k(Z)$ we can similarly define the norm $\|m\|_{[k;Z]}$. We have the following property of $\|m\|_{[k;Z]}$ (see [216]).

Lemma 3.2.10 (Composition and TT^*) *If $k_1, k_2 \geqslant 1$, m_1 and m_2 are functions on Z^{k_1} and Z^{k_2} respectively, then*

$$\|m_1(\xi_1, \ldots, \xi_{k_1}) m_2(\xi_{k_1+1}, \ldots, \xi_{k_1+k_2})\|_{[k_1+k_2;Z]}$$
$$\leqslant \|m_1(\xi_1, \ldots, \xi_{k_1})\|_{[k_1+1;Z]} \|m_2(\xi_1, \ldots, \xi_{k_2})\|_{[k_2+1;Z]}.$$

As a special case, for all functions $m : Z^k \to \mathbf{R}$, the following TT^* holds

$$\|m(\xi_1, \ldots, \xi_k)\overline{m(-\xi_{k+1}, \ldots, -\xi_{2k})}\|_{[2k;Z]} = \|m(\xi_1, \ldots, \xi_k)\|_{[k+1;Z]}^2.$$

Lemma 3.2.11 *The group* $\{S(t)\}_{-\infty}^{+\infty}$ *of the fractional Schrödinger equation satisfies*

$$\|D_x^{\alpha-\frac{1}{2}}S(t)u_0\|_{L_x^\infty L_t^2} \lesssim \|u_0\|_{L^2}, \quad \text{the local smoothing effect}$$

$$\|D_x^{-\frac{1}{4}}S(t)u_0\|_{L_x^4 L_t^\infty} \lesssim \|u_0\|_{L^2}, \quad \text{the maximal function estimate}$$

$$\|S(t)u_0\|_{L_x^4 L_t^4} \lesssim \|u_0\|_{L^2}, \quad \text{the Strichartz estimate}$$

$$\|D_x^{\frac{\alpha-1}{3}}S(t)u_0\|_{L_x^6 L_t^6} \lesssim \|u_0\|_{L^2}.$$

Lemma 3.2.12 *Let* $\mathcal{F}F_\rho(\xi,\tau) = \dfrac{f(\xi,\tau)}{(1+|\tau-\xi^{2\alpha}|)^\rho}$, *then*

$$\|D_x^{\alpha-\frac{1}{2}}F_\rho\|_{L_x^\infty L_t^2} \lesssim \|f\|_{L_\xi^2 L_\tau^2}, \ \rho > 1/2$$

$$\|D_x^{-\frac{1}{4}}F_\rho\|_{L_x^4 L_t^\infty} \lesssim \|f\|_{L_\xi^2 L_\tau^2}, \ \rho > 1/2$$

$$\|D_x^{\frac{\alpha-1}{3}}F_\rho\|_{L_x^6 L_t^6} \lesssim \|f\|_{L_\xi^2 L_\tau^2}, \ \rho > 1/2$$

$$\|D_x^{\frac{\alpha-1}{4}}F_\rho\|_{L_x^4 L_t^4} \lesssim \|f\|_{L_\xi^2 L_\tau^2}, \ \rho > 3/8$$

$$\|F_\rho\|_{L_x^4 L_t^4} \lesssim \|f\|_{L_\xi^2 L_\tau^2}, \ \rho > 1/2$$

$$\|D_x^{-1/2-}F_\rho\|_{L_x^\infty L_t^\infty} \lesssim \|f\|_{L_\xi^2 L_\tau^2}, \ \rho > 1/2$$

$$\|F_\rho\|_{L_x^q L_t^q} \lesssim \|f\|_{L_\xi^2 L_\tau^2}, \ \rho > \frac{2q-4}{2q}, 2 \leqslant q \leqslant 4$$

$$\|D_x^{-\frac{q-2}{2q}-}F_\rho\|_{L_x^q L_t^q} \lesssim \|f\|_{L_\xi^2 L_\tau^2}, \ \rho > \frac{q-2}{2q}, 2 \leqslant q < \infty.$$

Lemma 3.2.13 (Linear estimates [123, 124]) *Let* $s \in \mathbf{R}$, $\dfrac{1}{2} < b < 1$, *and* $0 < \delta < 1$, *then*

$$\|\psi_\delta(t)S(t)u_0\|_{X_{s,b}} \leqslant C\delta^{\frac{1}{2}-b}\|u_0\|_{H^s},$$

$$\left\|\psi_\delta(t)\int_0^t S(t-t')f(t')\mathrm{d}t'\right\|_{X_{s,b}} \leqslant C\delta^{\frac{1}{2}-b}\|f\|_{X_{s,b-1}},$$

$$\left\|\psi_\delta(t)\int_0^t S(t-t')f(t')\mathrm{d}t'\right\|_{H^s} \leqslant C\delta^{\frac{1}{2}-b}\|f\|_{X_{s,b-1}},$$

$$\|\psi_\delta(t)f\|_{X_{s,b-1}} \leqslant C\delta^{b'-b}\|f\|_{X_{s,b'-1}}.$$

Lemma 3.2.14 *If* $\dfrac{1}{4} < b < \dfrac{1}{2}$, *then there exists* $C > 0$ *such that*

$$\int_{\mathbf{R}} \frac{\mathrm{d}x}{\langle x-\alpha\rangle^{2b}\langle \tau-\beta\rangle^{2b}} \leqslant \frac{C}{\langle \alpha-\beta\rangle^{4b-1}}.$$

Lemma 3.2.15 *If f, f_1, f_2 and f_3 belong to the Schwartz space on $\mathbf{R}^2$, then*

$$\int_* \bar{\hat{f}}(\xi,\tau)\hat{f}_1(\xi_1,\tau_1)\hat{f}_2(\xi_2,\tau_2)\hat{f}_3(\xi_3,\tau_3)\mathrm{d}\delta = \int \bar{f}f_1f_2f_3(x,t)\mathrm{d}x\mathrm{d}t.$$

Lemma 3.2.16 *For arbitrary Schwartz functions u_1 and $\bar{u}_2$ with Fourier support in $|\xi_1| \sim R_1$ and $|\xi_2| \sim R_2$, respectively. If $\xi_1 \cdot \xi_2 < 0$ or $R_1 \ll R_2(R_2 \ll R_1)$, then*

$$\|u_1\bar{u}_2\|_{L_x^2 L_t^2} \lesssim \|u_1\|_{X_{0,\frac{1}{2}+}}\|u_2\|_{X_{0,\frac{1}{2}-}}.$$

Remark 3.2.1 *By multilinear expression, we have*

$$\left\|\frac{1}{\langle\sigma_1\rangle^{1/2+}\langle\bar{\sigma}_2\rangle^{1/2-}}\right\|_{[3,\mathbf{R}\times\mathbf{R}]} \lesssim 1.$$

Proof Define $\tau_2 = \tau - \tau_1$, $\xi_2 = \xi - \xi_1$ and $\sigma = \tau - |\xi|^{2\alpha}$. By symmetry, we can assume $|\xi_1| \geqslant |\xi_2|$.

Case 1). When $|\sigma_1| \gtrsim |\xi_1|^{2\alpha}$ or $|\bar{\sigma}_2| \gtrsim |\xi_2|^{2\alpha}$.

By symmetry we assume $|\sigma_1| \gtrsim |\xi_1|^{2\alpha}$, then using Lemma 3.2.12, we have

$$\|u_1\bar{u}_2\|_{L_x^2 L_t^2} \lesssim \|u_1\|_{L_x^{4+} L_t^{4+}}\|\bar{u}_2\|_{L_x^{4-} L_t^{4-}} \lesssim \|u_1\|_{X_{0,\frac{1}{2}+}}\|u_2\|_{X_{0,\frac{1}{2}-}}.$$

Case 2). When $|\sigma_1| \lesssim |\xi_1|^{2\alpha}$ and $|\bar{\sigma}_2| \lesssim |\xi_2|^{2\alpha}$.

Then from $\sigma - \sigma_1 - \bar{\sigma}_2 = -|\xi|^{2\alpha} + |\xi_1|^{2\alpha} - |\xi - \xi_1|^{2\alpha} \lesssim |\xi_1|^{2\alpha}$, it follows that $|\sigma| \lesssim |\xi_1|^{2\alpha}$. Let $f_1(\tau_1,\xi_1) = \langle\sigma_1\rangle^{1/2+}\hat{u}_1(\tau_1,\xi_1)$ and $f_2(\tau_2,\xi_2) = \langle\bar{\sigma}_2\rangle^{1/2-}\hat{\bar{u}}_2(\tau_2,\xi_2)$, then

$$\|u_1\bar{u}_2\|_{L_x^2 L_t^2} = \|\mathcal{F}(u_1\bar{u}_2)\|_{L_\xi^2 L_\tau^2} = \|(\hat{u}_1 * \hat{\bar{u}}_2)(\xi)\|_{L_\xi^2 L_\tau^2}$$

$$= \left\|\int\int \frac{f_1(\tau_1,\xi_1)f_2(\tau-\tau_1,\xi-\xi_1)}{\langle\sigma_1\rangle^{1/2+}\langle\bar{\sigma}_2\rangle^{1/2-}}\mathrm{d}\xi_1\mathrm{d}\tau_1\right\|_{L_\xi^2 L_\tau^2}$$

$$\leqslant \left\|\left(\int\int \frac{\mathrm{d}\xi_1\mathrm{d}\tau_1}{\langle\sigma_1\rangle^{1+}\langle\bar{\sigma}_2\rangle^{1-}}\right)^{\frac{1}{2}}\right.$$

$$\left(\int\int (f_1(\tau_1,\xi_1)f_2(\tau_2,\xi_2))^2\mathrm{d}\xi_1\mathrm{d}\tau_1\right)^{\frac{1}{2}}\Bigg\|_{L_\xi^2 L_\tau^2}$$

$$\leqslant \left\|\left(\int\int \frac{\mathrm{d}\xi_1\mathrm{d}\tau_1}{\langle\sigma_1\rangle^{1+}\langle\bar{\sigma}_2\rangle^{1-}}\right)^{\frac{1}{2}}\right\|_{L_\xi^\infty L_\tau^\infty}$$

$$\left\|\left(\int\int (f_1(\tau_1,\xi_1)f_2(\tau_2,\xi_2))^2\mathrm{d}\xi_1\mathrm{d}\tau_1\right)^{\frac{1}{2}}\right\|_{L_\xi^2 L_\tau^2}$$

$$\leqslant \left\|\left(\int\int \frac{\mathrm{d}\xi_1\mathrm{d}\tau_1}{\langle\sigma_1\rangle^{1+}\langle\bar{\sigma}_2\rangle^{1-}}\right)^{\frac{1}{2}}\right\|_{L_\xi^\infty L_\tau^\infty}\|f_1\|_{L_\xi^2 L_\tau^2}\|f_2\|_{L_\xi^2 L_\tau^2}.$$

It suffices to show that

$$\left\| \left(\int\int \frac{\mathrm{d}\xi_1 \mathrm{d}\tau_1}{\langle\sigma_1\rangle^{1+}\langle\bar\sigma_2\rangle^{1-}} \right)^{\frac{1}{2}} \right\|_{L_\xi^\infty L_\tau^\infty} \lesssim 1.$$

Using Lemma 3.2.14, when $\dfrac{1}{4} < b < 1/2$, we have

$$\int\int_{\mathbf{R}^2} \frac{\mathrm{d}\tau_1 \mathrm{d}\xi_1}{\langle\tau_1 - |\xi_1|^{2\alpha}\rangle^{2b}\langle\tau - \tau_1 + |\xi - \xi_1|^{2\alpha}\rangle^{2b}}$$
$$\leqslant C \int_{\mathbf{R}} \frac{\mathrm{d}\xi_1}{\langle\tau - |\xi_1|^{2\alpha} + |\xi - \xi_1|^{2\alpha}\rangle^{4b-1}}.$$

To integrate with respect to ξ_1, we change variable $\mu = \tau - |\xi_1|^{2\alpha} + |\xi - \xi_1|^{2\alpha}$. From $\xi_1(\xi - \xi_1) < 0$ or $|\xi - \xi_1| \ll |\xi_1|$, it follows that $\mathrm{d}\mu \sim |\xi_1|^{2\alpha-1}\mathrm{d}\xi_1$. Moreover,

$$\mu = \tau - |\xi|^{2\alpha} + |\xi|^{2\alpha} - |\xi_1|^{2\alpha} + |\xi - \xi_1|^{2\alpha} \lesssim |\xi_1|^{2\alpha}.$$

Taking $b = 1/2 - \varepsilon$ for small enough $\varepsilon > 0$, we have when $\alpha > 1/2$ that

$$\int_{\mathbf{R}} \frac{\mathrm{d}\xi_1}{\langle\tau - |\xi_1|^{2\alpha} + |\xi - \xi_1|^{2\alpha}\rangle^{4b-1}} \sim \frac{1}{|\xi_1|^{2\alpha-1}} \int_{\mathbf{R}} \frac{\mathrm{d}\mu}{\langle\mu\rangle^{4b-1}} \lesssim |\xi_1|^{1-2\alpha-\alpha\varepsilon} \lesssim 1.$$

This completes the proof of Lemma 3.2.16.

Below we consider trilinear estimates and prove the following

Theorem 3.2.4 *Assume* $\mathcal{F}u_1 = \hat{u}_1(\tau_1, \xi_1)$, $\mathcal{F}\bar{u}_2 = \hat{\bar{u}}_2(\tau_2, \xi_2)$ *and* $\mathcal{F}u_3 = \hat{u}_3(\tau_3, \xi_3)$ *are supported in* $\{(\xi_1, \tau_1) : |\xi_1| \leqslant 2\} \bigcup \{(\xi_2, \tau_2) : |\xi_2| \leqslant 2\} \bigcup \{(\xi_3, \tau_3) : |\xi_3| \leqslant 2\} \bigcup \{(\xi_1 + \xi_2 + \xi_3, \tau_1 + \tau_2 + \tau_3) : |\xi_1 + \xi_2 + \xi_3| \leqslant 6\}$, *then*

$$\|u_1\bar{u}_2 u_3\|_{X_{0,-1/2+}} \leqslant C\|u_1\|_{X_{0,1/2+}}\|u_2\|_{X_{0,1/2+}}\|u_3\|_{X_{0,1/2+}}.$$

Proof By duality and Plancherel identity, it suffices to show

$$\Gamma = \int_* \frac{\bar{f}(\tau, \xi)}{\langle\sigma\rangle^{1-b}} \mathcal{F}u_1(\tau_1, \xi_1)\mathcal{F}u_2(\tau_2, \xi_2)\mathcal{F}\bar{u}_3(\tau_3, \xi_3)\mathrm{d}\delta$$
$$= \int_* \frac{\bar{f}(\tau, \xi)f_1(\tau_1, \xi_1)f_2(\tau_2, \xi_2)f_3(\tau_3, \xi_3)\mathrm{d}\delta}{\langle\sigma\rangle^{1/2-}\langle\sigma_1\rangle^{1/2+}\langle\bar\sigma_2\rangle^{1/2+}\langle\sigma_3\rangle^{1/2+}}$$
$$\leqslant C\|f\|_{L_2} \prod_{j=1}^{3} \|f_j\|_{L_2}$$

for all $\bar{f} \in L_2, \bar{f} \geqslant 0$, where $f_1 = \langle\sigma_1\rangle^{1/2+}\widehat{u_1}$, $f_2 = \langle\bar\sigma_2\rangle^{1/2+}\widehat{u_2}$, $f_3 = \langle\sigma_3\rangle^{1/2+}\widehat{u_3}$. By multiple linear expressions, (3.2.3) is established only when

$$\left\| \frac{1}{\langle\sigma_1\rangle^{1/2+}\langle\bar\sigma_2\rangle^{1/2+}\langle\sigma_3\rangle^{1/2+}\langle\bar\sigma_4\rangle^{1/2-}} \right\|_{[4,\mathbf{R}\times\mathbf{R}]} \lesssim 1.$$

Let

$$\mathcal{F}F_\rho^j(\xi,\tau) = \frac{f_j(\xi,\tau)}{(1+|\tau-\xi^{2\alpha}|)^\rho}, \quad j=1,3; \quad \mathcal{F}F_\rho^2(\xi,\tau) = \frac{f_2(\xi,\tau)}{(1+|\tau+\xi^{2\alpha}|)^\rho},$$

$$\mathcal{F}F_\rho(\xi,\tau) = \frac{\bar{f}(\xi,\tau)}{(1+|\tau-\xi^{2\alpha}|)^\rho}.$$

By symmetry, we need only to consider the following two cases.

Case 1). When $|\xi| \lesssim 6$.

Using Lemma 3.2.12 and 3.2.15, the integral Γ restricted to this domain is estimated by

$$\int_* \frac{\bar{f}(\tau,\xi)f_1(\tau_1,\xi_1)f_2(\tau_2,\xi_2)f_3(\tau_3,\xi_3)\mathrm{d}\delta}{\langle\sigma\rangle^{1/2-}\langle\sigma_1\rangle^{1/2+}\langle\bar{\sigma}_2\rangle^{1/2+}\langle\sigma_3\rangle^{1/2+}}$$

$$=C\int \overline{F_{1/2-}} \cdot F_{1/2+}^1 \cdot F_{1/2+}^2 \cdot F_{1/2+}^3(x,t)\mathrm{d}x\mathrm{d}t$$

$$\leqslant C\|F_{1/2-}\|_{L_x^4 L_t^4}\|F_{1/2+}^1\|_{L_x^4 L_t^4}\|F_{1/2+}^2\|_{L_x^4 L_t^4}\|F_{1/2+}^3\|_{L_x^4 L_t^4}$$

$$\leqslant C\|f\|_{L_\xi^2 L_\tau^2}\|f_1\|_{L_\xi^2 L_\tau^2}\|f_2\|_{L_\xi^2 L_\tau^2}\|f_3\|_{L_\xi^2 L_\tau^2}.$$

Case 2). When $|\xi_1| \lesssim 2$.

Using Lemma 3.2.12 and 3.2.15, the integral Γ restricted to this area is bounded by

$$\int_* \frac{\bar{f}(\tau,\xi)f_1(\tau_1,\xi_1)f_2(\tau_2,\xi_2)f_3(\tau_3,\xi_3)\mathrm{d}\delta}{\langle\sigma\rangle^{1/2-}\langle\sigma_1\rangle^{1/2+}\langle\bar{\sigma}_2\rangle^{1/2+}\langle\sigma_3\rangle^{1/2+}}.$$

$$=C\int \overline{F_{1/2-}} \cdot F_{1/2+}^1 \cdot F_{1/2+}^2 \cdot F_{1/2+}^3(x,t)\mathrm{d}x\mathrm{d}t$$

$$\leqslant C\|F_{1/2-}\|_{L_x^3 L_t^3}\|F_{1/2+}^1\|_{L_x^6 L_t^6}\|F_{1/2+}^2\|_{L_x^4 L_t^4}\|F_{1/2+}^3\|_{L_x^4 L_t^4}$$

$$\leqslant C\|f\|_{L_\xi^2 L_\tau^2}\|f_1\|_{L_\xi^2 L_\tau^2}\|f_2\|_{L_\xi^2 L_\tau^2}\|f_3\|_{L_\xi^2 L_\tau^2}.$$

Thus we complete the proof of Lemma 3.2.16.

Theorem 3.2.5 (Trilinear estimates)   *If* $1/2 < \alpha < 1$, *then*

$$\|u_1\bar{u}_2 u_3\|_{X_{0,-1/2+}} \leqslant C\|u_1\|_{X_{0,1/2+}}\|u_2\|_{X_{0,1/2+}}\|u_3\|_{X_{0,1/2+}}.$$

Proof By duality and the Plancherel identity, it suffices to show that

$$\left\|m((\xi_1,\tau_1),...,(\xi_4,\tau_4))\right\|_{[4,\mathbf{R}\times\mathbf{R}]}$$

$$:=\left\|\frac{1}{\langle\sigma_1\rangle^{1/2+}\langle\bar{\sigma}_2\rangle^{1/2+}\langle\sigma_3\rangle^{1/2+}\langle\bar{\sigma}_4\rangle^{1/2-}}\right\|_{[4,\mathbf{R}\times\mathbf{R}]} \lesssim 1,$$

where

$$\xi_1 + \xi_2 + \xi_3 + \xi_4 = 0, \quad \tau_1 + \tau_2 + \tau_3 + \tau_4 = 0, \quad \xi = -\xi_4, \tau = -\tau_4,$$

$$\bar{\sigma}_4 = \tau_4 + |\xi_4|^{2\alpha}, \quad |\sigma_1 + \bar{\sigma}_2 + \sigma_3 + \bar{\sigma}_4| = |\xi_4|^{2\alpha} - |\xi_1|^{2\alpha} + |\xi_2|^{2\alpha} - |\xi_3|^{2\alpha}.$$

Define $N_i := |\xi_i|$, and adopt the notation that

$$1 \leqslant soprano, alto, tenor, baritone \leqslant 4$$

as the distinct indices such that

$$N_{soprano} \geqslant N_{alto} \geqslant N_{tenor} \geqslant N_{baritone}$$

are the highest, second highest, third highest and fourth highest values of the frequencies $N_1, \ldots, N_4$, respectively. Since $\xi_1 + \xi_2 + \xi_3 + \xi_4 = 0$, we have $N_{soprano} \sim N_{alto}$. Without loss of generality, we can assume that $N_{soprano} = N_1$ and $\xi_1 > 0$.

Case 1). Assume $N_2 = N_{alto}$.

This means that $\xi_1\xi_2 < 0$.

    Subcase 1-a). When $\xi_3\xi_4 < 0$. From Lemma 3.2.10 and Lemma 3.2.16, we know

$$\left\| m((\xi_1, \tau_1), \ldots(\xi_4, \tau_4)) \right\|_{[4, \mathbf{R} \times \mathbf{R}]}$$

$$\lesssim \left\| \frac{1}{\langle\sigma_1\rangle^{1/2+}\langle\bar{\sigma}_2\rangle^{1/2+}\langle\sigma_3\rangle^{1/2+}\langle\overline{\sigma}_4\rangle^{1/2-}} \right\|_{[4, \mathbf{R} \times \mathbf{R}]}$$

$$\lesssim \left\| \frac{1}{\langle\sigma_1\rangle^{1/2+}\langle\bar{\sigma}_2\rangle^{1/2+}} \right\|_{[3, \mathbf{R} \times \mathbf{R}]} \left\| \frac{1}{\langle\sigma_3\rangle^{1/2+}\langle\overline{\sigma}_4\rangle^{1/2-}} \right\|_{[3, \mathbf{R} \times \mathbf{R}]}$$

$$\lesssim 1.$$

    Subcase 1-b). When $\xi_3\xi_4 > 0$. Then it implies that $\xi_3 < 0$, $\xi_4 < 0$ and $|\xi_1 + \xi_2| = |\xi_3 + \xi_4| \geqslant \max\{|\xi_3|, |\xi_4|\}$.

    (1) If $N_3 = N_{tenor}$, then $|\xi_4|^{2\alpha} - |\xi_3|^{2\alpha} < 0$ and $-|\xi_1|^{2\alpha} + |\xi_2|^{2\alpha} < 0$. Using Taylor formula, we have

$$|\xi_3|^{2\alpha} - |\xi_4|^{2\alpha} \gtrsim 2\alpha N_4^{2\alpha-1} N_{12},$$

$$|\xi_1|^{2\alpha} - |\xi_2|^{2\alpha} \sim 2\alpha N_1^{2\alpha-1} N_{12},$$

and

$$|\xi_1|^{2\alpha} - |\xi_4|^{2\alpha} - |\xi_2|^{2\alpha} + |\xi_3|^{2\alpha} \gtrsim |\xi_2|^{2\alpha-1}|\xi_3|.$$

If $|\xi_4| \ll |\xi_3|$, similarly to *Subcase 1-a)*, we can obtain the result.

If $|\xi_4| \sim |\xi_3|$, then it implies that $|\xi| \sim |\xi_3|$. By symmetry, we assume $|\bar{\sigma}_4| = |\sigma| \gtrsim |\xi_2|^{2\alpha-1}|\xi_3| \geqslant |\xi_3|^{2\alpha}$. Similar to the proof of Theorem 3.2.4, using Lemma 3.2.12 and Lemma 3.2.15, we bound the integral Γ by

$$\int_* \frac{\bar{f}(\tau,\xi) f_1(\tau_1,\xi_1) f_2(\tau_2,\xi_2) f_3(\tau_3,\xi_3) d\delta}{\langle\sigma\rangle^{1/2-}\langle\sigma_1\rangle^{1/2+}\langle\bar{\sigma}_2\rangle^{1/2+}\langle\sigma_3\rangle^{1/2+}}$$

$$\leqslant \int_* \frac{\bar{f}(\tau,\xi) f_1(\tau_1,\xi_1) f_2(\tau_2,\xi_2) f_3(\tau_3,\xi_3) d\delta}{|\xi_3|^{\alpha-}\langle\sigma_1\rangle^{1/2+}\langle\bar{\sigma}_2\rangle^{1/2+}\langle\sigma_3\rangle^{1/2+}}$$

$$= C \int \overline{F_0} \cdot F_{1/2+}^1 \cdot F_{1/2+}^2 \cdot D_x^{-\alpha+} F_{1/2+}^3(x,t) dx dt$$

$$\leqslant C \|F_0\|_{L_x^2 L_t^2} \|F_{1/2+}^1\|_{L_x^4 L_t^4} \|F_{1/2+}^2\|_{L_x^4 L_t^4} \|D_x^{-\alpha+} F_{1/2+}^3\|_{L_x^\infty L_t^\infty}$$

$$\leqslant C \|f\|_{L_\xi^2 L_\tau^2} \|f_1\|_{L_\xi^2 L_\tau^2} \|f_2\|_{L_\xi^2 L_\tau^2} \|f_3\|_{L_\xi^2 L_\tau^2}.$$

(2) Suppose $N_4 = N_{tenor}$. Let $f(x) = (x+a)^{2\alpha} - a^{2\alpha} - x^{2\alpha}$ with $a, x > 0, 2\alpha > 1$. Then we have $f'(x) > 0$ for $x > 0$ and $f(x) \sim (x+a)\min\{x,a\}$. Thus we have

$$|\xi_1|^{2\alpha} - |\xi_4|^{2\alpha} - |\xi_2|^{2\alpha} + |\xi_3|^{2\alpha} = |\xi_2 + \xi_3 + \xi_4|^{2\alpha} - |\xi_4|^{2\alpha} - |\xi_2|^{2\alpha} + |\xi_3|^{2\alpha}$$

$$\gtrsim |\xi_2|^{2\alpha-1}|\xi_3|.$$

Similarly, we can obtain the result.

Case 2). Assume $N_3 = N_{alto}$.

It implies that $\xi_1 \xi_3 < 0$.

Subcase 2-a). If $\xi_2 < 0$, $\xi_4 > 0$, then similar to *Subcase 1-a)*, we can obtain the result.

Subcase 2-b). If $\xi_2 > 0$, $\xi_4 < 0$, then by Lemma 3.2.10 and Lemma 3.2.16, we have

$$\left\| m((\xi_1,\tau_1),...(\xi_4,\tau_4)) \right\|_{[4,\mathbf{R}\times\mathbf{R}]}$$

$$\lesssim \left\| \frac{1}{\langle\sigma_1\rangle^{1/2+}\langle\bar{\sigma}_2\rangle^{1/2+}\langle\sigma_3\rangle^{1/2+}\langle\bar{\sigma}_4\rangle^{1/2-}} \right\|_{[4,\mathbf{R}\times\mathbf{R}]}$$

$$\lesssim \left\| \frac{1}{\langle\sigma_1\rangle^{1/2+}\langle\bar{\sigma}_4\rangle^{1/2-}} \right\|_{[3,\mathbf{R}\times\mathbf{R}]} \left\| \frac{1}{\langle\sigma_3\rangle^{1/2+}\langle\bar{\sigma}_2\rangle^{1/2+}} \right\|_{[3,\mathbf{R}\times\mathbf{R}]}$$

$$\lesssim 1.$$

Subcase 2-c). If $\xi_2 < 0$, $\xi_4 < 0$, then it implies that $|\xi_1 + \xi_2| = |\xi_3 + \xi_4| \geqslant \max\{|\xi_3|, |\xi_4|\}$. Moreover, we have $|\xi_1|^{2\alpha} - |\xi_2|^{2\alpha} > 0$ and $|\xi_3|^{2\alpha} - |\xi_4|^{2\alpha} > 0$,

$$|\xi_1|^{2\alpha} - |\xi_4|^{2\alpha} - |\xi_2|^{2\alpha} + |\xi_3|^{2\alpha} \gtrsim |\xi_2|^{2\alpha-1}|\xi_3|.$$

Similar to the *Subcase 1-b)*, we can obtain the result. This completes the proof.

Moreover, we have the following energy inequality.

Lemma 3.2.17 *Let $u(t)$ be smooth solution of the Cauchy problem (3.2.34). Then*

$$\|u(t)\|_{L^2} \lesssim \|u_0\|_{L^2}.$$

Therefore, similar to [122, 124], using Lemma 3.2.13, Theorem 3.2.5 and Lemma 3.2.17, we can show that the Cauchy problem (3.2.34) is globally well-posed in L^2 with $1/2 < \alpha < 1$. Theorem 3.2.3 is hence established.

3.3   Fractional Ginzburg-Landau equation

In this section, we will focus on the fractional complex Ginzburg-Landau (FCGL) equation [184, 211]

$$u_t = Ru - (1 + \mathrm{i}\nu)(-\Delta)^\alpha u - (1 + \mathrm{i}\mu)|u|^{2\sigma}u, \qquad (3.3.1)$$

for $\alpha \in (0, 1)$, where $u(x, t)$ is a complex-valued function of t and x, R, μ, ν and σ are all real coefficients. If $\alpha = 1$, then (3.3.1) reduces to the classical Ginzburg-Landau equation [76, 105]. The main purpose of this section is to discuss the existence and uniqueness of solutions for (3.3.1) and its infinite-dimensional dynamical behaviors. For the sake of simplicity, we will discuss the periodic case on $\mathbf{T}^d = [0, 2\pi]^d$. What follows consists of three parts: the global existence of weak solutions, the global existence of strong solutions and the asymptotic behaviors of the infinite-dimensional dynamical systems.

3.3.1   Existence of weak solutions

In this section, we will discuss the existence of weak solutions of (3.3.1).

Theorem 3.3.1 *For any $\varphi \in L^2(\mathbf{T}^d)$, there exists a function*

$$u \in C([0, T]); w\text{-}L^2(\mathbf{T}^d)) \cap L^2([0, T]; H^\alpha(\mathbf{T}^d)) \cap L^{2\varsigma}([0, T]; L^{2\varsigma}(\mathbf{T}^d))$$

satisfying the FCGL equation in the weak sense

$$\begin{aligned}
\langle u(t), \phi^* \rangle - \langle \varphi, \phi^* \rangle =&R \int_0^t \langle u, \phi^* \rangle \mathrm{d}\tau - \int_0^t (1 + \mathrm{i}\nu)\langle \Lambda^\alpha u, \Lambda^\alpha \phi^* \rangle \mathrm{d}\tau \\
&- \int_0^t (1 + \mathrm{i}\mu)\langle |u|^{2\sigma} u, \phi^* \rangle \mathrm{d}\tau, \quad \forall \phi \in C^\infty(\mathbf{T}^d).
\end{aligned} \qquad (3.3.2)$$

Moreover, there holds the following energy inequality

$$\frac{1}{2}\|u(t)\|_{L^2}^2 + \int_0^t \|\Lambda^\alpha u\|_{L^2}^2 \mathrm{d}\tau + \int_0^t \|u\|_{L^{2\varsigma}}^{2\varsigma} \mathrm{d}\tau \leqslant \frac{1}{2}\|\varphi\|_{L^2}^2 + R \int_0^t \|u\|_{L^2}^2 \mathrm{d}\tau. \qquad (3.3.3)$$

By $u \in C([0,T]); w\text{-}L^2(\mathbf{T}^d))$, we mean $\langle u(t), \phi \rangle \in C([0,T])$ for arbitrary $\phi \in L^2(\mathbf{T}^d)$.

First, we will establish the following *a priori* estimates.

Lemma 3.3.1 *Let u be a smooth solution of FCGL equation with initial data φ, then*

$$\|u(t)\|_{L^2}^2 \leqslant e^{2Rt} \|\varphi\|_{L^2}^2, \tag{3.3.4}$$

and

$$\|u(t)\|_{L^2}^2 + 2\int_0^t \|\Lambda^\alpha u\|_{L^2}^2 d\tau + 2\int_0^t \|u\|_{L^{2\varsigma}}^{2\varsigma} d\tau \leqslant e^{2Rt} \|\varphi\|_{L^2}^2. \tag{3.3.5}$$

Proof Multiply FCGL equation by u^*, and integrate over $\mathbf{T}^d$ to get

$$\int_{\mathbf{T}^d} u_t u^* = R\int_{\mathbf{T}^d} uu^* dx - (1+i\nu)\int_{\mathbf{T}^d} (-\Delta)^\alpha uu^* dx - (1+i\mu)\int_{\mathbf{T}^d} |u|^{2\sigma} uu^* dx.$$

Adding this to its conjugate, and using the integration by parts formula, we have

$$\frac{d}{dt}\int_{\mathbf{T}^d} |u|^2 dx + 2\int_{\mathbf{T}^d} |\Lambda^\alpha u|^2 dx + 2\int_{\mathbf{T}^d} |u|^{2\varsigma} \leqslant 2R\int_{\mathbf{T}^d} |u|^2. \tag{3.3.6}$$

In particular,

$$\frac{d}{dt}\int_{\mathbf{T}^d} |u|^2 dx \leqslant 2R\int_{\mathbf{T}^d} |u|^2,$$

yielding by the Gronwall inequality that

$$\|u\|_{L^2}^2 \leqslant \|\varphi\|_{L^2}^2 e^{2Rt}.$$

Substituting this into (3.3.6) yields the estimate (3.3.5).

Lemma 3.3.2 *Let u be a smooth solution of the FCGL equation, then*

$$\left\|\frac{du}{dt}\right\|_{L^{\frac{2\varsigma}{2\varsigma-1}}(0,t;H^{-\beta})} \leqslant C, \quad \beta \geqslant \max\left\{\alpha, \frac{\sigma d}{2\varsigma}\right\}. \tag{3.3.7}$$

Proof Multiplying the function by $\phi^*(x)$ and integrating over $\mathbf{T}^d \times [0,t]$ yield

$$\int_0^t \left\langle \frac{du}{dt}, \phi^* \right\rangle = \int_0^t \langle Ru, \phi^* \rangle - \int_0^t (1+i\nu)\langle (-\Delta)^\alpha u, \phi^* \rangle - \int_0^t (1+i\mu)\langle |u|^{2\sigma} u, \phi^* \rangle.$$

Integrating by parts and applying the Hölder inequality, we have

$$\left|\int_0^t \langle Ru, \phi^* \rangle\right| \leqslant R\|u\|_{L^{2\varsigma}(\mathbf{T}^d \times [0,t])} \|\phi\|_{L^{2\varsigma}(\mathbf{T}^d \times [0,t])},$$

$$\left| \int_0^t (1+\mathrm{i}\nu)\langle(-\Delta)^\alpha u, \phi^*\rangle \right| \leqslant C\|u\|_{L^2(0,t;H^\alpha)}\|\phi\|_{L^2(0,t;H^\alpha)}$$

and

$$\left| \int_0^t (1+\mathrm{i}\mu)\langle|u|^{2\sigma}u, \phi^*\rangle \right| \leqslant C\|u\|_{L^{2\varsigma}(\mathbf{T}^d\times[0,t])}^{2\varsigma-1}\|\phi\|_{L^{2\varsigma}(\mathbf{T}^d\times[0,t])}.$$

Therefore, we know that for $\beta \geqslant \max\{\alpha, \sigma d/(2\sigma+2)\}$,

$$\left| \int_0^t \left\langle \frac{\mathrm{d}u}{\mathrm{d}t}, \phi^* \right\rangle \right| \leqslant C\|\phi\|_{L^{2\varsigma}(0,t;H^\alpha)}, \quad \forall \phi \in L^{2\varsigma}(0,t;H^\alpha).$$

It shows that $\dfrac{\mathrm{d}u}{\mathrm{d}t} \in L^{\frac{2\varsigma}{2\varsigma-1}}(0,t;H^{-\alpha})$, and the inequality (3.3.7) holds.

Further estimation can also be obtained. Let $I_\phi(t) = \langle u(t), \phi^*\rangle$.

Lemma 3.3.3 $I_\phi(t)$ *is a continuous function of t for arbitrary $\phi \in L^2(\mathbf{T}^d)$.*

Proof We first consider the case $\phi \in C^\infty(\mathbf{T}^d)$ and then use a density argument to extend to the general case for $\phi \in L^2(\mathbf{T}^d)$. Let $0 \leqslant t_1 < t - 2 \leqslant T$, by Hölder inequality,

$$
\begin{aligned}
|I_\phi(t_2) - I_\phi(t_1)| = & |R \int_{t_1}^{t_2} \langle u, \phi^*\rangle \mathrm{d}\tau - \int_{t_1}^{t_2}(1+\mathrm{i}\nu)\langle \Lambda^\alpha u, \Lambda^\alpha \phi^*\rangle \mathrm{d}\tau \\
& - \int_{t_1}^{t_2}(1+\mathrm{i}\mu)\langle|u|^{2\sigma}u, \phi_N^*\rangle \mathrm{d}\tau| \\
\leqslant & (|R|\|\phi\|_{L^\infty} + |1+\mathrm{i}\nu|\|\Lambda^{2\alpha}\phi^*\|_{L^\infty})\|u\|_{L^2(0,T;L^2)}|t_2-t_1|^{1/2} \\
& + |1+\mathrm{i}\mu|\|\phi^*\|_\infty\|u\|_{L^{2\varsigma}(0,T;L^{2\varsigma})}^{2\sigma-1}|t_2-t_1|^{\frac{1}{2\varsigma}} \\
\leqslant & C_\phi e^{RT}\|\varphi\|_{L^2}|t_2-t_1|^{1/2} + C_\phi(e^{2RT}\|\varphi\|_{L^2}^2)^{\frac{2\varsigma-1}{2\varsigma}}|t_2-t_1|^{\frac{1}{2\varsigma}}
\end{aligned}
$$
$$(3.3.8)$$

thanks to Lemma 3.3.1. The continuity of $I_\phi(t)$ then follows. Let $\varepsilon > 0$, and for arbitrary $\phi \in L^2(\mathbf{T}^d)$, choose $\phi_\varepsilon \in C^\infty(\mathbf{T}^d)$ such that $\|\phi_\varepsilon - \phi\|_{L^2(\mathbf{T}^d)} \leqslant \varepsilon$. Use Hölder inequality and triangle inequality to obtain

$$|I_\phi(t_2) - I_\phi(t_1)| \leqslant \varepsilon(\|u(t_2)\|_{L^2} + \|u(t_1)\|_{L^2}) + |I_{\phi_\varepsilon}(t_2) - I_{\phi_\varepsilon}(t_1)|. \quad (3.3.9)$$

Since $I_{\phi_\varepsilon}(t)$ is continuous in t, the second item on the right approaches to 0 as $t_1 \to t_2$. Noting that $\|u(t_2)\|_{L^2} + \|u(t_1)\|_{L^2}$ is independent of ε and ε is arbitrary, the continuity of $I_\phi(t)$ for $\phi \in L^2(\mathbf{T}^d)$ then follows.

The existence of weak solutions is proved by using the Fourier-Galerkin approximation method. Let $\{e_1, e_2, \cdots, e_N, \cdots\}$ be an orthonormal basis for

L^2 and $\mathscr{P}_N L^2$ be the orthogonal projection of L^2 into $span\{e_1, \cdots, e_N\}$. Consider the following approximation problem

$$\langle u_N(t), \phi^* \rangle - \langle \varphi_N, \phi^* \rangle = R \int_0^t \langle u_N, \phi^* \rangle d\tau - \int_0^t (1 + i\nu) \langle \Lambda^\alpha u_N, \Lambda^\alpha \phi^* \rangle d\tau$$
$$- \int_0^t (1 + i\mu) \langle |u_N|^{2\sigma} u_N, \phi_N^* \rangle d\tau,$$

(3.3.10)

where $\phi \in C^\infty(\mathbf{T}^d)$ and $\phi_N = \mathscr{P}_N \phi$. When $N \to \infty$, $\mathscr{P}_N \phi$ strongly converges to ϕ in $L^2(\mathbf{T}^d)$, and $\|\phi_N\|_{L^2} \leqslant \|\phi\|_{L^2}$ by the Parseval inequality.

Fix $T > 0$. First, Lemma 3.3.1-3.3.3 still hold for u_N, yielding

$$\|u_N(t)\|^2 + 2\int_0^t \|\Lambda^\alpha u_N\|_{L^2}^2 d\tau + 2\int_0^t \|u_N\|_{L^{2\varsigma}}^{2\varsigma} d\tau$$
$$= \|\varphi_N(t)\|^2 + 2R\int_0^t \|u_N\|_{L^2}^2 d\tau.$$

(3.3.11)

From above, we see $\{u_N\}$ is bounded in $L^2([0,T]; H^\alpha)$ and $L^{2\varsigma}([0,T]; L^{2\varsigma})$. Therefore, there exists a subsequence (still denoted by $\{u_N\}$) such that $u_N \to u$ weakly in $L^2([0,T]; H^\alpha)$ and $L^{2\varsigma}([0,T]; L^{2\varsigma})$. Moreover, (3.3.4) shows that for arbitrary $t \geqslant 0$, $\{u_N(t)\}$ is weakly compact in $L^2(\mathbf{T}^d)$ and Lemma 3.3.2 shows that du_N/dt is bounded in $L^{\frac{2\varsigma}{2\varsigma-1}}(0, T; H^{-\beta})$. Therefore, it follows that $\{u_N\}$ is compact in $L^2(0, T; L^2)$.

According to the interpolation inequality

$$\|\eta\|_{L^{2\varsigma-1}}^{2\varsigma-1} \leqslant \rho \|\eta\|_{L^{2\varsigma}}^{2\varsigma} + C(\rho)\|\eta\|_{L^2}^2, \qquad \forall \rho > 0,$$

it can be shown that $\{u_N\}$ is compact in the strong topology of $L^{2\varsigma-1}([0,T]; L^{2\varsigma-1})$. Let $\varepsilon > 0$ and $\eta_N = u_N - u$, then using the inequality above yields

$$\|\eta_N\|_{L^{2\varsigma-1}(\mathbf{T}^d \times [0,T])}^{2\varsigma-1} \leqslant \rho \|\eta_N\|_{L^{2\varsigma}(\mathbf{T}^d \times [0,T])}^{2\varsigma} + C(\rho)\|\eta_N\|_{L^2(\mathbf{T}^d \times [0,T])}^2.$$

Since $\eta_N \to 0$ weakly in $L^{2\varsigma}(\mathbf{T}^d \times [0,T])$, and strongly in $L^2(\mathbf{T}^d \times [0,T])$, we have

$$\limsup_{N\to\infty} \|\eta_N\|_{L^{2\varsigma-1}(\mathbf{T}^d \times [0,T])}^{2\varsigma-1} \leqslant \limsup_{N\to\infty} \rho \|\eta_N\|_{L^{2\varsigma}(\mathbf{T}^d \times [0,T])}^{2\varsigma} \leqslant \rho C < \varepsilon.$$

As $\varepsilon > 0$ is arbitrary, it shows that $u_N \to u$ strongly in $L^{2\varsigma-1}(\mathbf{T}^d \times [0,T])$, and hence $|u_N|^{2\sigma} u_N \to |u|^{2\sigma} u$ weakly in $L^1([0,T]; L^1(\mathbf{T}^d))$.

Similar to Lemma 3.3.3, $\{\langle u_N, \phi^* \rangle\}_N$ is a continuous function of t. Actually, as (3.3.8) and (3.3.9) are independent of N, then for arbitrary $\phi \in$

$L^2(\mathbf{T}^d)$, $\{\langle u_N, \phi^*\rangle\}_N$ is equicontinuous in $C([0,T])$. On the other hand, according to Lemma 3.3.1, $\{\langle u_N, \phi^*\rangle\}_N$ is uniformly bounded in $C([0,T])$. According to Arzelà-Ascoli theorem, $\{\langle u_N, \phi^*\rangle\}_N$ is compact in $C([0,T])$, i.e., $\{u_N\}$ is compact in $C([0,T]); w\text{-}L^2(\mathbf{T}^d))$.

Proof of 3.3.1 The starting point of the proof is the approximation problem (3.3.10). Let $\phi \in C^\infty(\mathbf{T}^d)$. Similar to Lemma 3.3.3, $\langle u_N(t), \phi \rangle$ is a continuous function of t, which converges to $\langle u(t), \phi \rangle$ as $N \to \infty$ for arbitrary $t \geqslant 0$. As $u_N \to u$ weakly in $L^2([0,T]; H^\alpha(\mathbf{T}^d))$, then

$$\int_0^t \langle u_N, \phi^*\rangle \mathrm{d}\tau \to \int_0^t \langle u, \phi^*\rangle \mathrm{d}\tau,$$

and

$$\int_0^t \langle \Lambda^\alpha u_N, \Lambda^\alpha \phi^*\rangle \mathrm{d}\tau \to \int_0^t \langle \Lambda^\alpha u, \Lambda^\alpha \phi^*\rangle \mathrm{d}\tau.$$

As $|u_N|^{2\sigma} u_N \to |u|^{2\sigma} u$ weakly in $L^1([0,T]; L^1(\mathbf{T}^d))$, and $\phi_N \to \phi$ uniformly in $\mathbf{T}^d$, then

$$\int_0^t \langle |u_N|^{2\sigma} u_N, \phi_N^*\rangle \mathrm{d}\tau \to \int_0^t \langle |u|^{2\sigma} u, \phi_N^*\rangle \mathrm{d}\tau.$$

Hence, the limit function u satisfies the fractional complex Ginzburg-Landau equation in the sense of (3.3.2). At last, (3.3.3) is obvious from (3.3.11) and Fatou Lemma.

Generally speaking, weak solution is not unique. However, the following uniquenes criterion holds.

Theorem 3.3.2 *Let* $\alpha \in \left(\dfrac{1}{2}, 1\right]$, $T > 0$, *and* $d < 4\alpha$. *Then there is at most one solution for the FCGL equation such that*

$$u \in L^\infty(0,T; L^2) \cap L^2(0,T; H^\alpha) \tag{3.3.12}$$

and

$$u \in L^{\frac{2\sigma}{1-\theta}}(0,T; L^{4\sigma}), \quad \theta = \frac{d}{4\alpha} \in (0,1). \tag{3.3.13}$$

Proof Let u_A and u_B be two solutions of the FCGL equation, then $w = u_A - u_B$ satisfies

$$w_t = Rw - (1+i\nu)(-\Delta)^\alpha w - (1+i\mu)(|u_A|^{2\sigma} u_A - |u_B|^{2\sigma} u_B).$$

Multiply the formula by w^* and integrate over $\mathbf{T}^d$ and take the real part to obtain

$$\frac{\mathrm{d}}{\mathrm{d}t}\|w\|_{L^2}^2 + 2\|\Lambda^\alpha w\|_{L^2}^2 = 2R\|w\|_{L^2}^2 - 2\int_{\mathbf{T}^d} |u_A|^{2\sigma}|w|^2$$
$$- \int_{\mathbf{T}^d} (|u_A|^{2\sigma} - |u_B|^{2\sigma})(u_B w' + u_B^* w). \tag{3.3.14}$$

For the second item on the right, using the following interpolation inequality

$$\|w\|_{L^4} \leqslant \|w\|_{L^2}^{1-\theta} \|w\|_{H^\alpha}^\theta, \qquad \theta = \frac{d}{4\alpha},$$

yields

$$\left| 2 \int_{\mathbf{T}^d} |u_A|^{2\sigma} |w|^2 \right| \leqslant \left(\int |u_A|^{4\sigma} \right)^{\frac{1}{2}} \left(\int |w|^4 \right)^{\frac{1}{2}}$$

$$\leqslant C \left(\int |u_A|^{4\sigma} \right)^{\frac{1}{2}} \|w\|_{L^2}^{2(1-\theta)} \|w\|_{H^\alpha}^{2\theta}$$

$$\leqslant C \|u_A\|_{L^{4\sigma}}^{2\sigma} \|w\|_{L^2}^2 + C \|u_A\|_{L^{4\sigma}}^{\frac{2\sigma}{1-\theta}} \|w\|_{L^2}^2 + \|\Lambda^\alpha w\|_{L^2}^2.$$

But since there is a constant $\epsilon \in (0,1)$ such that

$$\left| |u_A|^{2\sigma} - |u_B|^{2\sigma} \right| \leqslant 2\sigma \left| \epsilon |u_A| + (1-\epsilon)|u_B| \right|^{2\sigma-1} \left| |u_A| - |u_B| \right|$$

$$\leqslant C_\sigma \| |u_A| + |u_B| \|^{2\sigma-1} |w|,$$

the last item on the right can be bounded by

$$\leqslant C \left(\left(\int |u_A|^{4\sigma} \right)^{\frac{1}{2}} + \left(\int |u_B|^{4\sigma} \right)^{\frac{1}{2}} \right) \left(\int |w|^4 \right)^{\frac{1}{2}}$$

$$\leqslant C \|U\|_{L^{4\sigma}}^{2\sigma} \|w\|_{L^2}^2 + C \|U\|_{L^{4\sigma}}^{\frac{2\sigma}{1-\theta}} \|w\|_{L^2}^2 + \|\Lambda^\alpha w\|_{L^2}^2.$$

Here $|U| = |u_A| + |u_B|$. By using (3.3.14), we get

$$\frac{\mathrm{d}}{\mathrm{d}t} \|w\|_{L^2}^2 = 2R \|w\|_{L^2}^2 + C \|U\|_{L^{4\sigma}}^{2\sigma} \|w\|_{L^2}^2 + C \|U\|_{L^{4\sigma}}^{\frac{2\sigma}{1-\theta}} \|w\|_{L^2}^2.$$

It follows that $w = 0$ by (3.3.12), completing the proof.

3.3.2   Global existence of strong solutions

This section considers the global existence of fractional complex Ginzburg-Landau equation. Let $\mathcal{S}^\alpha(t) = \mathrm{e}^{-t(1+i\nu)(-\Delta)^\alpha + Rt}$, then the operator family $\mathcal{S}^\alpha(t)$ generates bounded linear operators in L^p for $p \in [1, \infty]$ (see [164]). We first consider the following linear equation

$$u_t = Ru - (1+i\nu)(-\Delta)^\alpha u, \quad u(0) = \varphi(x). \tag{3.3.15}$$

By taking the Fourier transform, we obtain

$$\frac{\mathrm{d}}{\mathrm{d}t} \hat{u}(t,\xi) = R\hat{u} - (1+i\nu)|\xi|^\alpha \hat{u}(t,\xi),$$

which yields by inverse Fourier transform

$$u(t) = \mathcal{S}^\alpha(t)\varphi = \mathcal{F}^{-1}\left(e^{-t(1+i\nu)|\xi|^{2\alpha}+Rt}\right) * \varphi = G_t^\alpha * \varphi.$$

For this, it suffices to study $\tilde{G}_t^\alpha := G_t^\alpha e^{-Rt}$. By a change of variable

$$\tilde{G}_t^\alpha(x) = \frac{1}{(2\pi)^d}\int_{\mathbf{R}^d} e^{ix\cdot\xi}e^{-t(1+i\nu)|\xi|^{2\alpha}}\,d\xi = t^{-\frac{d}{2\alpha}}\tilde{G}_1^\alpha\left(\frac{x}{t^{1/2\alpha}}\right),$$

which reduces to study the kernel function

$$\tilde{G}^\alpha(x) = \frac{1}{(2\pi)^d}\int_{\mathbf{R}^d} e^{ix\cdot\xi}e^{-(1+i\nu)|\xi|^{2\alpha}}.$$

Since $e^{-(1+i\nu)|\xi|^{2\alpha}} \in L^1(\mathbf{R}^d)$, from the Riemann-Lebesgue Lemma, we know $\tilde{G}^\alpha \in L^\infty(\mathbf{R}^d)\cap C(\mathbf{R}^d)$, and $\tilde{G}^\alpha(x) \to 0$ when $|x| \to \infty$. Then $\tilde{G}^\alpha \in C_0(\mathbf{R}^d)$, the space of continuous functions which tend to zero at infinite. Moreover, since for arbitrary $\beta > 0$, $|\xi|^{2\beta}e^{-(1+i\nu)|\xi|^{2\alpha}} \in L^1(\mathbf{R}^d)$, then $(-\Delta)^\beta\tilde{G}^\alpha \in C_0(\mathbf{R}^d)$.

The following two lemmas will be useful (see [164]).

Lemma 3.3.4 *Let $\alpha > 0$, then $\tilde{G}^\alpha \in L^p(\mathbf{R}^d)$ for all $p \in [1,\infty]$ and*

$$|\tilde{G}^\alpha(x)| \leqslant C(1+|x|)^{-d-2\alpha}, \qquad \forall x \in \mathbf{R}^d.$$

Lemma 3.3.5 *Let $\alpha > 0$, then $(-\Delta)^s\tilde{G}^\alpha \in L^p(\mathbf{R}^d)$ for all $p \in [1,\infty]$ and*

$$|(-\Delta)^s\tilde{G}^\alpha(x)| \leqslant C(1+|x|)^{-d-2s}, \qquad \forall s > 0, \forall x \in \mathbf{R}^d.$$

In particular, $\nabla\tilde{G}^\alpha \in L^p(\mathbf{R}^d)$ for arbitrary $p \in [1,\infty]$.

In the following, the semigroup method is used to establish the local and global existence of solutions for the equation. One may refer to [177] for the semigroup method. Consider an abstract evolution equation in Banach space X

$$u_t = Au + f(u), \qquad u(0) = \varphi \in X, \tag{3.3.16}$$

where A is the infinitesimal generator of a strong continuous semigroup $S(t)$ in Banach space X, while $f(u)$ can be viewed as a nonlinear perturbation in X.

Proposition 3.3.1 *Let $f : X \to X$ be a Lipshitz continuous function, then for arbitrary $\rho > 0$ there exists a time $T(\rho) > 0$ such that for arbitrary initial data $u(0) = \varphi \in X$ with $\|\varphi\|_X \leqslant \rho$, there exists a unique solution $u \in C([0,T];X)$ for (3.3.16) in the sense*

$$u(t) = S(t)\varphi + \int_0^t S(t-\tau)f(u(\tau))d\tau.$$

Moreover, u is a locally Lipshitz continuous function of φ.

For FCGL equation, let $Au = Ru - (1 + i\nu)\Lambda^{2\alpha}u$ and $f(u) = -(1 + i\mu)|u|^{2\sigma}u$. The semigroup $\mathcal{S}_{per}^{\alpha}(t)$ corresponding to the periodic FCGL equation can be written as $\mathcal{S}_{per}^{\alpha}(t) = G_{per,t}^{\alpha} * \varphi$, where

$$G_{per,t}^{\alpha}(x) = \sum_{n \in \mathbb{Z}^d} G_t^{\alpha}(x + n).$$

Next, we consider the integral form of the FCGL equation

$$u(t) = G_{per,t}^{\alpha} * \varphi + \int_0^t G_{per,t-\tau}^{\alpha} * f(u(\tau))\mathrm{d}\tau. \qquad (3.3.17)$$

By Lemma 3.3.4 and 3.3.5, applying the decay estimates gives $\|G_{per,t}^{\alpha}\|_{L^1} \leqslant Ce^{Rt}$ and by Young's inequality, we have

$$\|\mathcal{S}_{per}^{\alpha}(t)\varphi\|_{L^p} \leqslant \|G_{per,t}^{\alpha}\|_{L^1}\|\varphi\|_{L^p} \leqslant Ce^{Rt}\|\varphi\|_{L^p}, \qquad \forall p \in [1, \infty].$$

It follows that $\mathcal{S}_{per}^{\alpha}(t)$ is a strong continuous semigroup in $C(\mathbf{T}^d)$ and $L^p(\mathbf{T}^d)$ for $p \in [1, \infty)$.

We now prove the following local existence of strong solutions in $X = C(\mathbf{T}^d)$.

Theorem 3.3.3   *Let $\alpha \in (1/2, 1)$ and $\rho > 0$ be arbitrary. Then there is some $T = T(\rho) > 0$ such that there exists a unique solution for the FCGL equation*

$$u \in C([0, T]; C) \cap C((0, T]; C^2) \cap C^1((0, T]; C),$$

for every initial data $\varphi \in C(\mathbf{T}^d)$ with $\|\varphi\|_{L^{\infty}} \leqslant \rho$. Moreover, if the initial data $\varphi \in C^2(\mathbf{T}^d)$, then

$$u \in C([0, T]; C^2) \cap C^1([0, T]; C).$$

Proof Obviously, the nonlinear term $f(u)$ is locally Lipshitz function mapping $C(\mathbf{T}^d)$ into itself. By standard local existence theory, there exists a local mild solution for the FCGL equation on $[0, T]$ with T depending on $\|\varphi\|_{L^{\infty}}$, which is the limit of the sequence $\{u^{(n)}\}$ defined successively by

$$u^{(0)}(t) = G_t^{\alpha} * \varphi,$$
$$u^{(n+1)}(t) = G_t^{\alpha} * \varphi + \int_0^t G_{t-\tau}^{\alpha} * f(u^{(n)}(\tau))\mathrm{d}\tau. \qquad (3.3.18)$$

Using the standard bootstrapping argument, additional regularity can be obtained. The main ingredient is by observing the estimate

$$\|\boldsymbol{\nabla} G_{per,t}^{\alpha}\|_{L^1} \leqslant Ct^{-\frac{1}{2\alpha}}e^{Rt}. \qquad (3.3.19)$$

Indeed, applying the Gradient operator ∇ to (3.3.18), we obtain

$$\nabla u^{(n+1)}(t) = \nabla G^{\alpha}_{per,t} * \varphi + \int_0^t \nabla G^{\alpha}_{per,t-\tau} * f(u^{(n)}(\tau)) \mathrm{d}\tau.$$

Computing directly, we have

$$\|\nabla u^{(n+1)}(t)\|_{L^\infty}$$

$$\leqslant \|\nabla G^{\alpha}_{per,t}\|_{L^1} \|\varphi\|_{L^\infty} + \int_0^t \|\nabla G^{\alpha}_{per,t-\tau}\|_{L^1} \|f(u^{(n)}(\tau))\|_{L^\infty} \mathrm{d}\tau$$

$$\leqslant Ct^{-\frac{1}{2\alpha}} e^{Rt} \|\varphi\|_{L^\infty} + C \int_0^t (t-\tau)^{-\frac{1}{2\alpha}} e^{R(t-\tau)} \|f(u^{(n)}(\tau))\|_{L^\infty} \mathrm{d}\tau.$$

Since $\alpha \in (1/2, 1)$, this shows that $\nabla u^{(n)}$ is bounded in $C([0,T]; C(\mathbf{T}^d))$. On the other hand, since f is Lipshitz continuous, similar argument shows that $\nabla u^{(n)}$ is Cauchy in $C([0,T]; C(\mathbf{T}^d))$. Therefore $u \in C([0,T]; C^1(\mathbf{T}^d))$ and the singularity at $t = 0$ disappears. In this case, ∇u is a solution for the equation

$$\nabla u(t) = G^{\alpha}_{per,t} * \nabla\varphi + \int_0^t G^{\alpha}_{per,t-\tau} * [f'(u(\tau))\nabla u(\tau)] \mathrm{d}\tau \qquad (3.3.20)$$

where

$$f'(u(\tau))\nabla u(\tau) = -(1 + i\mu)[(\sigma + 1)|u|^{2\sigma}\nabla u + \sigma |u|^{2\sigma-2} u^2 \nabla u^*].$$

A repetition of the above procedure then shows that $u(t) \in C((0,T]; C^2(\mathbf{T}^d))$. Moreover, because the FGL equation trades the first time derivative to the 2α-order space derivative, the solution obtained must also be in $C^1((0,T]; C(\mathbf{T}^d))$. Therefore, the solution is indeed a classical one, completing the proof.

Generally, the above discussion cannot be repeated to show $u \in C((0,T]; C^3(\mathbf{T}^d))$, since further differentiation of the nonlinear term will introduce singularities at the zeros of u, yielding divergence of (3.3.20). But when σ is a positive integer, arbitrary differentiation of nonlinear term will not introduce singularities, thus one can show $u \in C((0,T]; C^\infty(\mathbf{T}^d))$. Similarity, one can also show that $u \in C^\infty((0,T]; C^\infty(\mathbf{T}^d))$ also holds.

When σ is not an integer, as long as the differentiation of nonlinear term does not introduce unbounded singularities at the zeros, we can still bootstrap to obtain higher regularity of the solution.

Theorem 3.3.4 (Local C^k solution)   *Let $\sigma \geqslant \dfrac{n}{2}$ for some positive integer n. Then for every $\rho > 0$, there exists $T(\rho) > 0$ such that for every initial data $\varphi \in C(\mathbf{T}^d)$ with $\|\varphi\|_\infty < \rho$, there exists a unique*

$\varphi \in C(\mathbf{T}^d)$ with $\|\varphi\|_\infty < \rho$, there exists a unique

$$u \in C([0,T]; C(\mathbf{T}^d)) \cap C((0,T]; C^{n+2}(\mathbf{T}^d)) \cap C^1((0,T]; C^n(\mathbf{T}^d))$$

satisfying the FCGL equation. Moreover, if $\varphi \in C^{n+2}(\mathbf{T}^d)$, then there holds $u \in C([0,T]; C^{n+2}(\mathbf{T}^d)) \cap C^1((0,T]; C^n(\mathbf{T}^d))$.

In the following, we consider the local existence of solutions in $X = L^p(\mathbf{T}^d)$. Assume that $\varphi \in L^p(\mathbf{T}^d)$ for $1 \leqslant p < \infty$. First, we have the L^r estimates

$$\|u(t)\|_{L^r} \leqslant \|G^\alpha_{per,t}\|_{L^q}\|\varphi\|_{L^p} + |1+\mathrm{i}\mu| \int_0^t \|G^\alpha_{per,t-\tau}\|_{L^s}\|u(\tau)\|_{L^r}^{2\sigma+1}\mathrm{d}\tau,$$

$$(3.3.21)$$

where p, q, r, s satisfy

$$1 + \frac{1}{r} = \frac{1}{p} + \frac{1}{q}, \quad and \quad 1 + \frac{1}{r} = \frac{1}{s} + \frac{2\sigma+1}{r}. \qquad (3.3.22)$$

Let

$$\|u(t)\|_{\widehat{L}^{r,q}} = \frac{\|u(t)\|_{L^r}}{\|G^\alpha_{per,t}\|_{L^q}},$$

and define the space $\Xi^{p,r}([0,T]; \mathbf{T}^d)$ to be the completion of $C([0,T]; L^r(\mathbf{T}^d))$ under the norm

$$\|u\|_{\Xi^{p,r}} := \sup_{t\in[0,T]} \{\|u(t)\|_{\widehat{L}^{r,q}}\}.$$

From (3.3.21), it is easy to get

$$\|u(t)\|_{\widehat{L}^{r,q}} \leqslant \|\varphi\|_{L^p} + \frac{|1+\mathrm{i}\mu|}{\|G^\alpha_{per,t}\|_{L^q}} \int_0^t \|G^\alpha_{per,t-\tau}\|_{L^s}\|G^\alpha_{per,\tau}\|_{L^q}^{2\sigma+1}\|u(\tau)\|_{\widehat{L}^{r,q}}^{2\sigma+1}\mathrm{d}\tau.$$

If the kernel satisfies the condition

$$\frac{1}{\|G^\alpha_{per,t}\|_{L^q}} \int_0^t \|G^\alpha_{per,t-\tau}\|_{L^s}\|G^\alpha_{per,\tau}\|_{L^q}^{2\sigma+1}\mathrm{d}\tau \to 0, \quad as\ \ t \to 0, \quad (3.3.23)$$

then there is a sufficiently small $T > 0$ such that iterative sequence (3.3.18) converges in the space Ξ. But by the definition of $G^\alpha_{per,t}$, it is easy to see that

$$|G^\alpha_{per,t}| \leqslant \sum_{n\in\mathbb{Z}^d} \mathrm{e}^{Rt}t^{-\frac{d}{2\alpha}}|\tilde{G}^\alpha\left(\frac{x}{t^{1/2\alpha}}\right)| \leqslant \mathrm{e}^{Rt}t^{-\frac{d}{2\alpha}}(a + bt^{\frac{d}{2\alpha}}),$$

where a, b are constants depending only on ν. By using interpolation inequality, there holds

$$\|G^{\alpha}_{per,t}\|_{L^q} \leqslant \|G^{\alpha}_{per,t}\|_{L^\infty}^{1/q^*} \|G^{\alpha}_{per,t}\|_{L^1}^{1/q} \leqslant \frac{Ce^{Rt}}{t^{d/(2\alpha q^*)}}(a + bt^{\frac{d}{2\alpha}})^{1/q^*}. \quad (3.3.24)$$

In particular, $\|G^{\alpha}_{per,t}\|_{L^q} = \mathcal{O}(t^{\frac{d}{2\alpha q^*}})$ when $t \to 0$.

So, if

A1. $d(2\sigma + 1) < 2\alpha q^*$, and

A2. $\dfrac{d}{2\alpha s^*} + \dfrac{\sigma d}{\alpha q^*} < 1$,

then the condition (3.3.23) holds. The condition A1 is equivalent to

$$\frac{1}{p} - \frac{2\alpha}{(2\sigma + 1)d} < \frac{1}{r} \quad (3.3.25)$$

By using (3.3.22), the condition A2 is equivalent to $\sigma d < \alpha p$.

Now we only assume

A. $\sigma d < \alpha p$,

and choose $r = (2\sigma + 1)p$, then (3.3.25) and (3.3.23) hold. Using contraction mapping principle in $\Xi^{p,r}([0,T]; \mathbf{T}^d)$ for $r = (2\sigma + 1)p$, then it is easy to know that FCGL equation has a unique local solution.

Let us consider the regularity of u. It is easy to show that $u \in C([0,T]; L^p(\mathbf{T}^d)) \cap C((0,T]; L^r(\mathbf{T}^d))$. Indeed, from the definition of $\Xi^{p,r}$, it is easy to know $\Xi^{p,r}([0,T]; \mathbf{T}^d) \subset C((0,T]; L^r(\mathbf{T}^d))$, thus to show that $u \in C([0,T]; L^p(\mathbf{T}^d))$ it suffices to prove the continuity of u at $t = 0$. Notice that

$$\|u(t) - \varphi\|_{L^p} \leqslant \|G^{\alpha}_{per,t} * \varphi - \varphi\|_{L^p}$$
$$+ |1 + i\mu| \int_0^t \|G^{\alpha}_{per,t-\tau}\|_{L^1} \|G^{\alpha}_{per,\tau}\|_{L^q}^{2\sigma+1} \|u(\tau)\|_{L^r/L^q_t}^{2\sigma+1} d\tau,$$

where $r = (2\sigma + 1)p$ and q is defined in (3.3.22). By the strong continuity of the operator semigroup in L^p, we have known that when $t \to 0$, the first item in the right approaches to 0. On the other side, according to $\sigma d < \alpha p$, it is known that $d(2\sigma + 1) < 2\alpha q^*$. Thus $\|G^{\alpha}_{per,\tau}\|_{L^q}^{2\sigma+1}$ is integrable near $\tau = 0$. Using the boundedness of u in $\Xi^{p,r}$ and (3.3.19), it's easy to see that when $t \to 0$, the second item in the right approaches to 0. Then u is continuous at $t = 0$, i.e. $u \in C([0,T]; L^p(\mathbf{T}^d))$.

Theorem 3.3.5 *If p satisfies $q \leqslant p$ and $\sigma d < \alpha p$, then for arbitrary $\rho > 0$, there is $T(\rho) > 0$ such that for arbitrary initial value $\varphi \in L^p(\mathbf{T}^d)$ with $\|\varphi\|_{L^p} \leqslant \rho$, there is a unique solution*

$$u \in \Xi^{p,r}([0,T]; \mathbf{T}^d) \cap C([0,T]; L^p(\mathbf{T}^d)),$$

in the sense of (3.3.17).

The mild solution here actually is the strong solution $u \in C((0,T]; C^2(\mathbf{T}^d)) \cap C^1((0,T]; C(\mathbf{T}^d))$. According to Theorem 3.3.3, we only need to prove $u \in C((0,T]; L^\infty)$. Therefore it suffices to prove that the solution with the L^p initial data belongs to $C((0,T]; L^r(\mathbf{T}^d))$ for $r > p$ satisfying (A1-A2) and (3.3.25). Indeed, if $p > \left(\sigma + \dfrac{1}{2}\right)d$, then for arbitrary $r \in [(2\sigma + 1)p, \infty]$, the conditions are all satisfied, then $u \in C((0,T]; L^\infty)$ holds. On the other side, if $\sigma d < p \leqslant \left(\sigma + \dfrac{1}{2}\right)d$, then $u \in C((0,T]; L^{p_1})$ for $p_1 = (2\sigma + 1)p$. When $p_1 \geqslant \left(\sigma + \dfrac{1}{2}\right)d$, by repeating the above reasoning, we can show $u \in C((0,T]; L^\infty)$. Otherwise, more generally, repeating the above reasoning we know that when $p_n = (2\sigma + 1)^n p \geqslant \left(\sigma + \dfrac{1}{2}\right)d$ for $n = 1, 2, \cdots$, then $u \in C((0,T]; L^{p_n}(\mathbf{T}^d))$, hence $u \in C((0,T]; L^\infty)$.

In particular, we have the following

Theorem 3.3.6 *Let $1 \leqslant p < \infty$ and $\sigma d < \alpha p$, then for arbitrary $\rho > 0$, there is $T(\rho) > 0$ such that for arbitrary initial value $\varphi \in L^p(\mathbf{T}^d)$ with $\|\varphi\|_{L^p} \leqslant \rho$, there exists a unique solution for the FCGL equation satisfying*

$$u \in C([0,T]; L^p(\mathbf{T}^d)) \cap C((0,T]; C^2(\mathbf{T}^d)) \cap C^1((0,T]; C(\mathbf{T}^d)).$$

A special situation is when $p = 2\sigma + 2$. Once $\sigma < \infty$ when $d = 1$ or $\dfrac{2\alpha}{d - 2\alpha}$ when $d \geqslant 2$, then there is a local strong solution according to the above theorem.

In the following, we extend the local solution to a global one by some *a priori* estimates. The purpose is to get the H^1-estimates of the solution, then using the Sobolev embedding theorem to get the L^p-estimates of the solution. Therefore it suffices to get the L^2 estimates of ∇u.

Theorem 3.3.7 (Global strong solution) *Let $\sigma \geqslant \dfrac{1}{2}$ and $d < 2 + 2/\sigma$. If*

$$\sigma \leqslant \frac{\sqrt{1 + \mu^2} + 1}{\mu^2}, \tag{3.3.26}$$

then the FCGL equation with C^2 initial data has a unique global strong solution.

Proof Multiply the equation by $-\Delta u^*$ and integrate over $\mathbf{T}^d$ to get

$$
\int_{\mathbf{T}^d} \partial_t u(-\Delta)u^* = R\int u(-\Delta)u^* - (1+\mathrm{i}\nu)\int (-\Delta)^\alpha u(-\Delta)u^*
$$
$$
- (1+\mathrm{i}\mu)\int |u|^{2\sigma}u(-\Delta)u^*,
$$

which yields

$$
\frac{\mathrm{d}}{\mathrm{d}t}\|\nabla u\|^2 = 2R\|\nabla u\|^2 - 2\|(-\Delta)^{\frac{1+\alpha}{2}}u\|^2 - \frac{1}{2}\int |u|^{2\sigma-2}[(1+2\sigma)|\nabla|u|^2|^2
$$
$$
- 2\mathrm{i}\mu\sigma\nabla|u|^2(u^*\nabla u - u\nabla u^*) + |u^*\nabla u - u\nabla u^*|^2]\mathrm{d}x.
\tag{3.3.27}
$$

If the matrix

$$
\begin{pmatrix} 1+2\sigma & -\sigma\mu \\ -\sigma\mu & 1 \end{pmatrix}
\tag{3.3.28}
$$

is non-negative definite, then the last item above in non-positive and hence

$$
\frac{\mathrm{d}}{\mathrm{d}t}\|\nabla u\|^2 \leqslant 2R\|\nabla u\|^2,
$$

yielding

$$
\|\nabla u(t)\| \leqslant \mathrm{e}^{RT}\|\nabla\varphi\|_{L^2}.
$$

The theorem is found to establish by the local existence. By the H^1 estimates of the solution, its L^p estimation can be obtained for

$$
1 \leqslant p < \begin{cases} \infty, & d=1,2 \\ \dfrac{2d}{d-2}, & d \geqslant 3. \end{cases}
\tag{3.3.29}
$$

In this case, besides (3.3.26) and (3.3.29), if moreover $\sigma d < \alpha p$, then the FCGL equation has a unique strong solution according to Theorem 3.3.5. But when $d < 2 + \dfrac{\alpha}{\sigma}$, we can select p such that all the above conditions hold, completing the proof.

3.3.3　Existence of attractors

This section will discuss the existence of attractor of fractional Ginzburg-Landau in L^2. For simplicity, we only consider the case when $d=1$ and $1/2 < \alpha \leqslant 1$. We let $\mathbf{T} = \mathbf{T}^1$ and prove the following

Theorem 3.3.8 *Suppose that* $\alpha \in \left(\dfrac{1}{2}, 1\right]$ *and* $d = 1$. *Then the solution operator* $S : S(t)\varphi = u(t)$ *for all* $t > 0$ *of the FGL equation well defines a semigroup in the space* $H = L^2$. *Moreover, the following statements hold:*

1. *For any* $t > 0$, $S(t)$ *is continuous in* H;
2. *For any* $\varphi \in H$, S *is continuous from* $[0, T]$ *to* H;
3. *For any* $t > 0$, $S(t)$ *is compact in* H;
4. *The semigroup* $\{S(t)\}_{t \geqslant 0}$ *possesses a global attractor* $\mathcal{A}$ *in* H. *Furthermore,* $\mathcal{A}$ *is compact and connected in* H, *and is the maximal bounded absorbing set and the minimal invariant set in* H *in the sense of set inclusion relation.*

First of all, we state the following theorem [120, 213].

Theorem 3.3.9 *Suppose that* H *is a metric space and the semigroup* $\{S(t)\}_{t \geqslant 0}$ *is a family of operators from* H *to itself such that*

1. *for any fixed* $t > 0$, $S(t) : H \to H$ *is continuous;*
2. *there is* $t_0 > 0$ *such that* $S(t_0)$ *is compact from* H *to itself;*
3. *there exists a bounded subset* $B_0 \subset H$ *and an open subset* $U \subset H$ *such that* $B_0 \subseteq U \subseteq H$, *and for arbitrary bounded subset* $B \subset U$, *there is a* $t_0 = t_0(B)$ *such that* $S(t)B \subset B_0$ *for all* $t > t_0(B)$.

Then $\mathcal{A} = \omega(B)$ *is a compact attractor which attracts all the bounded set of* U, *i.e.,*

$$\lim_{t \to +\infty} dist(S(t)x, \mathcal{A}) = 0, \qquad \forall x \in U.$$

$\mathcal{A}$ *is the maximal bounded absorbing set and minimal invariant set, in the sense that* $S(t)\mathcal{A} = \mathcal{A}$ *for all* $t \geqslant 0$.

Suppose in addition that H *is a Banach space,* U *is convex and*

4. *For all* $x \in H$, $S(t)x : R^+ \to H$ *is continuous.*

Then $\mathcal{A} := \omega(B)$ *is also connected.*

If $U = H$, $\mathcal{A}$ *is called the global attractor of the semigroup* $\{S(t)\}_{t \geqslant 0}$ *in* H.

Based on the preceding discussions, for $d = 1$ and $\alpha \in \left(\dfrac{1}{2}, 1\right]$, then for arbitrary $\varphi \in L^2(\mathbf{T})$, FCGL possesses a unique global solution u such that $u \in C([0, T]; L^2) \cap L^2(0, T; H^\alpha)$ for all $T < \infty$, and $S(t) : \varphi \to u(t)$ is a continuous mapping from L^2 to itself. To prove Theorem 3.3.8, it suffices to check the conditions in Theorem 3.3.9.

1. *Absorbing set in* $H = L^2$.

Taking the L^2 inner product of equation (3.3.1) with u^* over $\mathbf{T}$. Integrating by parts and taking the real part, we have

$$\frac{1}{2}\frac{\mathrm{d}}{\mathrm{d}t}\|u\|^2 + \|\Lambda^\alpha u\|^2 + \|u\|_{L^{2\sigma+2}}^{2\sigma+2} - R\|u\|^2 = 0. \tag{3.3.30}$$

When $R \leqslant 0$, this leads to trivial dynamical systems. Indeed, when $R < 0$, we have

$$\|u(t)\|_{L^2} \leqslant \|\varphi\|_{L^2}\mathrm{e}^{Rt},$$

from which it follows that for any initial data $\varphi \in L^2$, we have

$$\|u(t)\|_{L^2}^2 \to 0, \quad t \to \infty. \tag{3.3.31}$$

When $R = 0$, using Hölder inequality

$$\|u\|_{L^2}^2 \leqslant |\mathbf{T}|^{\frac{\sigma}{\sigma+1}}\|u\|_{L^{2\sigma+2}}^2,$$

we have from (3.3.30) that

$$\frac{\mathrm{d}}{\mathrm{d}t}\|u\|_{L^2}^2 + \frac{2}{(2\pi)^\sigma}\|u\|_{L^2}^{2\sigma+2} \leqslant 0,$$

which yields

$$\frac{1}{\|u(t)\|_{L^2}^2} \geqslant \frac{1}{\|\varphi\|^{2\sigma}} + \frac{2\sigma}{(2\pi)^\sigma}t.$$

Therefore, (3.3.31) still holds.

When $R > 0$, using Young's inequality we have $Ry^2 \leqslant \frac{1}{2}y^{2\sigma+2} + CR^{\frac{\sigma+1}{\sigma}}$, we have

$$\frac{1}{2}\|u\|_{L^{2\sigma+2}}^{2\sigma+2} - R\|u\|^2 \geqslant -2\pi CR^{\frac{\sigma+1}{\sigma}},$$

where C is a constant depending only on R and σ. One obtains from (3.3.30) that

$$\frac{\mathrm{d}}{\mathrm{d}t}\|u\|^2 + 2\|\Lambda^\alpha u\|^2 + \|u\|_{L^{2\sigma+2}}^{2\sigma+2} + R\|u\|_{L^2}^2 \leqslant 4\pi CR^{\frac{\sigma+1}{\sigma}}, \tag{3.3.32}$$

and hence by Gronwall inequality

$$\begin{aligned}
\|u(t)\|_{L^2}^2 &\leqslant \mathrm{e}^{-Rt}\left[\|\varphi\|_{L^2}^2 + 4\pi CR^{\frac{\sigma+1}{\sigma}}t\right]\\
&\leqslant \|\varphi\|_{L^2}^2\mathrm{e}^{-Rt} + 4\pi CR^{\frac{1}{\sigma}}(1 - \mathrm{e}^{-Rt}), \quad \forall t \geqslant 0.
\end{aligned} \tag{3.3.33}$$

Therefore,

$$\limsup_{t\to\infty}\|u(t)\|_{L^2}^2 \leqslant \rho_0^2, \quad \rho_0^2 = 4\pi CR^{\frac{1}{\sigma}}.$$

From (3.3.33), we can infer the existence of an absorbing ball in L^2. Indeed, the balls $B_H(0, \rho)$ with $\rho > \rho_0$ are positively invariants for the semigroup $S(t)$ associated with the FCGL equation, and these balls are absorbing for any $\rho > \rho_0$. Fix any $\rho_0' > \rho_0$ and denote $\mathscr{B}_0'$ the ball $B_H(0, \rho_0')$. Since any bounded set $\mathscr{B}$ can be included in a ball $B_H(0, \rho)$ of H centered at 0 of radius ρ, it follows that $S(t)\mathscr{B} \subset \mathscr{B}_0'$ for $t \geqslant t_0 = t_0(\mathscr{B}, \rho_0')$ with $t_0 = \dfrac{1}{R}\log\dfrac{\rho^2}{\rho_0'^2 - \rho_0^2}$.

Furthermore, by integrating (3.3.32) from t to $t + 1$, we have

$$\|u(t+1)\|_{L^2}^2 + 2\int_t^{t+1}\|\Lambda^\alpha u\|_{L^2}^2 ds + \int_t^{t+1}\|u\|_{L^4}^4 ds \leqslant \|u(t)\|_{L^2}^2 + 4\pi C R^2.$$

$$(3.3.34)$$

The inequalities (3.3.33) and (3.3.34) show that

$$2\int_t^{t+1}\|\Lambda^\alpha u\|_{L^2}^2 ds + \int_t^{t+1}\|u\|_{L^4}^4 ds \leqslant \rho_0'^2 + 4\pi C R^2.$$

Therefore, when $t \geqslant t_0$ (t_0 as above),

$$\int_t^{t+1}\|\Lambda^\alpha u\|_{L^2}^2 ds \leqslant a_1 \quad \text{and} \quad \int_t^{t+1}\|u\|_{L^4}^4 ds \leqslant a_2 \qquad (3.3.35)$$

are uniformly bounded independent of φ.

2. *Absorbing set in H^α.*

First of all, we state the uniform Gronwall Lemma [213].

Lemma 3.3.6   *Let g, h and y be three positive locally integrable functions on (t_0, ∞) such that y' is locally integrable on (t_0, ∞) and satisfy*

$$\frac{dy}{dt} \leqslant gy + h \quad \text{for } t > t_0,$$

$$\int_t^{t+r} g(s)ds \leqslant a_1, \quad \int_t^{t+r} h(s)ds \leqslant a_2, \quad \int_t^{t+r} y(s)ds \leqslant a_3, \quad \text{for } t \geqslant t_0,$$

where r, a_1, a_2, a_3 are positive constants, then

$$y(t+r) \leqslant \left(\frac{a_3}{r} + a_2\right)e^{a_1}, \quad \forall t \geqslant t_0.$$

The following will focus on the attractor set in H^α. Taking inner product of (3.3.1) with $(-\Delta)^\alpha u^*$, integrating by parts and then using Hölder inequality, we obtain

$$\frac{\mathrm{d}}{\mathrm{d}t}\|\Lambda^\alpha u\|_{L^2}^2 + 2\|\Lambda^{2\alpha}u\|_{L^2}^2 - 2R\|\Lambda^\alpha u\|_{L^2}^2 = -\Re\left[(1+i\mu)\int_{\mathbf{T}^d}|u|^{2\sigma}u\Lambda^{2\alpha}u^*\mathrm{d}x\right]$$

$$\leqslant \sqrt{1+\mu^2}\int_{\mathbf{T}^d}|u|^{2\sigma+1}|\Lambda^{2\alpha}u|\mathrm{d}x$$

$$\leqslant \frac{1}{2}\|\Lambda^{2\alpha}u\|_{L^2}^2 + \frac{\sqrt{1+\mu^2}}{2}\|u\|_{L^{2(2\sigma+1)}}^{2(2\sigma+1)}. \tag{3.3.36}$$

By using interpolation inequality, we have

$$\|u\|_{L^{2(2\sigma+1)}} \leqslant C_1\|u\|_{L^2}^{1-\rho}(\|u\|_{L^2}^2 + \|\Lambda^{2\alpha}u\|_{L^2}^2)^{\rho/2}, \quad \rho = \frac{2\sigma}{4\alpha(2\sigma+1)},$$

which follows from (3.3.36) that

$$\frac{\mathrm{d}}{\mathrm{d}t}\|\Lambda^\alpha u\|_{L^2}^2 + \frac{3}{2}\|\Lambda^{2\alpha}u\|_{L^2}^2 - 2R\|\Lambda^\alpha u\|_{L^2}^2 \leqslant \frac{\sqrt{1+\mu^2}}{2}\|u\|_{L^{2(2\sigma+1)}}^{2(2\sigma+1)}$$

$$\leqslant 2^{\rho(2\sigma+1)}C_1'C_\mu[\|u\|^{2(2\sigma+1)} + \|\Lambda^{2\alpha}u\|^{2\rho(2\sigma+1)}]$$

$$\leqslant 2^{\rho(2\sigma+1)}C_1'C_\mu\|u\|^{2(2\sigma+1)} + \frac{1}{2}\|\Lambda^{2\alpha}u\|_{L^2}^2 + C_2,$$

where $C_1' = C_1^{2(2\sigma+1)}$, $C_\mu = \sqrt{1+\mu^2}/2$, $C_2 = \dfrac{[(2\rho(2\sigma+1))^{\rho(2\sigma+1)}2^{\rho(2\sigma+1)}C_1'C_\mu]^q}{q}$,

and $q = \dfrac{1}{1-\rho(2\sigma+1)}$. Therefore,

$$\frac{\mathrm{d}}{\mathrm{d}t}\|\Lambda^\alpha u\|_{L^2}^2 + \|\Lambda^{2\alpha}u\|_{L^2}^2 \leqslant 2R\|\Lambda^\alpha u\|_{L^2}^2 + 2^{\rho(2\sigma+1)}C_1'C_\mu\|u\|^{2(2\sigma+1)} + C_2.$$

Let

$$y = \|\Lambda^\alpha u\|_{L^2}^2, \quad g = 2R, \quad h = 2^{\rho(2\sigma+1)}C_1'C_\mu\|u\|^{2(2\sigma+1)} + C_2,$$

it follows that from the uniform Gronwall inequality that

$$\|\Lambda^\alpha u\|^2 \leqslant (a_3 + a_2)e^{a_1}, \quad t \geqslant t_0 + 1, \tag{3.3.37}$$

where a_1, a_2, a_3 are constants. A careful inspection can assure us the existence of an absorbing ball of the solutions in the space H^α, which is similar to the L^2 case.

Let $\varphi \in \mathscr{B}$, a bounded set of H. Since $\mathscr{B}_1$ is bounded in H^α, and the embedding $H^\alpha \hookrightarrow L^2$ is compact, we infer that

$$\bigcup_{t\geqslant t_0+1} S(t)B \text{ is relatively compact in } L^2.$$

This observation proves 3 of Theorem 3.3.8.

Proof of Theorem 3.3.8 Theorem 3.3.8 is a direct consequence of Theorem 3.3.9. It suffices to check conditions 1 and 4 in Theorem 3.3.9, which are standard and obvious.

3.4 Fractional Landau-Lifshitz equation

In this section, we discuss the fractional Landau-Lifshitz equation. Classical Landau-Lifshitz equation has the following form

$$\frac{\partial u}{\partial t} = -\alpha u \times \left(u \times \frac{\delta E}{\delta u} \right) + \beta u \times \frac{\delta E}{\delta u},$$

where $u : \Omega \to \mathbf{R}^3$ is a three-dimensional vector in $\mathbf{R}^3$, $\alpha \geqslant 0, \beta > 0$ are constants and $\frac{\delta E}{\delta u}$ is the variation of the functional E with respect to u,

$$E(u) = \int_\Omega |\boldsymbol{\nabla} u|^2 \mathrm{d}x + \int_\Omega \phi(u) \mathrm{d}x + \int_{\mathbf{R}^3} |\nabla \Phi|^2 \mathrm{d}x,$$

where the three items in the right hand side are exchange energy, anisotropy energy and magnetostatic energy, respectively. When the magnetostatic energy cannot be neglected, the equation is nonlocal and complex. So, it's very important to consider some simplified models of the equation with magnetostatic energy. Therefore, we consider the following fractional Landau-Lifshitz equation,

$$\begin{cases} u_t = u \times (-\Delta)^\alpha u, & (x,t) \in \mathbf{R}^d \times (0,T), \\ u(x,0) = u_0, & x \in \mathbf{R}^d, \end{cases} \tag{3.4.1}$$

where $\alpha \in (0,1)$, Ω is a smooth domain of $\mathbf{R}^d$, $k_i \in \mathbf{Z}^d$ and e_i is an orthonomal basis in $\mathbf{R}^d$.

3.4.1 Vanishing viscosity method

In this section, we consider the periodic case and employ the vanishing viscosity method to prove the following theorem.

Theorem 3.4.1 *Let $0 < \alpha < 1$, $u_0 \in H_{per}^\alpha(\Omega)$ and $|u_0(x)| = 1$ for a.e. $x \in \mathbf{R}^d$. Then for any $T > 0$, there exists $u \in L^\infty(0,T; H^\alpha(\Omega))$ such that $|u(x,t)| = 1$ for a.e. $(x,t) \in \mathbf{R}^d \times [0,T]$ and satisfies (3.4.1) in the weak sense*

$$\int_{\Omega \times (0,T)} u \Phi_t \mathrm{d}x \mathrm{d}t + \int_\Omega u_0 \Phi(\cdot,0) \mathrm{d}x = \int_{\Omega \times (0,T)} (-\Delta)^{\frac{\alpha}{2}} u \times \Phi \cdot (-\Delta)^{\frac{\alpha}{2}} u \mathrm{d}x \mathrm{d}t,$$

for any $\Phi \in C^\infty(\mathbf{R}^d \times [0,T])$ with $\Phi(x,T) = 0$.

We first introduce the discrete Young's inequality. If $f \in L^2(\Omega)$, then f can be represented by Fourier series as $f = \sum_{n \in \mathbf{Z}^d} \hat{f}(n) \mathrm{e}^{2\pi i n \cdot x}$ for $\hat{f}(n) = \int_\Omega f(x) \mathrm{e}^{-2\pi i n \cdot x}$, and $n = (n_1, n_2, \cdots, n_d) \in \mathbf{Z}^d$ being a d-dimensional vector. For arbitrary non-negative multi-index $m = (m_1, m_2, \cdots, m_d) \in \mathbf{Z}^d (m_i \geqslant 0)$, one has at least formally

$$(-\Delta)^\alpha f = (2\pi)^\alpha \sum_{n \in \mathbf{Z}^d} |n|^{2\alpha} \hat{f}(n) \mathrm{e}^{2\pi i n \cdot x}.$$

Define

$$H_{per}^{2\alpha}(\Omega) = \left\{ f \,\Big|\, f \in L^2(\Omega) \text{or} \sum_{n \in \mathbf{Z}^d} |n|^{4\alpha} |\hat{f}(n)|^2 < \infty \right\},$$

with norm

$$\|f\|_{H_{per}^{2\alpha}(\Omega)} = \|f\|_2 + \|(-\Delta)^\alpha f\|_2.$$

If $f, g \in H_{per}^{2\alpha}(\Omega)$, by using the Parseval identity and integrating by parts, one has

$$\int_\Omega (-\Delta)^\alpha f \cdot g \mathrm{d}x = \int_\Omega (-\Delta)^{\alpha_1} f \cdot (-\Delta)^{\alpha_2} g \mathrm{d}x,$$

for arbitrary non-negative α_1 and α_2 such that $\alpha_1 + \alpha_2 = \alpha$.

Lemma 3.4.1 *If $\{f_n\} \in l^p$ and $\{g_n\} \in l^1$, then the "convolution"* $\left\{ \sum_{n_1+n_2=n} f_{n_1} g_{n_2} \right\}_{n=1}^\infty \in l^p$ *and*

$$\left\| \sum_{n_1+n_2=n} f_{n_1} g_{n_2} \right\|_{l^p} \leqslant \|f_n\|_p \|g_n\|_1.$$

Assume temporarily $u_0 \in H_{per}^1(\Omega)$ and consider the Ginzburg-Landau approximation

$$\begin{cases} u_t = \dfrac{u}{\max\{1, |u|\}} \times (-\Delta)^\alpha u - \beta \dfrac{u}{\max\{1, |u|\}} \times \Delta u + \varepsilon \Delta u \\ \qquad (x, t) \in \Omega \times (0, T), \\ u(x, 0) = u_0, \quad x \in \Omega, \end{cases} \qquad (3.4.2)$$

where β and ε are viscosity coefficients. Taking inner product of the approximation equation (3.4.2) with u, one has

$$\frac{1}{2} \frac{\mathrm{d}}{\mathrm{d}t} \int_\Omega |u|^2 \mathrm{d}x + \varepsilon \int_\Omega |\nabla u|^2 \mathrm{d}x = 0,$$

which yields by integrating with respect to time over $[0, t]$ that

$$\|u(\cdot, t)\|_2 \leqslant C, \qquad \forall 0 \leqslant t \leqslant T.$$

Take inner product of the approximation equation (3.4.2) with $\beta \Delta u$ to obtain

$$\beta \Delta u \cdot u_t = \beta \frac{u}{\max\{1, |u|\}} \times (-\Delta)^\alpha u \cdot \Delta u + \varepsilon \beta |\Delta u|^2,$$

and then with $(-\Delta)^\alpha u$ to obtain

$$(-\Delta)^\alpha u \cdot u_t = \varepsilon \Delta u \cdot (-\Delta)^\alpha u - \beta \frac{u}{\max\{1, |u|\}} \times \Delta u \cdot (\Delta)^{-\alpha} u.$$

Taking the difference of the two equations and then integrating over Ω, one has

$$-\frac{\beta}{2} \frac{\mathrm{d}}{\mathrm{d}t} \int_\Omega |\boldsymbol{\nabla} u|^2 \mathrm{d}x - \frac{1}{2} \frac{\mathrm{d}}{\mathrm{d}t} \int_\Omega |(-\Delta)^{\alpha/2} u|^2 \mathrm{d}x = \beta \varepsilon \int_\Omega |\Delta u|^2 \mathrm{d}x$$
$$-\varepsilon \int_\Omega \Delta u \cdot (-\Delta)^\alpha u \mathrm{d}x,$$

yielding

$$\beta \varepsilon \int_0^t \|\Delta u\|_2^2 \mathrm{d}t + \varepsilon \int_0^t \|(-\Delta)^{\frac{1+\alpha}{2}} u\|_2^2 + \frac{\beta}{2} \|\boldsymbol{\nabla} u\|_2^2$$
$$+ \frac{1}{2} \|(-\Delta)^{\alpha/2} u\|_2^2 = \frac{\beta}{2} \|\boldsymbol{\nabla} u_0\|_2^2 + \frac{1}{2} \|(-\Delta)^{\alpha/2} u_0\|_2^2. \tag{3.4.3}$$

Next we seek the approximation solution of (3.4.2) in the form

$$u_N(x, t) = \sum_{|n| \leqslant N} \varphi_n(t) e^{2\pi i n \cdot x},$$

such that for any $|n| \leqslant N$, there holds

$$\left\langle \frac{\partial u_N}{\partial t} - \frac{u_N}{\max\{1, |u_N|\}} \times (-\Delta)^\alpha u_N + \beta \frac{u_N}{\max\{1, |u_N|\} \times \Delta u_N} \right.$$
$$\left. -\varepsilon \Delta u_N, e^{2\pi i n \cdot x} \right\rangle = 0, \tag{3.4.4}$$

where φ_n is a vector in $\mathbf{R}^3$ and $\langle \cdot, \cdot \rangle$ represents the inner product in the space $L^2(\Omega)$. The initial value can be accordingly approximated in $H^1_{per}(\Omega)$ by

$$u_N(x, 0) = \sum_{i=1}^N \varphi_i(0) e_i(x) \to u_0.$$

We hence have obtained a system of ODEs for $\{\varphi_n(t)\}$ for $1 \leqslant |n| \leqslant N$. By the classical theory of ODEs, there exists a unique local solution $\{\varphi_n(t)\}$ satisfying (3.4.4). In what follows, we will obtain some useful *a priori* estimates to take the limit $N \to \infty$, and C will always denote some constant independent of β, ε and N. Multiplying the equation (3.4.4) by φ_n and then summing over $1 \leqslant |n| \leqslant N$, we have

$$\frac{1}{2}\frac{\mathrm{d}}{\mathrm{d}t}\int_\Omega |u_N|^2\mathrm{d}x + \varepsilon\int_\Omega |\nabla u_N|^2\mathrm{d}x = 0,$$

which yields by integrating over $[0,t]$ that $\|u_N(t)\|_2 \leqslant C$ for all $t \in [0,T]$. Similar to (3.4.3), we have

$$\beta\|\nabla u_N(t)\|_2^2 + \|(-\Delta)^{\alpha/2}u_N(t)\|_2^2 \leqslant C, \quad \forall t \in [0,T],$$

and

$$\beta\varepsilon\int_0^T \|\Delta u_N\|_2^2\mathrm{d}t \leqslant C.$$

The estimates of $\|u_{Nt}\|_2$ can also be obtained from (3.4.4). Let ε and β be fixed, and denote $Q_T = \Omega \times (0,T)$, then in view of the above bounded estimates, we have in the sense of a subsequence that

$$\Delta u_N \to \Delta u^{\beta,\varepsilon} \quad \text{weakly in } L^2(Q_T)$$
$$u_N \to u^{\beta,\varepsilon} \quad \text{weakly * in } L^\infty(0,T;H^1_{per}(\Omega))$$
$$u_N \to u^{\beta,\varepsilon} \quad \text{weakly and strongly in } L^2(Q_T) \text{ and a.e.}$$
$$u_{Nt} \to u_t^{\beta,\varepsilon} \quad \text{weakly in } L^2(Q_T).$$

Letting $N \to \infty$, then for any Fourier series ψ and smooth function $\varphi \in C^\infty[0,T]$, there holds

$$\int_{Q_T} u_t^{\beta,\varepsilon}\cdot\psi\varphi\mathrm{d}x\mathrm{d}t = \int_{Q_T}\left[\frac{u^{\beta,\varepsilon}}{\max\{1,|u^{\beta,\varepsilon}|\}}\times(-\Delta)^\alpha u^{\beta,\varepsilon}\cdot\psi\varphi\right.$$
$$\left.-\beta\frac{u^{\beta,\varepsilon}}{\max\{1,|u^{\beta,\varepsilon}|\}}\times\Delta u^{\beta,\varepsilon}\cdot\psi\varphi + \varepsilon\Delta u^{\beta,\varepsilon}\cdot\psi\varphi\right]\mathrm{d}x\mathrm{d}t.$$

It follows from density argument that for any $\phi \in L^2(Q_T)$,

$$\int_{Q_T} u_t^{\beta,\varepsilon}\cdot\phi\mathrm{d}x\mathrm{d}t = \int_{Q_T}\left[\frac{u^{\beta,\varepsilon}}{\max\{1,|u^{\beta,\varepsilon}|\}}\times(-\Delta)^\alpha u^{\beta,\varepsilon}\cdot\phi\right.$$
$$\left.-\beta\frac{u^{\beta,\varepsilon}}{\max\{1,|u^{\beta,\varepsilon}|\}}\times\Delta u^{\beta,\varepsilon}\cdot\phi + \varepsilon\Delta u^{\beta,\varepsilon}\cdot\phi\right]\mathrm{d}x\mathrm{d}t.$$
$$(3.4.5)$$

Lemma 3.4.2   If $u^{\beta,\varepsilon}$ satisfies (3.4.5), then $|u^{\beta,\varepsilon}| \leqslant 1$ for a.e. $(x,t) \in \Omega \times (0,T)$.

Proof We choose $\phi = u^{\beta,\varepsilon} - \min\{1, |u^{\beta,\varepsilon}|\} \dfrac{u^{\beta,\varepsilon}}{|u^{\beta,\varepsilon}|}$ in (3.4.5) to get

$$\frac{1}{2}\frac{d}{dt}\int_{|u^{\beta,\varepsilon}|\geqslant 1} |u^{\beta,\varepsilon}|^2\left(1 - \frac{1}{|u^{\beta,\varepsilon}|}\right)dx = \frac{1}{2}\int_{|u^{\beta,\varepsilon}|\geqslant 1}\frac{u^{\beta,\varepsilon}\cdot\partial_t u^{\beta,\varepsilon}}{|u^{\beta,\varepsilon}|}dx$$

$$-\varepsilon\int_{|u^{\beta,\varepsilon}|\geqslant 1}\frac{|u^{\beta,\varepsilon}\cdot\nabla u^{\beta,\varepsilon}|^2}{|u^{\beta,\varepsilon}|^3}dx - \varepsilon\int_{|u^{\beta,\varepsilon}|\geqslant 1}|\nabla u^{\beta,\varepsilon}|^2\left(1 - \frac{1}{|u^{\beta,\varepsilon}|}\right)dx.$$

Choose $\dfrac{\chi u^{\beta,\varepsilon}}{|u^{\beta,\varepsilon}|}$ as the test function, then

$$\frac{1}{2}\int_{|u^{\beta,\varepsilon}|\geqslant 1}\frac{u^{\beta,\varepsilon}\cdot\partial_t u^{\beta,\varepsilon}}{|u^{\beta,\varepsilon}|}dx = \frac{\varepsilon}{2}\int_{|u^{\beta,\varepsilon}|\geqslant 1}\frac{|u^{\beta,\varepsilon}\cdot\nabla u^{\beta,\varepsilon}|^2}{|u^{\beta,\varepsilon}|^3}dx$$

$$-\frac{\varepsilon}{2}\int_{|u^{\beta,\varepsilon}|\geqslant 1}\frac{|\nabla u^{\beta,\varepsilon}|^2}{|u^{\beta,\varepsilon}|}dx + \int_{|u^{\beta,\varepsilon}|=1}\frac{\partial u^{\beta,\varepsilon}}{\partial n}\cdot u^{\beta,\varepsilon}dS,$$

where χ is the characteristic function of the set $\{|u^{\beta,\varepsilon}| \geqslant 1\}$. Since $\dfrac{\partial u^{\beta,\varepsilon}}{\partial n}\cdot$ $u^{\beta,\varepsilon} = \dfrac{1}{2}\dfrac{|\partial u^{\beta,\varepsilon}|}{\partial n} \leqslant 0$ on the boundary $\{|u^{\beta,\varepsilon}| = 1\}$, then it follows that

$$\frac{1}{2}\frac{d}{dt}\int_{|u^{\beta,\varepsilon}|\geqslant 1}|u^{\beta,\varepsilon}|^2\left(1 - \frac{1}{|u^{\beta,\varepsilon}|}\right)dx \leqslant 0,$$

which implies that $|u^{\beta,\varepsilon}| \leqslant 1$ for a.e. $x \in \Omega \times (0,T)$.

Let β be fixed and $\varepsilon \to 0$, then (3.4.3) allows us to choose a subsequence from $\{u^{\beta,\varepsilon}\}$ such that $u^{\beta,\varepsilon} \to u^\beta$ weakly $*$ in $L^\infty(0,T;H^1(\Omega))$. We will show that

$$|u^\beta| = 1, \quad a.e.\ in\ \Omega \times (0,T). \tag{3.4.6}$$

In fact, for any $t > 0$, it is easy to know

$$\int_\Omega |u^{\beta,\varepsilon}(t)|^2 dx - \int_\Omega |u^{\beta,\varepsilon}(0)|^2 dx + \varepsilon\int_0^t\int_\Omega |\nabla u^{\beta,\varepsilon}|^2 dxdt = 0.$$

Letting $\varepsilon \to 0$, then $u^{\beta,\varepsilon} \to u^\beta$ strongly in $L^2(\Omega)$. Furthermore, since $\|\nabla u^{\beta,\varepsilon}\|_2$ is uniformly bounded and $|u^{\beta,\varepsilon}(0)| = 1$ always holds, letting $\varepsilon \to 0$ yields $\int_\Omega |u^\beta(t)|^2 - 1 dx = 0$, from which it follows (3.4.6) thanks to Lemma 3.4.2

We now consider the limit of (3.4.5) as $\varepsilon \to 0$. Let β be fixed, then for any $\phi \in C^\infty(\bar{Q}_T)$, ϕ with $\phi(\cdot, T) = 0$, then

$$\int_{Q_T} u_t^{\beta,\varepsilon} \cdot \phi \mathrm{d}x \mathrm{d}t = \int_{Q_T} [u^{\beta,\varepsilon} \times (-\Delta)^\alpha u^{\beta,\varepsilon} \cdot \phi - \beta u^{\beta,\varepsilon} \times \Delta u^{\beta,\varepsilon} \cdot \phi + \varepsilon \Delta u^{\beta,\varepsilon} \cdot \phi] \mathrm{d}x \mathrm{d}t.$$

$$(3.4.7)$$

The term on the left hand side can be written as

$$- \int_{Q_T} u^{\beta,\varepsilon} \cdot \phi_t \mathrm{d}x \mathrm{d}t - \int_\Omega u^{\beta,\varepsilon}(0) \cdot \phi(\cdot, 0) \mathrm{d}x,$$

which converges to

$$- \int_{Q_T} u^\beta \cdot \phi_t \mathrm{d}x \mathrm{d}t - \int_\Omega u_0 \cdot \phi(\cdot, 0) \mathrm{d}x, \qquad (3.4.8)$$

as $\varepsilon \to 0$. For the second term on the right of (3.4.7), we have

$$- \int_{Q_T} \beta u^{\beta,\varepsilon} \times \Delta u^{\beta,\varepsilon} \cdot \phi \mathrm{d}x \mathrm{d}t = \int_{Q_T} \beta u^{\beta,\varepsilon} \times \boldsymbol{\nabla} u^{\beta,\varepsilon} \cdot \boldsymbol{\nabla}\phi \mathrm{d}x \mathrm{d}t$$

$$\to \int_{Q_T} \beta u^\beta \times \boldsymbol{\nabla} u^\beta \cdot \boldsymbol{\nabla}\phi \mathrm{d}x \mathrm{d}t,$$

$$(3.4.9)$$

as $\varepsilon \to 0$. Similarly, the last term in (3.4.7) converges to 0.

Next we consider the limit when $\beta \to 0$. According to the *a priori* estimates (3.4.3), there is a subsequence $\{u^\beta\}$ such that

$$u^\beta \to u \qquad \text{weakly * in } L^\infty(0, T; H_{per}^\alpha(\Omega));$$
$$u^\beta \to u \qquad \text{strongly in } L^2(Q_T),$$

and $\beta \|\nabla u^\beta\|_2^2 \leqslant C$. Then if $\beta \to 0$, (3.4.8) converges to

$$- \int_{Q_T} u \cdot \phi_t \mathrm{d}x \mathrm{d}t - \int_\Omega u_0 \cdot \phi(\cdot, 0) \mathrm{d}x.$$

For (3.4.9), using Cauchy inequality, we have

$$\left| \int_{Q_T} \beta u^\beta \times \boldsymbol{\nabla} u^\beta \cdot \boldsymbol{\nabla}\phi \mathrm{d}x \mathrm{d}t \right| \leqslant \sqrt{\beta} \int_{Q_T} |\sqrt{\beta} \boldsymbol{\nabla} u^\beta| |\boldsymbol{\nabla}\phi| \mathrm{d}x \mathrm{d}t,$$

which converges to 0 as $\beta \to 0$.

Next we consider the convergence of the first term on the right of (3.4.7).

First of all, change this term into the form

$$\int_\Omega u^{\beta,\varepsilon} \times (-\Delta)^\alpha u^{\beta,\varepsilon} \cdot \phi \mathrm{d}x$$

$$= -\int_\Omega u^{\beta,\varepsilon} \times \phi \cdot (-\Delta)^\alpha u^{\beta,\varepsilon} \mathrm{d}x$$

$$= -\int_\Omega (-\Delta)^{\alpha/2}(u^{\beta,\varepsilon} \times \phi) \cdot (-\Delta)^{\alpha/2} u^{\beta,\varepsilon} \mathrm{d}x$$

$$= \int_\Omega \left[(-\Delta)^{\alpha/2}(u^{\beta,\varepsilon} \times \phi) - (-\Delta)^{\alpha/2} u^{\beta,\varepsilon} \times \phi\right] \cdot (-\Delta)^{\alpha/2} u^{\beta,\varepsilon} \mathrm{d}x =: \mathcal{I}^{\beta,\varepsilon}.$$

Define $\mathcal{L}u := (-\Delta)^{\alpha/2}(u\phi) - (-\Delta)^{\alpha/2} u\phi$, then by the following Proposition 3.4.1 $\mathcal{L} : H_{per}^\alpha \to L^2(\Omega)$ is compact. Therefore, we have from (3.4.3) that

$$\lim_{\beta \to 0} \lim_{\varepsilon \to 0} \mathcal{I}^{\beta,\varepsilon} = \int_\Omega \left[(-\Delta)^{\alpha/2}(u \times \phi) - (-\Delta)^{\alpha/2} u \times \phi\right] \cdot (-\Delta)^{\alpha/2} u \mathrm{d}x$$

$$= \int_\Omega (-\Delta)^{\alpha/2}(u \times \phi) \cdot (-\Delta)^{\alpha/2} u \mathrm{d}x.$$

Then the proof of Theorem 3.4.1 is complete.

Proposition 3.4.1 *Operator $\mathcal{L} : H_{per}^\alpha(\Omega) \to L^2(\Omega)$ is compact.*

Proof It suffices to prove that $\mathcal{L} : H_{per}^\alpha(\Omega) \to H_{per}^\alpha(\Omega)$ is a bounded linear operator, since $H_{per}^\alpha(\Omega)$ is compactly embedded into $L^2(\Omega)$. Obviously, by definition,

$$\|(-\Delta)^{\alpha/2}(u\phi)\|_{L^2}^2 = (2\pi)^\alpha \sum_{n \in \mathbf{Z}^d} |n|^{2\alpha} |\widehat{u\phi}(n)|^2,$$

and

$$\|(-\Delta)^{\alpha/2} u\phi\|_{L^2} \leqslant C\|u\|_{H_{per}^\alpha}.$$

Since $\widehat{u\phi}(n) = \sum_{n_1+n_2=n} \hat{u}(n_1)\hat{\phi}(n_2)$ and $|n_1 + n_2|^\alpha \leqslant C(|n_1|^\alpha + |n_2|^\alpha)$, then

$$|n|^\alpha |\widehat{u\phi}(n)| \leqslant |n|^\alpha \sum_{n_1+n_2=n} |\hat{u}(n_1)||\hat{\phi}(n_2)|$$

$$\leqslant C\left(\sum_{n_1+n_2=n} |n_1|^\alpha |\hat{u}(n_1)||\hat{\phi}(n_2)| + \sum_{n_1+n_2=n} |\hat{u}(n_1)||n_2|^\alpha |\hat{\phi}(n_2)|\right).$$

Applying the discrete Young's inequality, we have

$$\|(-\Delta)^{\alpha/2}(u\phi)\|_{L^2}^2 \leqslant C(\|(-\Delta)^{\alpha/2} u\|^2 \|\phi\|_1^2 + \|u\|_2^2 \|(-\Delta)^{\alpha/2}\phi\|_1^2).$$

We need also to estimate $\|(-\Delta)^{\alpha/2}(\mathcal{L}u)\|_{L^2}$. By definition, it suffices to prove that $\{|n|^\alpha \widehat{\mathcal{L}u}(n)\} \in l^2$. As $u \sim \sum_n \hat{u}(n)\mathrm{e}^{2\pi i n \cdot x}$ and $\phi \sim \sum_n \hat{\phi}(n)\mathrm{e}^{2\pi i n \cdot x}$, then

$u\phi \sim \sum\limits_{n} \sum\limits_{n_1+n_2=n} \hat{u}(n_1)\hat{\phi}(n_2)e^{2\pi i n \cdot x}$. By definition of $\mathcal{L}u$, we have

$$\widehat{\mathcal{L}u}(n) = \sum_{n_1+n_2=n} |n|^\alpha \hat{u}(n_1)\hat{\phi}(n_2) - \sum_{n_1+n_2=n} |n_1|^\alpha \hat{u}(n_1)\hat{\phi}(n_2),$$

yielding

$$|\widehat{\mathcal{L}u}(n)| \leqslant \sum_{n_1+n_2=n} |n_2|^\alpha |\hat{u}(n_1)||\hat{\phi}(n_2)|,$$

and

$$|n|^\alpha \widehat{\mathcal{L}u}(n) \leqslant \sum_{n_1+n_2=n} |n_1|^\alpha |\hat{u}(n_1)||n_2|^\alpha |\hat{\phi}(n_2)| + \sum_{n_1+n_2=n} |\hat{u}(n_1)||n_2|^{2\alpha}|\hat{\phi}(n_2)|.$$

Applying the discrete Young's inequality again, we obtain

$$\|(-\Delta)^{\alpha/2}\mathcal{L}u\|_{L^2}^2 \leqslant \|(-\Delta)^{\alpha/2}u\|_2^2\|(-\Delta)^{\alpha/2}\phi\|_1^2 + \|u\|_2^2\|(-\Delta)^\alpha \phi\|_1^2,$$

completing the proof.

3.4.2   Ginzburg-Landau approximation and asymptotic limit

In this section, we consider the Ginzburg-Landau approximation of the fractional Landau-Lifshitz equation and the limits when the coefficients tend to zero,

$$\begin{cases} \partial_t u = \nu u \times \Lambda^{2\alpha} u + \mu u \times (u \times \Lambda^{2\alpha} u), \\ u(0) = u_0 \text{ and } |u_0(x)| = 1, a.e. \ x \in \Omega, \end{cases} \tag{3.4.10}$$

where $\alpha \in (1/2, 1)$ and $u = (u_1, u_2, u_3)$ is the magnetization vector, $\Lambda = (-\Delta)^{1/2}$ represents fractional Laplace operator. For convenience, we let $\Omega = [0, 2\pi]$ be the one dimensional periodic domain, $\nu \in \mathbf{R}$ and $\mu \geqslant 0$ are physical parameters. Let $Q_T = (0, T) \times \Omega$.

When $\alpha = 1$, (3.4.10) reduces to the classical Landau-Lifshitz equation

$$\partial_t u = -\nu u \times \Delta u - \mu u \times (u \times \Delta u).$$

The equation was first proposed by Landau and Lifshitz [128], which was used to study the dispersion theory of permeability for ferromagnetic materials, then it was widely studied [102]. When $\nu = 0, \alpha = 1$, equation (3.4.10) reduces to the heat flow of harmonic maps [133]

$$u_t = \mu\Delta u + \mu|\nabla u|^2 u.$$

When $\alpha = 1$, $\nu = -1$ and $\mu = 0$, (3.4.10) reduces to the Schrodinger flow on the unit sphere, and was extensively studied [41, 173]. For this reason, when

$\nu = 0$, equation (3.4.10) is also called the fractional heat flow for harmonic maps. If $\mu = 0$, equation (3.4.10) reduces to the fractional Heisenberg equation [210]. It is easy to see, if $|u_0(x)| = 1$, then $|u(t,x)| = 1$ for any $t \geqslant 0$ and hence (3.4.10) is equivalent to the following Gilbert equation [91]

$$u_t = \frac{\nu^2 + \mu^2}{\nu} u \times \Lambda^{2\alpha} u + \frac{\mu}{\nu} u \times u_t. \tag{3.4.11}$$

The weak solution of this equation can be defined as follows.

Definition 3.4.1 *Let $u_0 \in H^\alpha$, $|u_0| = 1$ a.e., $u = (u_1, u_2, u_3)$ is called a weak solution of (3.4.11) if*
(i) for any $T > 0$, $u \in L^\infty(0, T; H^\alpha(\Omega))$, $u_t \in L^2(Q_T)$ and $|u| = 1$ a.e.;
(ii) for any three-dimensional vector $\varphi \in L^2(0, T; H^\alpha(\Omega))$, there holds

$$\frac{\mu}{\nu} \int_{Q_T} \left(u \times \frac{\partial u}{\partial t} \right) \cdot \varphi \mathrm{d}x\mathrm{d}t - \int_{Q_T} \frac{\partial u}{\partial t} \cdot \varphi \mathrm{d}x\mathrm{d}t$$
$$= \frac{\nu^2 + \mu^2}{\nu} \int_{Q_T} \Lambda^\alpha u \cdot \Lambda^\alpha(u \times \varphi) \mathrm{d}x\mathrm{d}t; \tag{3.4.12}$$

(iii) $u(0, x) = u_0(x)$ in the trace sense;
(iv) for any $T > 0$, there holds

$$\int_\Omega |\Lambda^\alpha u(T)|^2 \mathrm{d}x + \frac{2\mu}{1 + \mu^2} \int_{Q_T} \left| \frac{\partial u}{\partial t} \right|^2 \mathrm{d}x\mathrm{d}t \leqslant \int_\Omega |\Lambda^\alpha u_0|^2 \mathrm{d}x. \tag{3.4.13}$$

One of the main results in this section states that there exists at least one weak solution to the fractional Landau-Lifshitz equation, see Theorem 3.4.2. Since (3.4.10) is similar to the harmonic map heat flow equation but with one more nonlinear term $u \times \Lambda^\alpha u$, we generalize Chen's idea in [44] to the fractional case, where she proved the existence of weak solutions to the heat flow of harmonic maps by Ginzburg-Landau approximation. But since λ^α is a nonlocal operator, more is involved in the fractional case and the fractional calculus inequalities in Theorem 2.2.14 will play a critical role in the convergence of approximate solutions.

The fractional calculus inequality itself is not sufficient for the convergence, since the equation does not have any commutator structure or the divergence free condition as in the quasi-geostrophic equation in the subsequent sections. However, noting that cross product $a \times b$ of two three dimensional vectors is vertical to each of its components, we can introduce a commutator structure to get the convergence. The details are presented in the following and the commutator is defined as

$$[\Lambda^\alpha, \varphi]u := \Lambda^\alpha(\varphi \times u) - \varphi \times \Lambda^\alpha u.$$

Another aspect of this paper concerns the limiting behaviors as the Gilbert parameter μ varies. Formally, as $\mu \to 0$, (3.4.10) tends to the fractional Heisenberg equation and as $\nu \to \infty$, the solution of (3.4.10) tends to the weak solution of the fractional heat flow of harmonic maps after a scale transform. These observations will be justifies in this section, see Theorem 3.4.3 and Theorem 3.4.4 for detailed statements. All these results can be generalized to the case $\alpha = 1$, which is established in Guo-Ding [102] and Alougest-Soyeur [10].

If not specified, $\dot{H}^\alpha(\Omega)$ denotes the homogeneous fractional Sobolev space and $H^\alpha(\Omega)$ denotes the inhomogeneous one. The product functional spaces $(X)^3$ are all simplified to X. For example, $(L^\infty(0, T; H^\alpha(\Omega)))^3$ is simplified to $L^\infty(0, T; H^\alpha(\Omega))$.

Now we prove the existence of global weak solutions to the 1D periodic fractional Landau-Lifshitz equation (3.4.10). For simplicity we let $\nu = 1$, which will not affect the result essentially. We will prove the following.

Theorem 3.4.2 *Let $\alpha \in (1/2, 1)$ and $u_0 \in H^\alpha$ such that $|u_0| = 1$, a.e.. Then there exists at least a weak solution for the fractional Landau-Lifshitz equation in the sense of Definition 3.4.1.*

What follows focuses on the proof of this theorem. Before we doing so, we introduce a compact lemma due to Simon [201].

Lemma 3.4.3 *Assume B_0, B, B_1 are three Banach spaces and satisfy $B_0 \subset B \subset B_1$ with compact embedding $B_0 \hookrightarrow B$. Let W be bounded in $L^\infty(0, T; B_0)$ and $W_t := \{w_t; w \in W\}$ be bounded in $L^q(0, T; B_1)$ where $q > 1$. Then W is relatively compact in $C([0, T]; B)$.*

The proof can be found in Simon [201, Corollary 4, P.85]. Note that this lemma is an extension of the Aubin's compactness result, see [13, 135] for details. One may see also Theorem 3.2.7.

Inspired by Chen's work on the heat flow of harmonic maps, we consider the following Ginzburg-Landau approximation problem for u_ϵ

$$\mu \frac{\partial u_\epsilon}{\partial t} + u_\epsilon \times \frac{\partial u_\epsilon}{\partial t} + (1 + \mu^2) \left(\Lambda^{2\alpha} u_\epsilon - \frac{1}{\epsilon^2}(1 - |u_\epsilon|^2) u_\epsilon \right) = 0. \qquad (3.4.14)$$

Let $\{w_i\}_{i \in \mathbf{N}}$ be a complete orthonormal basis of $L^2(\Omega)$ consisting of eigenvectors of $\Lambda^{2\alpha}$

$$\Lambda^{2\alpha} w_j = \kappa_j w_j, \quad j = 1, 2, \cdots \qquad (3.4.15)$$

under periodic boundary conditions. The existence of such a basis can be proved as in Temam [213, §.2,Ch.II]. For fixed $\epsilon > 0$, we seek approximate solutions $\{u_N(t,x)\}$ for equation (3.4.14) of the form

$$u_N(t,x) = \sum_{i=1}^{N} \varphi_i(t)w_i(x),$$

where φ_i are $\mathbf{R}^3$-valued vectors, such that for $1 \leqslant i \leqslant n$ there holds

$$\mu \int_\Omega \frac{\partial u_N}{\partial t} w_i + \int_\Omega u_N \times \frac{\partial u_N}{\partial t} w_i + (1+\mu^2) \int_\Omega \Lambda^{2\alpha} u_N w_i$$
$$-\frac{1+\mu^2}{\epsilon^2} \int_\Omega (|u_N|^2 - 1)u_N w_i = 0, \quad 1 \leqslant i \leqslant n \tag{3.4.16}$$

with initial conditions

$$\int_\Omega u_N(0)w_i = \int_\Omega u_0 w_i$$

hold.

Since the coefficient matrix "$\mu + u_N \times$" before $\dfrac{\partial u_N}{\partial t}$ is anti-symmetric and hence invertible, from the standard ODE theory, there exists a local solution to the system (3.4.16) for $\{\varphi_i\}_{i=1}^N$. In the following, we make some *a priori* estimates to show these solutions exist at least on a common interval $[0,T]$. Multiplying the equality (3.4.16) by $\dfrac{d\varphi_i}{dt}$ and summing over $1 \leqslant i \leqslant n$ lead to

$$\frac{\mu}{1+\mu^2} \int_\Omega |\frac{\partial u_N}{\partial t}|^2 dx + \frac{1}{2}\frac{d}{dt} \int_\Omega |\Lambda^\alpha u_N|^2 dx + \frac{1}{4\epsilon^2}\frac{d}{dt} \int_\Omega (|u_N|^2 - 1)^2 dx = 0.$$

Then integrating the resulting equality over $[0,T]$ leads to

$$\frac{1}{2} \int_\Omega |\Lambda^\alpha u_N(T)|^2 dx + \frac{1}{4\epsilon^2} \int_\Omega (|u_N|^2 - 1)^2(T) dx + \frac{\mu}{1+\mu^2} \int_{Q_T} |\frac{\partial u_N}{\partial t}|^2 dx dt$$
$$= \frac{1}{2} \int_\Omega |\Lambda^\alpha u_{N0}|^2 dx + \frac{1}{4\epsilon^2} \int_\Omega (|u_{N0}|^2 - 1)^2 dx, \quad \forall t \in [0,T]. \tag{3.4.17}$$

Since $\alpha \in (1/2,1)$ and $H^\alpha(\Omega) \hookrightarrow L^4(\Omega)$, the right hand side is uniformly bounded. Furthermore, by Young's inequality

$$\int_\Omega |u_N|^2 dx \leqslant C + \frac{1}{2} \int_\Omega (|u_N|^2 - 1)^2 dx, \tag{3.4.18}$$

therefore for fixed $\epsilon > 0$, there holds

$$\{u_N\} \quad \text{is bounded in} L^\infty(0,T;H^\alpha(\Omega));$$

$$\{\frac{\partial u_N}{\partial t}\} \quad \text{is bounded in } L^2(0,T;L^2(\Omega));$$

$$\{|u_N|^2 - 1\} \quad \text{is bounded in } L^\infty(0,T;L^2(\Omega)).$$

These estimates imply that the solution can be extended to all time, and we can extract a subsequence (still denoted as $\{u_N(t)\}$) such that for any $1 < p < \infty$

$$u_N \to u_\epsilon \qquad weakly\ in \quad L^p(0, T; H^\alpha(\Omega));$$

$$u_N \to u_\epsilon \qquad strongly\ in \quad C([0, T]; H^\beta(\Omega))\ and\ a.e.\ for\ 0 \leqslant \beta < \alpha;$$

$$\frac{\partial u_N}{\partial t} \to \frac{\partial u_\epsilon}{\partial t} \qquad weakly\ in \quad L^2(0, T; L^2(\Omega));$$

$$|u_N|^2 - 1 \to \zeta \qquad weakly\ in\ L^p(0, T; L^2(\Omega)).$$

The second convergence is due to Lemma 3.4.3. On the other hand, since $u_N \to u_\epsilon$ a.e., it can be shown from [135, Lem1.3, Chap.1] that $\zeta = |u_\epsilon|^2 - 1$. Taking $N \to \infty$ in (3.4.16), we find a weak solution u_ϵ for the approximate problem (3.4.14), i.e., there holds

$$\mu \int_{Q_T} \frac{\partial u_\epsilon}{\partial t}\phi + \int_{\Omega_T} u_\epsilon \times \frac{\partial u_\epsilon}{\partial t}\phi + (1 + \mu^2) \int_{Q_T} \Lambda^\alpha u_\epsilon \Lambda^\alpha \phi$$
$$- \frac{1 + \mu^2}{\epsilon^2} \int_{Q_T} (|u_\epsilon|^2 - 1) u_\epsilon \phi = 0, \tag{3.4.19}$$

for any $\phi \in L^2(0, T; H^\alpha)$. Furthermore, passing $N \to \infty$ in (3.4.17), by Fatou lemma we have

$$\frac{1}{2} \int_\Omega |\Lambda^\alpha u_\epsilon(T)|^2 \mathrm{d}x + \frac{1}{4\epsilon^2} \int_\Omega (|u_\epsilon|^2 - 1)^2(T)\mathrm{d}x + \frac{\mu}{1 + \mu^2} \int_{Q_T} \left|\frac{\partial u_\epsilon}{\partial t}\right|^2 \mathrm{d}x \mathrm{d}t$$
$$\leqslant \frac{1}{2} \int_\Omega |\Lambda^\alpha u_0|^2 \mathrm{d}x, \qquad \forall t \in [0, T]. \tag{3.4.20}$$

In the following, we let $\epsilon \to 0$. From (3.4.20) and a similar inequality that leads to (3.4.18), we have

$$\{u_\epsilon\} \quad is\ bounded\ in\ L^\infty(0, T; H^\alpha(\Omega));$$

$$\left\{\frac{\partial u_\epsilon}{\partial t}\right\} \quad is\ bounded\ in\ L^2(0, T; L^2(\Omega));$$

$$\{|u_\epsilon|^2 - 1\} \quad is\ bounded\ in\ L^\infty(0, T; L^2(\Omega)).$$

Therefore, we can select a subsequence (still denoted as u_ϵ) such that for any $1 < p < \infty$ and for any $0 \leqslant \beta < \alpha$

$$u_\epsilon \to u \qquad weakly\ in \quad L^p(0, T; H^\alpha(\Omega));$$

$$u_\epsilon \to u \qquad strongly\ in \quad C([0, T]; H^\beta(\Omega))\ and\ a.e.,$$

$$\frac{\partial u_\epsilon}{\partial t} \to \frac{\partial u}{\partial t} \qquad weakly\ in \quad L^2(0, T; L^2(\Omega));$$

$$|u_\epsilon|^2 - 1 \to 0 \qquad strongly\ in \quad L^p(0, T; L^2(\Omega))\ and\ a.e..$$

It can be shown from the last convergence that $|u| = 1$, a.e.. On the other hand, since $H^\alpha(\Omega) \hookrightarrow L^\infty(\Omega)$, by Sobolev embedding inequality we have

$$u_\epsilon \text{ is bounded in } L^\infty(Q_T), \quad i.e. \; |u_\epsilon| \leqslant C \tag{3.4.21}$$

for some constant C depending only on the initial data and the Sobolev constant. In order to pass to the limit, we let $\varphi \in C^\infty([0,T] \times \Omega)$ and $\phi = u_\epsilon \times \varphi$ in (3.4.19). Applying the multiplicative estimates to u_ϵ and φ

$$\|\Lambda^\alpha(u_\epsilon \times \varphi)\|_{L^2} \leqslant C(\|u_\epsilon\|_{L^\infty}\|\varphi\|_{\dot{H}^{\alpha,2}} + \|u_\epsilon\|_{\dot{H}^{\alpha,2}}\|\varphi\|_{L^\infty}),$$

it can be shown $\phi \in L^2(0,T; H^\alpha(\Omega))$ (where $L^2(0,T; L^2(\Omega))$-norm is obvious), and hence

$$-\mu \int_{Q_T} \left(u_\epsilon \times \frac{\partial u_\epsilon}{\partial t} \right) \cdot \varphi + \int_{Q_T} |u_\epsilon|^2 \frac{\partial u_\epsilon}{\partial t} \cdot \varphi$$
$$-\int_{Q_T} \left(u_\epsilon \cdot \frac{\partial u_\epsilon}{\partial t} \right) u_\epsilon \cdot \varphi + (1+\mu^2) \int_{Q_T} \Lambda^\alpha u_\epsilon \cdot \Lambda^\alpha(u_\epsilon \times \varphi) = 0. \tag{3.4.22}$$

Taking $\epsilon \to 0$, it can be shown that

$$\int_{Q_T} |u_\epsilon|^2 \frac{\partial u_\epsilon}{\partial t} \cdot \varphi = \int_{Q_T} (|u_\epsilon|^2 - 1)\frac{\partial u_\epsilon}{\partial t} \cdot \varphi + \int_{Q_T} \frac{\partial u_\epsilon}{\partial t} \cdot \varphi$$
$$\to \int_{Q_T} \frac{\partial u}{\partial t} \cdot \varphi,$$

thanks to the strong convergence of $|u_\epsilon|^2 - 1$ to zero. For the third term, we have

$$\int_{Q_T} \left(u_\epsilon \cdot \frac{\partial u_\epsilon}{\partial t} \right) u_\epsilon \cdot \varphi - \int_{Q_T} u \cdot \frac{\partial u}{\partial t} u \cdot \varphi = \int_{Q_T} \left(u_\epsilon \cdot \frac{\partial u_\epsilon}{\partial t} \right)(u_\epsilon - u) \cdot \varphi$$
$$+ \int_{Q_T} \left(u_\epsilon \cdot \left(\frac{\partial u_\epsilon}{\partial t} - \frac{\partial u}{\partial t} \right) u \cdot \varphi \right.$$
$$+ \int_{Q_T} (u_\epsilon - u) \cdot \frac{\partial u}{\partial t} u \cdot \varphi$$
$$\to 0.$$

Finally, we consider the convergence of the last term in (3.4.22). This is by no means obvious since we encounter the fractional order derivatives and for this reason, the commutator estimate will be resorted to. Let $\mathcal{I}_\epsilon = -\int_{Q_T} \Lambda^\alpha u_\epsilon \cdot$

$\Lambda^\alpha(u_\epsilon \times \varphi)$ and accordingly $\mathcal{I} = -\int_{Q_T} \Lambda^\alpha u \cdot \Lambda^\alpha(u \times \varphi)$. Since $\int_{Q_T} \Lambda^\alpha u_\epsilon \cdot$

$$\Lambda^\alpha u_\epsilon \times \varphi = \int_{Q_T} \Lambda^\alpha u \cdot \Lambda^\alpha u \times \varphi = 0, \text{ we have}$$

$$\mathcal{I}_\epsilon = \int_{Q_T} \Lambda^\alpha u_\epsilon \cdot [\Lambda^\alpha, \varphi] u_\epsilon$$

and

$$\mathcal{I} = \int_{Q_T} \Lambda^\alpha u \cdot [\Lambda^\alpha, \varphi] u.$$

Now, using commutator estimate, it is shown

$$\|[\Lambda^\alpha, \varphi](u_\epsilon - u)\|_{L^2(\Omega)} \leqslant C(\|\nabla\varphi\|_{L^{p_1}} \|u_\epsilon - u\|_{\dot{H}^{\alpha-1,p_2}} + \|\varphi\|_{\dot{H}^{\alpha,p_3}} \|u_\epsilon - u\|_{L^{p_4}}).$$

Taking $p_1 = \dfrac{1}{1-\alpha}, p_2 = \dfrac{2}{2\alpha - 1}$ and for any $p_3, p_4 \in (2, \infty)$, we have

$$\|[\Lambda^\alpha, \varphi](u_\epsilon - u)\|_{L^2(Q_T)} \leqslant C(\|\nabla\varphi\|_{L^\infty(0,T;L^{p_1}(\Omega))} \|u_\epsilon - u\|_{L^2(Q_T)}$$
$$+ \|\varphi\|_{L^\infty(0,T;\dot{H}^{\alpha,p_3}(\Omega))} \|u_\epsilon - u\|_{L^2(0,T;H^\beta(\Omega))})$$
$$\to 0,$$

by the strong convergence of u_ϵ to u in $L^2(Q_T)$ and in $L^2(0, T; H^\beta(\Omega))$, where $\beta = \dfrac{1}{2} - \dfrac{1}{p_4} < \dfrac{1}{2} < \alpha$. On the other hand, following exactly the same commutator estimate, one can show that $[\Lambda^\alpha, \varphi]u \in L^2(Q_T)$. Therefore,

$$\mathcal{I}_\epsilon - \mathcal{I} = \int_{Q_T} \Lambda^\alpha u_\epsilon \cdot [\Lambda^\alpha, \varphi](u_\epsilon - u) + \int_{Q_T} \Lambda^\alpha(u_\epsilon - u) \cdot [\Lambda^\alpha, \varphi]u \to 0,$$

since u_ϵ is bounded in $L^2(0, T; H^\alpha(\Omega))$ and converges to u weakly in $L^2(0, T; H^\alpha(\Omega))$. This verifies the convergence of the last term in (3.4.22). Taking $\epsilon \to 0$, we have

$$\mu \int_{Q_T} \left(u \times \frac{\partial u}{\partial t}\right) \cdot \varphi \mathrm{d}x\mathrm{d}t - \int_{Q_T} \frac{\partial u}{\partial t} \cdot \varphi \mathrm{d}x\mathrm{d}t$$
$$= (1 + \mu^2) \int_{Q_T} \Lambda^\alpha u \cdot \Lambda^\alpha(u \times \varphi)\mathrm{d}x\mathrm{d}t,$$

and this relation holds for $\varphi \in L^2(0, T; H^\alpha(\Omega))$ by a standard density argument. Furthermore, the inequality (3.4.13) holds from estimates (3.4.20) and we finish the proof of Theorem 3.4.2.

Remark 3.4.1 *For technical reasons, the above analysis is confined to the 1-D case due to Sobolev embedding inequality in (3.4.21). Similar result is obtained through an auxiliary function g for the classical Landau-Lifshitz equation. See [10, pp.1079], but whose analysis breaks down in the fractional case.*

Remark 3.4.2   *When $\alpha = 1$, it can be shown that $|u_\epsilon| \leqslant 1$, see [10]. We would also like to mention that the advantage of the Ginzburg-Landau approximation lies in the fact that it can be shown $|u| = 1$ through the approximating process, which is difficulty to be accomplished by the usual Galerkin method.*

Remark 3.4.3   *When $\mu = 0$, by means of the Galerkin approximation, we can show that there exists at least a global weak solution for the fractional Heisenberg equation such that for all $\varphi \in L^2(0, T; H^\alpha(\Omega))$ there holds*

$$\int_{Q_T} \frac{\partial u}{\partial t} \cdot \varphi \mathrm{d}x\mathrm{d}t + \nu \int_{Q_T} \Lambda^\alpha u \cdot \Lambda^\alpha (u \times \varphi) \mathrm{d}x\mathrm{d}t = 0. \qquad (3.4.23)$$

On the other hand, it can be shown following the same steps as above that when $\nu = 0$, there exists a global weak solution for the fractional harmonic map heat flow such that for all $\varphi \in L^2(0, T; H^\alpha(\Omega))$, the following equality holds

$$\int_{Q_T} \left(u \times \frac{\partial u}{\partial t} \right) \cdot \varphi \mathrm{d}x\mathrm{d}t - \mu \int_{Q_T} \Lambda^\alpha u \cdot \Lambda^\alpha (u \times \varphi) \mathrm{d}x\mathrm{d}t = 0. \qquad (3.4.24)$$

In what follows, we discuss the relationships between the fractional Landau-Lifshitz equation and the fractional Heisenberg equation or the generalized fractional heat flow of harmonic maps. We will prove the following two theorems.

Theorem 3.4.3   *Let $\mu \to 0$, the weak solution obtained in Sect. 3 weakly converges, up to a subsequence, to a solution of the fractional Heisenburg equation in the sense of (3.4.23).*

Proof From the inequality (3.4.13), we know that u^μ is uniformly bounded in $L^p(0, T; H^\alpha)$ for any $1 \leqslant p \leqslant \infty$, and $\sqrt{\mu}\dfrac{\partial u^\mu}{\partial t}$ is bounded in $L^2(0, T; L^2(\Omega))$. By the calculus inequalities, we have

$$\|\Lambda^\alpha (u^\mu \times \varphi)\| \leqslant C(\|\Lambda^\alpha \varphi\| + \|\Lambda^\alpha u^\mu\|\|\varphi\|_{L^\infty(\Omega)})$$
$$\leqslant C(1 + \|\Lambda^\alpha u^\mu\|)\|\varphi\|_{H^\alpha}.$$

Therefore from (3.4.10)

$$\left| \int_{Q_T} \frac{\partial u^\mu}{\partial t} \cdot \varphi \right| \leqslant C\sqrt{\mu}\|\varphi\|_{L^2(Q_T)}$$
$$+ C(1 + \mu^2)(1 + \|\Lambda^\alpha u_0\|)\|\Lambda^\alpha u^\mu\|_{L^2(Q_T)}\|\varphi\|_{L^2(0,T;H^\alpha(\Omega))}$$
$$\leqslant C(1 + \mu^2)\|\varphi\|_{L^2(0,T;H^\alpha(\Omega))},$$

and hence $\left\{ \dfrac{\partial u^\mu}{\partial t} \right\}$ is bounded uniformly in $L^2(0, T; H^{-\alpha}(\Omega))$ as far as $\mu \leqslant 1$. We can then extract a subsequence (if necessary) such that for any $-\alpha \leqslant$

$\beta < \alpha$ and for any $1 < p < \infty$ there hold

$$u^\mu \to u \quad weakly\ in\ \ L^p(0,T;H^\alpha(\Omega));$$

$$u^\mu \to u \quad strongly\ in\ \ C([0,T];H^\beta(\Omega)).$$

Passing to the limit $\mu \to 0$ and taking advantage of the commutator estimate, we then find a weak solution of the fractional Heisenberg equation as a limit of u^μ in the sense of (3.4.23) and Theorem 3.4.3 is proved.

Theorem 3.4.4 *Let u^μ be weak solutions for the fractional Landau-Lifshitz equation and $\tilde{u}^\mu(t,x) = u^\mu(t/\mu,x)$. Then as $\mu \to \infty$, $\tilde{u}^\mu$ weakly converges, up to a subsequence, to a solution of the fractional heat flow of harmonic map equation in the sense of (3.4.24).*

Proof Taking the scale transform $\tilde{u}^\mu(t,x) = u^\mu(t/\mu,x)$, we have from (3.4.12) that

$$\int_{Q_T} \left(\tilde{u}^\mu \times \frac{\partial \tilde{u}^\mu}{\partial t}\right) \cdot \varphi \mathrm{d}x\mathrm{d}t - \frac{1}{\mu}\int_{Q_T}\frac{\partial \tilde{u}^\mu}{\partial t}\cdot \varphi\mathrm{d}x\mathrm{d}t$$
$$= \frac{1+\mu^2}{\mu^2}\int_{Q_T}\Lambda^\alpha\tilde{u}^\mu \cdot \Lambda^\alpha(\tilde{u}^\mu \times \varphi)\mathrm{d}x\mathrm{d}t. \tag{3.4.25}$$

Furthermore, we have the energy inequality

$$\int_\Omega |\Lambda^\alpha\tilde{u}^\mu(T)|^2\mathrm{d}x + \frac{2\mu^2}{1+\mu^2}\int_{Q_T}\left|\frac{\partial \tilde{u}^\mu}{\partial t}\right|^2\mathrm{d}x\mathrm{d}t \leqslant \int_\Omega |\Lambda^\alpha u_0|^2\mathrm{d}x.$$

Therefore, $\left\{\dfrac{\partial \tilde{u}^\mu}{\partial t}\right\}$ is uniformly bounded in μ in $L^2(Q_T)$ as soon as $\mu > 1$. We can then extract a subsequence if necessary such that for any $1 < p < \infty$ and for any $0 \leqslant \beta < \alpha$ there hold

$$\tilde{u}^\mu \to \tilde{u} \quad weakly\ in\ \ L^p(0,T;H^\alpha(\Omega));$$

$$\tilde{u}^\mu \to \tilde{u} \quad strongly\ in\ \ C([0,T];H^\beta(\Omega))\ and\ a.e.;$$

$$\frac{\partial \tilde{u}^\mu}{\partial t} \to \frac{\partial \tilde{u}}{\partial t} \quad weakly\ in\ \ L^2(0,T;L^2(\Omega)).$$

Taking $\mu \to \infty$ in (3.4.25) and taking advantage of the commutator estimate, we find that up to a subsequence, weak solutions for the fractional Landau-Lifshitz equation weakly converge as $\mu \to \infty$ to a weak solution for the fractional heat flow of harmonic map equation in the sense of (3.4.24). $\blacksquare$

3.4.3 Higher dimensional case—Galerkin approximation

Now, we consider the fractional Landau-Lifshitz equation in higher dimensions

$$\begin{cases} u_t = \gamma u \times \Lambda^{2\alpha}u + \lambda u \times (u \times \Lambda^{2\alpha}u) \\ u(0) = u_0 \in H^\alpha, \end{cases} \tag{3.4.26}$$

where $u(x,t)$ is still a three-dimensional vector, representing the magnetization vector of ferromagnetic materials, $\gamma, \lambda \geqslant 0$ and $\alpha \in (0,1)$ are real numbers. In the section, we still discuss the spatial periodic case as $\Omega = [0, 2\pi]^d$ and $d = 2, 3$. Let $Q_T = (0, T) \times \Omega$. When $\gamma = 0$, equation (3.4.26) is the fractional heat flow of harmonic maps

$$u_t = \lambda u \times (u \times \Lambda^{2\alpha} u). \qquad (3.4.27)$$

In what follows, we discuss the existence of global weak solutions of (3.4.26) and (3.4.27). We first make clear what we mean by a weak solution.

Definition 3.4.2 *Let $u_0 \in H^\alpha$, $|u_0| = 1$ a.e., we say that u is a weak solution of equation (3.4.26) if*
    (i) for all $T > 0$, $u \in L^\infty(0, T; H^\alpha(\Omega))$;
    (ii) for all $\varphi \in C^\infty(Q_T)$, there holds when $\lambda = 0$

$$\int_{Q_T} \frac{\partial u}{\partial t} \cdot \varphi = -\gamma \int_{Q_T} \Lambda^\alpha u \cdot \Lambda^\alpha (u \times \varphi) dxdt \qquad (3.4.28)$$

where $Q_T = (0, T) \times \Omega$; or when $\lambda > 0$

$$\int_{Q_T} \frac{\partial u}{\partial t} \cdot \varphi = \gamma \int_{Q_T} (u \times \Lambda^{2\alpha} u) \cdot \varphi dxdt - \int_{Q_T} \lambda(u \times \Lambda^{2\alpha} u) \cdot (u \times \varphi) dxdt. \quad (3.4.29)$$

When $\lambda > 0$, we will show below that $u \times \Lambda^{2\alpha} u$ makes sense in $L^2(Q_T)$, and for this reason, it will be clear that (3.4.29) makes sense.

Definition 3.4.3 *Let $u_0 \in H^\alpha$, $|u_0| = 1$ a.e., we say that u is a weak solution of equation (3.4.27) if*
    (i) for all $T > 0$, $u \in L^\infty(0, T; H^\alpha(\Omega))$, $\partial_t u \in L^2(0, +\infty; L^2(\Omega))$ and $|u| = 1$ a.e.;
    (ii) for all $\varphi \in C^\infty(Q_T)$, there holds

$$\int_{Q_T} \left(u \times \frac{\partial u}{\partial t} \right) \cdot \varphi = \lambda \int_{Q_T} \Lambda^\alpha u \cdot \Lambda^\alpha (u \times \varphi) dxdt; \qquad (3.4.30)$$

(iii) $u(0, x) = u_0(x)$ in the trace sense;
(iv) for all $T > 0$, there holds

$$\int_\Omega |\Lambda^\alpha u(T)|^2 dx + \frac{2}{\lambda} \int_{Q_T} \left| \frac{\partial u}{\partial t} \right|^2 dxdt \leqslant \int_\Omega |\Lambda^\alpha u_0|^2 dx.$$

This definition fully utilize the condition that u stays on the unit sphere as time evolves, and the readers are referred to Lemma 3.4.8 to see why we define a weak solution by the identity (3.4.30).

Consider the eigenvalue problem

$$
\begin{cases}
\Lambda^{2\alpha} u = \nu u, \\
\text{with periodic boundary conditions.}
\end{cases}
\tag{3.4.31}
$$

Since $\Lambda^{-2\alpha}$ is a compact self-adjoint operator in $L^2(\Omega)$, there exists a complete orthonormal family of $L^2(\Omega)$, $\{w_j\}_{j\in\mathbf{N}}$, made of eigenvectors of $\Lambda^{-2\alpha}$

$$
\Lambda^{-2\alpha} w_j = \mu_j w_j, \quad \forall j \in \mathbf{N},
$$

where the sequence μ_j is decreasing and tends to 0. It is clear that $w_j \in D(\Lambda^{2\alpha})$ for all $j \in \mathbf{N}$, and setting $\nu_j = \mu_j^{-1}$ we obtain

$$
\begin{cases}
\Lambda^{2\alpha} w_j = \nu_j w_j, \quad j = 1, 2 \cdots \\
0 < \nu_1 \leqslant \nu_2 \leqslant \cdots, \quad \nu_j \to \infty \ (\text{as } j \to \infty).
\end{cases}
$$

The family $\{w_j\}$ satisfies

$$
\begin{cases}
(w_j, w_k) = \delta_{jk}, \text{ the Kronecker symbol,} \\
\langle \Lambda^{2\alpha} w_j, w_k \rangle = \nu_j \delta_{jk}, \quad \forall j, k.
\end{cases}
$$

What follows is dedicated to constructing the global weak solutions for (3.4.26) via the Galerkin method. In particular, the following global existence theorem for the fractional Landau-Lifshitz-Gilbert equation (3.4.26) will be proved. We set $\gamma = 1$ in the following.

Theorem 3.4.5 *Let $\alpha \in (0, 1)$. Then for all $u_0 \in H^\alpha(\Omega)$, $|u_0| = 1$ a.e., there exists at least a weak solution for the fractional Landau-Lifshitz-Gilbert equation (3.4.26) such that*
(i) when $\lambda = 0$,

$$
u(x, t) \in L^\infty(0, T; H^\alpha(\Omega)) \bigcap C^{0, \frac{\alpha}{\alpha+s}}(0, T; L^2(\Omega))
$$

where $s > \alpha + \dfrac{d}{2}$;
(ii) when $\lambda > 0$,

$$
u(x, t) \in L^\infty(0, T; H^\alpha(\Omega)) \bigcap C^{0, \frac{r-1}{r}}(0, T; L^r(\Omega)),
$$

where $r < 2$ and $1 \leqslant r \leqslant r^* = \dfrac{d}{d - \alpha}$, or $r = 2$, $d = 1$, and $\alpha > 1/2$.

The proof of this theorem is divided into two major parts. First, we show some a priori estimates, and then some compactness arguments which insure the existence of global weak solutions.

Let $\{w_n(x)\}_{n=1}^{\infty}$ be the normalized eigenfunctions of (3.4.31), and let $\lambda_1, \lambda_2, \cdots$ be the corresponding eigenvalues. Then $\{w_n\}$ are smooth over Ω and form a basis of $H^{\alpha}(\Omega)$. Define the orthogonal projection

$$\mathscr{P}_N : H^{\alpha}(\Omega) \to \mathcal{S}_N := \operatorname{span}\{w_1, w_2, \cdots, w_N\} \subset H^{\alpha}(\Omega).$$

We look for approximate solutions $\{u_N(t, x)\}$ for equation (3.4.26) under the form

$$u_N(t, x) = \sum_{s=1}^{N} \varphi_{sN}(t) w_s(x),$$

where φ_{sN} are three dimensional vector valued functions and are chosen such that

$$\int_{\Omega} \frac{\partial u_N}{\partial t} \cdot w_s - \int_{\Omega} u_N \times \Lambda^{2\alpha} u_N \cdot w_s - \lambda \int_{\Omega} u_N \times (u_N \times \Lambda^{2\alpha} u_N) \cdot w_s = 0, \quad 1 \leqslant s \leqslant N \tag{3.4.32}$$

with the initial conditions

$$\int_{\Omega} u_N(x, 0) \cdot w_s(x) = \int_{\Omega} u_0(x) \cdot w_s(x), \quad 1 \leqslant s \leqslant N. \tag{3.4.33}$$

The local in time existence of solutions

$$(\varphi_{sN}^1, \varphi_{sN}^2, \varphi_{sN}^3), \quad 1 \leqslant s \leqslant N$$

to problem (3.4.32)-(3.4.33) follows from the standard Picard's theorem, which can be found in a standard ODE textbook. In order to take the limit $N \to \infty$, we need to make sure that all the functions φ_{sN} are defined at least in a common interval $[0, T]$, and this is a consequence of the following lemmas.

Lemma 3.4.4 *Let $u_0 \in H^{\alpha}(\Omega)$. Then for any $0 < T < \infty$, for the solutions u_N to the approximating system (3.4.32)-(3.4.33), there holds the estimate*

$$\sup_{0 \leqslant t \leqslant T} \|u_N\|_{H^{\alpha}}^2 + \lambda \int_0^T \|u_N \times \Lambda^{2\alpha} u_N\|_{L^2}^2 dt \leqslant K_1. \tag{3.4.34}$$

If $p < \infty$ and $2 \leqslant p \leqslant p^* = \dfrac{2d}{d - 2\alpha}$, or $p = \infty, d = 1$ and $\alpha > \dfrac{1}{2}$, then there holds

$$\sup_{0 \leqslant t \leqslant T} \|u_N\|_{L^p}^2 \leqslant CK_1.$$

Furthermore, for $r < 2$ and $1 \leqslant r \leqslant r^* = \dfrac{d}{d - \alpha}$, or $r = 2, d = 1$ and $\alpha > \dfrac{1}{2}$, there holds

$$\|u_N \times \Lambda^{\alpha} u_N\|_{L^r(\Omega)} \leqslant CK_2.$$

In particular, the constants C, K_1 and K_2 are independent of N.

Proof Multiplying the equation (3.4.32) with φ_{sN} and summing up the resulting product for $s = 1, 2, \cdots, N$, we have

$$\frac{\mathrm{d}}{\mathrm{d}t} \int_\Omega |u_N(x, t)|^2 \mathrm{d}x = 0.$$

Then we have

$$\|u_N\|^2_{L^2(\Omega)} \leqslant \|\mathscr{P}u_0\|^2_{L^2(\Omega)} \leqslant K_0, \tag{3.4.35}$$

where K_0 depends only on the initial data $\|u_0\|_{L^2(\Omega)}$ and is independent of N. Multiplying the equation (3.4.32) with $\nu_s \varphi_{sN}$ and summing up to N, we have

$$\int_\Omega \frac{\partial u_N}{\partial t} \cdot \Lambda^{2\alpha} u_N - \lambda \int_\Omega u_N \times (u_N \times \Lambda^{2\alpha} u_N) \cdot \Lambda^{2\alpha} u_N = 0,$$

i.e.

$$\frac{1}{2} \frac{\mathrm{d}}{\mathrm{d}t} \int_\Omega |\Lambda^\alpha u_N|^2 + \lambda \int_\Omega |u_N \times \Lambda^{2\alpha} u_N|^2 = 0.$$

Integrating over $[0, T]$, we have

$$\sup_{0 \leqslant t \leqslant T} \|\Lambda^\alpha u_N(t)\|^2_{L^2(\Omega)} + \lambda \|u_N \times \Lambda^{2\alpha} u_N\|^2_{L^2(0, T; L^2(\Omega))} \leqslant K_1, \tag{3.4.36}$$

where the constant K_1 depends only on the initial data $\|\Lambda^\alpha u_0\|_{L^2(\Omega)}$. By Sobolev embedding, we have for any $p < \infty$ such that $2 \leqslant p \leqslant p^* = \dfrac{2d}{d - 2\alpha}$, there holds

$$\sup_{0 \leqslant t \leqslant T} \|u_N(t)\|_{L^p(\Omega)} \leqslant CK_1.$$

In particular, when $d = 1$, there holds for any $\alpha > 1/2$ that

$$\sup_{0 \leqslant t \leqslant T} \|u_N(t)\|_{L^\infty(\Omega)} \leqslant CK_1. \tag{3.4.37}$$

Finally, by Hölder inequality, for $r < 2$ such that $1 \leqslant r \leqslant r^*$ there holds

$$\left(\int_\Omega |u_N \times \Lambda^\alpha u_N|^r \mathrm{d}x \right)^{1/r} \leqslant \left(\int_\Omega |\Lambda^\alpha u_N|^2 \mathrm{d}x \right)^{1/2} \left(\int_\Omega |u_N|^{\frac{2r}{2-r}} \mathrm{d}x \right)^{\frac{2-r}{2r}}, \tag{3.4.38}$$

where $r^* = \dfrac{d}{d - \alpha}$. In this case, since $\dfrac{2r}{2 - r} \leqslant p^*$ for $r \leqslant r^*$, we have

$$\{u_N \times \Lambda^\alpha u_N\}_{N \geqslant 1} \text{ are uniformly bounded in } L^r(\Omega). \tag{3.4.39}$$

On the other hand, when $r = 2$, from the Hölder inequality and inequality (3.4.37) there holds

$$\left(\int_\Omega |u_N \times \Lambda^\alpha u_N|^2 \mathrm{d}x \right)^{1/2} \leqslant |u_N|_{L^\infty(\Omega)} \left(\int_\Omega |\Lambda^\alpha u_N|^2 \mathrm{d}x \right)^{1/2},$$

for $d = 1$ and $\alpha > 1/2$. In particular, all these constants are independent of N.

Lemma 3.4.5 *Let u_N be solutions for (3.4.32). Then under the conditions of Lemma 3.4.4, there hold the following estimates*
  (i) when $\lambda = 0$,

$$\sup_{0 \leqslant t \leqslant T} \left\| \frac{\partial u_N}{\partial t} \right\|_{H^{-s}(\Omega)} \leqslant K_2, \quad \forall s > \alpha + \frac{d}{2}; \tag{3.4.40}$$

  (ii) when $\lambda > 0$, for r as in Lemma 3.4.4,

$$\left\| \frac{\partial u_N}{\partial t} \right\|_{L^r(Q_T)} \leqslant K_3, \tag{3.4.41}$$

where the constants K_2, K_3 are independent of N.

Proof Consider the case $\lambda = 0$. For any three dimensional vector $\varphi \in H^s(\Omega)$, φ can be represented as

$$\varphi = \varphi_N + \bar{\varphi}_N,$$

where

$$\varphi_N(x) = \sum_{s=1}^{N} \beta_s w_s(x), \quad \bar{\varphi}_N(x) = \sum_{s=N+1}^{\infty} \beta_s w_s(s).$$

Then by Lemma 3.4.4, we have

$$\int_\Omega \frac{\partial u_N}{\partial t} \varphi = \int_\Omega \frac{\partial u_N}{\partial t} \varphi_N = \int_\Omega u_N \times \Lambda^{2\alpha} u_N \cdot \varphi_N$$

$$= -\int_\Omega \Lambda^\alpha u_N \cdot \Lambda^\alpha (u_N \times \varphi_N).$$

Using the calculus inequality in Theorem 2.2.14, for $\frac{1}{2} = \frac{1}{p} + \frac{1}{q}, q < \infty$ we have

$$\left| \int_\Omega \frac{\partial u_N}{\partial t} \varphi \right| \leqslant \|\Lambda^\alpha u_N\|_{L^2(\Omega)} \|\Lambda^\alpha (u_N \times \varphi_N)\|_{L^2(\Omega)}$$

$$\leqslant \|\Lambda^\alpha u_N\|_{L^2(\Omega)} \{\|u_N\|_{L^p} \|\varphi_N\|_{\dot{H}^{\alpha,q}} + \|u_N\|_{\dot{H}^{\alpha,2}} \|\varphi_N\|_\infty\}.$$

Taking $2 < p < p^*$, using Sobolev embedding, we have

$$\|u_N\|_{L^p} \leqslant \|u_N\|_{H^\alpha}.$$

Therefore

$$\left| \int_\Omega \frac{\partial u_N}{\partial t} \varphi \right| \leqslant \|u_N\|_{\dot{H}^{\alpha,2}}^2 \|\varphi_N\|_{H^{s,2}}$$

for any $s > \alpha + \frac{d}{2}$ and (3.4.40) is proved.

Consider the case $\lambda > 0$. We have from (3.4.26) that

$$\left|\int_{Q_T} \frac{\partial u_N}{\partial t}\varphi\right| \leqslant \left|\int_{Q_T} (u_N \times \Lambda^{2\alpha} u_N) \cdot \varphi\right|$$
$$+ \lambda \left|\int_{Q_T} [u_N \times (u_N \times \Lambda^{2\alpha} u_N)] \cdot \varphi\right|$$
$$\leqslant \|u_N \times \Lambda^{2\alpha} u_N\|_{L^2(Q_T)} \|\varphi\|_{L^2(Q_T)}$$
$$+ \lambda \|u_N \times \Lambda^{2\alpha} u_N\|_{L^2(Q_T)} \|u_N\|_{L^p(Q_T)} \|\varphi\|_{L^q(Q_T)}$$
$$\leqslant K_3 \|\varphi\|_{L^q(Q_T)}$$

for $\dfrac{1}{2} = \dfrac{1}{p} + \dfrac{1}{q}$. Let p and r be as in Lemma 3.4.4, then we have $\|\partial u_N/\partial t\|_{L^r(Q_T)}$
$\leqslant K_3$, completing the proof.

Lemma 3.4.6 *Under the conditions in Lemma 3.4.4, for the solution $u_N(t,x)$
for (3.4.32) the following estimates hold:*

(i) when $\lambda = 0$, for $s > \alpha + \dfrac{d}{2}$

$$\|u_N(t_1) - u_N(t_2)\|_{L^2(\Omega)} \leqslant K_4 |t_1 - t_2|^{\frac{\alpha}{\alpha+s}};$$

(ii) when $\lambda > 0$, for r as in Lemma 3.4.4 and $r > 1$

$$\|u_N(t_1) - u_N(t_2)\|_{L^r(\Omega)} \leqslant K_4 |t_1 - t_2|^{\frac{r-1}{r}},$$

where the constant K_4 is independent of N.

Proof When $\lambda = 0$, by Sobolev embedding theorem and the interpolation
inequalities, we have from Lemma 3.4.5

$$\|u_N(t_1) - u_N(t_2)\|_{L^2(\Omega)} \leqslant C \|u_N(t_1) - u_N(t_2)\|_{H^{-s}(\Omega)}^{\frac{\alpha}{\alpha+s}} \|u_N(t_1) - u_N(t_2)\|_{H^\alpha(\Omega)}^{\frac{s}{\alpha+s}}$$
$$\leqslant C \left\|\int_{t_1}^{t_2} \frac{\partial u_N}{\partial t} dt\right\|_{H^{-s}(\Omega)}^{\frac{\alpha}{\alpha+s}}$$
$$\leqslant C \sup_{0\leqslant t\leqslant T} \left\|\frac{\partial u_N}{\partial t}\right\|_{H^{-s}(\Omega)}^{\frac{\alpha}{\alpha+s}} |t_2 - t_1|^{\frac{\alpha}{\alpha+s}}$$
$$\leqslant C |t_2 - t_1|^{\frac{\alpha}{\alpha+s}},$$

where in the last inequality we have used the estimate (3.4.40).

Consider $\lambda > 0$. Let $r > 1$ be as in Lemma 3.4.4, then from Young's inequality and Hölder inequality, we have

$$
\|u_N(t_1) - u_N(t_2)\|_{L^r(\Omega)} = \left\| \int_{t_1}^{t_2} \frac{\partial u_N}{\partial t} dt \right\|_{L^r(\Omega)}
$$

$$
\leqslant \int_{t_1}^{t_2} \left\| \frac{\partial u_N}{\partial t} \right\|_{L^r(\Omega)} dt
$$

$$
\leqslant |t_2 - t_1|^{\frac{r-1}{r}} \left(\int_{Q_T} \left| \frac{\partial u_N}{\partial t} \right|^r \right)^{1/r}
$$

$$
\leqslant C |t_2 - t_1|^{\frac{r-1}{r}},
$$

where in the last inequality we have used the estimate (3.4.41). The proof is complete.

Based on the above *a priori* estimates, we have the following.

Lemma 3.4.7 *Let N and T be arbitrarily fixed. Then under the conditions of Lemma 3.4.4, the initial value problem for the ordinary differential equations (3.4.32)-(3.4.33) has at least one continuous differentiable and global solution $\{\varphi_{sN}(t)\}$ for $s = 1, 2, \cdots, N$ and $t \in [0, T]$.*

In the following, we will take $N \to \infty$ to get a global weak solutions for the fractional Landau-Lifshitz-Gilbert equation. We first consider the convergence for the case of $\lambda = 0$.

It follows from these *a priori* estimates that $\{u_N(t, x)\}_{N \geqslant 1}$ is uniformly bounded in the space

$$
\mathbb{G}_0 = L^\infty(0, T; H^\alpha(\Omega)) \cap W^{1, \infty}(0, T; H^{-s}(\Omega)).
$$

Applying the compactness lemma, there exists some $u \in L^\infty(0, T; H^\alpha(\Omega))$ such that up to a subsequence

$$
u_{Nt} \rightharpoonup u_t, \qquad \text{weakly in } L^p(0, T; H^{-s}(\Omega)),
$$

$$
u_N \to u, \qquad \text{strongly in } L^p(0, T; H^\beta(\Omega)),
$$

where $1 < p < \infty$ and $-s < \beta < \alpha$. In particular, $u_N \to u$ strongly in $L^2(0, T; H^\beta(\Omega))$.

Then the convergence of the first term is obvious

$$
\int_{Q_T} \frac{\partial u_N}{\partial t} \cdot \varphi \to \int_{Q_T} \frac{\partial u}{\partial t} \cdot \varphi, \quad \forall \varphi \in C^\infty(Q_T). \tag{3.4.42}
$$

However the convergence for the nonlinear nonlocal term is not obvious at all, since we only have compactness in H^β for $\beta < \alpha$. In the following, we show

$$\int_{Q_T} \Lambda^\alpha u_N \cdot \Lambda^\alpha(u_N \times \varphi)\mathrm{d}x\mathrm{d}t \to \int_{Q_T} \Lambda^\alpha u \cdot \Lambda^\alpha(u \times \varphi)\mathrm{d}x\mathrm{d}t. \qquad (3.4.43)$$

As a first check, we show that the r.h.s. of (3.4.43) makes sense. Indeed, for any $u \in H^\alpha$, we have

$$\left| \int_\Omega \Lambda^\alpha u \cdot \Lambda^\alpha(u \times \varphi)\mathrm{d}x \right| \leqslant \|\Lambda^\alpha u\| \|\Lambda^\alpha(u \times \varphi)\|$$
$$\leqslant C\|\Lambda^\alpha u\|(\|\Lambda^\alpha u\|\|\varphi\|_\infty + \|u\|_{L^p}\|\Lambda^\alpha \varphi\|_{L^q}) \qquad (3.4.44)$$
$$\leqslant C\|\Lambda^\alpha u\|^2 \|\varphi\|_{H^s},$$

for $\dfrac{1}{p} + \dfrac{1}{q} = \dfrac{1}{p} + \dfrac{\alpha}{d} = \dfrac{1}{2}$ and $s > \max\left\{\alpha + \dfrac{d}{p}, \dfrac{d}{2}\right\}$.

Now the special structure of this equation plays an important role in the convergence. Indeed, let $\mathcal{C}_\varphi(u) = \Lambda^\alpha(u \times \varphi) - \Lambda^\alpha u \times \varphi$. Since

$$\Lambda^\alpha u \cdot (\Lambda^\alpha u \times \varphi) = 0,$$

it suffices to prove

$$\int_{Q_T} \Lambda^\alpha u_N \cdot \mathcal{C}_\varphi(u_N - u) + \int_{Q_T} \Lambda^\alpha(u_N - u) \cdot \mathcal{C}_\varphi(u) \to 0.$$

Applying the commutator estimate and Sobolev embedding, we have

$$\|\mathcal{C}_\varphi(u_N - u)\|_{L^2(\Omega)}$$
$$\leqslant c\left(\|\nabla\varphi\|_{L^{p_1}(\Omega)}\|u_N - u\|_{\dot{H}^{\alpha-1,p_2}(\Omega)} + \|\varphi\|_{\dot{H}^{\alpha,p_3}(\Omega)}\|u_N - u\|_{L^{p_4}(\Omega)}\right)$$
$$\leqslant c\left(\|\nabla\varphi\|_{L^{p_1}(\Omega)}\|u_N - u\|_{L^2(\Omega)} + \|\varphi\|_{\dot{H}^{\alpha,p_3}(\Omega)}\|u_N - u\|_{H^\beta(\Omega)}\right)$$
$$\leqslant C\|\varphi\|_{H^s(\Omega)}\|u_N - u\|_{H^\beta(\Omega)},$$

where $p_2, p_3 \in (1, +\infty)$ are such that

$$\frac{1}{2} = \frac{1}{p_1} + \frac{1}{p_2} \quad \text{and} \quad \frac{\alpha-1}{d} + \frac{1}{2} = \frac{1}{p_2},$$

$$\frac{1}{2} = \frac{1}{p_3} + \frac{1}{p_4} \quad \text{and} \quad \frac{\beta}{d} + \frac{1}{p_4} = \frac{1}{2}, \ 0 < \beta < \alpha,$$

and

$$s > \frac{d}{2} + 1.$$

Note that in this case, $s > \alpha + \dfrac{d}{2} - \dfrac{d}{p_3}$ holds automatically.

Again by Hölder inequality,

$$\left| \int_{Q_T} \Lambda^\alpha u_N \cdot \mathcal{C}_\varphi(u_N - u) \mathrm{d}x\mathrm{d}t \right|$$

$$\leqslant c\|\varphi\|_{H^s(\Omega)} \|\Lambda^\alpha u_N\|_{L^2(Q_T)} \|u_N - u\|_{L^2(0,T;H^\beta(\Omega))}$$

$$\rightarrow 0 \qquad \text{as } N \rightarrow \infty.$$

On the other hand, since $\mathcal{C}_\varphi(u) \in L^2(Q_T)$ and $u_N \rightarrow u$ in $L^2(0,T;H^\alpha(\Omega))$ weakly, the convergence

$$\int_{Q_T} \Lambda^\alpha(u_N - u) \cdot \mathcal{C}_\varphi(u) \rightarrow 0$$

is obvious and (3.4.43) is proved. Therefore, letting $N \rightarrow \infty$, we have from (3.4.42) and (3.4.44) that

$$\int_{Q_T} \frac{\partial u}{\partial t} \cdot \varphi \mathrm{d}x\mathrm{d}t = - \int_{Q_T} \Lambda^\alpha u \cdot \Lambda^\alpha(u \times \varphi) \mathrm{d}x\mathrm{d}t,$$

and the global existence of weak solutions for the fractional Heisenberg equation ($\lambda = 0$ in (3.4.26)) is proved.

Next we consider the convergence for the case of $\lambda > 0$. From the estimates established before, we have that $\{u_N\}_{N \geqslant 1}$ is uniformly bounded in

$$\mathbb{G}_\lambda = L^\infty(0,T;H^\alpha(\Omega)) \bigcap W^{1,r}(0,T;L^r(\Omega)),$$

for $r > 1$ as in Lemma 3.4.4. Therefore, from the compactness lemma, there exists some $u \in L^\infty(0,T;H^\alpha(\Omega))$ such that

$$u_N \rightharpoonup u \quad \text{weakly in } L^p(0,T;H^\alpha(\Omega)) \text{ for } 1 < p < \infty;$$

$$u_N \rightharpoonup u \quad \text{strongly in } L^p(0,T;H^\beta(\Omega)) \text{ for } 1 < p < \infty, \ 0 \leqslant \beta < \alpha;$$

$$u_N \rightharpoonup u \quad \text{weakly in } L^p(Q_T) \text{ for } 1 < p < \infty \text{ as in Lemma 3.4.4;}$$

$$\frac{\partial u_N}{\partial t} \rightharpoonup \frac{\partial u}{\partial t} \quad \text{weakly in } L^r(Q_T) \text{ for } r > 1 \text{ as in Lemma 3.4.4.}$$

$$(3.4.45)$$

The handling of the case $\lambda > 0$ is much trickier since it involves one more nonlinear nonlocal term of the highest order derivative and the usual (integral order) Leibniz formula does not hold anymore. Even worse, in this case, the former commutator estimates method cannot be applied to get the desired global existence result directly. Observing the fact that from (3.4.36)

$$\|u_N \times \Lambda^{2\alpha} u_N\|_{L^2(0,T;L^2(\Omega))}^2 \leqslant \frac{K_1}{\lambda},$$

there exists some element ζ in $L^2(0, T; L^2(\Omega))$ such that

$$u_N \times \Lambda^{2\alpha} u_N \rightharpoonup \zeta \qquad \text{weakly in} \quad L^2(0, T; L^2(\Omega)). \qquad (3.4.46)$$

In the following, we show

$$u \times \Lambda^{2\alpha} u = \zeta \in L^2(0, T; L^2(\Omega)). \qquad (3.4.47)$$

In fact, let $\varphi \in H^s(\Omega)$ for $s > \alpha + \dfrac{d}{2}$,

$$\int_{Q_T} u_N \times \Lambda^{2\alpha} u_N \cdot \varphi = -\int_{Q_T} \Lambda^\alpha u_N \cdot \Lambda^\alpha (u_N \times \varphi)$$

$$= -\int_{Q_T} \Lambda^\alpha u_N \cdot C_\varphi(u_N).$$

On the other hand, using commutator estimate together with the same reasonings that lead to (3.4.43), we have

$$\int_{Q_T} \Lambda^\alpha u_N \cdot C_\varphi(u_N) \to \int_{Q_T} \Lambda^\alpha u \cdot C_\varphi(u)$$

$$= \int_{Q_T} \Lambda^\alpha u \cdot \Lambda^\alpha (u \times \varphi)$$

$$= -\int_{Q_T} u \times \Lambda^{2\alpha} u \cdot \varphi,$$

and therefore (3.4.47) is proved. In particular, we have

$$u_N \times \Lambda^{2\alpha} u_N \rightharpoonup u \times \Lambda^{2\alpha} u \qquad \text{weakly in} \quad L^2(0, T; L^2(\Omega)).$$

From (3.4.46) and (3.4.47), we know that for any $\varphi \in C^\infty(Q_T)$

$$\int_{Q_T} (u_N \times \Lambda^{2\alpha} u_N) \cdot \varphi \mathrm{d}x \mathrm{d}t \to \int_{Q_T} (u \times \Lambda^{2\alpha} u) \cdot \varphi \mathrm{d}x \mathrm{d}t.$$

Furthermore, since $u_N \to u$ strongly in $L^2(Q_T)$, the following convergence also holds for any $\varphi \in C^\infty(Q_T)$

$$\int_{Q_T} (u_N \times \Lambda^{2\alpha} u_N) \cdot (u_N \times \varphi) \mathrm{d}x \mathrm{d}t \to \int_{Q_T} (u \times \Lambda^{2\alpha} u) \cdot (u \times \varphi) \mathrm{d}x \mathrm{d}t,$$

and the r.h.s. term makes sense. Then the global existence of weak solutions for the fractional Landau-Lifshitz equation (3.4.26) is proved in the sense of (3.4.29) in Definition 3.4.2.

Theorem 3.4.5 is now completely proved.

Next we prove the global existence of weak solutions for equation (3.4.27). We will prove

Theorem 3.4.6 *Let $\alpha \in (0, 1)$ be such that $\alpha > \dfrac{d}{4}$. Then for any $u_0 \in H^\alpha$, $|u_0| = 1$ a.e., there exists at least one weak solution for (3.4.27) in the sense of Definition 3.4.3.*

First we prove the following

Lemma 3.4.8 *A map $u : \Omega \times \mathbf{R}^+ \to \mathbb{S}^2$, with $\Lambda^\alpha u \in L^\infty(\mathbf{R}^+; L^2(\Omega))$ and $\partial_t u \in L^2(\Omega \times \mathbf{R}^+)$, is a weak solution of (3.4.27) if and only if*

$$u \times u_t = -\lambda u \times \Lambda^{2\alpha} u, \tag{3.4.48}$$

holds in the sense of Definition 3.4.3.

Proof If u weakly solves (3.4.27), then for any three dimensional vector $\phi \in C^\infty(\Omega)$, multiplying (3.4.27) with $(u \times \phi)$ and then integrating give

$$\int_\Omega u_t \cdot (u \times \phi) = \lambda \int_\Omega (u \cdot \Lambda^{2\alpha}) u \cdot (u \times \phi) - \lambda \int_\Omega \Lambda^{2\alpha} u \cdot (u \times \phi)$$

$$= -\lambda \int_\Omega \Lambda^{2\alpha} u \cdot (u \times \phi),$$

since $u \cdot (u \times \phi) = 0$. Note also that

$$\int_\Omega u_t \cdot (u \times \phi) = -\int_\Omega (u \times u_t) \cdot \phi.$$

Hence u weakly solves (3.4.48) in the sense of (3.4.30) in Definition 3.4.3.
Conversely, if u weakly solves equation (3.4.48), then we have

$$(\partial_t u + \lambda \Lambda^{2\alpha} u) \times u = 0.$$

Hence there exists a multiplier $m : \Omega \times \mathbf{R}^+ \to \mathbf{R}$ such that

$$\partial_t u + \lambda \Lambda^{2\alpha} u = mu.$$

Multiplying it by $u\phi$ and using $\partial_t u \cdot u = 0$, we obtain that for any three dimensional vector $\phi \in C^\infty(\Omega)$,

$$\int_\Omega m\phi = \lambda \int_\Omega \Lambda^{2\alpha} u \cdot u\phi. \tag{3.4.49}$$

Therefore u weakly solves

$$\partial_t u + \lambda \Lambda^{2\alpha} u = \lambda (u \cdot \Lambda^{2\alpha} u) u,$$

i.e. u weakly solves (3.4.27).

Remark 3.4.4 *Strictly speaking, one needs to show that the integral on the right hand side of (3.4.49) makes sense for any $u \in H^\alpha(\Omega)$. Indeed, this is the case. For any $\phi \in C^\infty(\Omega)$, and for any u such that $u(\cdot, t) \in H^\alpha(\Omega)$,*

$$\left| \int_\Omega \Lambda^{2\alpha} u \cdot (u\phi) \right| = \left| \int_\Omega \Lambda^\alpha u \cdot \Lambda^\alpha(u\phi) \right|$$
$$\leqslant \|\Lambda^\alpha u\| \|\Lambda^\alpha(u\phi)\|,$$

while

$$\|\Lambda^\alpha(u\phi)\| \leqslant C(\|\Lambda^\alpha u\|_{L^2} |\phi|_\infty + |u|_\infty \|\Lambda^\alpha \phi\|_{L^2}),$$

thanks to the calculus inequality in Lemma 3.4.4. On the other hand, this lemma explains why we define a weak solutions for (3.4.27) as in Definition 3.4.3. This observation is crucial for us to get the convergence in the following proofs.

In the following, we consider the Ginzburg-Landau approximation. For $k \geqslant 1$ integer, consider the problem for maps $u^\varepsilon : \Omega \times \mathbf{R}^+ \to \mathbf{R}^3$:

$$\partial_t u^\varepsilon + \lambda \Lambda^{2\alpha} u^\varepsilon = \frac{\lambda}{\varepsilon^2}(1 - |u^\varepsilon|^2)u^\varepsilon. \tag{3.4.50}$$

We seek approximate solutions $\{u_n(t, x)\}$ for equation (3.4.50) under the form

$$u_n(t, x) = \sum_{i=1}^n \varphi_i(t) w_i(x),$$

where φ_i are $\mathbf{R}^3$-valued vectors, such that for $1 \leqslant i \leqslant n$ there holds

$$\int_\Omega \frac{\partial u_n}{\partial t} w_i + \lambda \int_\Omega \Lambda^{2\alpha} u_n w_i + \frac{\lambda}{\varepsilon^2} \int_\Omega (|u_n|^2 - 1)u_n w_i = 0, \tag{3.4.51}$$

with the initial conditions

$$\int_\Omega u_n(0)w_i = \int_\Omega u_0 w_i.$$

This is a system of ODEs for φ_i's and from standard ODEs theory, one can easily show the existence of local solutions. Now, we show some *a priori* estimates. Multiplying the equality (3.4.51) by $\frac{d\varphi_i}{dt}$ and summing over $1 \leqslant i \leqslant n$ lead to

$$\int_\Omega \left| \frac{\partial u_n}{\partial t} \right|^2 dx + \frac{\lambda}{2} \frac{d}{dt} \int_\Omega |\Lambda^\alpha u_n|^2 dx + \frac{\lambda}{4\varepsilon^2} \frac{d}{dt} \int_\Omega (|u_n|^2 - 1)^2 dx = 0,$$

from which, after integration in time, we have

$$\frac{1}{2}\int_\Omega |\Lambda^\alpha u_n(t)|^2 dx + \frac{1}{4\varepsilon^2}\int_\Omega (|u_n|^2 - 1)^2(t)dx + \frac{1}{\lambda}\int_{Q_T}\left|\frac{\partial u_n}{\partial t}\right|^2 dxdt$$

$$=\frac{1}{2}\int_\Omega |\Lambda^\alpha u_{n0}|^2 dx + \frac{1}{4\varepsilon^2}\int_\Omega (|u_{n0}|^2 - 1)^2 dx, \qquad \text{for all } t \in [0, T].$$

$$(3.4.52)$$

Since the initial data $u_0 \in H^\alpha$ which is embedded into $L^4(\Omega)$ for $\alpha \geqslant \frac{d}{4}$, the right hand side is uniformly bounded. Furthermore by Young's inequality

$$\int_\Omega |u_n|^2 \leqslant C + \frac{1}{2}\int_\Omega (|u_n|^2 - 1)^2,$$

we know that $\{u_n\}$ is bounded in $L^\infty(0, T; L^2(\Omega))$. Then we deduce that

$$\{u_n\} \text{ is bounded in } L^\infty(0, T; H^\alpha(\Omega));$$

$$\left\{\frac{\partial u_n}{\partial t}\right\} \text{ is bounded in } L^2(0, T; L^2(\Omega));$$

$$\{|u_n|^2 - 1\} \text{ is bounded in } L^2(0, T; L^2(\Omega)).$$

These estimates imply that the solution can be extended to all time, and we can extract a subsequence (still denoted as $\{u_n(t)\}$) such that

$$u_n \rightarrow u^\varepsilon \qquad \text{weakly in } L^2(0, T; H^\alpha(\Omega)),$$

$$\frac{\partial u_n}{\partial t} \rightarrow \frac{\partial u^\varepsilon}{\partial t} \qquad \text{weakly in } L^2(Q_T),$$

$$u_n \rightarrow u^\varepsilon \qquad \text{strongly in } L^2(0, T; H^\beta(\Omega)) \text{ for } 0 \leqslant \beta < \alpha \text{ and a.e.,}$$

$$|u_n|^2 - 1 \rightarrow \chi \qquad \text{weakly in } L^2(Q_T).$$

From [135, Lemma 1.3, Ch1], it is easy to show that $\chi = |u^\varepsilon|^2 - 1$. Passing to the limit ($n \rightarrow \infty$), we find a global weak solution u^ε to the approximate equation (3.4.50), i.e. for any $\tilde{\varphi} \in L^2(0, T; H^\alpha(\Omega))$, there holds

$$\int_{Q_T}\frac{\partial u^\varepsilon}{\partial t} \cdot \tilde{\varphi} dxdt + \lambda\int_{Q_T}\Lambda^\alpha u^\varepsilon \cdot \Lambda^\alpha \tilde{\varphi} dxdt$$

$$+ \frac{\lambda}{\varepsilon^2}\int_{Q_T}(|u^\varepsilon|^2 - 1)u^\varepsilon \cdot \tilde{\varphi} dxdt = 0.$$

$$(3.4.53)$$

Furthermore, applying Fatou lemma to (3.4.52) we have the estimates for u^ε

$$\frac{1}{2}\int_\Omega |\Lambda^\alpha u^\varepsilon(t)|^2 dx + \frac{1}{4\varepsilon^2}\int_\Omega (|u^\varepsilon|^2 - 1)^2(t)dx$$

$$+ \frac{1}{\lambda}\int_{Q_T}\left|\frac{\partial u^\varepsilon}{\partial t}\right|^2 dxdt \leqslant \frac{1}{2}\int_\Omega |\Lambda^\alpha u_0|^2 dx.$$

$$(3.4.54)$$

From (3.4.54), we have

$$\{u^\varepsilon\} \text{ is bounded in } L^2(0,T;H^\alpha(\Omega)),$$

$$\left\{\frac{\partial u^\varepsilon}{\partial t}\right\} \text{ is bounded in } L^2(Q_T),$$

$$\text{and} \quad |u^\varepsilon|^2 - 1 \to 0 \text{ in } L^2(Q_T) \text{ as } \varepsilon \to 0,$$

and therefore, up to a subsequence, we have

$$
\begin{aligned}
u^\varepsilon \to u & \qquad \text{weakly in } L^2(0,T;H^\alpha(\Omega)), \\
\frac{\partial u^\varepsilon}{\partial t} \to \frac{\partial u}{\partial t} & \qquad \text{weakly in } L^2(Q_T), \\
u^\varepsilon \to u & \qquad \text{strongly in } L^2(0,T;H^\beta(\Omega)) \text{ for } 0 \leqslant \beta < \alpha \text{ and a.e.}, \\
|u^\varepsilon|^2 - 1 \to 0 & \qquad \text{strongly in } L^2(Q_T) \text{ and a.e..}
\end{aligned}
$$

$$(3.4.55)$$

Let $\tilde\varphi = (u^\varepsilon \times \varphi)$ in (3.4.53), we have

$$\int_{Q_T} \frac{\partial u^\varepsilon}{\partial t} \cdot (u^\varepsilon \times \varphi)\mathrm{d}x\mathrm{d}t + \lambda \int_{Q_T} \Lambda^\alpha u^\varepsilon \cdot \Lambda^\alpha (u^\varepsilon \times \varphi)\mathrm{d}x\mathrm{d}t$$

$$+ \frac{\lambda}{\varepsilon^2} \int_{Q_T} (|u^\varepsilon|^2 - 1)u^\varepsilon \cdot (u^\varepsilon \times \varphi)\mathrm{d}x\mathrm{d}t = 0. \quad (3.4.56)$$

The third term on the left is zero since $a \cdot (a \times b) = 0$ and by (3.4.55),

$$\int_{Q_T} \frac{\partial u^\varepsilon}{\partial t} \cdot (u^\varepsilon \times \varphi)\mathrm{d}x\mathrm{d}t \to \int_{Q_T} \frac{\partial u}{\partial t} \cdot (u \times \varphi)\mathrm{d}x\mathrm{d}t.$$

Finally, following exactly the same steps that lead to (3.4.43), we have

$$\int_{Q_T} \Lambda^\alpha u^\varepsilon \cdot \Lambda^\alpha (u^\varepsilon \times \varphi)\mathrm{d}x\mathrm{d}t \to \int_{Q_T} \Lambda^\alpha u \cdot \Lambda^\alpha (u \times \varphi)\mathrm{d}x\mathrm{d}t,$$

as $\varepsilon \to 0$, where the r.h.s. makes sense following the same remarks after Lemma 3.4.8.

Taking $\varepsilon \to 0$ in (3.4.56), we have

$$\int_{Q_T} \frac{\partial u}{\partial t} \cdot (u \times \varphi)\mathrm{d}x\mathrm{d}t + \lambda \int_{Q_T} \Lambda^\alpha u \cdot \Lambda^\alpha (u \times \varphi)\mathrm{d}x\mathrm{d}t = 0.$$

This is exactly the expression in (3.4.30), and we finish the proof of Theorem 3.4.6.

3.4.4 Local well-posedness

In what follows, we consider the local smooth solutions for the fractional Landau-Lifshitz equation (3.4.10) for $\nu = 1$ and $\mu = 0$ on the d-dimensional torus with $d \leqslant 3$:

$$\partial_t u = u \times \Lambda^{2\alpha} u \tag{3.4.57}$$

with initial data

$$u(\cdot, 0) = u_0(\cdot). \tag{3.4.58}$$

The approach is based on the vanishing viscosity method and Kato's method on local in time existence for quasi-linear equations [121]. We first consider the approximate system

$$u_t = \varepsilon \Delta u + u \times \Lambda^{2\alpha} u, \tag{3.4.59}$$

with smooth initial data $u(x, 0) = u_0$. We will show that the viscous equation (3.4.59) has a unique global classical solution. We first show [185]

Theorem 3.4.7 *Let $d \leqslant 3$, $\varepsilon > 0$ be fixed and $0 < \alpha \leqslant 1/2$. Assume that $u_0 \in H^{4+\alpha}$. Then there exists a $T > 0$ depending only on the initial data u_0, such that (3.4.59) possesses a unique solution*

$$u \in \mathcal{C}([0, T]; H^{4+\alpha}) \cap \mathcal{C}^1([0, T]; H^\alpha).$$

Proof The proof is based on Kato's method for evolutionary equations, see [121, Sect. 7]. For that purpose, let $X = H^\alpha$, $Y = H^{4+\alpha}$ and $S = (I - \Delta)^2$. We choose W as the ball in Y with center 0 and radius R and define the operator

$$A(y) \cdot = -\varepsilon \Delta \cdot - y \times \Lambda^{2\alpha} \cdot, \quad \text{for } y \in W.$$

It is known that X and Y are both reflexive Banach spaces with $Y \hookrightarrow X$ continuously and densely and $S = (I - \Delta)^2$ is an isomorphism of Y onto X. Finally, we denote by $G(X, M, \beta)$ the set of all linear operators $\mathcal{L}$ in X such that $-\mathcal{L}$ generates a C_0 semigroup $\{e^{-t\mathcal{L}}\}$ with $\|e^{-t\mathcal{L}}\| \leqslant Me^{\beta t}$ for $0 \leqslant t < \infty$.

We check the following properties one by one.

(A1) *$A(\cdot)$ is a function of W into $G(X, 1, \beta)$, where W is an open ball in Y and β is a real number.*

For $y \in Y$, we will check that if u is a solution of

$$u_t = \varepsilon \Delta u + y \times \Lambda^{2\alpha} u, \tag{3.4.60}$$

then there exists some constants $\beta > 0$ such that

$$\|u\|_{H^\alpha} \leqslant e^{\beta t}\|u_0\|_{H^\alpha}. \tag{3.4.61}$$

Multiplying (3.4.60) with u and integration by parts, we have

$$\begin{aligned}
\frac{1}{2}\frac{d}{dt}\|u\|_{L^2}^2 + \varepsilon\|\nabla u\|_{L^2}^2 &= (\Lambda^\alpha u, \Lambda^\alpha(y \times u)) \\
&\leqslant \|\Lambda^\alpha u\|_{L^2}^2 + \|\Lambda^\alpha(y \times u)\|_{L^2}^2 \\
&\leqslant \|\Lambda^\alpha u\|_{L^2}^2 + C(\|\Lambda^\alpha u\|_{L^2}^2\|y\|_{L^\infty}^2 + \|\Lambda^\alpha y\|_{L^p}^2\|u\|_{L^q}^2) \\
&\leqslant C\|u\|_{H^\alpha}^2,
\end{aligned} \tag{3.4.62}$$

where $\dfrac{1}{p} + \dfrac{1}{q} = \dfrac{1}{2}$, and $q \leqslant \dfrac{2d}{d - 2\alpha}$ for $d > 2\alpha$ and $1 < q < \infty$ for $d = 1$ and $\alpha = 1/2$. Similarly, multiplying the equation with $\Lambda^{2\alpha}u$ and integration by parts, we have

$$\frac{1}{2}\frac{d}{dt}\|\Lambda^\alpha u\|_{L^2}^2 + \varepsilon\|\nabla\Lambda^\alpha u\|_{L^2}^2 \leqslant 0. \tag{3.4.63}$$

Adding together (3.4.62) and (3.4.63), we have

$$\frac{d}{dt}\|u\|_{H^\alpha}^2 + 2\varepsilon\|\nabla u\|_{H^\alpha}^2 \leqslant C\|u\|_{H^\alpha}^2,$$

from which we know that

$$\|u(t)\|_{H^\alpha} \leqslant e^{Ct}\|u_0\|_{H^\alpha}, \quad t \in [0, \infty), \ y \in W.$$

Let $\beta = C$, (3.4.61) is proved and

$$\|e^{-tA(y)}\|_{H^\alpha} \leqslant e^{\beta t}, \quad t \in [0, \infty), \ y \in W.$$

Therefore, $A(\cdot)$ maps W into $G(X, 1, \beta)$.

(A2) *For each $y \in W$, we have*

$$SAS^{-1} = A(y) + B(y),$$

and $B(y) \in \mathcal{L}(X, X)$ with $\|B(y)\|_X \leqslant K$ for some constant $K > 0$.

Indeed, by direct computation, we have for $y \in Y$ and $w \in X$,

$$\begin{aligned}
B(y)w &= [S, y \times \Lambda^{2\alpha}]S^{-1}w \\
&= -2\Delta y \times \Lambda^{2\alpha}S^{-1}w - 4\nabla \times \Lambda^{2\alpha}\nabla S^{-1}w + \Delta^2 y \times \Lambda^{2\alpha}S^{-1}w \\
&\quad + 4\nabla^3 y \times \Lambda^{2\alpha}\nabla S^{-1}w + 6\Delta y \times \Lambda^{2\alpha}\Delta S^{-1}w + 4\nabla y \times \Lambda^{2\alpha}\nabla^3 S^{-1}w \\
&=: I + II + III + IV + V + VI.
\end{aligned}$$

For I, we have since Λ^α and S commutes,

$$\|\mathcal{J}^\alpha I\|_{L^2} \leqslant C(\|\mathcal{J}^\alpha \Delta y\|_{L^p}\|\Lambda^{2\alpha} S^{-1} w\|_{L^q}$$
$$+ \|\Delta y\|_{L^\infty}\|\mathcal{J}^\alpha \Lambda^\alpha S^{-1}\Lambda^\alpha w\|_{L^2})$$
$$\leqslant C\|y\|_{H^{4+\alpha}}\|w\|_{H^\alpha},$$

for any $2 < p < \infty$ with $\dfrac{1}{p} + \dfrac{1}{q} = \dfrac{1}{2}$.

The treatment for II is similar, and we have

$$\|\mathcal{J}^\alpha II\|_{L^2} \leqslant C(\|\mathcal{J}^\alpha \boldsymbol{\nabla} y\|_{L^p}\|\Lambda^\alpha S^{-1}\boldsymbol{\nabla}\Lambda^\alpha w\|_{L^q}$$
$$+ \|\boldsymbol{\nabla} y\|_{L^\infty}\|\mathcal{J}^\alpha \Lambda^\alpha S^{-1}\boldsymbol{\nabla}\Lambda^\alpha w\|_{L^2})$$
$$\leqslant C\|y\|_{H^{4+\alpha}}\|w\|_{H^\alpha}.$$

For III, we have by Sobolev embedding

$$\|\mathcal{J}^\alpha III\| \leqslant C(\|\mathcal{J}^\alpha \Delta^2 y\|_{L^2}\|\Lambda^\alpha S^{-1}\Lambda^\alpha w\|_{L^\infty}$$
$$+ \|\Delta^2 y\|_{L^p}\|\mathcal{J}^\alpha \Lambda^\alpha S^{-1}\Lambda^\alpha w\|_{L^q})$$
$$\leqslant C\|y\|_{H^{4+\alpha}}\|w\|_{H^\alpha},$$

for any $p \in \left(2, \dfrac{2d}{d-2\alpha}\right)$ and q with $\dfrac{1}{p} + \dfrac{1}{q} = \dfrac{1}{2}$.

For IV, we have

$$\|\mathcal{J}^\alpha IV\| \leqslant C(\|\Lambda^\alpha \boldsymbol{\nabla}^3 y\|_{L^{p_1}}\|\mathcal{J}^\alpha \boldsymbol{\nabla} S^{-1}\Lambda^\alpha w\|_{L^{q_1}}$$
$$+ \|\boldsymbol{\nabla}^3 y\|_{L^{p_2}}\|\mathcal{J}^\alpha \Lambda^\alpha \boldsymbol{\nabla} S^{-1}\Lambda^\alpha w\|_{L^{q_2}})$$
$$\leqslant C\|y\|_{H^{4+\alpha}}\|w\|_{H^\alpha},$$

where we choose $p_1 \in \left(2, \dfrac{2d}{d-2}\right)$ and q_1 with $\dfrac{1}{p_1} + \dfrac{1}{q_1} = \dfrac{1}{2}$ and $p_2 \in$ $\left(2, \dfrac{2d}{d-2(1+\alpha)}\right)$ and q_2 with $\dfrac{1}{p_1} + \dfrac{1}{q_1} = \dfrac{1}{2}$.

For V, we have

$$\|\mathcal{J}^\alpha V\| \leqslant C(\|\mathcal{J}^\alpha \Delta y\|_{L^p}\|\Lambda^\alpha \Delta S^{-1}\Lambda^\alpha w\|_{L^q}$$
$$+ \|\Delta y\|_{L^\infty}\|\mathcal{J}^\alpha \Lambda^\alpha \Delta S^{-1}\Lambda^\alpha w\|_{L^2})$$
$$\leqslant C\|y\|_{H^{4+\alpha}}\|w\|_{H^\alpha},$$

for all $p, q \in (2, \infty)$ such that $\dfrac{1}{p} + \dfrac{1}{q} = \dfrac{1}{2}$.

The term VI can be handled in the following way,

$$\|\mathcal{J}^\alpha VI\| \leqslant C(\|\mathcal{J}^\alpha \boldsymbol{\nabla} y\|_{L^{p_1}}\|\Lambda^\alpha \boldsymbol{\nabla}^3 S^{-1}\Lambda^\alpha w\|_{L^{q_1}}$$
$$+ \|\boldsymbol{\nabla} y\|_{L^{p_2}}\|\mathcal{J}^\alpha \Lambda^\alpha \boldsymbol{\nabla}^3 S^{-1}\Lambda^\alpha w\|_{L^{q_2}})$$
$$\leqslant C\|y\|_{H^{4+\alpha}}\|w\|_{H^\alpha},$$

for any $q_1 \in \left(2, \dfrac{2d}{d-2(1-\alpha)}\right)$ for $\alpha \leqslant \dfrac{1}{2}$, and $\dfrac{1}{p_1} + \dfrac{1}{q_1} = \dfrac{1}{2}$ and for $p_2 \in$

$(2, \infty]$ and $q_2 \in [2, \infty)$ such that $\dfrac{1}{2} - \dfrac{1-2\alpha}{d} \leqslant \dfrac{1}{q_2}$ and $\dfrac{1}{p_2} + \dfrac{1}{q_2} = \dfrac{1}{2}$. Note

here $q_2 \geqslant 2$, we require $\alpha \leqslant \dfrac{1}{2}$ in the last inequality.

Therefore, we have shown that

$$\|\mathscr{J}^\alpha B(y)w\| \leqslant C\|y\|_{H^{4+\alpha}}\|w\|_{H^\alpha}.$$

In other words, that $B(y)$ is a bounded operator form X to X. In particular, there exists some positive constant $K = C\|y\|_{H^{4+\alpha}}$ such that $\|B(y)\| \leqslant K$.

(A3) *For each $y \in W$, we have $A(y) \in \mathcal{L}(Y, X)$ and the function $y \to A(y)$ is Lipshitz continuous.*

Indeed, since $y \in H^{4+\alpha}$, we have for $w \in H^{4+\alpha}$

$$
\begin{aligned}
\|(A(y) - A(z))w\|_{H^\alpha} &\leqslant C\|\mathscr{J}^\alpha\{(y-z) \times \Lambda^{2\alpha}w\}\| \\
&\leqslant C(\|\mathscr{J}^\alpha(y-z)\|_{L^2}\|\Lambda^{2\alpha}w\|_{L^\infty} \\
&\quad + \|y-z\|_{L^p}\|\mathscr{J}^\alpha\Lambda^{2\alpha}w\|_{L^q}) \\
&\leqslant C\|y-z\|_{H^\alpha}\|w\|_{H^{4+\alpha}},
\end{aligned}
$$

for $p = \dfrac{2d}{d-2\alpha}$, and $\dfrac{1}{p} + \dfrac{1}{q} = \dfrac{1}{2}$. This implies that

$$\|A(y) - A(z)\|_{\mathcal{L}(Y,X)} \leqslant C\|y-z\|_X, \quad \forall y, z \in W.$$

(A4) *Let y_0 be the center of W, then $A(y)y_0 \in Y$ for all $y \in W$ with*

$$\|A(y)y_0\|_Y \leqslant K_2, \quad y \in W.$$

This is obvious since $y_0 = 0$ and $A(y)$ is linear.

Applying Kato's local existence theorem [121, Theorem 6], (A1)-(A4) imply the local existence of classical solutions for the viscous problem (3.4.59) with initial data Z_0, and we complete the proof of this theorem.

Remark 3.4.5 *In the proof, the choice of the fractional space $X = H^\alpha$ as our working space is vital. It seems difficulty to get the well-posedness for any other choice of working space such as H^s ($s \neq k + \alpha, k$ nonnegative integers) as we have tried.*

In the following two lemmas, we give some *a priori* estimates, which will lead to the global existence of smooth solutions for the approximate equation (3.4.60).

Lemma 3.4.9 *Suppose that $u_0 \in \mathbb{S}^2$ is smooth and u is a smooth solution to (3.4.59) on $[0, T]$ with initial data u_0, then*

$$\|u(t)\|_{L^p} \leqslant \|u_0\|_{L^p}, \quad \forall t \in [0, T] \text{ and } p \in [2, \infty].$$

Proof Multiplying the equation with $p|u|^{p-2}u$ for $p \geqslant 2$, integrating over Ω and integrating by parts, we have

$$\frac{\mathrm{d}}{\mathrm{d}t}\|u\|_{L^p}^p = \varepsilon p \int_\Omega |u|^{p-2}u \cdot \Delta u \mathrm{d}x$$

$$= -\varepsilon p(p-1) \int_\Omega |u|^{p-2}|\nabla u|^2 \mathrm{d}x \leqslant 0.$$

Therefore, one easily obtains that

$$\|u(t)\|_{L^p} \leqslant \|u_0\|_{L^p}, \quad \forall p \in [2, \infty).$$

Letting $p \to \infty$ then completes the proof.

Lemma 3.4.10 *Let $d \leqslant 3$, $\alpha \in (0, 1/2]$ and $\varepsilon > 0$ be fixed. Let $u_0 \in \mathbb{S}^2$ be smooth and u be a smooth solution to (3.4.59) on $[0, T]$ with initial data u_0, then there exists constant $C > 0$ such that*

$$\sup_{0 \leqslant t \leqslant T} \|u\|_{H^{m+\alpha}} \leqslant C\|u_0\|_{H^{m+\alpha}}, \tag{3.4.64}$$

for all nonnegative integers $m = 0, 1, 2, \cdots$.

Proof When $m = 0$, by multiplying the equation (3.4.67) by $\Lambda^{2\alpha}u$ and integrating over Ω, we can show that for any $T > 0$, there exists a constant $C > 0$ independent of T such that

$$\sup_{0 \leqslant t \leqslant T} \|u\|_{H^\alpha} + \varepsilon \int_0^T \|u\|_{H^{1+\alpha}}^2 \leqslant C. \tag{3.4.65}$$

When $m = 1$, multiplying the equation (3.4.59) with $\Delta\Lambda^{2\alpha}u$ and integration by parts, we have

$$\frac{\mathrm{d}}{\mathrm{d}t}\|\nabla\Lambda^\alpha u\|_{L^2}^2 + 2\varepsilon\|\Delta\Lambda^\alpha u\|_{L^2}^2 = 2(u \times \Lambda^{2\alpha}u, \Delta\Lambda^{2\alpha}u). \tag{3.4.66}$$

For the right hand side term, we have

$$(u \times \Lambda^{2\alpha}u, \Delta\Lambda^{2\alpha}u) = (\Lambda^\alpha(u \times \Lambda^{2\alpha}u), \Delta\Lambda^\alpha u)$$

$$\leqslant \delta\|\Delta\Lambda^\alpha u\|_{L^2}^2 + C\|\Lambda^\alpha(u \times \Lambda^{2\alpha}u)\|_{L^2}$$

$$\leqslant \delta\|\Delta\Lambda^\alpha u\|_{L^2}^2 + C\left(\|\Lambda^\alpha u\|_{L^{p_1}}^2\|\Lambda^{2\alpha}u\|_{L^{q_1}}^2 + \|u\|_{L^{p_2}}^2\|\Lambda^{3\alpha}u\|_{L^{q_2}}^2\right)$$

$$\leqslant \delta\|\Delta\Lambda^\alpha u\|_{L^2}^2 + C(\|u\|_{H^{1+\alpha}}^2\|u\|_{H^{1+\alpha}}^2 + \|u\|_{H^{1+\alpha}}^2\|u\|_{H^{1+\alpha}}^2), \tag{3.4.67}$$

where we have used the calculus inequality in Theorem 2.2.14 and Sobolev embedding $H^{1+\alpha} \hookrightarrow L^\infty$ for $d = 1, 2$ for $\alpha > 0$. The constants here p_1, q_1 satisfy

$$\begin{cases} \dfrac{1}{p_1} + \dfrac{1}{q_1} = \dfrac{1}{2}; \\ p_1 = \infty, \quad q_1 = 2, & \text{for } d = 1; \\ p_1 < \infty, \quad q_1 \leqslant \dfrac{2d}{d - 2(1 - \alpha)}, & \text{for } d = 2, 3, \end{cases}$$

and the constants p_2, q_2 satsify

$$\begin{cases} \dfrac{1}{p_2} + \dfrac{1}{q_2} = \dfrac{1}{2}; \\ p_2 = \infty, \quad q_2 = 2, & \text{for } d = 1, 2; \\ p_2 \leqslant \dfrac{6}{1 - 2\alpha}, q_2 < \infty, & \text{for } d = 3, 0 < \alpha < 1/2. \end{cases} \qquad (3.4.68)$$

From (3.4.66)-(3.4.67), and choosing δ sufficiently small such that $2\delta < \varepsilon$, then we have

$$\frac{\mathrm{d}}{\mathrm{d}t} \|\nabla \Lambda^\alpha u\|_{L^2}^2 + \varepsilon \|\Delta \Lambda^\alpha u\|_{L^2}^2 \leqslant C \|u\|_{H^{1+\alpha}}^2 \|u\|_{H^{1+\alpha}}^2,$$

which implies

$$\sup_{0 \leqslant t \leqslant T} \|u\|_{H^{1+\alpha}} + \varepsilon \int_0^T \|u\|_{H^{2+\alpha}}^2 \leqslant C, \qquad (3.4.69)$$

thanks to the Gronwall inequality and the integrability of $\|u\|_{H^{1+\alpha}}^2$ in (3.4.65).

The only case that fails here is when $d = 3$ and $\alpha = 1/2$ because of the failure of embedding $H^{3/2} \hookrightarrow L^\infty$ when $d = 3$. Indeed, in this case, since $\alpha = 1/2$, the index q_2 can be only chosen to be $q_2 = 2$ in (3.4.68), but p_2 can not be $p_2 = \infty$ when applying Theorem 2.2.14, hence contradicting the condition $\dfrac{1}{p_2} + \dfrac{1}{q_2} = \dfrac{1}{2}$.

When $d = 3$ and $\alpha = 1/2$, we start by bounding the term $\|u\|_{L^{p_2}}^2 \|\Lambda^{3\alpha} u\|_{L^{q_2}}^2$ in (3.4.67) using the result in Lemma 3.4.9. We choose $p_2 = \infty$ and $q_2 = 2$

$$\|u\|_{L^\infty}^2 \|\Lambda^{3/2} u\|_{L^2}^2 \leqslant \|u_0\|_{L^\infty}^2 \|u\|_{H^{3/2}}^2.$$

Repeating the Gronwall inequality and using the fact $u_0 \in \mathbb{S}^2$, we obtain the estimate (3.4.69) for $d = 3$ and $\alpha = 1/2$ and we complete the case $m = 1$.

Repeating the above arguments will lead to estimates (3.4.64) for $m = 2, \cdots$, and therefore complete the proof of Lemma 3.4.10.

Combining the local existence result and the global *a priori* estimates, we indeed show that the solution is globally smooth when u_0 is smooth. We state this fact in the following theorem.

Theorem 3.4.8 *Let $d \leqslant 3$, $\alpha \in (0, 1/2]$ and $\varepsilon > 0$ be fixed. Then there exists a unique global smooth solution $u^\varepsilon \in C([0, T]; H^{m+\alpha})$ of the approximate equation (3.4.59) for any initial data $u_0 \in H^{m+\alpha}$.*

Now, we consider the local existence of classical solutions of (3.4.57). We will prove

Theorem 3.4.9 *Let $d \leqslant 3$, $\alpha \in (0, 1/2]$ and $u_0 \in H^{m+\alpha}$ with $m \geqslant 4$. Then there exists a $T_* > 0$ depending only on u_0 such that (3.4.57) possesses a unique classical solution u, which remains in $C([0, T_*]; H^{m+\alpha} \cap C^2)$.*

Proof Let $\varepsilon > 0$ be a small parameter. Consider the regularized equation for u^ε

$$\partial_t u^\varepsilon = u^\varepsilon \times \Lambda^{2\alpha} u^\varepsilon + \varepsilon \Delta u^\varepsilon, \qquad (3.4.70)$$

with initial data $u^\varepsilon(x, 0) = u_0(x)$. For such a system, the previous results imply that (3.4.70) possesses a global classical solution $u^\varepsilon \in C([0, T]; H^{m+\alpha})$ when $\varepsilon > 0$ is fixed. We now show that there exists $T_* > 0$ over which u^ε is regular and converges to a classical solution u of (3.4.57). It suffices to establish certain *a priori* bounds for u^ε independent of ε.

More specifically, we will prove that u^ε is uniformly bounded in $H^{m+\alpha}$ and $\partial_t u^\varepsilon$ in H^s for some $d/2 < s < m + \alpha$. By integration by parts, we have

$$\frac{1}{2} \frac{d}{dt} \|\nabla^m \Lambda^\alpha u^\varepsilon\|_{L^2}^2 + \varepsilon \|\nabla^{m+1} \Lambda^\alpha u^\varepsilon\|_{L^2}^2 = (u^\varepsilon \times \Lambda^{2\alpha} u^\varepsilon, \Delta^m \Lambda^{2\alpha} u^\varepsilon).$$

For the right hand side term, we have

$$\begin{aligned}
&(u^\varepsilon \times \Lambda^{2\alpha}, \Delta^m \Lambda^{2\alpha} u^\varepsilon) \\
=&(\nabla^m (u^\varepsilon \times \Lambda^{2\alpha} u^\varepsilon), \nabla^m \Lambda^{2\alpha} u^\varepsilon) \\
=&(\nabla^m u^\varepsilon \times \Lambda^{2\alpha} u^\varepsilon, \nabla^m \Lambda^{2\alpha} u^\varepsilon) + \cdots + (u^\varepsilon \times \nabla^m \Lambda^{2\alpha} u^\varepsilon, \nabla^m \Lambda^{2\alpha} u^\varepsilon).
\end{aligned}$$

By the elementary property of cross product, the last term vanishes, and we need only to consider the remaining terms.

Let us consider the first term. Since ∇ commutes with $\Lambda^{2\alpha}$, the first term can be written as

$$(\nabla^m u^\varepsilon \times \Lambda^{2\alpha} u^\varepsilon, \nabla^m \Lambda^{2\alpha} u^\varepsilon) = (\Lambda^\alpha (\nabla^m u^\varepsilon \times \Lambda^{2\alpha} u^\varepsilon), \nabla^m \Lambda^\alpha u^\varepsilon).$$

By calculus inequality, we have

$$\begin{aligned}
\|\Lambda^\alpha (\nabla^m u^\varepsilon \times \Lambda^{2\alpha} u^\varepsilon)\|_{L^2} \leqslant & C(\|\nabla^m \Lambda^\alpha u^\varepsilon\|_{L^2} \|\Lambda^{2\alpha} u^\varepsilon\|_{L^\infty} \\
& + \|\nabla^m u^\varepsilon\|_{L^p} \|\Lambda^{3\alpha} u^\varepsilon\|_{L^q})
\end{aligned}$$

for some $2 < p, q < \infty$ such that $\dfrac{1}{p} + \dfrac{1}{q} = \dfrac{1}{2}$. Choosing $p = \dfrac{2d}{d - 2\alpha} \in (2, \infty)$

for $d \geqslant 2$ and any $p \in (2, \infty)$ for $d = 1$, we have by Sobolev embedding

$$\|\boldsymbol{\nabla}^m u^\varepsilon\|_{L^p}\|\Lambda^{3\alpha} u^\varepsilon\|_{L^q} \leqslant \|u^\varepsilon\|_{H^{m+\alpha}}\|u^\varepsilon\|_{H^{m+\alpha}},$$

when $m > 2\alpha + \dfrac{d}{2}$. On the other hand, for any $j = 0, 1, \cdots, m-1$, we can always select $(p_1, q_1) \in [2, \infty) \times (2, \infty]$ and $(p_2, q_2) \in (2, \infty] \times [2, \infty)$ that may depend on j such that

$$
\begin{aligned}
\|\Lambda^\alpha(\boldsymbol{\nabla}^j u^\varepsilon \times \boldsymbol{\nabla}^{m-j}\Lambda^{2\alpha} u^\varepsilon)\|_{L^2} &\leqslant C(\|\boldsymbol{\nabla}^j \Lambda^\alpha u^\varepsilon\|_{L^{p_1}}^2 \|\boldsymbol{\nabla}^{m-j}\Lambda^{2\alpha} u^\varepsilon\|_{L^{q_1}} \\
&\quad + \|\boldsymbol{\nabla}^j u^\varepsilon\|_{L^{p_2}}\|\boldsymbol{\nabla}^{m-j}\Lambda^{2\alpha} u^\varepsilon\|_{L^{q_2}}) \\
&\leqslant C\|u^\varepsilon\|_{H^{m+\alpha}}^2.
\end{aligned}
$$

Therefore, we deduce that there exists some constant $C > 0$ such that

$$\frac{d}{dt}\|u^\varepsilon\|_{H^{m+\alpha}} \leqslant C\|u^\varepsilon\|_{H^{m+\alpha}}^2,$$

and therefore, there exists $0 < T_* < (C\|u_0\|_{H^{m+\alpha}})^{-1}$ such that for all ε and $0 < T \leqslant T_*$,

$$\sup_{0 \leqslant t \leqslant T} \|u^\varepsilon\|_{H^{m+\alpha}} \leqslant \frac{\|u_0\|_{H^{m+\alpha}}}{1 - CT\|u_0\|_{H^{m+\alpha}}}. \tag{3.4.71}$$

This inequality implies that u^ε is uniformly bounded in $C([0, T]; H^{m+\alpha})$, provided that $T \leqslant T_*$.

On the other hand, since H^s is an algebra for $s > \dfrac{d}{2}$, there holds

$$\|fg\|_{H^s} \leqslant C\|f\|_{H^s}\|g\|_{H^s}, \quad \forall f, g \in H^s.$$

Using this inequality and expressing $\partial_t u^\varepsilon$ in terms of the other terms in (3.4.70), we get

$$\sup_{t \in [0, T]} \|\partial_t u^\varepsilon\|_{H^s} \leqslant C, \quad \forall t \in [0, T_*].$$

Finally, using the Lions-Aubin compactness theorem [152, 222] (see for example, Lemma 10.4, Chapter 10 in [152]), $\{u^\varepsilon\}$ is compact in $C([0, T_*]; H^s)$, which is also compact in $C([0, T_*]; C^2)$ by Sobolev embedding theorem. Therefore, we can pass to the limit $\varepsilon \to 0$ to obtain a local classical solution $u \in L^\infty([0, T_*]; H^{m+\alpha})$ of the equation (3.4.57).

In addition, u is continuous in the weak topology of $H^{m+\alpha}$, i.e., $u \in C_W([0, T]; H^{m+\alpha})$. Let $\varphi \in H^{-(m+\alpha)}$, since $u^\varepsilon \to u$ in $C([0, T_*]; H^s)$ for $s < m + \alpha$, it follows that $\langle u^\varepsilon, \varphi \rangle \to \langle u, \varphi \rangle$ uniformly on $[0, T]$ for any $\varphi \in H^{-s}$. Using uniform boundedness of $\|u^\varepsilon\|$ in $H^{m+\alpha}$ and the fact that H^{-s} is dense in $H^{-(m+\alpha)}$, we have $\langle u^\varepsilon, \varphi \rangle \to \langle u, \varphi \rangle$ uniformly on $[0, T]$ for any $\varphi \in H^{-(m+\alpha)}$ by means of $\varepsilon/3$ argument. This shows $u \in C_W([0, T]; H^{m+\alpha})$ and hence

$$\liminf_{t \to 0^+} \|u(t, \cdot)\|_{H^{m+\alpha}} \geqslant \|u_0\|_{H^{m+\alpha}}.$$

On the other hand, estimate (3.4.71) implies that

$$\sup_{0\leqslant t\leqslant T}\|u^{\varepsilon}\|_{H^{m+\alpha}} \leqslant \|u_0\|_{H^{m+\alpha}} + \frac{C\|u_0\|^2_{H^{m+\alpha}}T}{1-CT\|u_0\|_{H^{m+\alpha}}}, \quad \forall 0 < T \leqslant T_*,$$

and hence

$$\limsup_{t\to 0^+}\|u(t,\cdot)\|_{H^{m+\alpha}} \leqslant \|u_0\|_{H^{m+\alpha}}.$$

In particular, we get the right continuity of $\|u(t,\cdot)\|_{H^{m+\alpha}}$ at $t=0$,

$$\lim_{t\to 0^+}\|u(t,\cdot)\|_{H^{m+\alpha}} = \|u_0\|_{H^{m+\alpha}}.$$

The left continuity at $t=0$ can be similarly deduced and gives us the strong continuity at $t=0$. For the continuity at $t' \in [0,T_*]$, we can take $u(t')$ as initial data and repeat the argument to show the continuity of $\|u(t)\|_{H^{m+\alpha}}$ at $t' \in [0,T_*]$. We therefore complete the proof of the theorem.

A natural question is whether the unique local solution exists globally or will it develop singularities in finite time. This question is not easy to answer since there is no regularizing effects in this equation. However we can give some regularity criteria to show that it can be extended globally if we know before hand that u is in a reasonably regular functional space.

By the construction of the local classical solution of (3.4.57), the solution can be continued in time provided that $\|u\|_{H^{m+\alpha}}$ remains bounded. That is, T be the maximal time of the existence of smooth solutions $u \in C([0,T);H^{m+\alpha})$ if and only if $\lim_{t\to T}\|u\|_{H^{m+\alpha}} = \infty$. Although we cannot show directly that T can be chosen to be ∞, we can give some necessary condition for such a maximal time. I.e., we can give the following regularity criteria, involving some integrability of higher order derivatives. Hereafter, we denote the space $L^{r,s}$ of all functions v such that the quantity

$$\|v\|_{L^{r,s}} := \begin{cases} \left(\displaystyle\int_0^T \|v(\cdot,\tau)\|^r_{L^s}d\tau\right)^{1/r}, & \text{if } 1 \leqslant p < \infty; \\ \text{ess sup}_{0<\tau<T}\|v(\cdot,\tau)\|_{L^s}, & \text{if } r = \infty \end{cases}$$

is finite, where $\|v(\cdot,\tau)\|_{L^s}$ is the usual Lebesgue norm.

Lemma 3.4.11 *Suppose that $u_0 \in \mathbb{S}^2$ is smooth and u is a smooth solution to (3.4.57), then*

$$\|u(t)\|_{L^p} \leqslant \|u_0\|_{L^p}, \quad \forall p \in [2,\infty].$$

Proof The proof is the same as that in Lemma 3.4.9.

Theorem 3.4.10   *Let $d \leqslant 3$, $\alpha \in (0, 1/2]$ and $u_0 \in H^{m+\alpha}$ with $m \geqslant 4$, so that there exists a classical solution $u \in C([0, T); H^{m+\alpha} \cap C^2)$ to the fractional Landau-Lifshitz equation (3.4.57). Then for any $0 < T < \infty$, if when $\alpha = 1/2$ that*

$$\int_0^T \|\nabla u\|_{L^\infty} dt < \infty, \quad \int_0^T \|\Lambda u\|_{L^\infty} dt < \infty, \tag{3.4.72}$$

or when $0 < \alpha < 1/2$ that

$$\int_0^T \|\nabla u\|_{L^s} dt < \infty, \quad \int_0^T \|\Lambda^{2\alpha} u\|_{L^\infty} dt < \infty, \tag{3.4.73}$$

for some $s > 1$ satisfying $2\alpha + \frac{d}{s} \leqslant 1$, then the solution u exists globally in time, i.e, $u \in C([0, \infty); H^{1+\alpha})$.

Proof From the *a priori* estimates, we know that for any $T > 0$, we have $u \in L^\infty(0, T; H^\alpha)$ and

$$\sup_{t \in [0, T]} \|u\|_{H^\alpha} \leqslant \|u_0\|_{H^\alpha}.$$

Multiplying the equation with $\Delta \Lambda^{2\alpha} u$ and by integration by parts, we have

$$\begin{aligned}
\frac{1}{2} \frac{d}{dt} \|\nabla \Lambda^\alpha u\|_{L^2}^2 &= (\nabla u \times \Lambda^{2\alpha} u, \nabla \Lambda^{2\alpha} u) \\
&= (\Lambda^\alpha (\nabla u \times \Lambda^{2\alpha} u), \nabla \Lambda^\alpha u) \\
&\leqslant \|\nabla \Lambda^\alpha u\|_{L^2} \|\Lambda^\alpha (\nabla u \times \Lambda^{2\alpha} u)\|_{L^2}.
\end{aligned} \tag{3.4.74}$$

By calculus inequality, we have

$$\|\Lambda^\alpha (\nabla u \times \Lambda^{2\alpha} u)\|_{L^2} \leqslant C \left(\|\Lambda^\alpha \nabla u\|_{L^2} \|\Lambda^{2\alpha} u\|_{L^\infty} + \|\nabla u\|_{L^p} \|\Lambda^{3\alpha} u\|_{L^q} \right) \tag{3.4.75}$$

where $\dfrac{1}{p} + \dfrac{1}{q} = \dfrac{1}{2}$, with $p, q \geqslant 2$ and $q \neq \infty$.

When $\alpha = 1/2$, we can only let $q = 2$ and $p = \infty$ and in this case we obtain

$$\frac{d}{dt} \|\nabla \Lambda^{1/2} u\|_{L^2}^2 \leqslant C \|\nabla \Lambda^{1/2} u\|_{L^2}^2 \left(\|\Lambda u\|_{L^\infty} + \|\nabla u\|_{L^\infty} \right),$$

which implies that the local solution can be extended to $[0, T]$ under assumption (3.4.72) and remains in $L^\infty([0, T]; H^{1+\alpha})$.

When $1/3 \leqslant \alpha < \dfrac{1}{2}$, we have

$$\|\nabla u\|_{L^p} \|\Lambda^{3\alpha} u\|_{L^q} \leqslant C \|\nabla \Lambda^\alpha u\|_{L^2}^{\theta+\delta} \|\nabla u\|_{L^s}^{2} \cdot {}^{(\theta \mid \delta)} \tag{3.4.76}$$

where we have used the Gagliardo-Nirenberg inequality for the fractional Sobolev spaces [203]

$$\|\nabla u\|_{L^p} \leqslant C\|\nabla\Lambda^\alpha u\|_{L^2}^\theta \|\nabla u\|_{L^s}^{1-\theta};$$
$$\|\Lambda^{3\alpha} u\|_{L^q} \leqslant C\|\nabla\Lambda^\alpha u\|_{L^2}^\delta \|\nabla u\|_{L^s}^{1-\delta}.$$

Here the constants $2 \leqslant p, q \leqslant \infty$, $q \neq \infty$, $0 \leqslant \theta, \delta \leqslant 1$ satisfy

$$
\begin{cases}
\dfrac{1}{p} + \dfrac{1}{q} = \dfrac{1}{2}, \\[2mm]
1 - \dfrac{d}{p} = \theta\left(1 + \alpha - \dfrac{d}{2}\right) + (1-\theta)\left(1 - \dfrac{d}{s}\right), \\[2mm]
3\alpha - \dfrac{d}{p} = \delta\left(1 + \alpha - \dfrac{d}{2}\right) + (1-\delta)\left(1 - \dfrac{d}{s}\right).
\end{cases}
\tag{3.4.77}
$$

This system has many solutions, among which one solution to (3.4.77) can be written as

$$
p = \frac{2(d-\alpha)}{d - 2\alpha}, \quad
q = \frac{2(d-\alpha)}{\alpha}, \quad
\theta = \frac{\dfrac{d}{s} - \dfrac{d}{p}}{\alpha - \dfrac{d}{2} + \dfrac{d}{s}}, \quad
\delta = \frac{3\alpha - 1 + \dfrac{d}{s} - \dfrac{d}{q}}{\alpha - \dfrac{d}{2} + \dfrac{d}{s}}.
$$

Note in this case,

$$
\theta + \delta = \frac{3\alpha - 1 + \dfrac{2d}{s} - \dfrac{d}{2}}{\alpha - \dfrac{d}{2} + \dfrac{d}{s}} \leqslant 1;
$$

provided that $2\alpha + \dfrac{d}{s} \leqslant 1$. Moreover, for any p, q, θ, δ satisfy (3.4.77), we have from the assumption $2\alpha + \dfrac{d}{s} \leqslant 1$ that $\theta + \delta \leqslant 1$. Note also that the larger $\theta + \delta$ is, the less regularity of $\|\nabla u\|_{L^{r,s}}$ is required. Hence we choose $\theta + \delta = 1$ in the following.

Putting (3.4.75) and (3.4.76) into (3.4.74), we obtain

$$\frac{d}{dt}\|\nabla\Lambda^{2\alpha}u\|_{L^2}^2 \leqslant \|\Lambda^\alpha\nabla u\|_{L^2}^2(\|\Lambda^{2\alpha}u\|_{L^\infty} + \|\nabla u\|_{L^s}). \tag{3.4.78}$$

Therefore, by applying the Gronwall inequality, we have

$$\sup_{0\leqslant t\leqslant T} \|\nabla\Lambda^{2\alpha}u\|_{L^2}^2 \leqslant \|\nabla\Lambda^\alpha u_0\|_{L^2}^2 \exp\left\{\int_0^T \|\Lambda^{2\alpha}u(\cdot,t)\|_{L^\infty} + \|\nabla u(\cdot,t)\|_{L^s}dt\right\}.$$

Under the assumption (3.4.73), this inequality implies that

$$u \in L^\infty(0,T; H^{1+\alpha}).$$

When $0 < \alpha < 1/3$, we choose $p = 2d/(d - 2\alpha)$, and $q = d/\alpha$, we have

$$\|\nabla u\|_{L^p} \leqslant C\|\nabla \Lambda^\alpha u\|_{L^2}$$

and for $2\alpha + \dfrac{d}{s} \leqslant 1$,

$$\|\Lambda^{3\alpha} u\|_{L^q} \leqslant C\|\nabla u\|_{L^s},$$

thanks to Lemma 3.4.11. Again, we recover the inequality (3.4.78). Therefore, we complete the proof for all $0 < \alpha \leqslant 1/2$.

Scaling analysis. Let $u(x,t)$ be a solution of the equation (3.4.57), then the scaling $u_\lambda(x,t) = u(\lambda x, \lambda^{2\alpha} t)$ is also a solution. Motivated by the work of Caffarelli, Kohn and Nirenberg [33] (see also [229]) for the Navier-Stokes equation, we call the norm $\|\Lambda^\beta u\|_{L^r(0,T;L^s)}$ is of dimension zero if

$$\|\Lambda^\beta u_\lambda\|_{L^r(0,T;L^s)} = \|\Lambda^\beta u\|_{L^r(0,T;L^s)}$$

holds for any $\lambda > 0$, see also [229]. It is easy to see this holds if and only if

$$\beta = \frac{2\alpha}{r} + \frac{d}{s}. \tag{3.4.79}$$

Therefore, the regularity criteria may involve the finiteness of $\|\Lambda^\beta u\|_{L^{r,s}}$ for β, r, s satisfying (3.4.79). Note that condition (3.4.72) satisfy this relationship with $r = 1, s = \infty$ and $\beta = 1$ when $\alpha = \dfrac{1}{2}$. Therefore, we expect the following regularity criteria concerned with $\|\Lambda^\beta u\|_{L^{r,s}}$.

Theorem 3.4.11   *Let $d \leqslant 3$, $\alpha \in (0, 1/2]$ and $u_0 \in H^{m+\alpha}$ with $m \geqslant 4$, so that there exists a classical solution $u \in C([0,T); H^{m+\alpha} \cap C^2)$ to the fractional Landau-Lifshitz equation (3.4.57). Then for any $0 < T < \infty$, if*

$$\int_0^T \|\Lambda^{2\alpha} u(t)\|_{L^\infty} dt < \infty,$$

and

$$\int_0^T \|\Lambda^\beta u(\cdot, t)\|_{L^s} dt < \infty,$$

for some $\beta \geqslant 2\alpha + \dfrac{d}{s}$ and $1 < s < \infty$, then the local solution can be extended into a global classical solution and remains in $L^\infty(0,T; H^{m+\alpha})$.

Remark 3.4.6   *When $\alpha < 1/2$, we can choose s sufficiently large such that we can choose $\beta < 1/2$ to reduce the differentiability requirement of Z as required in Theorem 3.4.10, therefore the theorem is meaningful.*

Proof Repeating the arguments used in Theorem 3.4.10, we easily prove that

$$\sup_{0 \leqslant t \leqslant T} \|u\|_{H^{1+\alpha}} \leqslant C\|u_0\|_{H^{1+\alpha}}.$$

Now we go to the global estimates for $\|\Delta\Lambda^\alpha u\|_{L^2}$.

Multiply the equation (3.4.57) with $\Delta^2\Lambda^{2\alpha}Z$ to obtain

$$
\begin{aligned}
\frac{1}{2}\frac{d}{dt}\|\Delta\Lambda^\alpha u\|_{L^2}^2 &= (\Delta(u \times \Lambda^{2\alpha}u, \Delta\Lambda^{2\alpha}u)) \\
&= (\Delta u \times \Lambda^{2\alpha}u + 2\nabla u \times \nabla\Lambda^{2\alpha}u, \Delta\Lambda^{2\alpha}u) \\
&= (\Lambda^\alpha(\Delta u \times \Lambda^{2\alpha}u) + 2\Lambda^\alpha(\nabla u \times \nabla\Lambda^{2\alpha}u), \Delta\Lambda^\alpha u) \\
&=: I + II.
\end{aligned}
\tag{3.4.80}
$$

The following proof is divided into three cases, according to $2\alpha < \beta \leqslant 3\alpha$ and $\beta > 3\alpha$.

Case 1. $2\alpha < \beta \leqslant 3\alpha$.

For I, we have

$$
\begin{aligned}
|I| &= |(\Lambda^\alpha(\Delta u \times \Lambda^{2\alpha}u), \Delta\Lambda^\alpha u)| \\
&\leqslant C\|\Delta\Lambda^\alpha u\|_{L^2} \left(\|\Delta\Lambda^\alpha u\|_{L^2}\|\Lambda^{2\alpha}u\|_{L^\infty} + \|\Delta u\|_{L^p}\|\Lambda^{3\alpha}u\|_{L^q}\right) \\
&\leqslant C\|\Delta\Lambda^\alpha u\|_{L^2} \left(\|\Delta\Lambda^\alpha u\|_{L^2}\|\Lambda^{2\alpha}u\|_{L^\infty} + \|\nabla\Lambda^\alpha u\|_{L^2}^{\theta+\delta}\|\Lambda^\beta u\|_{L^s}^{2-(\theta+\delta)}\right),
\end{aligned}
$$

where we have used the Gagliardo-Nirenberg interpolation inequality and the constants $2 \leqslant p, q \leqslant \infty$, $q \neq \infty$, $0 \leqslant \theta, \delta \leqslant 1$ satisfy

$$
\begin{cases}
\dfrac{1}{p} + \dfrac{1}{q} = \dfrac{1}{2}, \\[2mm]
1 - \dfrac{d}{p} = \theta\left(1+\alpha-\dfrac{d}{2}\right) + (1-\theta)\left(\beta-\dfrac{d}{s}\right), \\[2mm]
3\alpha - \dfrac{d}{p} = \delta\left(1+\alpha-\dfrac{d}{2}\right) + (1-\delta)\left(\beta-\dfrac{d}{s}\right).
\end{cases}
$$

Since $2\alpha + \dfrac{d}{s} < \beta$, we find $\theta + \delta \leqslant 1$. Choosing $\theta + \delta = 1$, we have

$$|I| \leqslant C\|\Delta\Lambda^\alpha u\|_{L^2}^2(\|\Lambda^{2\alpha}u\|_{L^\infty} + \|\Lambda^\beta u\|_{L^s}).$$

For the second term II, by Gagliardo-Nirenberg interpolation inequality for the fractional Sobolev spaces, we have

$$|II| \leqslant C\|\Delta\Lambda^\alpha u\|_{L^2}^2\|\Lambda^\beta u\|_{L^s}.$$

Therefore, combining the estimates for I and II, we have

$$\frac{1}{2}\frac{d}{dt}\|\Delta\Lambda^\alpha u\|_{L^2}^2 \leqslant C\|\Delta\Lambda^\alpha u\|_{L^2}^2(\|\Lambda^{2\alpha}u\|_{L^\infty} + \|\Lambda^\beta u\|_{L^s}),$$

which implies that

$$\sup_{0\leqslant t\leqslant T}\|\Delta\Lambda^\alpha u(t)\|_{L^2}^2 \leqslant C\|\|\Delta\Lambda^\alpha u_0\|_{L^2}^2\|e^{\int_0^T(\|\Lambda^{2\alpha}u\|_{L^\infty}+\|\Lambda^\beta u\|_{L^s})dt}.$$

Case 2. $\beta > 3\alpha$.

For the first term I, we have

$$\begin{aligned}|I| =&|(\Lambda^\alpha(\Delta u \times \Lambda^{2\alpha}u), \Delta\Lambda^\alpha u)\\ \leqslant&\|\Delta\Lambda^\alpha u\|_{L^2}\left(\|\Delta\Lambda^\alpha u\|_{L^2}\|\Lambda^{2\alpha}u\|_{L^\infty} + \|\Delta u\|_{L^p}\|\Lambda^{3\alpha}u\|_{L^q}\right)\end{aligned}$$

where $\dfrac{1}{p}+\dfrac{1}{q}=\dfrac{1}{2}$ and $q \in [2,\infty)$. Choosing $p = \dfrac{2d}{d-2\alpha} > 2$ and $q = \dfrac{d}{\alpha} < \infty$, we have

$$\|\Delta u\|_{L^p} \leqslant C\|\Delta\Lambda^\alpha u\|_{L^2},$$

and

$$\|\Lambda^{3\alpha}u\|_{L^q} \leqslant C\|\Lambda^\beta u\|_{L^s}$$

thanks to the condition $\beta \geqslant 2\alpha + \dfrac{d}{s}$ and Lemma 3.4.11. Hence

$$|I| \leqslant C\|\Delta\Lambda^\alpha u\|_{L^2}^2\left(\|\Lambda^{2\alpha}u\|_{L^\infty} + \|\Lambda^\beta u\|_{L^s}\right).$$

For the second term,

$$\begin{aligned}|II| \leqslant& C|(\Lambda^\alpha(\nabla u \times \nabla\Lambda^{2\alpha}u), \Delta\Lambda^\alpha u)|\\ \leqslant& C\|\Delta\Lambda^\alpha u\|_{L^2}\left(\|\nabla\Lambda^\alpha u\|_{L^{p_1}}\|\nabla\Lambda^{2\alpha}u\|_{L^{q_1}} + \|\nabla u\|_{L^{p_2}}\|\nabla\Lambda^{3\alpha}u\|_{L^{q_2}}\right)\\ =&:C\|\Delta\Lambda^\alpha u\|_{L^2}(II_1 + II_2).\end{aligned}$$

The terms in the parentheses $(\cdots)$ can be handled similarly, by using the Gagliardo-Nirenberg interpolation inequality and adjusting the index p_1, q_1, p_2 and q_2 appropriately. We then obtain

$$|II| \leqslant C\|\Delta\Lambda^\alpha u\|_{L^2}^2\left(\|\Lambda^{2\alpha}u\|_{L^\infty} + \|\Lambda^\beta u\|_{L^s}\right).$$

Therefore, in the case $\beta > 3\alpha$, we get the bound

$$\sup_{0\leqslant t\leqslant T}\|\Delta\Lambda^\alpha u\|_{L^2} \leqslant C.$$

Higher order estimates can be obtained by induction. By interpolation between $m-1+\alpha$ and $m+\alpha$,

$$\|u\|_{H^m} \leqslant C\|u\|_{H^{m-1+\alpha}}^\alpha\|u\|_{H^{m+\alpha}}^{1-\alpha},$$

we recover all the integer order bound for the solution. The proof is complete.

It will be very interesting if one can prove the global existence of smooth solutions when the regularity criteria is dropped.

3.5   Fractional QG equations

This section considers the following inviscid two-dimensional quasigeostrophic (QG) equation

$$\theta_t + u \cdot \nabla\theta = 0 \tag{3.5.1}$$

as well as the two-dimensional viscous QG equation

$$\theta_t + u \cdot \nabla\theta + \kappa(-\Delta)^\alpha\theta = 0, \tag{3.5.2}$$

where $\theta = \theta(x,t)$ is a real valued function of x and t, $0 \leqslant \alpha < 1$ and $\kappa > 0$ are real numbers. As mentioned in Chapter 1, θ represents potential temperature, u represents the fluid velocity, which can be represented by the stream function ψ

$$u = (u_1, u_2) = \left(-\frac{\partial\psi}{\partial x_2}, \frac{\partial\psi}{\partial x_1}\right), \qquad (-\Delta)^{1/2}\psi = -\theta, \tag{3.5.3}$$

and ψ is identified with the pressure. In what follows, the spatial domain is either periodic when $x \in \mathbf{T}^2$ or the whole space when $x \in \mathbf{R}^2$. Hereinafter, we usually denote $\Lambda = (-\Delta)^{1/2}$. Introducing the Riesz operator, u can be expressed as

$$u = (\partial_{x_2}\Lambda^{-1}\theta, -\partial_{x_1}\Lambda^{-1}\theta) = (-\mathcal{R}_2\theta, \mathcal{R}_1\theta) =: \mathcal{R}^\perp\theta,$$

where $R_j, j = 1, 2$ are Riesz operators

$$\widehat{\mathcal{R}_j f}(k) = -\mathrm{i}\frac{k_j}{|k|}\hat{f}(k), \quad k \in \mathbf{Z}^2\backslash\{0\},$$

$$\widehat{\mathcal{R}_j f}(\xi) = -\mathrm{i}\frac{\xi_j}{|\xi|}\hat{f}(\xi), \quad \xi \in \mathbf{R}^2\backslash\{0\}.$$

In case of $\mathbf{R}^2$, the Riesz operator can be expressed in terms of the singular integral

$$\mathcal{R}_j f(x) = C \mathrm{P.V.} \int_{\mathbf{R}^2} \frac{f(x-y)y_j}{|y|^3}dy, \quad j = 1, 2,$$

where C is a constant. By Calderon-Zygmund theory of singular integral, there exists a constant $C = C_p$ for any $p \in (1, \infty)$ such that

$$\|u\|_{L^p} \leqslant C_p\|\theta\|_{L^p}. \tag{3.5.4}$$

In what follows, we will also consider the nonhomogeneous QG equation

$$\theta_t + u \cdot \nabla\theta + \kappa(-\Delta)^\alpha\theta = f, \tag{3.5.5}$$

where f is a known function.

In what follows, we will present some of the recent results for the two-dimensional fractional QG equation, such as the existence and uniqueness of its solutions, inviscid limit, and long-time behavior. For more details, readers are referred to [22, 34, 39, 40, 52–56, 58, 119, 120, 125, 190, 197, 220, 221, 223] and the references therein. In particular, [39, 54] proved existence of smooth solutions for subcritical case, and [52, 58] proved global existence of solutions with small initial data in L^∞. Recently, [125] proved the global well-posedness for the critical 2D dissipative quasi geostrophic equation based on a non-local maximum principle involving appropriate moduli of continuity, and [34] showed that solutions of the quasi-geostrophic equation with initial L^2 data and critical diffusion $(-\Delta)^{1/2}$ are locally smooth for any space dimension based on the De Giorgi iteration idea. Under the supercritical situation, [40] proved global existence for small initial data in a scale invariant Besov space.

3.5.1   Existence and uniqueness of solutions

By a weak solution of (3.5.5), we mean a function $\theta \in L^\infty(0, T; L^2(\mathbf{T}^2)) \cap L^2(0, T; H^\alpha(T^2))$ such that

$$
\int_{\mathbf{T}^2} \theta(T)\varphi \mathrm{d}x - \int_{\mathbf{T}^2} \theta_0\varphi \mathrm{d}x - \int_0^T \int_{\mathbf{T}^2} \theta u \cdot \nabla \varphi \mathrm{d}x\mathrm{d}t
$$
$$
+\kappa \int_0^T \int_{\mathbf{T}^2} \Lambda^\alpha\theta \cdot \Lambda^\alpha\varphi \mathrm{d}x\mathrm{d}t = \int_0^T \int_{\mathbf{T}^2} f\varphi \mathrm{d}x\mathrm{d}t,
$$

for any $\varphi \in C^\infty(\mathbf{T}^2)$.

Theorem 3.5.1 *Let $T > 0$ be arbitrary, $\theta_0 \in L^2(\mathbf{T}^2)$ and $f \in L^2(0, T; L^2(\mathbf{T}^2))$, then there exists at least one weak solution $\theta \in L^\infty(0, T; L^2(\mathbf{T}^2)) \cap L^2(0, T; H^\alpha(T^2))$ for two-dimensional QG equation.*

By Faedo-Galerkin method, we can construct approximate solutions θ_n such that $\|\theta_n(t)\|_{L^2} \leqslant C\|\theta_0\|_{L^2}$. By selecting a subsequence if necessary, θ_n converges to a weak solution of (3.5.5). Indeed, it suffices to consider the convergence for the nonlinear terms

$$
\int_{\mathbf{T}^2} \theta_n u_n \cdot \nabla\varphi \mathrm{d}x = -\frac{(-1)^{j+1}}{2} \int_{\mathbf{T}^2} \sum_{j=1}^2 \mathcal{R}_{\{j\}}(\theta_n) \left[\Lambda, \frac{\partial\varphi}{\partial x_j}\right] (\Lambda^{-1}\theta_n)\mathrm{d}x,
$$

$$(3.5.6)$$

where $\left[\Lambda, \dfrac{\partial\varphi}{\partial x_j}\right] = \Lambda\dfrac{\partial\varphi}{\partial x_j} - \dfrac{\partial\varphi}{\partial x_j}\Lambda$ denotes the commutator and $\{j\} = 2$ when $j - 1$ and $\{j\} - 1$ when $j = 2$. Then the right hand side of (3.5.6) can be rewritten as

$$\int_{\mathbf{T}^2} \sum_{j=1}^{2} \mathcal{R}_{\{j\}}(\theta_n) K_j(\theta_n) \mathrm{d}x.$$

It can be shown that K_j is a compact operator depending on the test function φ, from which it follows the convergence of the nonlinear terms.

In general, weak solution of the equation is not unique. However, one can show that weak solution is unique among strong solutions.

Theorem 3.5.2 *Suppose that $\alpha \in (1/2, 1]$, $T > 0$ and p, q satisfy relations*

$$p \geqslant 1, \quad q > 0, \quad \frac{1}{p} + \frac{\alpha}{q} = \alpha - \frac{1}{2}.$$

Then for any $\theta_0 \in L^2$, the two-dimensional QG equation admits at most one solution $\theta \in L^\infty(0, T; L^2) \cap L^2(0, T; H^\alpha)$ such that $\theta \in L^p(0, T; L^q)$.

Proof Suppose that θ_1 and θ_2 are two different weak solutions with the same initial data. Then $\theta = \theta_1 - \theta_2$ satisfies

$$\partial_t \theta + u \cdot \nabla \theta_1 + u_2 \cdot \nabla \theta + \kappa \Lambda^{2\alpha} \theta = 0,$$

where $u = u_1 - u_2 = \mathcal{R}^\perp \theta_1 - \mathcal{R}^\perp \theta_2$. Taking inner product of this equation with $\psi = -\Lambda^{-1}\theta$ and using $\int_{\mathbf{T}^2} \psi u \cdot \nabla \theta_1 = 0$ yield

$$\left| \int_{\mathbf{T}^2} \theta u_2 \cdot \nabla \psi \right| \leqslant \kappa \|\psi\|_{H^{\alpha + \frac{1}{2}}}^2 + C(\kappa) \|\theta_2\|_{L^p}^{\frac{1}{1-\beta}} \|\psi\|_{H^{1/2}}^2,$$

where $\beta = \frac{1}{\alpha}\left(\frac{1}{2} + \frac{1}{p}\right)$ and $C(\kappa) = C\kappa^{-\frac{\beta}{1-\beta}}$. Then

$$\frac{\mathrm{d}}{\mathrm{d}t} \|\psi\|_{H^{1/2}}^2 \leqslant C(\kappa) \|\theta_2\|_{L^p}^{\frac{1}{1-\beta}} \|\psi\|_{H^{1/2}}^2,$$

from which it follows that $\psi = 0$ and hence $\theta = 0$.

By applying the energy method, the existence and uniqueness of local smooth solution for the inviscid QG equation can be proved, as it did in the monographs [151, 152].

Theorem 3.5.3 *Let $\kappa = 0$, and $\theta_0 \in H^k(\mathbf{R}^2)$ with $k \geqslant 3$ being an integer, then there exists $T_* > 0$ such that there exists a unique local smooth solution $\theta \in H^k(\mathbf{R}^2)$ for the two-dimensional QG equation on $[0, T_*)$. Furthermore, if $T_* < \infty$, then $\|\theta(\cdot, t)\|_{H^k} \to \infty$ as $t \nearrow T_*$.*

To give blow up criteria similar to the Beale-Kato-Majda criteria for the Euler equation [18], we introduce $\alpha(x,t) = \mathcal{D}(x,t)\xi \cdot \xi$, and $\alpha^*(t) = \max_{\xi \in \mathbf{R}^2} \alpha(x,t)$ where $\xi(x,t) = \dfrac{\nabla^\perp \theta}{|\nabla^\perp \theta|}$ represents the direction vector of $\nabla^\perp \theta$ and $\mathcal{D} = \dfrac{1}{2}((\nabla u) + (\nabla u)^t)$ is the symmetric part of the velocity gradient.

Theorem 3.5.4 *Suppose that $\theta = \theta(x,t)$ is the unique smooth solution of the two-dimensional inviscid QG equation with initial data $\theta_0 \in H^k(k \geqslant 3)$, then the following statements are equivalent:*

(1) $0 \leqslant t < T_* < \infty$ is maximum existence interval of solution H^k;

(2) when $T \nearrow T_*$,

$$\int_0^T \|\nabla \theta\|_{L^\infty}(s)\mathrm{d}s \to \infty, \quad T \to T_*;$$

(3) $\alpha^*(t)$ satisfy

$$\int_0^T \alpha^*(s)\mathrm{d}s \to \infty, \quad T \to T_*. \tag{3.5.7}$$

Proof Similar to [18], it is easy to show that (1) and (2) are equivalent. It then suffices to show that (1) and (3) are equivalent. Since from Riesz operator, $\hat{u}(\xi) = \dfrac{\mathrm{i}(-\xi_2, \xi_1)}{|\xi|}\hat{\theta}(\xi)$, one can show by Sobolev theorem that

$$\alpha^*(t) \leqslant C\|\nabla u(t)\|_{L^\infty(\mathbf{R}^2)} \leqslant C\|\nabla u(t)\|_{H^{k-1}}$$
$$\leqslant C\|u\|_{H^k} \leqslant C\|\theta(t)\|_{H^k}, \quad \forall k \geqslant 3, \tag{3.5.8}$$

where C is a constant. Hence, if (3.5.7) holds, then integrating (3.5.8) over $[0,T]$ yields

$$\int_0^T \|\theta(s)\|_{H^k}\mathrm{d}s \to \infty, \quad T \to T_*.$$

Therefore $[0, T_*)$ is the maximal existence interval of $\theta(x,t)$.

On the contrary, if

$$\int_0^{T_*} \alpha^*(s)\mathrm{d}s \leqslant M < \infty,$$

then it follows

$$\int_0^{T_*} \|\nabla^\perp \theta(s)\|_{L^\infty}\mathrm{d}s \leqslant e^M \|\nabla^\perp \theta_0\|_{L^\infty} < \infty. \tag{3.5.9}$$

In fact, based on (1.3.7) and definition of α^*, one obtains

$$\frac{\mathrm{d}}{\mathrm{d}t}\|\boldsymbol{\nabla}^\perp\theta(t)\|_{L^\infty} \leqslant \alpha^*(t)\|\boldsymbol{\nabla}^\perp\theta(t)\|_{L^\infty}.$$

(3.5.9) then follows from the Gronwall inequality. The proof is complete.

Theorem 3.5.5 *Let $\alpha \in (0,1)$, $\kappa > 0$, $\Omega = \mathbf{R}^2$ and $\theta_0 \in \dot{H}^s$, then the following assertions hold.*

(1) If $s = 2 - 2\alpha$, then there exists constant C_0 such that for any weak solution of (3.5.2) satisfying $\|\Lambda^s\theta_0\|_{L^2} \leqslant \kappa/C_0$ there holds $\|\Lambda^s\theta(t)\|_{L^2} \leqslant \|\Lambda^s\theta_0\|_{L^2}$ for all $\forall t > 0$ and $\theta \in L^2(0,\infty;\dot{H}^{s+\alpha})$. The solution is unique if $\theta_0 \in L^2$.

(2) If $s \in (2-2\alpha, 2-\alpha]$, then there exists $T > 0$ depending on κ and $\|\Lambda^s\theta_0\|_{L^2}$, such that for any weak solution of (3.5.2), there holds $\theta \in L^\infty(0,T;\dot{H}^s) \cap L^2(0,T;\dot{H}^{s+\alpha})$ and θ is unique if $\theta_0 \in L^2$.

(3) If $s > 2-\alpha$, there exists $T > 0$ depending on κ, $\|\theta_0\|_{L^2}$ and $\|\Lambda^s\theta_0\|_{L^2}$ such that any weak solution θ of (3.5.2) belongs to $L^\infty(0,T;H^s) \cap L^2(0,T; \dot{H}^{s+\alpha})$ if $\theta_0 \in H^s$ and is unique.

(4) If $s > 2-2\alpha$, then there exists a constant $C_0 > 0$ such that if

$$\|\theta_0\|_{L^2}^{\frac{s-2(1-\alpha)}{s}} \|\Lambda^s\theta_0\|_{L^2}^{\frac{2(1-\alpha)}{s}} \leqslant \kappa/C_0, \qquad (3.5.10)$$

then weak solution of equation (3.5.2) is unique and

$$\|\Lambda^s\theta(t)\|_{L^2} \leqslant \|\Lambda^s\theta_0\|_{L^2}.$$

Furthermore, if (3.5.10) holds strictly, then $\theta \in L^2(0,\infty;\dot{H}^{s+\alpha})$.

Proof First, we note $(u \cdot \nabla(\Lambda^s\theta), \Lambda^s\theta) = 0$ since $\nabla \cdot u = 0$. Taking L^2 inner product of (3.5.2) with θ, one has

$$\frac{1}{2}\frac{\mathrm{d}}{\mathrm{d}t}\|\theta\|_{L^2}^2 + \kappa\|\Lambda^\alpha\theta\|_{L^2}^2 \leqslant 0.$$

Integration then yields $\theta \in L^\infty(0,\infty;L^2) \cap L^2(0,\infty;\dot{H}^\alpha)$. Taking L^2 inner product of (3.5.2) with $\Lambda^{2s}\theta$ yields

$$\frac{1}{2}\frac{\mathrm{d}}{\mathrm{d}t}\|\Lambda^s\theta\|_{L^2}^2 + \kappa\|\Lambda^{s+\alpha}\theta\|_{L^2}^2 = -(\Lambda^s(u \cdot \nabla\theta) - u \cdot \nabla(\Lambda^s\theta), \Lambda^s\theta). \quad (3.5.11)$$

Since Λ^s and ∇ commute, one has from commutator estimates and (3.5.4) that

$$
\begin{aligned}
|(\Lambda^s(u \cdot \nabla\theta) - u \cdot \nabla(\Lambda^s\theta), \Lambda^s\theta)| &= |(\Lambda^s(u \cdot \nabla\theta) - u \cdot (\Lambda^s\nabla\theta), \Lambda^s\theta)| \\
&\leqslant C\|\Lambda^s(u \cdot \nabla\theta) - u \cdot (\Lambda^s\nabla\theta)\|_{L^2}\|\Lambda^s\theta\|_{L^2} \\
&\leqslant C\big(\|\nabla u\|_{L^{p_1}}\|\Lambda^s\theta\|_{L^{p_2}} \\
&\quad + \|\Lambda^s u\|_{L^{p_2}}\|\nabla\theta\|_{L^{p_1}}\big)\|\Lambda^s\theta\|_{L^2} \\
&\leqslant C\|\Lambda\theta\|_{L^{p_1}}\|\Lambda^s\theta\|_{L^{p_2}}\|\Lambda^s\theta\|_{L^2},
\end{aligned}
$$

$$(3.5.12)$$

where $p_1, p_2 > 2$ and $\dfrac{1}{p_1} + \dfrac{1}{p_2} = \dfrac{1}{2}$. In particular by selecting $p_1 = \dfrac{2}{\alpha}$ and $p_2 = \dfrac{2}{1-\alpha}$, one has from (3.5.11) and (3.5.12)

$$\frac{1}{2}\frac{d}{dt}\|\Lambda^s\theta\|_{L^2}^2 + \kappa\|\Lambda^{s+\alpha}\theta\|_{L^2}^2 \leqslant C\|\Lambda^{2-\alpha}\theta\|_{L^2}\|\Lambda^{s+\alpha}\theta\|_{L^2}\|\Lambda^s\theta\|_{L^2}, \quad (3.5.13)$$

thanks to the Sobolev embedding

$$\|\Lambda\theta\|_{L^{p_1}} \leqslant C\|\Lambda^{2-\alpha}\theta\|_{L^2}, \quad \|\Lambda^s\theta\|_{L^{p_2}} \leqslant C\|\Lambda^{s+\alpha}\theta\|_{L^2}.$$

Case 1. When $s = 2 - 2\alpha$.

In this case,

$$\frac{1}{2}\frac{d}{dt}\|\Lambda^s\theta\|_{L^2}^2 + \kappa\|\Lambda^{s+\alpha}\theta\|_{L^2}^2 \leqslant C\|\Lambda^s\theta\|_{L^2}^2\|\Lambda^{s+\alpha}\theta\|_{L^2}^2.$$

If $\|\Lambda^s\theta_0\|_{L^2} \leqslant \kappa/C$, then for any $t \geqslant 0$, there holds for all $t > 0$

$$\|\Lambda^s\theta(t)\|_{L^2} \leqslant \|\Lambda^s\theta_0\|_{L^2} \leqslant \frac{\kappa}{C}.$$

Hence, θ exists in the $\dot{H}^s$ for all $t > 0$ and is uniformly bounded. Furthermore, if $\|\Lambda^s\theta_0\|_{L^2} < \kappa/C$ strictly, then $\theta \in L^2(0, +\infty; \dot{H}^{s+\alpha})$.

Case 2. When $s \in (2(1-\alpha), 2-\alpha]$.

Recall the Gagliardo-Nirenberg inequality

$$\|\Lambda^{2-\alpha}\theta\|_{L^2} \leqslant C\|\Lambda^{s+\alpha}\theta\|_{L^2}^\beta\|\Lambda^s\theta\|_{L^2}^{1-\beta}, \quad \beta = \frac{2-\alpha-s}{\alpha} \in [0, 1).$$

Then by using (3.5.13) and Young's inequality, we obtain

$$\begin{aligned}
\frac{1}{2}\frac{d}{dt}\|\Lambda^s\theta\|_{L^2}^2 + \kappa\|\Lambda^{s+\alpha}\theta\|_{L^2}^2 &\leqslant C\|\Lambda^{s+\alpha}\theta\|_{L^2}^{1+\beta}\|\Lambda^s\theta\|_{L^2}^{2-\beta} \\
&\leqslant \frac{\kappa}{2}\|\Lambda^{s+\alpha}\theta\|_{L^2}^2 + C(\kappa)\|\Lambda^s\theta\|_{L^2}^{\frac{2(2-\beta)}{1-\beta}}.
\end{aligned} \quad (3.5.14)$$

Therefore, one has

$$\frac{d}{dt}\|\Lambda^s\theta\|_{L^2}^2 + \kappa\|\Lambda^{s+\alpha}\theta\|_{L^2}^2 \leqslant C(\kappa)\|\Lambda^s\theta\|_{L^2}^{\frac{2(3\alpha+s-2)}{2\alpha+s-2}} \quad (3.5.15)$$

and by ignoring the positive term on the LHS and direct integration

$$\|\Lambda^s\theta\|_{L^2}^2 \leqslant \|\Lambda^s\theta_0\|_{L^2}^2 \left[1 - \frac{tC(\kappa)\alpha}{s-2+2\alpha}\|\Lambda^s\theta_0\|_{L^2}^{-\frac{s-2+2\alpha}{\alpha}}\right].$$

This inequality shows local existence in the $\dot{H}^s$ for any initial data $\theta_0 \in \dot{H}^s$. Furthermore, (3.5.15) implies within the interval of existence that

$$\int_0^t \|\Lambda^{s+\alpha}\theta(s)\|_{L^2}^2 ds \leqslant \frac{1}{\kappa}\|\Lambda^s\theta_0\|_{L^2}^2 + \frac{C(\kappa)}{\kappa}\int_0^t \|\Lambda^s\theta(s)\|_{L^2}^{\frac{2(3\alpha+s-2)}{2\alpha+s-2}} ds < \infty.$$

It is standard to derive global existence when the initial data is small. Since $\alpha < 1$, then $1 + \beta + (2 - \beta)\dfrac{s}{s+\alpha} > 2$. Let $\gamma \in (0, 2 - \beta)$ then

$$1+\beta+\frac{s\gamma}{s+\alpha} = 2, \text{ i.e., } \gamma = \frac{(s+\alpha)(2\alpha-2+s)}{s(s+\alpha)} = \frac{2\alpha-2+s}{s} \text{ and } 2-\beta-\gamma > 0.$$

From the interpolation inequality

$$\|\Lambda^s\theta\|_{L^2} \leqslant C\|\Lambda^{s+\alpha}\theta\|_{L^2}^{\frac{s}{s+\alpha}}\|\theta\|_{L^2}^{\frac{\alpha}{s+\alpha}},$$

it follows that

$$\|\Lambda^s\theta\|_{L^2}^{2-\beta} = \|\Lambda^s\theta\|_{L^2}^{\gamma}\|\Lambda^s\theta\|_{L^2}^{2-\beta-\gamma} \leqslant C\|\Lambda^{s+\alpha}\theta\|_{L^2}^{\frac{s\gamma}{s+\alpha}}\|\theta\|_{L^2}^{\frac{\alpha\gamma}{s+\alpha}}\|\Lambda^s\theta\|_{L^2}^{2-\beta-\gamma}.$$
$$(3.5.16)$$

Using (3.5.14) and (3.5.16), one has

$$\frac{1}{2}\frac{d}{dt}\|\Lambda^s\theta\|_{L^2}^2 + \kappa\|\Lambda^{s+\alpha}\theta\|_{L^2}^2 \leqslant C\|\Lambda^{s+\alpha}\theta\|_{L^2}^2\|\theta\|_{L^2}^{\frac{\alpha\gamma}{s+\alpha}}\|\Lambda^s\theta\|_{L^2}^{2-\beta-\gamma}$$

$$= C\|\Lambda^{s+\alpha}\theta\|_{L^2}^2\|\theta\|_{L^2}^{\frac{s-2(1-\alpha)}{s}}\|\Lambda^s\theta\|_{L^2}^{\frac{2(1-\alpha)}{s}}.$$

Therefore if initial values satisfies (3.5.10), then global existence in $\dot{H}^s$ follows.

Case 3. When $s > 2 - \alpha$.

It follows from the Gagliardo-Nirenberg inequality that

$$\|\Lambda^{2-\alpha}\theta\|_{L^2} \leqslant C\|\Lambda^s\theta\|_{L^2}^{\frac{2-\alpha}{s}}\|\theta\|_{L^2}^{\frac{s+2-\alpha}{s}}.$$

Then from (3.5.13) and Young's inequality,

$$\frac{1}{2}\frac{d}{dt}\|\Lambda^s\theta\|_{L^2}^2 + \kappa\|\Lambda^{s+\alpha}\theta\|_{L^2}^2 \leqslant C\|\Lambda^{s+\alpha}\theta\|_{L^2}\|\Lambda^s\theta\|_{L^2}^{\frac{s+2-\alpha}{s}}\|\theta\|_{L^2}^{\frac{s-2+\alpha}{s}}$$

$$\leqslant \frac{\kappa}{2}\|\Lambda^{s+\alpha}\theta\|_{L^2}^2 + \frac{C}{\kappa}\|\Lambda^s\theta\|_{L^2}^{\frac{2(s+2-\alpha)}{s}}\|\theta\|_{L^2}^{\frac{2(s-2+\alpha)}{s}}.$$
$$(3.5.17)$$

If $\theta_0 \in L^2$, then

$$\frac{d}{dt}\|\Lambda^s\theta\|_{L^2}^2 + \kappa\|\Lambda^{s+\alpha}\theta\|_{L^2}^2 \leqslant \frac{C}{\kappa}\|\theta_0\|_{L^2}^{\frac{2(s-2+\alpha)}{s}}\|\Lambda^s\theta\|_{L^2}^{\frac{2(s+2-\alpha)}{s}}. \qquad (3.5.18)$$

Ignoring the positive term on the LHS and integrating over $[0, T]$ then yield

$$\|\Lambda^s\theta\|_{L^2}^2 \leqslant \|\Lambda^s\theta_0\|_{L^2}^2 \left[1 - \frac{Ct(2-\alpha)}{s\kappa}\|\theta_0\|_{L^2}^{\frac{2(s-2+\alpha)}{s}}\|\Lambda^s\theta_0\|_{L^2}^{\frac{2-2\alpha}{s}}\right]^{-\frac{s}{2-\alpha}}.$$

Therefore, one obtains the local existence of solutions in H^s for any given initial data $\theta_0 \in H^s$ for $s > 2 - \alpha$. Furthermore, $\theta \in L^2(0, T; \dot{H}^{s+\alpha})$ follows by integrating (3.5.18) over $[0, T]$. Similar to Case 2, it is standard that small initial data implies global existence in H^s. The details are omitted for clarity.

So far, the proof is formal since we only provide the *a priori* estimates. In order to give a rigorous proof, we can make use of the standard method of retard mollification (cf. [33]) to first obtain as above the uniform bounds for the mollified solutions and then pass to the limit to obtain the same bounds for the weak solution θ. The approximate solution can be constructed as follows. Let θ_n satisfy

$$\partial_t\theta_n + u_n \cdot \nabla\theta_n + \Lambda^{2\alpha}\theta_n = 0, \tag{3.5.19}$$

where $u_n = S_{\delta_n}(\theta_n)$ is

$$S_{\delta_n}(\theta_n) = \int_0^\infty \phi(\tau)\mathcal{R}^\perp\theta_n(t - \delta_n\tau)\mathrm{d}\tau,$$

and $\delta_n \to 0$. The function ϕ is smooth and non-negative, supported on $[1, 2]$ and $\int_0^\infty \phi(t)\mathrm{d}t = 1$. For any n, (3.5.19) is linear and $u_n(t)$ only depends on θ_n in $[t - 2\delta_n, t - \delta_n]$.

To complete the proof, we need to show uniqueness. First we state the following proposition, whose proof will be postponed.

Proposition 3.5.1 *Let $\kappa > 0$, $\alpha > 0$ and θ is a weak solution of the two-dimensional QG equation (3.5.2) with initial data $\theta_0 \in L^2$. If moreover for some $\varepsilon \in (0, \alpha]$ and $q < \infty$, there holds*

$$\int_0^T \|\Lambda^{1-\alpha+\varepsilon}\theta(s)\|_{L^p}^q\mathrm{d}s < \infty, \quad \frac{1}{p} + \frac{\alpha}{q} = \frac{\alpha+\varepsilon}{2},$$

then the weak solution is unique on $[0, T]$.

Using the proposition, we immediately know the uniqueness of the weak solution in the following cases.

1. When $\kappa > 0$, $\alpha \in (0, 1)$, $s = 2(1-\alpha)$, $\theta_0 \in L^2$ and $\|\Lambda^s\theta_0\|_{L^2} < \kappa/C_0$ holds strictly. In this case, the solution $\theta \in L^2(0, \infty; H^{s+\alpha}) \cap \theta \in L^\infty(0, \infty; H^s)$ is global. By interpolation, θ satisfy the criterion in Proposition 3.5.1 and uniqueness follows.

2. When $\kappa > 0$, $\alpha \in (0,1)$, $s \geqslant 2(1-\alpha)$ and $\theta_0 \in L^2$. Similarly, from interpolation uniqueness follows.

3. When $\kappa > 0$, $\alpha \in (0,1)$, $s > 2(1-\alpha)$ and $\theta_0 \in L^2$. In this case, one has $\theta \in L^\infty(0,\infty; H^s)$. Since $s > 2(1-\alpha)$, one can choose $q < \infty$ such that $H^s \hookrightarrow H^{1-\alpha+\varepsilon,p}$. Therefore, $\theta \in L^q(0,T; H^{1-\alpha+\varepsilon,p})$ and uniqueness follows.

Proof of Proposition 3.5.1 Suppose that θ_1, θ_2 are two solutions of (3.5.2) and $u_i = \mathcal{R}^\perp \theta_i$, $i = 1, 2$. Let $\theta = \theta_1 - \theta_2$ and $u = u_1 - u_2$, then

$$\theta_t + u_1 \cdot \nabla\theta + u \cdot \nabla\theta_2 + \kappa\Lambda^{2\alpha}\theta = 0.$$

Therefore, by taking inner product with φ, one has

$$(\theta_t, \varphi) + \kappa(\Lambda^\alpha\theta, \Lambda^\alpha\varphi) = -(u_1 \cdot \nabla\theta, \varphi) - (u \cdot \nabla\theta_2, \varphi).$$

Since u is divergence free, letting $\varphi = \theta$ then yields

$$\frac{1}{2}\frac{\mathrm{d}}{\mathrm{d}t}\|\theta\|_{L^2}^2 + \kappa\|\Lambda^\alpha\theta\|_{L^2}^2 = -(u \cdot \nabla\theta_2, \varphi) \leqslant C\|\nabla\theta_2\|_{L^\infty}\|\theta\|_{L^2}^2,$$

from which it follows by Gronwall inequality that

$$\|\theta\|_{L^2}^2 \leqslant C\|\theta_0\|_{L^2}^2 \exp\left\{\int_0^t \|\nabla\theta_2\|_{L^\infty}\mathrm{d}\tau\right\}. \tag{3.5.20}$$

Let $\alpha > 0$ and $\varepsilon \in (0, \alpha]$, then since $\nabla \cdot u = 0$, one has

$$-(u \cdot \nabla\theta_2, \theta) = -(\Lambda^{\alpha-\varepsilon}(\theta u), \Lambda^{-\alpha+\varepsilon}\nabla\theta_2) \leqslant \|\Lambda^{1-\alpha+\varepsilon}\theta_2\|_{L^{p_1}}\|\Lambda^{\alpha-\varepsilon}(\theta u)\|_{L^{p_1'}},$$

where by calculus inequalities

$$\|\Lambda^{\alpha-\varepsilon}(\theta u)\|_{L^{p_1'}} \leqslant C\|\theta\|_{L^{q_1}}\|u\|_{\dot{H}^{\alpha-\varepsilon,q_2}} + C\|u\|_{L^{q_1}}\|\theta\|_{\dot{H}^{\alpha-\varepsilon,q_2}}$$

$$\leqslant C\|\theta\|_{L^{q_1}}\|\theta\|_{\dot{H}^{\alpha-\varepsilon,q_2}},$$

where $\dfrac{1}{q_1} + \dfrac{1}{q_2} = \dfrac{1}{p_1'} = 1 - \dfrac{1}{p_1}$. Therefore

$$\frac{1}{2}\frac{\mathrm{d}}{\mathrm{d}t}\|\theta\|_{L^2}^2 + \kappa\|\theta\|_{\dot{H}^\alpha}^2 \leqslant C\|\theta\|_{L^{q_1}}\|\theta\|_{\dot{H}^{\alpha-\varepsilon,q_2}}\|\theta_2\|_{\dot{H}^{1-\alpha+\varepsilon,p_1}}.$$

Furthermore, by Gagliardo-Nirenberg inequality

$$\|\theta\|_{L^{q_1}} \leqslant C\|\theta\|_{L^2}^{1-\beta}\|\theta\|_{\dot{H}^\alpha}^\beta, \qquad \beta = \frac{1}{\alpha}\left(1 - \frac{2}{q_1}\right) \in (0,1);$$

$$\|\theta\|_{\dot{H}^{\alpha-\varepsilon,q_2}} \leqslant C\|\theta\|_{\dot{H}^\alpha}^\gamma\|\theta\|_{L^2}^{1-\gamma}, \qquad \gamma = \frac{1}{\alpha}\left(1 + \alpha - \varepsilon - \frac{2}{q_2}\right) \in (0,1),$$

which yields

$$\frac{1}{2}\frac{\mathrm{d}}{\mathrm{d}t}\|\theta\|_{L^2}^2 + \kappa\|\theta\|_{\dot{H}^\alpha}^2 \leqslant C\|\theta_2\|_{\dot{H}^{1-\alpha+\varepsilon,p_1}}\|\theta\|_{L^2}^{2-(\beta+\gamma)}\|\theta\|_{\dot{H}^\alpha}^{\beta+\gamma}.$$

Let $p_2 = 2/(\beta+\gamma)$, then by Young inequality

$$\frac{1}{2}\frac{\mathrm{d}}{\mathrm{d}t}\|\theta\|_{L^2}^2 + \kappa\|\theta\|_{\dot{H}^\alpha}^2 \leqslant \frac{\kappa}{2}\|\theta\|_{\dot{H}^\alpha}^2 + C(\kappa)\|\theta_2\|_{\dot{H}^{1-\alpha+\varepsilon,p_1}}^{p_2'}\|\theta\|_{L^2}^2,$$

where $p_2' = p_2/(p_2-1) = 2/(2-(\beta+\gamma))$. Thus

$$\frac{\mathrm{d}}{\mathrm{d}t}\|\theta\|_{L^2}^2 + \kappa\|\theta\|_{\dot{H}^\alpha}^2 \leqslant C\|\theta_2\|_{\dot{H}^{1-\alpha+\varepsilon,p_1}}^{p_2'}\|\theta\|_{L^2}^2,$$

from which it follows from Gronwall inequality that

$$\|\theta\|_{L^2}^2 \leqslant C\|\theta_0\|_{L^2}^2 \exp\left\{\int_0^T \|\theta_2\|_{\dot{H}^{1-\alpha+\varepsilon,p_1}}^{p_2'}\,\mathrm{d}\tau\right\}.$$

We complete the proof by identifying $p_1 = p$, $p_2' = q$ and noting $\beta+\gamma = \frac{1}{\alpha}\left(\alpha + \varepsilon + \frac{2}{p_1}\right)$.

Remark 3.5.1 *By (3.5.20), the solution is unique if $\displaystyle\int_0^T \|\boldsymbol{\nabla}\theta\|_{L^\infty}\mathrm{d}t < \infty$, which is the BKM blow up criterion for (3.5.2). Refer to Theorem 3.5.4.*

In particular, when $\alpha \in (1/2, 1]$, we have

Theorem 3.5.6 *Let $\alpha \in (1/2, 1]$ and $s \geqslant 0$ satisfy $s + 2\alpha > 2$, then if $\theta_0 \in H^s(\mathbf{T}^2)$, there holds for solution of the two dimensional EQ equation (3.5.2) that*

$$\|\Lambda^s\theta(t)\|_{L^2} \leqslant C, \qquad \forall t \leqslant T,$$

where C is constant depending only on T and $\|\theta_0\|_{H^s}$.

Proof Taking inner product of (3.5.1) and $\Lambda^{2s}\theta$ yields

$$\frac{1}{2}\frac{\mathrm{d}}{\mathrm{d}t}\|\Lambda^s\theta\|_{L^2}^2 + \kappa\|\Lambda^{s+\alpha}\theta\|_{L^2}^2 = -\left((u\cdot\boldsymbol{\nabla}\theta), \Lambda^{2s}\theta\right).$$

By maximum principle, Lemma 3.5.1 and multiplicative estimates,

$$\left|\left((u\cdot\boldsymbol{\nabla}\theta), \Lambda^{2s}\theta\right)\right| \leqslant \frac{\kappa}{2}\|\Lambda^{s+\alpha}\theta\|_{L^2}^2 + C(\kappa)\|\Lambda^s\theta\|_{L^2}^2.$$

The result follows from Gronwall inequality.

For later applications, we make the L^p-estimates in the following

Lemma 3.5.1   *Let θ be a solution of the 2D QG equation, then for any $p \in (1, \infty)$, there holds*

$$\|\theta(t)\|_{L^p} \leqslant \|\theta_0\|_{L^p}.$$

Proof Recall that for any $p \in (1, \infty)$, there holds

$$\int |\theta|^{p-2}\theta\Lambda^s\theta\mathrm{d}x \geqslant \frac{1}{p}\int |\Lambda^{s/2}\theta^{p/2}|^2\mathrm{d}x.$$

Therefore, multiplying the 2D QG equation by $p|\theta|^{p-2}\theta$, integrating over $x \in \mathbf{T}^2$ and noting $\nabla \cdot u = 0$, one has

$$\frac{\mathrm{d}}{\mathrm{d}t}\|\theta\|_{L^p}^p + \kappa\int |\Lambda^\alpha\theta^{p/2}|^2\mathrm{d}x \leqslant 0,$$

completing the proof since $\kappa \geqslant 0$.

3.5.2   Inviscid limit

This subsection considers the inviscid limit of the 2D QG equation when $\kappa \to 0$. In Theorem 3.5.4, we have obtained local existence of smooth solutions for $\kappa \geqslant 0$. In what follows, we denote the solutions by $(\theta_\kappa, u_\kappa)$ when $\kappa > 0$, and simply by (θ, u) when $\kappa = 0$.

Theorem 3.5.7   *Let $\Omega = \mathbf{R}^2$ or $\mathbf{T}^2$, θ and θ_κ are solutions of two-dimensional QG equations (3.5.1) and (3.5.2) with the same initial data $\theta_0 \in H^s$ with $s \geqslant 3$, respectively. If $[0, T_*)$ is maximal time interval of existence, then for any $t < T_*$,*

$$\|\theta(t) - \theta_\kappa(t)\|_{L^2} \leqslant C\kappa,$$

where C is constant depending on θ_0 and T_* only. In particular, C does not depends on κ.

Proof Let $\Theta = \theta_\kappa - \theta$ and $U = u_\kappa - u$. Then Θ satisfy

$$\partial_t\Theta + u_\kappa \cdot \nabla\Theta + U \cdot \nabla\theta + \kappa\Lambda^{2\alpha}(\Theta + \theta) = 0.$$

Taking inner product with Θ yields

$$\frac{1}{2}\frac{\mathrm{d}}{\mathrm{d}t}\|\Theta(t)\|_{L^2}^2 + \kappa\|\Lambda^\alpha\Theta\|_{L^2}^2 = (u_\kappa \cdot \nabla\Theta, \Theta) + (U \cdot \nabla\theta, \Theta) + \kappa(\Lambda^{2\alpha}\theta, \Theta)$$
$$= I_1 + I_2 + I_3.$$

It follows that $I_1 = 0$ by integration by parts. For I_2 and I_3, we have

$$|I_2| \leqslant \|\nabla\theta\|_{L^\infty}\|U\|_{L^2}\|\Theta\|_{L^2} = \|\nabla\theta\|_{L^\infty}\|\Theta\|_{L^2}^2,$$
$$|I_3| \leqslant \frac{\kappa^2}{2}\|\Lambda^{2\alpha}\theta\|_{L^2}^2 + \frac{1}{2}\|\Theta\|_{L^2}^2.$$

It then follows that

$$\frac{\mathrm{d}}{\mathrm{d}t}\|\Theta\|_{L^2}^2 + \kappa\|\Lambda^\alpha\Theta\|_{L^2}^2 \leqslant (2\|\nabla\theta\|_{L^\infty} + 1)\|\Theta\|_{L^2}^2 + \kappa^2\|\theta\|_{H^{2\alpha}}^2.$$

By Gronwall inequality, we then have

$$\|\Theta\|_{L^2}^2 \leqslant e^{\int_0^t (2\|\nabla\theta\|_{L^\infty}+1)\mathrm{d}s}\|\Theta_0\|_{L^2}^2 + \kappa^2 \int_0^t e^{\int_\tau^t (2\|\nabla\theta\|_{L^\infty}+1)\mathrm{d}s}\|\theta\|_{H^{2\alpha}}^2 \mathrm{d}\tau.$$

Since $\Theta_0 = 0$, it then follows from Theorem 3.5.4 that $\|\Theta\|_{L^2} \leqslant C\kappa$, completing the proof.

Next, we state the inviscid limit result in space $H^m(\mathbf{R}^2)$. For this we first note

Lemma 3.5.2 *Let* $m \geqslant 2$, $\theta \in H^{m+1}$ *and* $u \in H^m$ *with* $\nabla \cdot u = 0$, *then*

$$\|u \cdot \nabla\theta\|_{H^m} \leqslant C\|u\|_{H^m}\|\theta\|_{H^{m+1}}, \tag{3.5.21}$$

$$|(u \cdot \nabla\theta, \theta)_m| \leqslant C\|u\|_{H^m}\|\theta\|_{H^m}^2, \tag{3.5.22}$$

$$|(u \cdot \nabla\theta, \theta)_2| \leqslant C\|u\|_{H^3}\|\theta\|_{H^2}^2, \tag{3.5.23}$$

where $(\cdot, \cdot)$ *represents scalar product in* H^m.

Proof It suffices to note that H^m, $m \geqslant 2$ is an algebra.

Theorem 3.5.8 *Let* $\alpha \in \left(\dfrac{1}{2}, 1\right]$, $\theta_0 \in H^m$, $m \geqslant 3$, *then*

(1) there exists $0 < T_0 \leqslant T$ *depending on* $\|\theta_0\|_{H^m}$ *but independent of* κ, *such that there exists a unique solution of* (3.5.2)

$$\theta_\kappa \in C([0, T_0]; H^m) \cap AC([0, T_0]; H^{m-1}) \cap L^1(0, T_0; H^{m+\alpha}). \tag{3.5.24}$$

Furthermore, $\{\theta_\kappa\}_{\kappa \geqslant 0}$ *is uniformly bounded in* $C([0, T_0]; H^m)$.

(2) the limit $\theta(t) := \lim\limits_{\kappa \to 0} \theta_\kappa(t)$ *exists strongly in* H^{m-1} *and weakly in* H^m *uniformly for* $t \in [0, T_0]$. *And* θ *is a solution of* (3.5.1) *such that*

$$\theta \in C([0, T_0]; H^m) \cap AC([0, T_0]; H^{m-1}).$$

Proof The existence of smooth solutions have already been shown in Theorem 3.5.5 and 3.5.6.

Now, we show that $\{\theta_\kappa\}$ is uniformly bounded. Taking H^m inner product with θ_κ, one has

$$\frac{1}{2}\frac{\mathrm{d}}{\mathrm{d}t}\|\theta_\kappa\|_{H^m}^2 + \kappa(\Lambda^{2\alpha}\theta_\kappa(t), \theta_\kappa(t))_m \leqslant C\|\theta_\kappa\|_{H^m}^3. \tag{3.5.25}$$

From the positivity of the second term on the LHS, it follows

$$\frac{d}{dt}\|\theta_\kappa\|_{H^m} \leqslant C\|\theta_\kappa\|_{H^m}^2.$$

By comparison principle, we have $\|\theta_\kappa(t)\|_{H^m} \leqslant \varphi(t)$, where $\varphi(t)$ is solution of the ODE

$$\frac{d\varphi}{dt} = C\varphi^2(t) \tag{3.5.26}$$

with initial data $\varphi(0) = \|\theta_0\|_{H^m}$. From classical theory of ODEs, there exist $T_0 > 0$ and φ such that φ is absolutely continuous in $[0, T_0]$ and satisfy (3.5.26). In particular, it follows the local existence of smooth solutions by Faedo-Galerkin approximation and such *a priori* estimates. It can be seen that φ and T_0 are independent of κ. From (3.5.25), there exists a continuous function ψ on $[0, T_0]$ depending only on $\|\theta_0\|_{H^m}$ such that

$$\kappa \int_0^t (\Lambda^{2\alpha}\theta_\kappa(t), \theta_\kappa(t))_m ds \leqslant \psi(t), \quad \forall t \in [0, T_0]. \tag{3.5.27}$$

Let $\kappa_1 < \kappa_2$, $\Theta = \theta_{\kappa_1} - \theta_{\kappa_2}$ and $U = u_{\kappa_1} - u_{\kappa_2}$, then

$$\frac{\partial \Theta_k}{\partial t} + \kappa_1 \Lambda^{2\alpha}\Theta + (\kappa_1 - \kappa_2)\Lambda^{2\alpha}\theta_{\kappa_2} = -U \cdot \nabla\theta_{\kappa_1} = u_{\kappa_2} \cdot \nabla\Theta.$$

Taking H^{m-1} inner product with Θ, and using (3.5.4) and Lemma 3.5.2, one obtains

$$\frac{1}{2}\frac{d}{dt}\|\Theta\|_{H^{m-1}}^2 \leqslant (\kappa_2 - \kappa_1)(\Lambda^{2\alpha}\theta_{\kappa_2}, \Theta)_{m-1} + C_2 C(\|\theta_{\kappa_1}\|_{H^m} + \|\theta_{\kappa_2}\|_{H^m})\|\Theta\|_{m-1}^2.$$

Since $\|\theta_\kappa\|_{H^m}$ is uniformly bounded in $[0, T_0]$, we have

$$\frac{d}{dt}\|\Theta\|_{H^{m-1}} \leqslant \kappa_2\|\Lambda^{2\alpha}\theta_{\kappa_2}\|_{H^{m-1}} + K\|\Theta\|_{H^{m-1}},$$

where $K = C\varphi(T_0)$ is a constant independent of κ_1 and κ_2. It follows

$$\|\Theta\|_{H^{m-1}} \leqslant \kappa_2 e^{Kt} \int_0^t \|\Lambda^{2\alpha}\theta_{\kappa_2}(s)\|_{H^{m-1}} ds$$

$$\leqslant \sqrt{\kappa_2 t}\, e^{Kt} \left(\kappa_2 \int_0^t \|\Lambda^{2\alpha}\theta_{\kappa_2}(s)\|_{H^{m-1}}^2 ds\right)^{1/2},$$

where we have used $\Theta_0 = 0$.

Since

$$\|\Lambda^{2\alpha}\theta_{\kappa_2}\|_{H^{m-1}}^2 \sim \|(I + \Lambda^2)^{-1/2}\Lambda^{2\alpha}\theta_{\kappa_2}\|_{H^m}^2 \leqslant \|\Lambda^{2\alpha-1}\theta_{\kappa_2}\|_{H^m}^2,$$

then for $2\alpha - 1 \leqslant \alpha$,

$$\|\Lambda^{2\alpha}\theta_{\kappa_2}\|_{H^{m-1}}^2 \leqslant \|\Lambda^\alpha\theta_{\kappa_2}\|_{H^m}^2 = (\Lambda^{2\alpha}\theta_{\kappa_2}, \theta_{\kappa_2})_m.$$

It then follows from (3.5.27) that

$$\|\Theta\|_{H^{m-1}} \leqslant \sqrt{\kappa_2}te^{Kt}\psi(t)^{1/2},$$

yielding

$$\lim_{\kappa_2 \to 0} \|\Theta(t)\|_{H^{m-1}} = 0.$$

Therefore, the limit $\theta(x,t) = \lim_{\kappa \to 0}\theta_\kappa(x,t)$ exists strongly in H^{m-1}, and is uniform in $[0, T_0]$. Thus $\theta(\cdot, t) \in H^{m-1}$ is a continuous function of time. Moreover, $\|\theta_\kappa\|_{H^m}$ is bounded in $[0, T_0]$ thanks to (3.5.26), hence $\theta(t) \in H^m$ for any $t \in [0, T_0]$, $\theta_\kappa \rightharpoonup \theta(t)$ weakly in H^m and uniformly in $[0, T_0]$ and $\theta(t)$ is weakly continuous in H^m.

We have already shown that θ_κ belongs to the class of functions in (3.5.24), hence $u_\kappa \cdot \nabla\theta_\kappa(t) \rightharpoonup u \cdot \nabla\theta(t)$ weakly in H^{m-1} and uniformly in $[0, T_0]$. Therefore, $u \cdot \nabla\theta(t)$ is weakly continuous in H^{m-1}. Next we show that θ is a solution of (3.5.1). Fix $\kappa > 0$, integrating (3.5.1) over $[t_1, t_2]$ then yields

$$\theta_\kappa(t_2) - \theta_\kappa(t_1) = -\int_{t_1}^{t_2} \kappa\Lambda^{2\alpha}\theta_\kappa + u_\kappa \cdot \nabla\theta_\kappa \mathrm{d}\tau.$$

Let $\zeta \in H^{m-1}$ be a smooth function, then taking inner product of the above equation with ζ in H^{m-1} yields

$$(\theta(t_2) - \theta(t_1), \zeta)_{m-1} = -\int_{t_1}^{t_2} (u \cdot \nabla\theta, \zeta)_{m-1}\mathrm{d}\tau.$$

When $\kappa \to 0$, obviously $\kappa(\Lambda^{2\alpha}\theta_\kappa, \zeta)_{m-1} = \kappa(\theta_\kappa, \Lambda^{2\alpha}\zeta)_{m-1} \to 0$. Therefore,

$$\theta(t_2) - \theta(t_1) = -\int_{t_1}^{t_2} u \cdot \nabla\theta\mathrm{d}\tau.$$

It then follows that θ is a solution of (3.5.1) belonging to

$$\theta \in L^\infty(0, T_0; H^m) \cap AC([0, T_0]; H^{m-1}).$$

Uniqueness follows from Proposition 3.5.1. Finally, since θ is weakly continuous in H^m and $\limsup_{t \to 0} \|\theta(t)\|_{H^m} \leqslant \|\theta_0\|_{H^m}$, it is easy to show θ is right continuous at $t = 0$ and hence right continuous at any $t \in [0, T_0]$. By time reverse, the left continuity of $\theta(t) \in H^m$ is obtained. The proof is complete.

Remark 3.5.2 *1. Theorem 3.5.8 still holds if the 2D QG equation has an external force term $f \in L^1(0, T; H^m)$, $m \geqslant 3$.*

2. Theorem 3.5.8 still holds in the periodic case.

The following theorem concerns the inviscid limit of weak solutions with initial data in L^2.

Theorem 3.5.9   *Let* $\theta_0 \in L^2(\mathbf{T}^2)$, θ, θ_κ *are weak solutions of equation* (3.5.1) *and* (3.5.2) *with the same initial data* θ_0, *respectively. Then for any* $T > 0$ *and any* $\varphi \in L^2(\mathbf{T}^2)$,

$$\limsup_{\kappa \to 0}(\theta_\kappa(\cdot, t) - \theta(\cdot, t), \varphi) = 0, \quad \forall t \leqslant T.$$

Proof Consider the Galerkin approximate sequence of solutions $\theta^n \in S_n$ and $\theta_\kappa^n \in S_n$, where $S_n = span\{e^{imx}\}$, $0 < |m| \leqslant n$. Then there are suitable subsequences such that $\theta_\kappa^n \to \theta_\kappa$ and $\theta^n \to \theta$ weakly in $L^2(\mathbf{T}^2)$. Therefore when n is large enough

$$\begin{aligned}
|(\theta_\kappa(\cdot, t) - \theta(\cdot, t), \varphi)| &\leqslant \varepsilon + |(\theta_\kappa^n - \theta^n, \varphi)| \\
&\leqslant \varepsilon + \|\varphi\|_{L^2}\|\theta_\kappa^n - \theta^n\|_{L^2} \\
&\leqslant \varepsilon + C_n \kappa,
\end{aligned}$$

where in the last step, we have used the inviscid limit of smooth solutions in Theorem 3.5.7. The proof is complete.

3.5.3   Decay and approximation

This subsection considers the decay and approximation of the solutions of the 2D QG equation.

Theorem 3.5.10   *Let* $\alpha \in (0, 1]$, $\theta_0 \in L^1(\mathbf{R}^2) \cap L^2(\mathbf{R}^2)$. *Then there exists a weak solution* θ *of equation* (3.5.2) *such that*

$$\|\theta(\cdot, t)\|_{L^2(\mathbf{R}^2)} \leqslant C(1 + t)^{-\frac{1}{2\alpha}},$$

where C *is a constant depending on* $\|\theta_0\|_{L^1}$ *and* $\|\theta_0\|_{L^2}$.

Proof Taking Fourier transform of the equation to obtain

$$\partial_t \hat{\theta} + |\xi|^{2\alpha}\theta = -\widehat{u \cdot \nabla\theta}.$$

Since $\nabla \cdot u = 0$, $|\widehat{u \cdot \nabla\theta}| \leqslant |\xi|\|\theta\|_{L^2}^2$ by Hölder inequality, and hence it follows from the Gronwall inequality that

$$|\hat{\theta}(\xi, t)| \leqslant |\hat{\theta}_0(\xi)| + |\xi| \int_0^t \|\theta\|_{L^2}^2 d\tau \leqslant \|\theta_0\|_{L^1} + |\xi|\|\theta_0\|_{L^2}^2 t. \quad (3.5.28)$$

Taking inner product of equation (3.5.2) with θ to obtain

$$\frac{1}{2}\frac{d}{dt}\int_{\mathbf{R}^2}|\theta|^2 + \int_{\mathbf{R}^2}|\Lambda^\alpha\theta|^2 = 0,$$

which implies by Plancherel identity that

$$\frac{\mathrm{d}}{\mathrm{d}t}\int_{\mathbf{R}^2}|\hat{\theta}|^2 + 2\int_{\mathbf{R}^2}|\xi|^{2\alpha}|\hat{\theta}|^2 = 0.$$

For the second term, one has

$$\int_{\mathbf{R}^2}|\xi|^{2\alpha}|\hat{\theta}|^2 \geqslant \int_{B(t)^c}|\xi|^{2\alpha}|\hat{\theta}|^2 \geqslant g^{2\alpha}(t)\int_{B(t)^c}|\hat{\theta}|^2$$

$$= g^{2\alpha}(t)\int_{\mathbf{R}^2}|\hat{\theta}|^2 - g^{2\alpha}(t)\int_{B(t)}|\hat{\theta}|^2,$$

where $g \in C([0,\infty); \mathbf{R}^+)$ is to be determined, $B(t) = \{\xi \in \mathbf{R}^2 : |\xi| < g(t)\}$ and $B(t)^c$ is its complement. From (3.5.28) it then follows

$$\frac{\mathrm{d}}{\mathrm{d}t}\int_{bfR^2}|\hat{\theta}|^2 + 2g^{2\alpha}(t)\int_{\mathbf{R}^2}|\hat{\theta}|^2$$

$$\leqslant 2\pi g^{2\alpha}(t)\int_0^{g(t)}\left[\|\theta_0\|_{L^1} + r\int_0^t\|\theta(\tau)\|_{L^2}^2\mathrm{d}\tau\right]^2 r\mathrm{d}r,$$

which yields by integrating on $[0,t]$

$$e^{2\int_0^t g^{2\alpha}(\tau)\mathrm{d}\tau}\int_{\mathbf{R}^2}|\hat{\theta}|^2$$

$$\leqslant \|\theta_0\|_{L^2}^2 + \int_0^t e^{2\int_0^s g^{2\alpha}(\tau)\mathrm{d}\tau}\left[C_1 g^{2\alpha+2}(s) + C_2 s g^{2\alpha+4}(s)\int_0^s\|\theta(\tau)\|_{L^4}^4\mathrm{d}\tau\right]\mathrm{d}s,$$

where $C_1 = 2\pi\|\theta_0\|_{L^1}^2$ and $C_2 = \pi$. Let $g^{2\alpha}(t) = \left(\frac{1}{2}+\frac{1}{2\alpha}\right)[(e+t)\ln(e+t)]^{-1}$, then $e^{2\int_0^t g^{2\alpha}(\tau)\mathrm{d}\tau} = [\ln(e+t)]^{1+\frac{1}{\alpha}}$, and hence

$$\|\theta\|_{L^2}^2 \leqslant C[\ln(e+t)]^{-1-\frac{1}{\alpha}}.$$

Let $g^{2\alpha} = \frac{1}{2\alpha(t+1)}$, then $e^{2\int_0^t g^{2\alpha}(\tau)\mathrm{d}\tau} = (1+t)^{1/\alpha}$, thus

$$\|\theta(t)\|_{L^2}^2 \leqslant C(1+t)^{-1/\alpha} + C(1+t)^{1-\frac{2}{\alpha}}\int_0^t\|\theta(s)\|_{L^2}^2[\ln(e+s)]^{-1-\frac{1}{\alpha}}\mathrm{d}s.$$

It then follows from Gronwall inequality that

$$\|\theta(t)\|_{L^2}^2 \leqslant C(1+t)^{-1/\alpha}, \quad \alpha \leqslant 1,$$

where C depends on $\|\theta_0\|_{L^1}$ and $\|\theta_0\|_{L^2}$. So far, we have proved the theorem formally. It can be made rigorous by the retard mollification method in the proof of Theorem 3.5.5.

Similarly, we have

Theorem 3.5.11 *Let $\alpha \in (0,1]$, $\theta_0 \in L^1(\mathbf{R}^2) \cap L^2(\mathbf{R}^2)$. If $f \in L^1([0,\infty); L^2)$, and there exists constant C such that*

$$\|f(\cdot,t)\|_{L^2} \leqslant C(1+t)^{-1-\frac{1}{\alpha}}, \qquad |\hat{f}(\xi,t)| \leqslant C|\xi|^\alpha, \qquad (3.5.29)$$

then there exists a weak solution θ of the QG equation such that

$$\|\theta(\cdot,t)\|_{L^2} \leqslant C(1+t)^{-\frac{1}{2\alpha}}.$$

We will establish derivative estimates of solutions.

Theorem 3.5.12 *Let $\alpha \in (1/2,1]$, $\beta \geqslant \alpha$ and $\dfrac{2}{2\alpha-1} < q < \infty$. Suppose that $\theta_0 \in L^1 \cap L^2$, $\Lambda^\beta \theta_0 \in L^2$ and $f \in L^1([0,\infty]; L^q \cap L^2)$ satisfies (3.5.29) and $\Lambda^{\beta-\alpha} f \in L^2((0,\infty); L^2)$. Then the solution θ of (3.5.5) satisfies*

$$\|\Lambda^\beta \theta(t)\|_{L^2} \leqslant C_0(1+t)^{-\frac{1}{2\alpha}} + C_1 \left(\int_0^t \|\Lambda^{\beta-\alpha} f(s)\|_{L^2}^2 \mathrm{d}s \right)^{1/2}, \qquad \forall t \geqslant 0,$$

$$(3.5.30)$$

where C_0, C_1 depend only on θ_0 and f.

Proof We only give the formal proof. Take the inner product of (3.5.5) with $\Lambda^{2\beta}\theta(t)$ to obtain

$$\frac{1}{2}\frac{\mathrm{d}}{\mathrm{d}t}\int_{\mathbf{R}^2} |\Lambda^\beta\theta(t)|^2\mathrm{d}x + \kappa\int_{\mathbf{R}^2} |\Lambda^{\alpha+\beta}\theta(t)|^2\mathrm{d}x$$

$$= -\int_{\mathbf{R}^2}(u\cdot\nabla\theta)\Lambda^{2\beta}\theta\mathrm{d}x + \int_{\mathbf{R}^2} f\Lambda^{2\beta}\theta\mathrm{d}x = I_1 + I_2.$$

$$(3.5.31)$$

For the term I_2, we have

$$|I_2| \leqslant \frac{\kappa}{8}\int_{\mathbf{R}^2} |\Lambda^{\alpha+\beta}\theta(t)|^2\mathrm{d}x + \frac{2}{\kappa}\int_{\mathbf{R}^2} |\Lambda^{\beta-\alpha}f|^2\mathrm{d}x. \qquad (3.5.32)$$

Next we will show that for I_1, we have

$$|I_1| \leqslant \frac{\kappa}{8}\|\Lambda^{\alpha+\beta}\theta(t)\|_{L^2}^2 + C_0(\theta_0, \kappa, f)\|\Lambda^{s+1-\frac{2}{p}}\theta\|_{L^2}^2, \qquad (3.5.33)$$

where $s = \beta - \alpha + 1$, $\dfrac{1}{p} + \dfrac{1}{q} + \dfrac{1}{2}$. Indeed, since $u = (u_1, u_2)$ and $\nabla \cdot u = 0$, then $u \cdot \nabla\theta = \nabla \cdot (u\theta)$. Using Plancherel theorem and Hölder inequality, we obtain

$$\left| \int_{\mathbf{R}^2} (u \cdot \nabla \theta) \Lambda^{2\beta} \theta dx \right| = \left| \int_{\mathbf{R}^2} (\xi_1 \widehat{\theta u_1}(\xi) + \xi_2 \widehat{\theta u_2}(\xi)) |\xi|^{2\beta} \hat{\theta}(\xi) d\xi \right|$$

$$\leqslant \sum_{i=1}^{2} \int_{\mathbf{R}^2} |\xi|^{\beta - \alpha + 1} |\widehat{\theta u_i}(\xi)| |\xi|^{\alpha + \beta} |\hat{\theta}(\xi)| d\xi \qquad (3.5.34)$$

$$\leqslant \|\Lambda^{\beta - \alpha + 1}(\theta u)\|_{L^2} \|\Lambda^{\alpha + \beta} \theta\|_{L^2}$$

$$\leqslant \frac{\kappa}{8} \|\Lambda^{\alpha + \beta} \theta\|_{L^2}^2 + \frac{2}{\kappa} \|\Lambda^{s}(\theta u)\|_{L^2}^2,$$

where $s = \beta - \alpha + 1$. By multiplicative estimates, we obtain

$$\|\Lambda^{s}(\theta u)\|_{L^2} \leqslant C(\|u\|_{L^q} \|\Lambda^{s}\theta\|_{L^p} + \|\theta\|_{L^q} \|\Lambda^{s}u\|_{L^p}).$$

By the boundedness of Riesz operator, for any $p \in (2, \infty)$ there exists constant C such that $\|\Lambda^{s}u\|_{L^p} \leqslant C\|\Lambda^{s}\theta\|_{L^p}$ and $\|u\|_{L^q} \leqslant C\|\theta\|_{L^q}$. Therefore,

$$\|\Lambda^{s}(\theta u)\|_{L^2} \leqslant C\|\theta\|_{L^q} \|\Lambda^{s}\theta\|_{L^p}, \quad i = 1, 2. \qquad (3.5.35)$$

On the other hand, multiplying (3.5.5) with $q|\theta|^{q-2}\theta$ and integrating over $\mathbf{R}^2$, one has

$$\frac{d}{dt} \|\theta\|_{L^q}^q \leqslant q \left(\int |\theta|^{q-2}\theta f dx - \int |\theta|^{q-2}\theta (u \cdot \nabla \theta) dx - \kappa \int |\theta|^{q-2}\theta(-\Delta)^{\alpha}\theta \right).$$

Since $\nabla \cdot u = 0$, the second item on right hand side is zero, and by the positivity of the fractional Laplacian $(-\Delta)^{\alpha}$, the third term on the right hand side is nonnegative, we obtain

$$\frac{d}{dt} \|\theta\|_{L^q}^q \leqslant q \int |\theta|^{q-2}\theta f dx \leqslant q\|f\|_{L^q} \|\theta\|_{L^q}^{q-1}.$$

This leads to the L^p estimates of the solution,

$$\|\theta\|_{L^q} \leqslant \|\theta_0\|_{L^q} + \int_0^t \|f(\tau)\|_{L^q} d\tau. \qquad (3.5.36)$$

Inserting this into (3.5.35) and noting the hypothesis $f \in L^1(0, \infty; L^q)$, one obtains

$$\|\Lambda^{s}(\theta u)\|_{L^2} \leqslant C(\theta_0, f)\|\Lambda^{s}\theta\|_{L^p} \leqslant C(\theta_0, f)\|\Lambda^{s+1-\frac{2}{p}}\theta\|_{L^2}.$$

Inserting this into (3.5.34), one then obtains (3.5.33), where $C(\kappa, \theta_0, f) = \frac{4}{\kappa}C(\theta_0, f)^2$. Finally, from (3.5.31), (3.5.32) and (3.5.33), we obtain

$$\frac{1}{2}\frac{d}{dt} \|\Lambda^{\beta}\theta\|_{L^2}^2 dx + \frac{3\kappa}{4} \|\Lambda^{\alpha+\beta}\theta\|_{L^2}^2 dx \leqslant C_0 \|\Lambda^{\prime}\theta\|_{L^2}^2 + \frac{2}{\kappa} \|\Lambda^{\beta-\alpha}f\|_{L^2}^2, \quad (3.5.37)$$

where $C_0 = C(\kappa, \theta, f)$ and $\gamma = s + 1 - \dfrac{2}{p} = \beta - \alpha + 2\left(1 - \dfrac{1}{p}\right)$.

Let $B_M = \{\xi : |\xi|^2 \leqslant M\}$, where $M > 0$ is to be determined. Choose $\dfrac{2}{2\alpha - 1} < q < \infty$, then $\dfrac{d}{2} > \dfrac{1}{p} = \dfrac{1}{2} - \dfrac{1}{q} > 1 - \alpha$ and $\dfrac{1}{p} + \alpha - 1 > 0$. In this case, $\gamma = \alpha + \beta - 2\left(\dfrac{1}{p} + \alpha - 1\right) < \alpha + \beta$ and hence

$$\|\Lambda^\gamma \theta\|_{L^2}^2 = \int_{B_M} |\xi|^{2\gamma} |\hat{\theta}(t)|^2 \mathrm{d}\xi + \int_{B_M^c} |\xi|^{2\gamma} |\hat{\theta}(t)|^2 \mathrm{d}\xi$$

$$\leqslant M^{2\gamma} \|\theta(t)\|_{L^2}^2 + M^{-4\left(\frac{1}{p} + \alpha - 1\right)} \|\Lambda^{\alpha+\beta} \theta(t)\|_{L^2}^2.$$

Lete M be sufficiently large such that $M^{-4\left(\frac{1}{p} + \alpha - 1\right)} < \dfrac{\kappa}{4C_0}$ then yields

$$C_0 \|\Lambda^\gamma \theta(t)\|_{L^2}^2 \leqslant \frac{\kappa}{4} \|\Lambda^{\alpha+\beta} \theta(t)\|_{L^2}^2 + C_0 M^{2\gamma} \|\theta(t)\|_{L^2}^2. \tag{3.5.38}$$

Furthermore, since

$$\|\Lambda^{\alpha+\beta} \theta\|_{L^2}^2 \geqslant \int_{B_M^c} |\xi|^{2(\alpha+\beta)} |\hat{\theta}|^2 \mathrm{d}\xi \geqslant M^{2\alpha} \int_{B_M^c} |\xi|^{2\beta} |\hat{\theta}|^2 \mathrm{d}\xi$$

$$= M^{2\alpha} \|\Lambda^\beta \theta\|_{L^2}^2 - M^{2\alpha} \int_{B_M} |\xi|^{2\beta} |\hat{\theta}|^2 \mathrm{d}\xi,$$

then we obtain

$$\|\Lambda^{\alpha+\beta} \theta(t)\|_{L^2}^2 \geqslant M^{2\alpha} \|\Lambda^\beta \theta\|_{L^2}^2 - M^{2(\alpha+\alpha)} \|\theta(t)\|_{L^2}^2. \tag{3.5.39}$$

From Theorem 3.5.11, (3.5.37), (3.5.38) and (3.5.39), we have

$$\frac{\mathrm{d}}{\mathrm{d}t} \|\Lambda^\beta \theta(t)\|_{L^2}^2 + \kappa M \|\Lambda^\beta \theta(t)\|_{L^2}^2 \leqslant \tilde{C}_0 M^c (1+t)^{-\frac{1}{\alpha}} + \frac{2}{\kappa} \|\Lambda^{\beta-\alpha} f(t)\|_{L^2}^2, \tag{3.5.40}$$

where $\tilde{C}_0$ is a constant depending on f, θ_0 and κ and $c = \max\{2\gamma, 2\alpha + 2\beta\}$. Let $\nu = \kappa M^{2\alpha}$, then multiplying (3.5.40) with $e^{\nu t}$ yields

$$\|\Lambda^\beta \theta(t)\|_{L^2}^2 \leqslant e^{-\nu t} \|\Lambda^\beta \theta_0\|_{L^2}^2 + \tilde{C}_0 M^c \int_0^t e^{-\nu(t-s)} (s+1)^{-\frac{1}{\alpha}} \mathrm{d}s$$

$$+ \frac{2}{\kappa} \int_0^t e^{-\nu(t-s)} \|\Lambda^{\beta-\alpha} f(s)\|_{L^2}^2 \mathrm{d}s.$$

By observing

$$\int_0^t e^{-\nu(t-s)} (s+1)^{-\frac{1}{\alpha}} \mathrm{d}s \leqslant C(1+t)^{-\frac{1}{\alpha}}$$

and

$$\int_0^t e^{-\nu(t-s)} \|\Lambda^{\beta-\alpha} f(s)\|_{L^2}^2 \mathrm{d}s \leqslant \int_0^t \|\Lambda^{\beta-\alpha} f(s)\|_{L^2}^2 \mathrm{d}s,$$

we thus complete the proof of Theorem 3.5.12.

Corollary 3.5.1 *Let* $\alpha \in \left(\dfrac{1}{2}, 1\right]$, $m \geqslant \alpha$, *and suppose that* θ *is a smooth solution of* (3.5.2) *with initial data* $\theta_0 \in L^1(\mathbf{R}^2) \cap H^m(\mathbf{R}^2)$, *then for any* $t \geqslant 0$, *there holds*

$$\|\theta(t)\|_{H^m} \leqslant C(1+t)^{-\frac{1}{2\alpha}}, \qquad \|u(t)\|_{H^m} \leqslant C(1+t)^{-\frac{1}{2\alpha}},$$

where C *only depends on the initial data. Furthermore, if* $m \geqslant 1$ *and* $r \in [2, \infty)$, *then*

$$\|\Lambda^\gamma \theta(t)\|_{L^r} \leqslant C_r(1+t)^{-\frac{1}{2\alpha}}, \ \|\Lambda^\gamma u(t)\|_{L^r} \leqslant C_r(1+t)^{-\frac{1}{2\alpha}}, \qquad 0 \leqslant \gamma \leqslant \beta - 1,$$

where C_r *is constant depending on initial data and* r.

Corollary 3.5.2 *Let* $\beta > 1$, *then under the assumptions of Theorem 3.5.12,*

$$\|\theta(t)\|_{L^\infty} \leqslant C,$$

where constant C *depends on* f *and* θ_0.

Proof It suffices to show that $\hat{\theta}(t) \in L^1$ and $\|\hat{\theta}(t)\|_{L^1}$ is uniformly bounded. In fact, if $\theta \in H^\beta$, then

$$\int_{\mathbf{R}^2} |\hat{\theta}(\xi)| \mathrm{d}\xi \leqslant C \left(\int_{\mathbf{R}^2} (1 + |\xi|^2)^\beta |\hat{\theta}(\xi)|^2 \mathrm{d}\xi \right)^{\frac{1}{2}},$$

where $C^2 = \int_{\mathbf{R}^2} (1 + |\xi|^2)^{-\beta} \mathrm{d}\xi < \infty$.

The following theorem shows that when $\beta = 1$, we can get L^∞ estimates of the solution. We need only to provide L^1 bound of its Fourier transform.

Lemma 3.5.3 *Let* $\beta = 1$, *then under the assumptions of Theorem 3.5.11, if moreover* $\hat{\theta}_0 \in L^1$ *and* $\hat{f} \in L^1(0, \infty; L^1)$,

$$\|\hat{\theta}(t)\|_{L^1} \leqslant C, \qquad \forall t \geqslant 0$$

for some constant $C > 0$.

Proof From Theorem 3.5.11, there exists constant $C \geqslant 0$ such that $\|\nabla \theta(t)\|_{L^2}^2 = \|\Lambda \theta(t)\|_{L^2} \leqslant C$ for all $t \geqslant 0$. Using Fourier transform, $\hat{\theta}$ can be expressed as

$$\hat{\theta} = \mathrm{e}^{-\kappa |\xi|^{2\alpha} t} \hat{\theta}_0 - \int_0^t \mathrm{e}^{-\kappa |\xi|^{2\alpha}(t-s)} \widehat{u \cdot \nabla \theta} \mathrm{d}s + H(t),$$

where $H(t) = \displaystyle\int_0^t \mathrm{e}^{-\kappa |\xi|^{2\alpha}(t-s)} \hat{f}(s) \mathrm{d}s$. By the assumption of f, $\|H(t)\|_{L^1} \leqslant C$

is uniformly bounded and

$$\|\hat{\theta}(t)\|_{L^1} \leqslant \|\hat{\theta}_0\|_{L^1} + \int_0^t \|e^{-\kappa|\xi|^{2\alpha}(t-s)} \widehat{u \cdot \nabla\theta}\|_{L^1} ds + C. \tag{3.5.41}$$

In the following, we will show that the second term on the RHS is uniformly bounded. For this we let $\varepsilon > 0$ to be determined and

$$I = \int_0^{t-\varepsilon} \|\widehat{u \cdot \nabla\theta}\|_{L^1} ds, \quad II = \int_{t-\varepsilon}^t \|e^{-\kappa|\xi|^{2\alpha}(t-s)} \widehat{u \cdot \nabla\theta}\|_{L^1} ds, \quad \text{if } \varepsilon \leqslant t;$$

$$I = 0, \qquad II = \int_{t-\varepsilon}^t \|\widehat{u \cdot \nabla\theta}\|_{L^1} ds, \qquad \text{if } \varepsilon > t \geqslant 0.$$

We first estimate II. When $t > \varepsilon$, then

$$II \leqslant \int_{t-\varepsilon}^t \|e^{-\kappa|\xi|^{2\alpha}(t-s)}\|_{L^2} \|\widehat{u \cdot \nabla\theta}\|_{L^2} ds$$

$$\leqslant C \int_{t-\varepsilon}^t \frac{1}{(t-s)^{\frac{1}{2\alpha}}} \|\nabla\theta\|_{L^2} \|u\|_{L^\infty} ds$$

$$\leqslant C \sup_{t \geqslant 0} \|\nabla\theta\|_{L^2} \sup_{0 \leqslant s \leqslant t} \|\hat{\theta}(s)\|_{L^1} \varepsilon^{1-\frac{1}{2\alpha}},$$

where we have used the fundamental estimate $\|u(t)\|_{L^\infty} \leqslant C\|\hat{u}(t)\|_{L^1} \leqslant C\|\hat{\theta}(t)\|_{L^1}$. Since $\|\nabla\theta(t)\|_{L^2}$ is bounded in time, we can select $\varepsilon > 0$ such that

$$II \leqslant \frac{1}{2} \sup_{0 \leqslant s \leqslant t} \|\hat{\theta}(s)\|_{L^1}, \quad \forall t \geqslant \varepsilon. \tag{3.5.42}$$

When $t < \varepsilon$, similarly we obtain

$$II \leqslant C \sup_{t \geqslant 0} \|\nabla\theta\|_{L^2} \sup_{0 \leqslant s \leqslant t} \|\hat{\theta}(s)\|_{L^1} \int_0^t (t-s)^{-\frac{1}{2\alpha}} ds \leqslant C \sup_{0 \leqslant s \leqslant t} \|\hat{\theta}(s)\|_{L^1} \varepsilon^{1-\frac{1}{2\alpha}}.$$

Thus (3.5.42) still holds.

We next estimate I. When $t \geqslant \varepsilon$, then for any $s \geqslant 0$, $\|u(s)\|_{L^2} = \|\theta(s)\|_{L^2} \leqslant C(1+s)^{-\frac{1}{2\alpha}} \leqslant C$ and $\|\nabla\theta(s)\|_{L^2} \leqslant C$, thus

$$I \leqslant \int_0^{t-\varepsilon} \|e^{-\kappa|\xi|^{2\alpha}(t-s)}\|_{L^1} \|\widehat{u \cdot \nabla\theta}\|_{L^\infty} ds \leqslant C \int_0^{t-\varepsilon} \frac{1}{(t-s)^{1/\alpha}} \|\widehat{u \cdot \nabla\theta}\|_{L^\infty} ds$$

$$\leqslant C \int_0^{t-\varepsilon} \frac{1}{(t-s)^{1/\alpha}} \|u\|_{L^2} \|\nabla\theta\|_{L^2} ds \leqslant C \int_0^{t-\varepsilon} \frac{1}{(t-s)^{1/\alpha}} ds$$

$$\leqslant \begin{cases} C\varepsilon^{1-\frac{1}{\alpha}}, & \alpha < 1; \\ C\log(1/\varepsilon), & \alpha = 1. \end{cases}$$

Since ε is fixed, we have $I \leqslant C$ either when $\alpha < 1$ or $\alpha = 1$, i.e., I is uniformly bounded in time. By (3.5.42) and (3.5.41), we obtain

$$\|\hat{\theta}(t)\|_{L^1} \leqslant C + \frac{1}{2} \sup_{0 \leqslant s \leqslant t} \|\hat{\theta}(s)\|_{L^1}, \quad \forall t \geqslant 0,$$

where C is independent of t. Therefore, we complete the proof.

Next we consider the approximation to the QG equation by linear equations. Let $\Theta(t)$ satisfy the linear equation

$$\partial_t \theta + \Lambda^{2\alpha}\theta = 0, \quad \theta|_{t=0} = \theta_0.$$

Theorem 3.5.13 *Let $\alpha \in (0,1]$, $\theta_0 \in L^1(\mathbf{R}^2) \cap L^2(\mathbf{R}^2)$ and θ is a weak solution of 2D QG equation with initial data θ_0. Then there exists some constant $C > 0$ depending only on $\|\theta_0\|_{L^1}$ and $\|\theta_0\|_{L^2}$ such that*

$$\|\theta(t) - \Theta(t)\|_{L^2(\mathbf{R}^2)} \leqslant C(1+t)^{\frac{1}{2}-\frac{1}{\alpha}}.$$

Proof Let $w = \theta - \Theta$, then w satisfies

$$\partial_t w + \Lambda^{2\alpha}w = -u \cdot \nabla\theta. \tag{3.5.43}$$

Take the L^2 inner product of this equation with w to obtain

$$\frac{\mathrm{d}}{\mathrm{d}t} \int |w|^2 + 2 \int |\Lambda^\alpha w|^2 = \int \Theta(u \cdot \nabla\theta)\mathrm{d}x, \tag{3.5.44}$$

where we have used the fact $\int_{\mathbf{R}^2}(u \cdot \nabla\theta)\theta = 0$. From Proposition 3.1.2 and Theorem 3.5.10, the right hand side of (3.5.44) is bounded by

$$\left| \int \Theta(u \cdot \nabla\theta)\mathrm{d}x \right| \leqslant \|\nabla\Theta\|_{L^\infty} \|\theta\|_{L^2}^2 \leqslant C(1+t)^{-\frac{2}{\alpha}}.$$

Similar to the proof of Theorem 3.5.10, we obtain

$$\frac{\mathrm{d}}{\mathrm{d}t} \int |\hat{w}|^2 + 2g^{2\alpha}(t) \int |\hat{w}|^2 \leqslant 2g^{2\alpha}(t) \int_{|\xi| \leqslant g(t)} |\hat{w}|^2 + C(1+t)^{-\frac{2}{\alpha}}, \tag{3.5.45}$$

where $g(t)$ is to be determined. On the other hand, since $\alpha \leqslant 1$, by taking Fourier transform of (3.5.43), and by analogy of the proof of Theorem 3.5.10, one obtains

$$|w(\xi,t)| \leqslant |\xi| \int_0^t \|\theta(s)\|_{L^2}^2 \mathrm{d}s \leqslant |\xi| \int_0^t (1+s)^{-\frac{1}{\alpha}} \mathrm{d}s \leqslant C|\xi|.$$

Let $g^{2\alpha}(t) = \dfrac{\beta}{2(1+t)}$, then we get by integrating (3.5.45)

$$(1+t)^\beta \int |\hat{w}|^2 \leqslant C \left[(1+s)^{\beta-\frac{2}{\alpha}} \mathrm{d}s + \int_0^t (1+s)^\beta g^4(s)\mathrm{d}s \right].$$

Therefore, we obtain
$$\|w\|_{L^2}^2 \leqslant C(1+t)^{1-\frac{2}{\alpha}},$$
completing the proof.

3.5.4   Existence of attractors

This subsection considers the existence of attractors for the two-dimensional QG equation. We first introduce some general concepts of attractors. For more details of attractors in infinite dimensional dynamical systems, readers may refer to Temam [213].

Definition 3.5.1 *Let (W, d) be a metric space. A semi-flow in (W, d) is defined by a family of mappings $S(t) : W \to W$, $t \geqslant 0$ such that*
   (1) for any fixed $t \geqslant 0$, $S(t)$ is continuous in W,
   (2) for any fixed $w \in W$, $S(0)w = w$, and
   (3) for any $w \in W$ and $s, t \in [0, \infty)$, $S(s)S(t)w = S(s+t)w$.

Definition 3.5.2 *Let $S(t)$ be a semi-flow in the metric space (W, d), a set $\mathscr{A} \subset W$ is called a global attractor if*
   (1) $\mathscr{A}$ is a non-empty compact subset,
   (2) $\mathscr{A}$ is invariant, i.e., $S(t)\mathscr{A} = \mathscr{A}$ for any $t \geqslant 0$, and
   (3) $\lim_{t \to +\infty} d(S(t)B, \mathscr{A}) = 0$ for any bounded set $B \subset W$, where $d(A, B)$
$:= \sup_{x \in A} \inf_{y \in B} d(x, y)$.

Definition 3.5.3 (absorbing set). *A set $\mathscr{B} \subset W$ is called absorbing or an absorbing set, if for any bounded subset $B_0 \subset W$, there exists $t_1(B_0)$ such that $S(t)B_0 \subset \mathscr{B}$ when $t \geqslant t_1$.*

Uniformly compact The operators $S(t)$ are uniformly compact for t large, if for every bounded set $\mathscr{B}$ there exists t_0 depending possibly on $\mathscr{B}$ such that $\cup_{t \geqslant t_0} S(t)\mathscr{B}$ is relatively compact in W.

Theorem 3.5.14 *Let $S(t)$ be a uniformly compact semi-flow in a metric space W, and there exists a bounded absorbing set $\mathscr{B}$. Then the ω-limit set of $\mathscr{B}$, $\mathscr{A} = \omega(\mathscr{B})$ is a maximal, compact attractor, where the ω-limit set of $\mathscr{B} \subset W$ are defined as*

$$\omega(\mathscr{B}) := \bigcap_{\tau \geqslant 0} \overline{\bigcup_{t \geqslant \tau} S(t)\mathscr{B}}.$$

In applications, $x \in \omega(\mathscr{B})$-limit is and only if there exist sequences $x_n \in \mathscr{B}$ and $t_n \to \infty$ such that $S(t_n)x_n \to x$ when $n \to \infty$.

For completeness, we next introduce some concepts of weak attractors (cf. [23, 198]). The purpose of introducing weak attractors is to deal with the case when semi-flows is not uniformly compact with respect to d. Suppose that there exists another metric δ in W. Roughly speaking, the weak attractor is bounded in the d-topology, compact in the δ-topology, and the dynamic maps d-bounded sets into sets that are d-bounded and δ-compact.

Definition 3.5.4 *The semi-flow $S(t)$ is called d/δ uniformly compact, if for arbitrary d-bounded set $\mathscr{B} \subset W$, there exists t_0 possibly depending on $\mathscr{B}$ such that $\cup_{t \geq t_0} S(t)\mathscr{B}$ is relatively compact in the δ-topology of W.*

A set $\mathscr{B} \subset W$ is called a d-absorbing set if $\mathscr{B}$ is d-bounded and for any d-bounded set $B_0 \subset W$ there exists $t_1(B_0)$ such that $S(t)B_0 \subset \mathscr{B}$ when $t \geq t_1$.

For any $B \subset W$, its weak ω-limit set $\omega^\delta(B)$ can be defined as

$$\omega^\delta(\mathscr{B}) := \bigcap_{\tau \geq 0} \overline{\bigcup_{t \geq \tau} S(t)\mathscr{B}}^\delta,$$

where the closure is taken in the δ-topology.

Similarly, $x \in \omega^\delta(B)$ if and only if there exist sequences $x_n \in B$ and $t_n \to \infty$ such that $\delta(S(t_n)x_n, x) \to 0$ when $n \to \infty$.

Definition 3.5.5 *A set $\mathscr{A} \subset W$ is called a global d/δ weak attractor, if*
(1) $\mathscr{A}$ is non-empty, d-bounded and δ-compact,
(2) $\mathscr{A}$ is invariant, i.e., $S(t)\mathscr{A} = \mathscr{A}$ for all $t \geq 0$, and
(3) for any d-bounded set $B \subset W$, $\lim_{t \to +\infty} \delta(S(t)B, \mathscr{A}) = 0$.

Theorem 3.5.15 *Let $S(t)$ be a semi-flow in the metric space (W, d) and δ be another metric in W such that $S(t) : W \to W$ is δ-continuous $\forall t \geq 0$. If there exists a d-bounded absorbing set $\mathscr{B}$ and $S(t)$ is d/δ uniformly compact, then $\omega^\delta(\mathscr{B})$ is a global weak attractor.*

Proof Denote $\mathscr{A} = \omega^\delta(\mathscr{B})$.

(1) By definition of uniformly compact, there exists $t_0(\mathscr{B})$ such that $\cup_{t \geq t_0} S(t)\mathscr{B}$ is relatively δ-compact in W. That is, $\overline{\bigcup_{t \geq \tau} S(t)\mathscr{B}}^\delta$ is δ-compact when $\tau \geq t_0(\mathscr{B})$. By definition, $\mathscr{A}$ is the intersection of a family of non-empty, decreasing, δ-compact sets, and thus is non-empty and δ-compact.

(2) Next we show $S(t)\mathscr{A} = \mathscr{A}$. Let $x \in S(t)\mathscr{A}$, then there exists $y \in \mathscr{A}$ such that $x = S(t)y$. By definition, there exist sequences $\{y_n\}$ and $t_n \to \infty$ such that $\delta(S(t_n)y_n, y) \to 0$ when $n \to \infty$. By the semigroup property and δ-continuity of $S(t)$, one has

$$S(t + t_n)y_n = S(t)S(t_n)y_n \xrightarrow{\delta} S(t)y = x,$$

as $n \to \infty$. Therefore, $x \in \mathscr{A}$.

We also need to show $S(t)\mathscr{A} \supset \mathscr{A}$. Let $x \in \mathscr{A}$, then there exist sequences $\{y_n\}$ and $t_n \to \infty$ such that $S(t_n)y_n \overset{\delta}{\to} x$. When $t_n \geqslant t$, we have $S(t_n)y_n = S(t)S(t_n - t)y_n$. Since $S(t)$ is d/δ-uniformly compact, we can select a subsequence t_{n_k} such that $S(t_{n_k} - t)y_{n_k} \overset{\delta}{\to} \tilde{y} \in W$. As $t_{n_k} - t \to \infty$, it follows $\tilde{y} \in \mathscr{A}$ and hence

$$S(t_{n_k})y_{n_k} = S(t)S(t_{n_k} - t)y_{n_k} \overset{\delta}{\to} S(t)\tilde{y} = x,$$

by δ-continuity of $S(t)$. Therefore $x \in S(t)\mathscr{A}$.

(3) We prove by contradiction. Suppose that there exists a d-bounded set $B_0 \subset W$ such that $\delta(S(t)B_0, \mathscr{A})$ does not tend to zero as $t \to \infty$. That is there exist $\alpha > 0$, $t_n \to \infty$ and $u_n \in B_0$ such that $\delta(S(t_n)u_n, \mathscr{A}) \geqslant \alpha > 0$. Since $\mathscr{B}$ absorbs B_0, there exists $\tau = \tau(B_0)$ such that $v_n := S(\tau)u_n \in \mathscr{B}$ and thus there exist sequences $s_n = t_n - t \to \infty$ and $v_n \in \mathscr{B}$ such that

$$\delta(S(s_n)v_n, \mathscr{A}) \geqslant \alpha > 0. \tag{3.5.46}$$

On the other side, since $S(t)$ is d/δ-uniformly compact $S(s_n)v_n$ has a δ-limit in W, which belongs to $\mathscr{A}$ by definition of weak ω-limit set. This contradicts with (3.5.46), completing the proof.

Here we are interested in existence of strong attractors of two-dimensional QG equation (3.5.5). Readers who are interested in weak attractors can refer to [22]. Let $\Omega = [0, 2\pi]^2$ and suppose that θ and f have mean zero over Ω without loss of generality. That is, $\bar{\theta} := \frac{1}{|\Omega|} \int_\Omega \theta dx = 0$ and $\bar{f} := \frac{1}{|\Omega|} \int_\Omega f dx = 0$. Otherwise, we can consider (3.5.5) with θ replaced by $\theta - \bar{\theta}$ and f replaced by $f - \bar{f}$. The main result is as follows.

Theorem 3.5.16 Let $\alpha \in \left(\frac{1}{2}, 1\right]$, $\kappa > 0$, $s > 2(1 - \alpha)$ and $f \in H^{s-\alpha} \cap L^p$ do not depend on time. Then the solution operator $S: S(t)\theta_0 = \theta(t), \forall t > 0$ well defines a semigroup in H^s, and

(1) for any fixed $t > 0$, $S(t)$ is continuous in H^s;

(2) for any $\theta_0 \in H^s$, $S: [0, t] \to H^s$ is continuous;

(3) for any $t > 0$, $S(t)$ is a compact operator in H^s;

(4) $\{S(t)\}_{t \geqslant 0}$ has a global attractor $\mathscr{A}$ in H^s. $\mathscr{A}$ is compact and connected in H^s, is the maximal bounded absorbing set and minimal invariant set in H^s and attracts all bounded subsets in H^s in the norm of H^s for any $s > 2(\alpha - 1)$;

(5) if $\alpha > 2/3$, then $\mathscr{A}$ attracts all bounded subsets of all periodic functions in the space L^2 in the H^s-norm for any $s > 2(\alpha - 1)$.

Proof The rest of this section is dedicated to the proof of Theorem 3.5.16. The proof is divided into three parts.

1. *A priori* estimates and proof of (3).

Firstly we provide some useful *a priori* estimates. Let $s > 2(1-\alpha)$. Take L^2 inner product of (3.5.5) with θ to obtain

$$\frac{1}{2}\frac{d}{dt}\|\theta\|_{L^2}^2 + \kappa\|\Lambda^\alpha\theta\|_{L^2}^2 = (f,\theta) \leqslant \frac{\kappa}{2}\|\Lambda^\alpha\theta\|_{L^2}^2 + \frac{1}{2\kappa}\|\Lambda^{-\alpha}f\|_{L^2}^2,$$

from which it follows

$$\frac{d}{dt}\|\theta\|_{L^2}^2 + \kappa\|\Lambda^\alpha\theta\|_{L^2}^2 \leqslant \frac{1}{\kappa}\|\Lambda^{-\alpha}f\|_{L^2}^2. \tag{3.5.47}$$

Let λ_1 denote the eigenvalue of Λ, then since θ has mean zero over Ω,

$$\frac{d}{dt}\|\theta\|_{L^2}^2 + \kappa\lambda_1^{2\alpha}\|\theta\|_{L^2}^2 \leqslant \frac{1}{\kappa\lambda_1^{2\alpha}}\|f\|_{L^2}^2.$$

By Gronwall inequality, one has

$$\|\theta(t)\|_{L^2}^2 \leqslant \left(\|\theta_0\|_{L^2}^2 - \frac{F^2}{\mu_1}\right)e^{-\mu_1 t} + \frac{F^2}{\mu_1}, \tag{3.5.48}$$

where $\mu_1 = \kappa\lambda_1^{2\alpha}$ and $F = \|f\|_{L^2}$. It follows that there exists an absorbing set in L^2 with radius $\|\theta_0\|_{L^2}^2 + F^2/\mu_1$. Furthermore, by integrating (3.5.47) over $[t, t+1]$, one has

$$\|\theta(t+1)\|_{L^2}^2 + \kappa\int_t^{t+1}\|\Lambda^\alpha\theta(s)\|_{L^2}^2 ds \leqslant \|\theta(t)\|_{L^2}^2 + \frac{1}{\kappa}\|\Lambda^{-\alpha}f\|_{L^2}^2$$

$$\leqslant \left(\|\theta_0\|_{L^2}^2 - \frac{F^2}{\mu_1}\right)e^{-\mu_1 t} \tag{3.5.49}$$

$$+ \frac{F^2}{\mu_1} + \frac{1}{\kappa}\|\Lambda^{-\alpha}f\|_{L^2}^2.$$

Therefore there exists $t_* = t_*(\|\theta_0\|_{L^2}^2)$ such that $\int_t^{t+1}\|\Lambda^\alpha\theta(s)\|_{L^2}^2 ds$ independent of the initial data θ_0 when $t \geqslant t_*$.

Suppose $p \geqslant 2$, multiply (3.5.5) by $p|\theta|^{p-2}\theta$ and integrate over Ω to obtain

$$\frac{d}{dt}\|\theta\|_{L^p}^p + \kappa\|\Lambda^\alpha\theta^{p/2}\|_{L^2}^2 \leqslant p\int_\Omega f|\theta|^{p-2}\theta dx,$$

thanks to the divergence free condition $\nabla \cdot u = 0$. Therefore

$$\frac{d}{dt}\|\theta\|_{L^p}^p + \kappa\lambda^{2\alpha}\|\theta\|_{L^p}^p \leqslant p\|f\|_{L^p}\|\theta^{p-1}\|_{L^{p'}} = p\|f\|_{L^p}\|\theta\|_{L^p}^{p-1},$$

following

$$\frac{d}{dt}\|\theta\|_{L^p} + \frac{\kappa\lambda^{2\alpha}}{p}\|\theta\|_{L^p} \leqslant \|f\|_{L^p}.$$

Thus

$$\|\theta(t)\|_{L^p} \leqslant \left(\|\theta_0\|_{L^p} - \frac{p\|f\|_{L^p}}{\kappa\lambda_1^{2\alpha}}\right) e^{-\frac{\kappa\lambda_1^{2\alpha}}{p}t} + \frac{p\|f\|_{L^p}}{\kappa\lambda_1^{2\alpha}}. \tag{3.5.50}$$

It follows that $\|\theta\|_{L^p}$ is uniformly bounded and there exists an absorbing ball in L^p for any $\theta_0 \in L^p$, for any $p \in [2, \infty)$.

Next we consider uniform *a priori* estimate in H^s when $s > 2(1 - \alpha)$. Let $\alpha \in (1/2, 1)$ and $\theta_0 \in H^s$. We let $r = s$ if $s \in (2(1 - \alpha), 1)$ and let r be arbitrary in $(2(1 - \alpha), 1)$ if $s \in [1, \infty)$. Then $\theta_0 \in H^s \subset H^r \subset L^p$, where $\frac{1}{p} = \frac{1-r}{2} < \alpha - \frac{1}{2}$. Therefore $\theta, u \in L^\infty(0, +\infty; L^p)$ thanks to (3.5.50). On the other hand, by taking H^s inner product with θ, one has

$$\frac{d}{dt}\|\Lambda^s\theta\|_{L^2}^2 + \kappa\|\Lambda^{s+\alpha}\theta\|_{L^2}^2 \leqslant \frac{1}{\kappa}\|\Lambda^{s-\alpha}f\|_{L^2}^2 + C_0(\|\theta\|_{L^p} + \|u\|_{L^p})\|\Lambda^{s+\beta}\theta\|_{L^2}^2$$

$$\leqslant \frac{1}{\kappa}\|\Lambda^{s-\alpha}f\|_{L^2}^2 + C\|\Lambda^{s+\beta}\theta\|_{L^2}^2,$$

where $s > 0$, $p \in [2, \infty)$ and $\beta = \frac{1}{2} + \frac{1}{p} < \alpha$. By Gagliardo-Nirenberg inequality,

$$\|\Lambda^{s+\beta}\theta\|_{L^2} \leqslant C\|\Lambda^{s+\alpha}\theta\|_{L^2}^{\frac{\beta}{\alpha}}\|\Lambda^s\theta\|_{L^2}^{1-\frac{\beta}{\alpha}}$$

$$\leqslant \frac{\kappa}{2}\|\Lambda^{s+\alpha}\theta\|_{L^2}^2 + \frac{C}{\kappa}\|\Lambda^s\theta\|_{L^2}^2,$$

for any $\frac{1}{p} \in \left[0, \alpha - \frac{1}{2}\right)$. Therefore,

$$\frac{d}{dt}\|\Lambda^s\theta\|_{L^2}^2 + \frac{\kappa}{2}\|\Lambda^{s+\alpha}\theta\|_{L^2}^2 \leqslant \frac{1}{\kappa}\|\Lambda^{s-\alpha}f\|_{L^2}^2 + \frac{C}{\kappa}\|\Lambda^s\theta\|_{L^2}^2.$$

When $s \leqslant \alpha$, by uniform Gronwall inequality and (3.5.49), we know that $\|\Lambda^s\theta\|_{L^2}$ is uniformly bounded with respect to $\|\theta_0\|_{H^s}$ and there exists an absorbing set in H^s. Furthermore, by integrating this inequality over $[t, t+1]$, we get

$$\int_0^T \|\Lambda^{s+\alpha}\theta(t)\|_{L^2}^2 dt < \infty, \tag{3.5.51}$$

and is uniformly bounded with respect to $\|\theta_0\|_{H^s}$. It follows the uniform boundedness of $\|\Lambda^s\theta\|_{L^2}$ when $s > 2(1 - \alpha)$ by uniform Gronwall inequality with a bootstrapping argument and the existence of an absorbing set in H^s for $s > 2(1 - \alpha)$. Since the embedding $H^{s_1} \hookrightarrow\hookrightarrow H^{s_1}$ is compact for all $s_2 > s_1$, and there exists an absorbing set H^s when $s > 2(1 - \alpha)$, we know that $S(t)$ is compact in H^s for all $t > 0$.

2. Proof of (2).

For any fixed $\theta_0 \in H^s$, we regard $S(t)\theta_0$ as a function from $\mathbf{R}^+$ to H^s. We will show that this map is continuous. Since we have already shown $\theta \in L^2(0, T; H^{s+\alpha})$, to show $\theta \in C([0, T]; H^s)$ it suffices to show that $\Lambda^s \theta_t \in L^2(0, T; H^{-\alpha})$. Let $\varphi \in H^\alpha$ be arbitrary, then

$$(\Lambda^s \theta_t, \varphi) = -(\Lambda^s (u \cdot \nabla \theta), \varphi) - (\Lambda^{s+2}\theta, \varphi) + (\Lambda^s f, \varphi).$$

Therefore

$$|(\Lambda^s \theta_t, \varphi)| \leqslant \left(\|\Lambda^{s-\alpha}(u \cdot \nabla \theta)\|_{L^2} + \|\Lambda^{s+\alpha}\theta\|_{L^2} + \|\Lambda^{s-\alpha} f\|_{L^2} \right) \|\Lambda^\alpha \varphi\|_{L^2},$$

from which it follows

$$\|\Lambda^s \theta_t\|_{H^{-\alpha}} \leqslant \|\Lambda^{s-\alpha}(u \cdot \nabla \theta)\|_{L^2} + \|\Lambda^{s+\alpha}\theta\|_{L^2} + \|\Lambda^{s-\alpha} f\|_{L^2}. \qquad (3.5.52)$$

Since $\nabla \cdot u = 0$, we have

$$\|\Lambda^{s-\alpha}(u \cdot \nabla \theta)\|_{L^2} = \|\Lambda^{s-\alpha} \nabla (\theta u)\|_{L^2} \leqslant \|\Lambda^{1+s-\alpha}(\theta u)\|_{L^2}. \qquad (3.5.53)$$

Let $\alpha \in (1/2, 1)$ and $\theta_0 \in H^s$. We let $r = s$ if $s \in (2(1-\alpha), 1)$ and let r be arbitrary in $(2(1-\alpha), 1)$ if $s \in [1, \infty)$. Then $\theta_0 \in H^s \subset H^r \subset L^p$, where $\frac{1}{p} = \frac{1-r}{2} < \alpha - \frac{1}{2} \leqslant 1/2$. Therefore θ, $u \in L^\infty(0, +\infty; L^p)$ by (3.5.50). Moreover, by multiplicative estimates

$$\begin{aligned}
\|\Lambda^{1+s-\alpha}(\theta u)\|_{L^2} &\leqslant C(\|\theta\|_{L^p}\|\Lambda^{1+s-\alpha}u\|_{L^q} + \|u\|_{L^p}\|\Lambda^{1+s-\alpha}\theta\|_{L^q}) \\
&\leqslant C\|\theta\|_{L^p}\|\Lambda^{1+s-\alpha}\theta\|_{L^q},
\end{aligned} \qquad (3.5.54)$$

where $\frac{1}{p} + \frac{1}{q} = \frac{1}{2}$. Denote $q^* = \frac{1}{1-\alpha}$, then $q = \frac{2}{r} < q^*$. Since $\frac{1}{q^*} + \frac{(s+\alpha) - (1+s-\alpha)}{2} = \frac{1}{2}$, then

$$\|\Lambda^{1+s-\alpha}\theta\|_{L^q} \leqslant C\|\Lambda^{1+s-\alpha}\theta\|_{L^{q^*}} \leqslant C\|\Lambda^{s+\alpha}\theta\|_{L^2}. \qquad (3.5.55)$$

It follows from (3.5.52)-(3.5.55)

$$\|\Lambda^s \theta_t\|_{H^{-\alpha}} \leqslant (C\|\theta\|_{L^p} + 1)\|\Lambda^{s+\alpha}\theta\|_{L^2} + \|\Lambda^{s-\alpha} f\|_{L^2}.$$

Therefore, $\int_0^T \|\Lambda^s \theta_t(s)\|_{H^{-\alpha}}^2 ds < \infty$ thanks to (3.5.50) and (3.5.51).

3. Proof of (1).

Finally, we show that for any fixed $t > 0$, the solution operator $S(t)$ is continuous from H^s to itself. For this purpose, we let θ and η are two solutions of the 2D QG equation with initial data θ_0 and η_0, respectively. Let $\zeta = \theta - \eta$, $w = u - v$ for $u = \mathcal{R}^\perp \theta$ and $v = \mathcal{R}^\perp \eta$, then $\nabla \cdot w = 0$ and

$$(u \cdot \nabla \theta, \varphi) - (v \cdot \eta, \varphi) = (u \cdot \nabla \theta, \varphi) - (u \cdot \nabla \eta, \varphi) + (u \cdot \nabla \eta, \varphi) - (v \cdot \nabla \eta, \varphi)$$
$$= (u \cdot \nabla \zeta, \varphi) + (w \cdot \nabla \eta, \varphi),$$

which implies

$$(\zeta_t, \varphi) + \kappa (\Lambda^\alpha \zeta, \Lambda^\alpha \varphi) = -(u \cdot \nabla \zeta, \varphi) + (w \cdot \nabla \eta, \varphi). \tag{3.5.56}$$

Since $\nabla \cdot u = 0$, $(u \cdot \nabla \zeta, \zeta) = 0$ and letting $\varphi = \zeta$ and using Gagliardo-Nirenberg inequality yield

$$\frac{1}{2} \frac{d}{dt} \|\zeta\|_{L^2}^2 + \kappa \|\Lambda^\alpha \zeta\|_{L^2}^2 = -(w \cdot \nabla \eta, \varphi)$$
$$\leqslant C \|\Lambda \eta\|_{L^{p_1}} \|\zeta\|_{L^q} \|w\|_{L^q} \leqslant C \|\Lambda \eta\|_{L^{p_1}} \|\zeta\|_{L^q}^2$$
$$\leqslant C \|\eta\|_{W^{1,p_1}} \|\zeta\|_{L^2}^{2(1-\beta)} \|\zeta\|_{H^\alpha}^{2\beta}$$
$$\leqslant \frac{\kappa}{2} \|\zeta\|_{H^\alpha}^2 + C \|\eta\|_{W^{1,p_1}}^{p_2'} \|\zeta\|_{L^2}^2,$$

where $\dfrac{1}{p_1} + \dfrac{2}{q} = 1$ and $\beta = \dfrac{1}{\alpha p_1} \in (0,1)$. Therefore,

$$\frac{d}{dt} \|\zeta\|_{L^2}^2 + \kappa \|\zeta\|_{H^\alpha}^2 \leqslant C \|\eta\|_{W^{1,p_1}}^{p_2'} \|\zeta\|_{L^2}^2,$$

where $p_2 = 1/\beta = \alpha p_1$ and $p_2' = p_2/(p_2 - 1) = \dfrac{1}{1-\beta}$. From Gronwall inequality, it follows

$$\|\zeta\|_{L^2}^2 \leqslant C \|\zeta(0)\|_{L^2} e^{\int_0^T \|\eta(s)\|_{W^{1,p_1}}^{p_2'} ds}.$$

We let $r = s$ when $s \in (2(1-\alpha), 2-\alpha)$ and let r be arbitrary in $(2(1-\alpha), 2-\alpha)$ when $s \in [2-\alpha, +\infty)$, then $H^s \subset H^r$. Let $p_1 = 2/(2 - r - \alpha) > 1$, then $p_2' = 2\alpha/(3\alpha + s - 2) \in (1, 2]$ and hence

$$L^{p_2'}(0, T; W^{1,p_1}) \subset L^{p_2'}(0, T; H^{r+\alpha}) \subset L^2(0, T; H^{r+\alpha}) \subset L^2(0, T; H^{s+\alpha}).$$

Denote $C_1(\eta, T) := \int_0^T \|\eta(s)\|_{W^{1,p_1}}^{p_2'} ds$, then

$$C_1(\eta, T) \leqslant \int_0^T \|\eta(s)\|_{H^{s+\alpha}}^2 ds < +\infty.$$

Therefore,

$$\kappa \int_0^T \|\zeta(s)\|_{H^\alpha}^2 ds \leqslant \|\zeta(0)\|_{L^2}^2 \left\{ 1 + C C_1(\eta, T) e^{C_1(\eta, T)} \right\}.$$

By Riesz Lemma from real analysis, we have $\|\zeta(t)\|_{H^\alpha} \to 0$ for a.e. $t > 0$ if $\|\zeta(0)\|_{L^2}$ goes to zero. By the continuity of $\|\zeta(t)\|_{H^\alpha}$ in t, we know $\|\zeta(t)\|_{H^\alpha} \to 0$ for every t. Therefore, $\|S(t)\|$ is continuous in H^s when $s \in (2(1-\alpha), \alpha]$.

When $s > \alpha$, to prove (1) in Theorem 3.5.16 we need only to show the Lipshitz continuity of the solution operator in H^s. We only treat the case when $\alpha \in \left(\dfrac{1}{2}, 1\right)$ while the case when $\alpha = 1$ can be treated similarly to the 2D Navier-Stokes equations. Letting $\varphi = \Lambda^{2\alpha}\zeta$ (3.5.56) then yields

$$\frac{1}{2}\frac{d}{dt}\|\Lambda^s\zeta\|_{L^2}^2 + \kappa\|\Lambda^{s+\alpha}\zeta\|_{L^2}^2 = (\Lambda^s(u\cdot\nabla\zeta) - u\cdot\nabla(\Lambda^s\zeta), \Lambda^s\zeta)$$
$$- (\Lambda^{s-\alpha}(w\cdot\nabla\eta), \Lambda^{s+\alpha}\zeta) = I_1 + I_2.$$

For the term I_2, one has

$$|I_2| = |(\Lambda^{s-\alpha}(w\cdot\nabla\eta), \Lambda^{s+\alpha}\zeta)| \leqslant C\|\Lambda^{s-\alpha}(w\cdot\nabla\eta)\|_{L^2}^2 + \frac{\kappa}{4}\|\Lambda^{s+\alpha}\zeta\|_{L^2}^2.$$

By multiplicative estimates,

$$\|\Lambda^{s-\alpha}(w\cdot\nabla\eta)\|_{L^2} \leqslant C(\|\Lambda^{s-\alpha}w\|_{L^{p_1}}\|\nabla\eta\|_{L^{p_2}} + \|w\|_{L^{q_1}}\|\Lambda^{s-\alpha+1}\eta\|_{L^{q_2}})$$
$$\leqslant C(\|\Lambda^{s-\alpha}\zeta\|_{L^{p_1}}\|\Lambda\eta\|_{L^{p_2}} + \|\zeta\|_{L^{q_1}}\|\Lambda^{s-\alpha+1}\eta\|_{L^{q_2}}),$$

where $p_1, p_2, q_1, q_2 > 2$, $\dfrac{1}{p_1} + \dfrac{1}{p_2} = \dfrac{1}{2}$ and $\dfrac{1}{q_1} + \dfrac{1}{q_2} = \dfrac{1}{2}$. Selecting $p_1 = \dfrac{2}{1-\alpha}$, $p_2 = \dfrac{2}{\alpha}$, $q_1 = \dfrac{1}{1-\alpha}$ and $q_2 = \dfrac{2}{2\alpha - 1}$, we have

$$\|\Lambda^{s-\alpha}\zeta\|_{L^{p_1}} \leqslant C\|\Lambda^s\zeta\|_{L^2}, \quad \|\Lambda\eta\|_{L^{p_2}} \leqslant C\|\Lambda^{2-\alpha}\eta\|_{L^2} \leqslant C\|\Lambda^{s+\alpha}\eta\|_{L^2}$$
$$\|\zeta\|_{L^{q_1}} \leqslant C\|\Lambda^{2\alpha-1}\zeta\|_{L^2} \leqslant C\|\Lambda^\alpha\zeta\|_{L^2} \leqslant C\|\Lambda^s\zeta\|_{L^2},$$
$$\|\Lambda^{s-\alpha+1}\eta\|_{L^{q_2}} \leqslant C\|\Lambda^{s+\alpha}\eta\|_{L^2}.$$

Therefore

$$|I_2| \leqslant C\|\Lambda^{s+\alpha}\eta\|_{L^2}^2\|\Lambda^s\zeta\|_{L^2}^2 + \frac{\kappa}{4}\|\Lambda^{s+\alpha}\zeta\|_{L^2}^2.$$

For the term I_1, we have

$$|I_1| = |(\Lambda^s(u\cdot\nabla\zeta) - u\cdot(\Lambda^s\nabla\zeta), \Lambda^s\zeta)|$$
$$\leqslant C\|\Lambda^s(u\cdot\nabla\zeta) - u\cdot(\Lambda^s\nabla\zeta)\|_{L^2}\|\Lambda^s\zeta\|_{L^2}^2,$$

since ∇ and Λ commute. By commutator estimates, we obtain

$$\|\Lambda^s(u\cdot\nabla\zeta) - u\cdot(\Lambda^s\nabla\zeta)\|_{L^2} \leqslant C(\|\nabla u\|_{L^{p_1}}\|\Lambda^s\zeta\|_{L^{p_2}} + \|\Lambda^s u\|_{L^{q_1}}\|\nabla\zeta\|_{L^{q_2}})$$
$$\leqslant C(\|\Lambda\theta\|_{L^{p_1}}\|\Lambda^s\zeta\|_{L^{p_2}} + \|\Lambda^s\theta\|_{L^{q_1}}\|\nabla\zeta\|_{L^{q_2}}),$$

where $p_1, p_2, q_1, q_2 > 2$, $\dfrac{1}{p_1} + \dfrac{1}{p_2} = \dfrac{1}{2}$ and $\dfrac{1}{q_1} + \dfrac{1}{q_2} = \dfrac{1}{2}$. Selecting $p_1 = \dfrac{2}{\alpha}$, $p_2 = \dfrac{2}{1-\alpha}$, $q_1 = \dfrac{2}{1-\alpha}$ and $q_2 = \dfrac{2}{\alpha}$, we get

$$\|\Lambda\theta\|_{L^{p_1}} \leqslant C\|\Lambda^{2-\alpha}\theta\|_{L^2} \leqslant C\|\Lambda^{s+\alpha}\theta\|_{L^2}, \quad \|\Lambda^s\zeta\|_{L^{p_2}} \leqslant C\|\Lambda^{s+\alpha}\zeta\|_{L^2}$$
$$\|\Lambda^s\theta\|_{L^{q_1}} \leqslant C\|\Lambda^{s+\alpha}\theta\|_{L^2}, \quad \|\Lambda\zeta\|_{L^{q_2}} \leqslant C\|\Lambda^{2-\alpha}\zeta\|_{L^2} \leqslant C\|\Lambda^{s+\alpha}\zeta\|_{L^2}.$$

It follows that

$$|I_1| \leqslant C\|\Lambda^{s+\alpha}\theta\|_{L^2}\|\Lambda^{s+\alpha}\zeta\|_{L^2}\|\Lambda^s\zeta\|_{L^2}$$
$$\leqslant C\|\Lambda^{s+\alpha}\theta\|_{L^2}^2\|\Lambda^s\zeta\|_{L^2}^2 + \frac{\kappa}{4}\|\Lambda^{s+\alpha}\zeta\|_{L^2}^2.$$

Therefore

$$\frac{\mathrm{d}}{\mathrm{d}t}\|\Lambda^s\zeta\|_{L^2}^2 + \kappa\|\Lambda^{s+\alpha}\zeta\|_{L^2}^2 \leqslant C(\|\Lambda^{s+\alpha}\theta\|_{L^2}^2 + \|\Lambda^{s+\alpha}\eta\|_{L^2}^2)\|\Lambda^s\zeta\|_{L^2}^2,$$

which implies by integration in time that

$$\|\Lambda^s\zeta(t)\|_{L^2}^2 \leqslant C\|\Lambda^s\zeta(0)\|_{L^2}^2 e^{\int_0^t(\|\Lambda^{s+\alpha}\theta\|_{L^2}^2 + \|\Lambda^{s+\alpha}\eta\|_{L^2}^2)\mathrm{d}s}.$$

Notice that

$$\int_0^t (\|\Lambda^{s+\alpha}\theta\|_{L^2}^2 + \|\Lambda^{s+\alpha}\eta\|_{L^2}^2)\mathrm{d}s < \infty.$$

We complete the proof of continuity in (1).

Finally, item (4) can be proved by (1), (2), (3) and Theorem 3.3.8. For item (5), we note that $\alpha > 2(1-\alpha)$ since $\alpha > \dfrac{2}{3}$, and thus for any $\theta_0 \in L^2$, there holds for any $T > 0$ that

$$\int_0^T \|\Lambda^\alpha\theta\|_{L^2} < \infty$$

and

$$\|\theta(t+1)\|_{L^2}^2 + \kappa\int_t^{t+1} \|\Lambda^\alpha\theta(s)\|_{L^2}^2\mathrm{d}s \leqslant \|\theta(t)\|_{L^2}^2 + \frac{1}{\kappa}\|\Lambda^{-\alpha}f\|_{L^2}^2.$$

The proof is complete.

3.6   Fractional Boussinesq approximation

The study of flows in the Earth's mantle consists of thermal convection in a highly viscous fluid. For a description of dynamics of flows of an incompressible fluid in processes where the thermal effects play an essential role, the Boussinesq approximation is a reasonable model to present essential phenomena of such flows. In this section, we study the fractional Boussinesq approximation for the non-Newtonian fluids [109]

$$u_t + u \cdot \nabla u - \nabla \cdot \tau(e(u)) = -\nabla \pi + \eta \theta, \qquad (3.6.1)$$

$$\nabla \cdot u = 0, \qquad (3.6.2)$$

$$\theta_t + u \cdot \nabla \theta + \kappa \Lambda^{2\alpha} \theta = 0, \qquad (3.6.3)$$

subject to the initial values

$$u(0) = u_0, \quad \theta(0) = \theta_0,$$

with the periodic boundary conditions

$$u(x,t) = u(x + L\chi_i, t), \ \theta(x,t) = \theta(x + L\chi_i, t), \ t > 0, \ x \in \Omega, \qquad (3.6.4)$$

where $\Omega = [0, L]^2$, $L > 0$ and $\{\chi_i\}_{i=1}^2$ is the natural basis of R^2. The unknown vector function u denotes the velocity of the fluid, the scalar function π represents the pressure and $\eta = (0, 1)$ is a unit vector in R^2. θ is the scalar temperature, $\Lambda^{2\alpha}(0 < \alpha < 1)$ is the power of the square root of the Laplacian $\Lambda = (-\Delta)^{\frac{1}{2}}$ and $\kappa > 0$ is the thermometric conductivity. $\tau_{ij}(e(u))$ is a symmetric stress tensor

$$\tau_{ij}(e(\cdot)) = 2\mu_0(\epsilon + |e|^2)^{\frac{p-2}{2}} e_{ij} - 2\mu_1 \Delta e_{ij}, \ \epsilon > 0, \ i, j = 1, 2, \qquad (3.6.5)$$

$$e_{ij}(\cdot) = \frac{1}{2}\left(\frac{\partial u_i}{\partial x_j} + \frac{\partial u_j}{\partial x_i}\right) \ , \ |e(u)|^2 = \sum_{i,j=1}^2 |e_{ij}(u)|^2,$$

where $\mu_0, \mu_1 > 0$ are constants. There are many fluid materials, for example, liquid foams, polymeric fluids such as oil in water, blood, etc., satisfying such constitutive relation. If $\mu_0\mu_1 \neq 0$ in the constitutive relation (3.6.5), the fluids are called bipolar. While $\mu_0 \neq 0, \mu_1 = 0$, the fluids are said to be monopolar, because only the first derivative of the velocity is involved in the stress tensor, such as the Ladyzhenskaya's model. While $p = 2$, $\mu_0 \neq 0, \mu_1 = 0$, equation (3.6.1) turns out to be the famous Navier-Stokes equation. The fluids are shear thinning in the case of $1 < p < 2$, and shear thickening in the case of $p > 2$. For definiteness, we consider the case $1 < p < 2$ in the following.

The objective of this section is to study the existence, uniqueness and the long time behavior of weak solutions. First of all, we give the definition of weak solution as follows.

Definition 3.6.1 *Let u_0, θ_0 be given. A couple (u, θ) is called a weak solution of fractional Boussinesq approximate* (3.6.1)-(3.6.4), *if for all $T > 0$ and $1 < p < 2$,*

$$u \in L^\infty(0, T; \dot{H}_\sigma) \cap L^p(0, T; \dot{H}^{1,p}) \cap L^2(0, T; \dot{V}_\sigma), \ u_t \in L^2(0, T; V'),$$

$$\theta \in L^\infty(0, T; \dot{H}) \cap L^2(0, T; \dot{H}^\alpha),$$

and satisfy

$$-\int_0^T \int_\Omega u \frac{\partial \psi}{\partial t} dx + \int_0^T \int_\Omega u_i \frac{\partial u_j}{\partial x_i} \psi_j dx dt + \mu_1 \int_0^T \int_\Omega \Delta u \Delta \psi dx dt$$

$$+ 2\mu_0 \int_0^T \int_\Omega (\epsilon + |e(u)|^2)^{\frac{p-2}{2}} e_{ij}(u) e_{ij}(\psi) dx$$

$$= \int_0^T \int_\Omega \theta \psi_2 dx dt + \int_\Omega u_0 \psi(0) dx,$$

for every $\psi = (\psi_1, \psi_2) \in L^\infty(0, T; \dot{V}_\sigma) \cap W^{1,p}(0, T; \dot{H}^{1,p})$ with $\psi(T) = 0$, and

$$-\int_0^T \int_\Omega \theta \varphi_t dx + \int_0^T \int_\Omega u_i \frac{\partial \varphi}{\partial x_i} \theta dx + \kappa \int_0^T \int_\Omega (\Lambda^\alpha \theta)(\Lambda^\alpha \varphi) dx dt = \int_\Omega \theta_0 \varphi(0) dx,$$

for $\varphi \in L^\infty(0, T; \dot{H}) \cap W^{1,2}(0, T; \dot{H}^\alpha)$ with $\varphi(T) = 0$.

The spaces appeared in the above definition are defined as follows. First we let $\Omega = [0, L]^2 (L > 0)$ denote the periodic domain. $L^q(\Omega)$ denotes the Lebesgue space with norm $\| \cdot \|_{L^q}$ for $q \in [1, \infty]$. $\mathscr{C}(I, X)$ denotes the space of continuous functions from the interval I to X and $L^q(0, T; X)$ denotes the space of all measurable functions u on $[0, T]$ valued into X, with the norm $\|u\|_{L^q(0,T;X)}^q = \int_0^T \|u(t)\|_X^q dt$, for $1 \leqslant q < \infty$ and when $\|u\|_{L^\infty(0,T;X)} = ess\sup_{t \in [0,T]} \|u(t)\|_X$ when $q = \infty$. H^s and $\dot{H}^s$ denote the Sobolev spaces as before. We also define a space of smooth periodic functions with zero integral as follows:

$$\mathscr{V} := \{v \in C_{per}^\infty(\Omega) : \int_\Omega v dx = 0\},$$

where *per* represents that v is periodic with respect to x. Let H_{per}^s be the completion of $C_{per}^\infty(\Omega)$ in the H^s norm. Actually, if $v \in H_{per}^s$, we can deduce $v \in \dot{H}^s$ for $s \geqslant 0$. In the following, we always assume $v \in \dot{H}^s$ and the zero mean condition $\int_\Omega v dx = 0$ is included in $\dot{H}^s$. The H_{per}^s-norm is equivalent to H^s-norm under the zero integral condition. In this case, the Sobolev embedding theorems are also valid [192]. The space H^s satisfying the divergence free is denoted by

$$H_\sigma^s := \{u \in H^s | \nabla \cdot u = 0\},$$

and the space $\dot{H}^s$ satisfying the divergence free is denoted by

$$\dot{H}^s_\sigma := \{u \in \dot{H}^s | \ \nabla \cdot u = 0\}.$$

In particular, when $s = 0$, $\dot{H} = \dot{H}^0$, $H_\sigma = H^0_\sigma$, $\dot{H}_\sigma = \dot{H}^0_\sigma$, respectively. We also let $(\cdot, \cdot)$ denote the inner product of $\dot{H}_\sigma$. When $s = 2$, $\dot{V}_\sigma = \dot{H}^2_\sigma$, and V'_σ is the dual space of $\dot{V}_\sigma$.

We also define a continuous trilinear form as follows,

$$b(u,v,w) = \int_\Omega u_i \frac{\partial v_j}{\partial x_i} w_j \mathrm{d}x, \quad u,v,w \in \dot{H}^1,$$

which has the properties : $b(u,v,w) = -b(u,w,v)$, and $b(u,v,v) = 0$. In particular,

$$(B(u),w) = b(u,u,w) = \int_\Omega u_i \frac{\partial u_j}{\partial x_i} w_j \mathrm{d}x \quad u,w \in \dot{H}^1.$$

For $u \in \dot{V}_\sigma$, the operator $A_p(\cdot) : \dot{V}_\sigma \to V'$ is defined by

$$(A_p(u),v) = \int_\Omega \gamma(u) e_{ij}(u) e_{ij}(v) \mathrm{d}x, \quad u,v \in \dot{V}_\sigma,$$

where $\gamma(u) = (\epsilon + |e(u)|^2)^{\frac{p-2}{2}}$.

Consider the following eigenvalue problem $-\Delta u = \lambda u$, with periodic boundary conditions. Let $A = -\Delta$, according to Rellich theorem, A^{-1} is compact in $\dot{H}$, then

$$A\omega_n = \lambda_n \omega_n, \quad \omega_n \in D(A), \tag{3.6.6}$$

where $\{\omega_n\}_{n=1}^\infty$ are the eigenfunctions and also are basis of $\dot{V}$, $\lambda_n > 0$ and $\lambda_n \to \infty$, when $n \to \infty$. On the other hand, for the fractional diffusion operator $\Lambda^{2\alpha}(0 < \alpha < 1)$, we have

$$\Lambda^{2\alpha}\omega_n = \lambda_n^\alpha \omega_n.$$

Now, we apply Galerkin method to construct weak solution. We start with some useful a priori estimates, where the commutator estimates play an important role. Then by the compactness method, we can take the limit for the approximating solutions, whose limit is a weak solution of Boussinesq approximation. Finally, uniqueness is also established.

Lemma 3.6.1 *Assume that $u \in L^\infty(0,T;\dot{H}_\sigma) \cap L^2(0,T;\dot{V}_\sigma)$, then the function $B(u(t))$ defined by*

$$(B(u(t)),\phi) = b(u(t),u(t),\phi), \quad \forall \phi \in \dot{V}_\sigma$$

belongs to $L^2(0, T; V')$, and the function $A_p(u(t))$ defined by,

$$(A_p(u(t)), \phi) = 2\mu_0 \int_\Omega (\epsilon + |e(u)|^2)^{\frac{p-2}{2}} e_{ij}(u) e_{ij}(\phi) \mathrm{d}x$$

belongs to $L^2(0, T; V')$.

For the detailed proof, one can refer to [27].

Let $\{\omega_n\}$ be the normalized eigenfunctions defined as (3.6.6). Then $\{\omega_n\}$ forms an orthonormal basis of $\dot{H}_\sigma$, which is also a basis of $\dot{H}^\alpha$. Let

$$u_m(t) = \sum_{i=1}^m h_{im}(t) \omega_i(x), \quad \theta_m(t) = \sum_{i=1}^m j_{im}(t) \omega_i(x).$$

We consider the following abstract approximating equation

$$(u_{mt}, \omega_i) + (u_m \cdot \nabla u_m, \omega_i) + 2\mu_0((\epsilon + |e(u_m)|^2)^{\frac{p-2}{2}} e(u_m), e(\omega_i))$$
$$+ \mu_1(\Delta u_m, \Delta \omega_i) = (\eta \theta_m, \omega_i), \tag{3.6.7}$$

$$(\theta_{mt}, \omega_i) + (u_m \cdot \nabla \theta_m, \omega_i) + \kappa(\Lambda^\alpha \theta_m, \Lambda^\alpha \omega_i) = 0, \tag{3.6.8}$$

with the initial conditions

$$(u_m(0), \omega_i) = (u_{0m}, \omega_i), \quad (\theta_m(0), \omega_i) = (\theta_{0m}, \omega_i), \tag{3.6.9}$$

where $u_{0m} \to u_0$ in $\dot{H}_\sigma$, $\theta_{0m} \to \theta_0$ in $\dot{H}^s \cap L^2$.

From the local existence and uniqueness theory of ODEs, the local in time existence and uniqueness of solutions to equations (3.6.7)-(3.6.9) are obtained. In order to prove the global solution, we will show some *a priori* estimates independent of m.

Lemma 3.6.2 *Suppose that $\alpha \in (0, 1)$, $u_0 \in \dot{H}_\sigma$, $\theta_0 \in \dot{H}^s \cap L^2$, with $2(1 - \alpha) \leqslant s \leqslant 2 - \alpha$, for any $0 < T < \infty$, and the approximating solution (u_m, θ_m) to (3.6.7)-(3.6.9), there holds the following estimates*

$$\sup_{0 \leqslant t \leqslant T} \|u_m(t)\| \leqslant C, \quad \int_0^T \|u_m(\sigma)\|_2^2 \mathrm{d}\sigma \leqslant C,$$

$$\sup_{0 \leqslant t \leqslant T} \|\theta_m(t)\| \leqslant C, \quad \sup_{0 \leqslant t \leqslant T} \|\theta_m(t)\|_s \leqslant C, \quad \int_0^T \|\theta_m(\sigma)\|_{s+\alpha}^2 \mathrm{d}\sigma \leqslant C.$$

Proof Multiplying (3.6.8) with j_{im} and summing up the equation, we get

$$\frac{1}{2} \frac{\mathrm{d}}{\mathrm{d}t} \|\theta_m\|^2 + \kappa \|\Lambda^\alpha \theta_m\|^2 = 0,$$

where we have used the divergence free condition $\nabla \cdot u_m = 0$. Therefore, for any $T > 0$,

$$\|\theta_m(t)\|^2 + 2\kappa \int_0^T \|\Lambda^\alpha \theta_m(s)\|^2 ds \leqslant \|\theta_0\|^2,$$

which gives the basic uniform boundedness of θ_m in L^2, and the property $\theta_m \in L^2(0, +\infty; \dot{H}^\alpha)$, which are independent of m.

Multiplying (3.6.7) with h_{im} and summing up the equation, we have

$$\frac{1}{2}\frac{d}{dt}\|u_m\|^2 + 2\mu_0 \int_\Omega (\epsilon + |e(u_m)|^2)^{\frac{p-2}{2}}$$
$$e_{ij}(u_m)e_{ij}(u_m)dx + \mu_1\|\Delta u_m\|^2 = (\eta\theta_m, u_m),$$

where we have used the divergence free condition. Noting that the second term in the left hand side is nonnegative, we drop it in the following computation. Applying the Young's inequality yields

$$\frac{1}{2}\frac{d}{dt}\|u_m\|^2 + \mu_1\|\Delta u_m\|^2 \leqslant \frac{\|\theta_m\|^2}{2} + \frac{1}{2}\|u_m\|^2.$$

For any $T > 0$, we have from Gronwall inequality that for any $t \in [0, T]$

$$\|u_m(t)\|^2 \leqslant e^t\|u_0\|^2 + \int_0^t e^{t-s}\|\theta_m(s)\|^2 ds$$
$$\leqslant e^T\left(\|u_0\|^2 + \int_0^T \|\theta_m(s)\|^2 ds\right),$$

which gives the basic uniform boundedness of u_m in $\dot{H}_\sigma$, and the boundedness of u_m in $L^2(0, \infty; \dot{V}_\sigma)$, which are independent of m.

Multiplying (3.6.8) with $\lambda_i^s j_{im}$ and summing up the equation, we obtain

$$\frac{1}{2}\frac{d}{dt}\|\Lambda^s\theta_m\|^2 + \kappa\|\Lambda^{s+\alpha}\theta_m\|^2 = -(\Lambda^s(u_m \cdot \nabla\theta_m) - u_m \cdot \nabla(\Lambda^s\theta_m), \Lambda^s\theta_m),$$

where we have used the condition $\nabla \cdot u_m = 0$, and $(u_m \cdot \nabla(\Lambda^s\theta_m), \Lambda^s\theta_m) = 0$. Next we estimate the right hand side of the above equality. Noting that Λ^s and ∇ are commutable, then

$$|(\Lambda^s(u_m \cdot \nabla\theta_m) - u_m \cdot \nabla(\Lambda^s\theta_m), \Lambda^s\theta_m)|$$
$$= |(\Lambda^s(u_m \cdot \nabla\theta_m) - u_m \cdot (\Lambda^s\nabla\theta_m), \Lambda^s\theta_m)|$$
$$\leqslant C\|\Lambda^s(u_m \cdot \nabla\theta_m) - u_m \cdot (\Lambda^s\nabla\theta_m)\|\|\Lambda^s\theta_m\|.$$

Since for any $p_1, p_2 > 2$ with $\dfrac{1}{p_1} + \dfrac{1}{p_2} = \dfrac{1}{2}$, we have

$$\|\Lambda^s(u_m \cdot \nabla\theta_m) - u_m \cdot (\Lambda^s\nabla\theta_m)\|$$
$$\leqslant C\|\nabla u_m\|_{L^{p_1}}\|\Lambda^s\theta_m\|_{L^{p_2}} + \|\Lambda^s u_m\|_{L^{p_2}}\|\nabla\theta_m\|_{L^{p_1}}.$$

We can select $p_1 = \dfrac{2}{\alpha}$, $p_2 = \dfrac{2}{1-\alpha}$ such that

$$\|\nabla u_m\|_{L^{p_1}} \leqslant C\|\Lambda^{2-\alpha}u_m\|, \quad \|\nabla\theta_m\|_{L^{p_1}} \leqslant C\|\Lambda^{2-\alpha}\theta_m\|,$$

$$\|\Lambda^s u_m\|_{L^{p_2}} \leqslant C\|\Lambda^{s+\alpha}u_m\|, \quad \|\Lambda^s\theta_m\|_{L^{p_2}} \leqslant C\|\Lambda^{s+\alpha}\theta_m\|.$$

Thus, we have

$$\begin{aligned}
\|\Lambda^s(u_m \cdot \nabla\theta_m) &- u_m \cdot (\Lambda^s\nabla\theta_m)\| \\
&\leqslant C(\|\Lambda^{2-\alpha}u_m\|\|\Lambda^{s+\alpha}\theta_m\| + \|\Lambda^{s+\alpha}u_m\|\|\Lambda^{2-\alpha}\theta_m\|) \\
&= C(\|u_m\|_{2-\alpha}\|\Lambda^{s+\alpha}\theta_m\| + \|u_m\|_{s+\alpha}\|\Lambda^{2-\alpha}\theta_m\|).
\end{aligned}$$

From the assumption $2(1-\alpha) \leqslant s \leqslant 2-\alpha$, we can apply the Sobolev embedding $\dot{H}^2 \hookrightarrow \dot{H}^{2-\alpha}$, and $\dot{H}^2 \hookrightarrow \dot{H}^{s+\alpha}$, then

$$\|\Lambda^s(u_m \cdot \nabla\theta_m) - u_m \cdot (\Lambda^s\nabla\theta_m)\| \leqslant C(\|u_m\|_2\|\Lambda^{s+\alpha}\theta_m\| + \|u_m\|_2\|\Lambda^{2-\alpha}\theta_m\|),$$

and

$$\begin{aligned}
|(\Lambda^s(u_m \cdot \nabla\theta_m) &- u_m \cdot \nabla(\Lambda^s\theta_m), \Lambda^s\theta_m)| \hspace{2cm} (3.6.10) \\
&\leqslant C\|u_m\|_2\|\Lambda^s\theta_m\|(\|\Lambda^{s+\alpha}\theta_m\| + \|\Lambda^{2-\alpha}\theta_m\|).
\end{aligned}$$

From the ε-Young's inequality,

$$C\|u_m\|_2\|\Lambda^s\theta_m\|\|\Lambda^{s+\alpha}\theta_m\| \leqslant \frac{\kappa}{4}\|\Lambda^{s+\alpha}\theta_m\|^2 + \frac{C^2}{\kappa}(\|u_m\|_2^2\|\Lambda^s\theta_m\|^2). \quad (3.6.11)$$

For the second term in the right hand side of (3.6.10), noticing the assumed condition $2-\alpha \leqslant s+\alpha$, if $2-\alpha = s+\alpha$, the estimate is similar to (3.6.11), if $2-\alpha < s+\alpha$, we can apply the Gagliardo-Nirenberg inequality,

$$\|\Lambda^{2-\alpha}\theta_m\| \leqslant C\|\theta_m\|^{1-\beta}\|\Lambda^{s+\alpha}\theta_m\|^\beta, \quad 0 < \beta = \frac{2-\alpha}{s+\alpha} < 1.$$

Thus

$$\begin{aligned}
\|u_m\|_2\|\Lambda^{2-\alpha}\theta_m\|\|\Lambda^s\theta_m\| &\leqslant C\|u_m\|_2\|\Lambda^s\theta_m\|\|\theta_m\|^{1-\beta}\|\Lambda^{s+\alpha}\theta_m\|^\beta \\
&\leqslant C\|u_m\|_2^2\|\Lambda^s\theta_m\|^2 + \|\Lambda^{s+\alpha}\theta_m\|^{2\beta} \\
&\leqslant C\|u_m\|_2^2\|\Lambda^s\theta_m\|^2 + \frac{\kappa}{4}\|\Lambda^{s+\alpha}\theta_m\|^2 + C,
\end{aligned}$$

where the last inequality is due to the ϵ-Young inequality.

Then

$$\frac{1}{2}\frac{d}{dt}\|\Lambda^s\theta_m\|^2 + \frac{\kappa}{2}\|\Lambda^{s+\alpha}\theta_m\|^2 \leqslant C(\|u_m\|_2^2\|\Lambda^s\theta_m\|^2) + C,$$

noticing that $\int_0^T \|u_m(s)\|_2^2 ds \leqslant C$, then

$$\|\Lambda^s \theta_m(t)\|^2 \leqslant C \|\Lambda^s \theta_{m0}\|^2 \exp^{C \int_0^t \|u_m(s)\|_2^2 ds} + CT \leqslant C,$$

and

$$\int_0^T \|\Lambda^{s+\alpha} \theta_m(\sigma)\|^2 d\sigma \leqslant C,$$

where the constant is independent of m.

Now, we can prove

Theorem 3.6.1 *Suppose that $\alpha \in (0,1)$, $u_0 \in \dot{H}_\sigma$, $\theta_0 \in \dot{H}^s \cap L^2$, with $2(1-\alpha) \leqslant s \leqslant 2 - \alpha$, then there exists a unique couple weak solution (u, θ) to equations (3.6.1)-(3.6.4), such that*

$$u \in L^\infty(0, T; \dot{H}_\sigma) \cap L^p(0, T; \dot{H}^{1,p}) \cap L^2(0, T; \dot{V}_\sigma),$$
$$\theta \in L^\infty(0, T; \dot{H}^s) \cap L^2(0, T; \dot{H}^{s+\alpha}), \quad \forall T > 0.$$

Proof **Existence:** From the above *a priori* estimates, then we can extract a subsequence still denoted by u_m such that for any $T > 0$,

$$\{u_m\} \text{ converges to } u \text{ weakly star in } L^\infty(0, T; \dot{H}_\sigma),$$
$$\{u_m\} \text{ converges to } u \text{ weakly in } L^2(0, T; \dot{V}_\sigma),$$
$$\{\theta_m\} \text{ converges to } \theta \text{ weakly star in } L^\infty(0, T; \dot{H}^s),$$
$$\{\theta_m\} \text{ converges to } \theta \text{ weakly in } L^2(0, T; \dot{H}^{s+\alpha}).$$

Combining Lemma 3.6.1, we have $\{u_{mt}\}$ is uniformly bounded in $L^2(0, T; V')$, thus $\{u_m\}$ converges to u strongly in $L^2(0, T; \dot{H}^1)$. Then by a standard procedure we can pass the limit $m \to \infty$, the above convergences are sufficient to pass the limit in the linear term.

The difficulties in the limiting process, lie in the nonlinear terms. We notice that

$$\int_0^T \int_\Omega u_m \cdot \nabla u_m \, \phi \, dx dt = -\int_0^T \int_\Omega u_m \cdot \nabla \phi \, u_m dx dt,$$
$$\int_0^T \int_\Omega u_m \cdot \nabla \theta_m \, \phi \, dx dt = -\int_0^T \int_\Omega u_m \cdot \nabla \phi \, \theta_m dx dt.$$

Thus, both convective terms are handled by the strong convergence of u_m in $L^2(0, T; \dot{H})$.

On the other hand, for $1 < p < 2$, $\phi \in \mathscr{C}^\infty$,

$$\int_0^t \int_\Omega [(\epsilon + |e(u_m(s))|^2)^{\frac{p-2}{2}} e_{ij}(u_m(s))$$
$$- (\epsilon + |e(u(s))|^2)^{\frac{p-2}{2}} e_{ij}(u(s))]e_{ij}(\phi)dxds$$
$$\leqslant |\int_0^t \int_\Omega (\epsilon + |e(u_m)|^2)^{\frac{p-2}{2}} e_{ij}(u_m - u)e_{ij}(\phi)dxds| \qquad (3.6.12)$$
$$+ |\int_0^t \int_\Omega [(\epsilon + |e(u_m)|^2)^{\frac{p-2}{2}} - (\epsilon + |e(u)|^2)^{\frac{p-2}{2}}]e_{ij}(u)e_{ij}(\phi)dxds|$$
$$= \mathcal{A}_1 + \mathcal{A}_2.$$

Obviously,

$$\mathcal{A}_1 \leqslant \epsilon^{\frac{p-2}{2}} \int_0^t \|e(u_m - u)\| \|e(\phi)\| ds \leqslant C \int_0^t \|u_m - u\|_1 \|\phi\|_1 ds.$$

If $|e(u_m)| < |e(u)|$, by mean value theorem, there exists ξ such that $|e(u_m)| < \xi < |e(u)|$, and

$$(\epsilon + |e(u_m)|^2)^{\frac{p-2}{2}} - (\epsilon + |e(u)|^2)^{\frac{p-2}{2}}$$
$$= \frac{p-2}{2}(\epsilon + \xi^2)^{\frac{p-4}{2}}(|e(u_m)|^2 - |e(u)|^2)$$
$$\leqslant \frac{p-2}{2}(\epsilon + |e(u)|^2)^{\frac{p-4}{2}}(2|e(u)|)(|e(u_m)| - |e(u)|)$$
$$\leqslant (p-2)(\epsilon + |e(u)|^2)^{\frac{p-4}{2}}|e(u)||e(u_m) - e(u)|,$$

then

$$\mathcal{A}_2 \leqslant (p-2)\int_0^t \int_\Omega (\epsilon + |e(u)|^2)^{\frac{p-2}{2}}|e(u_m - u)||e(\phi)|dxds$$
$$\leqslant (p-2)\epsilon^{\frac{p-2}{2}}\int_0^t \|e(u_m - u)\| \|e(\phi)\| ds$$
$$\leqslant C \int_0^t \|u_m - u\|_1 \|\phi\|_1 ds.$$

For the case of $|e(u_m)| > |e(u)|$, the result can be obtained similarly.

From the above estimates, we deduce that

$$\int_0^t \int_\Omega [(\epsilon + |e(u_m(s))|^2)^{\frac{p-2}{2}} e_{ij}(u_m(s)) - (\epsilon + |e(u(s))|^2)^{\frac{p-2}{2}} e_{ij}(u(s))]e_{ij}(\phi)dxds$$
$$\leqslant C \int_0^t \|u_m(s) - u(s)\|_1 \|\phi\|_1 ds$$
$$\leqslant C \left(\int_0^t \|u_m(s) - u(s)\|_1^2 ds \right)^{\frac{1}{2}} \left(\int_0^t \|\phi\|_1^2 ds \right)^{\frac{1}{2}},$$

where the constant is independent of m. The convergence of (3.6.12) can be deduced from the strong convergence of u_m in $L^2(0, T; \dot{H}^1)$. Finally, by a standard procedure, we can show (u, θ) is a weak solution to equations (3.6.1)-(3.6.4).

Uniqueness: Now suppose that (u_1, θ_1), (u_2, θ_2) are two solutions of the fractional Boussinesq approximation with the same initial data. Let $u = u_1 - u_2$, $\theta = \theta_1 - \theta_2$, then

$$u_t + u_1 \cdot \nabla u_1 - u_2 \cdot \nabla u_2 - 2\mu_0 \nabla \cdot [(\epsilon + |e(u_1)|^2)^{\frac{p-2}{2}} e(u_1)$$
$$- (\epsilon + |e(u_2)|^2)^{\frac{p-2}{2}} e(u_2)] + \mu_1 \Delta^2 u = \eta\theta, \tag{3.6.13}$$

$$\theta_t + u_1 \cdot \nabla\theta_1 - u_2 \cdot \nabla\theta_2 + \kappa\Lambda^{2\alpha}\theta = 0. \tag{3.6.14}$$

Multiplying (3.6.13) with u and taking the inner product in L^2, then

$$\frac{1}{2}\frac{d}{dt}\|u\|^2 + \mu_1\|u\|_2^2 \leqslant |(\eta\theta, u)| + |(u \cdot \nabla u_2, u)|, \tag{3.6.15}$$

where we have used the divergence free condition $\nabla \cdot u = 0$, and the property

$$(-\nabla \cdot [(\epsilon + |e(u_1)|^2)^{\frac{p-2}{2}} e(u_1) - (\epsilon + |e(u_2)|^2)^{\frac{p-2}{2}} e(u_2)], u_1 - u_2) \geqslant 0.$$

Obviously,

$$|(\eta\theta, u)| \leqslant \frac{1}{2}\|u\|^2 + \frac{1}{2}\|\theta\|^2,$$

and

$$|(u \cdot \nabla u_2, u)| \leqslant \|u\|_{L^\infty}\|\nabla u_2\|\|u\|$$
$$\leqslant C\|u\|_2\|u_2\|_2\|u\|$$
$$\leqslant \frac{\mu_1}{4}\|u\|_2^2 + C\|u_2\|_2^2\|u\|^2.$$

Multiplying (3.6.14) with θ and taking the inner product in L^2, we have

$$(u_1 \cdot \nabla\theta_1, \theta) - (u_2 \cdot \nabla\theta_2, \theta) = (u_1 \cdot \nabla\theta_1, \theta) - (u_1 \cdot \nabla\theta_2, \theta)$$
$$+ (u_1 \cdot \nabla\theta_2, \theta) - (u_2 \cdot \nabla\theta_2, \theta)$$
$$= (u_1 \cdot \nabla\theta, \theta) + (u \cdot \nabla\theta_2, \theta).$$

From the divergence free condition $\nabla \cdot u = 0$,

$$\frac{1}{2}\frac{d}{dt}\|\theta\|^2 + \kappa\|\Lambda^\alpha\theta\|^2 \leqslant |(u \cdot \nabla\theta_2, \theta)|, \tag{3.6.16}$$

and from the Hölder inequality and Sobolev embedding, we know

$$|(u \cdot \nabla\theta_2, \theta)| \leqslant \|u\|_{L^\infty}\|\nabla\theta_2\|\|\theta\| \leqslant C\|u\|_2\|\nabla\theta_2\|\|\theta\|.$$

Noticing $2(1 - \alpha) \leqslant s \leqslant 2 - \alpha$, we have

$$|(u \cdot \nabla \theta_2, \theta)| \leqslant C\|u\|_2\|\theta_2\|_{s+\alpha}\|\theta\|$$
$$\leqslant \frac{\mu_1}{4}\|u\|_2^2 + C\|\theta_2\|_{s+\alpha}^2\|\theta\|^2.$$

Combining the above estimates together, we have

$$\frac{1}{2}\frac{d}{dt}(\|u\|^2 + \|\theta\|^2) + \left(\frac{\mu_1}{2}\|u\|_2^2 + \kappa\|\Lambda^\alpha\theta\|^2\right)$$
$$\leqslant \left(\frac{1}{2} + C\|\theta_2\|_{s+\alpha}^2\right)\|\theta\|^2 + \left(\frac{1}{2} + C\|u_2\|_2^2\right)\|u\|^2.$$

Taking the maximum of $\left\{\dfrac{1}{2} + C\|u_2\|_2^2, \dfrac{1}{2} + C\|\theta_2\|_{s+\alpha}^2\right\} = \mathscr{A}(t)$, we have

$$\|u(t)\|^2 + \|\theta(t)\|^2 \leqslant \exp^{\int_0^t 2\mathscr{A}(s)ds}(\|u(0)\|^2 + \|\theta(0)\|^2).$$

From above proof, we know $\displaystyle\int_0^t \mathscr{A}(s)$ is bounded, which completes proof of uniqueness.

Remark 3.6.1 *The a priori estimates obtained in this section are independent of the periodic domain Ω. Thus let $|\Omega| \to \infty$, suppose that $\alpha \in (0,1)$, $u_0 \in H_\sigma$, $\theta_0 \in H^s$, with $2(1 - \alpha) \leqslant s \leqslant 2 - \alpha$, we directly obtain the existence and uniqueness of weak solution to the initial value problem.*

In the rest of this section, we consider the decay of velocity and temperature. For simplicity, we set the constant $\kappa = 1$. Now, we consider the following equations with dissipative condition in the whole space R^2:

$$u_t + u \cdot \nabla u - \nabla \cdot \left(2\mu_0(\epsilon + |e(u)|^2)^{\frac{p-2}{2}}e(u) - 2\mu_1\Delta e(u)\right) = -f(u) + \eta\theta, \quad (3.6.17)$$

$$\nabla \cdot u = 0, \quad (3.6.18)$$

$$\theta_t + u \cdot \nabla\theta + \Lambda^{2\alpha}\theta = 0, \quad (3.6.19)$$

where $f(u)$ satisfies the condition

$$(f(u), u) \geqslant l\|u\|^2, \quad \text{for some } l > 0.$$

The conditions of $f(u)$ are the dissipative conditions, which play an important role in proving the decay of solution. To obtain the decay of temperature, we first show

Lemma 3.6.3 *Let $u_0 \in H_\sigma$, $\theta_0 \in L^2 \cap L^1$. Then*

$$|\hat{\theta}(\xi,t)| \leqslant \|\theta_0\|_{L^1} + C|\xi|t.$$

Proof The Fourier transform of the temperature satisfies the following equation:

$$\hat{\theta}_t + \mathcal{F}[(u \cdot \nabla)\theta] + |\xi|^{2\alpha}\hat{\theta} = 0,$$

and

$$\mathcal{F}[(u \cdot \nabla)\theta] = \mathcal{F}[\nabla \cdot (\theta u)] \leqslant \sum_i \int_\Omega |u_i||\theta||\xi_i| dx \leqslant |\xi|\|u\|\|\theta\|.$$

Thus, we have

$$\hat{\theta}_t + |\xi|^{2\alpha}\hat{\theta} \leqslant |\xi|\|u\|\|\theta\|.$$

Now using $\exp(t|\xi|^{2\alpha})$ as a multiplier, then

$$\frac{d}{dt}(\hat{\theta}\exp^{t|\xi|^{2\alpha}}) \leqslant \exp^{t|\xi|^{2\alpha}}(|\xi|\|u\|\|\theta\|).$$

Integrating in time over $[0, t]$, it follows that

$$\hat{\theta}(\xi, t) \leqslant \exp^{-t|\xi|^{2\alpha}}\hat{\theta}_0(\xi) + \int_0^t \exp^{-|\xi|^{2\alpha}(t-s)}(|\xi|\|u\|\|\theta\|) ds$$

$$\leqslant \|\theta_0\|_{L^1} + \int_0^t |\xi|\|u\|\|\theta\| ds$$

$$\leqslant \|\theta_0\|_{L^1} + C|\xi|t,$$

where the last inequality is due to the uniform boundedness of velocity and temperature.

Now, we present the main decay results of temperature and velocity.

Theorem 3.6.2 *Let $(u(t), \theta(t))$ be a solution of equations (3.6.17)-(3.6.19), assume initial value $u_0 \in H_\sigma$, $\theta_0 \in L^2 \cap L^1$, $2(1 - \alpha) \leqslant s \leqslant 2 - \alpha$, $(f(u), u) \geqslant l\|u\|^2$, for some $l > 0$. Then*
Case 1: *when $0 < \alpha < \dfrac{1}{2}$,*

$$\|u(t)\|^2 \leqslant \exp^{-lt}\|u_0\|^2 + \frac{C}{l^2}\left(\exp^{-\frac{lt}{2}} + \frac{1}{\left(\frac{t}{2} + 1\right)^2}\right),$$

$$\|\theta(t)\|^2 \leqslant \frac{C}{(1+t)^2}.$$

Case 2: *when* $\alpha = \dfrac{1}{2}$,

$$\|u(t)\|^2 \leqslant \exp^{-lt} \|u_0\|^2 + \frac{C}{l^2}\left(\exp^{-\frac{lt}{2}} + \frac{1}{\left(\dfrac{t}{2}+1\right)} \right),$$

$$\|\theta(t)\|^2 \leqslant \frac{C}{(1+t)}.$$

Case 3: *when* $\dfrac{1}{2} < \alpha < 1$,

$$\|u(t)\|^2 \leqslant \exp^{-lt} \|u_0\|^2 + \frac{C}{l^2}\left(\exp^{-\frac{lt}{2}} + \frac{1}{\left(\dfrac{t}{2}+1\right)^{\frac{2}{\alpha}-2}} \right),$$

$$\|\theta(t)\|^2 \leqslant \frac{C}{(1+t)^{\frac{2}{\alpha}-2}}.$$

Proof First, we consider the L^2 decay of temperature, then the decay of velocity can be obtained by the decay of the temperature. Taking inner product of (3.6.19) with θ in L^2, we obtain

$$\frac{d}{dt}\|\theta\|^2 + 2\|\Lambda^\alpha \theta\|^2 = 0.$$

This energy equality is the starting point of the Fourier splitting method. The idea is to obtain an ordinary differential inequality for the energy norm of the temperature. This is obtained by working in the frequency space and splitting the space into two appropriately chosen time dependent subspaces.

By Parseval's equality,

$$\frac{d}{dt}\int_{R^2} |\hat{\theta}(\xi,t)|^2 d\xi + 2\int_{R^2} |\xi|^{2\alpha}|\hat{\theta}(\xi,t)|^2 d\xi = 0. \tag{3.6.20}$$

Multiplying (3.6.20) with $(1+t)^2$, then

$$\frac{d}{dt}\left[(1+t)^2 \int_{R^2} |\hat{\theta}(\xi,t)|^2 d\xi\right] + 2(1+t)^2 \int_{R^2} |\xi|^{2\alpha}|\hat{\theta}(\xi,t)|^2 d\xi$$

$$= 2(1+t)\int_{R^2} |\hat{\theta}(\xi,t)|^2 d\xi. \tag{3.6.21}$$

Let $B(t) = \{\xi \in R^2 | (1+t)|\xi|^{2\alpha} \leqslant 1\}$, then

$$(1+t)\int_{R^2} |\xi|^{2\alpha}|\hat{\theta}(\xi,t)|^2 d\xi \geqslant (1+t)\int_{B(t)^c} |\xi|^{2\alpha}|\hat{\theta}(\xi,t)|^2 d\xi$$

$$\geqslant \int_{R^2} |\hat{\theta}(\xi,t)|^2 d\xi - \int_{B(t)} |\hat{\theta}(\xi,t)|^2 d\xi.$$

From (3.6.21), we can deduce that

$$\frac{d}{dt}[(1+t)^2 \int_{R^2} |\hat{\theta}(\xi,t)|^2 d\xi] \leqslant 2(1+t) \int_{B(t)} |\hat{\theta}(\xi,t)|^2 d\xi. \qquad (3.6.22)$$

To obtain decay of the temperature, we need an intermediate estimate for the Fourier transform of the temperature for frequency values in $B(t)$. From (3.6.22) and Lemma 3.6.3

$$\frac{d}{dt}[(1+t)^2 \int_{R^2} |\hat{\theta}(\xi,t)|^2 d\xi] \leqslant 2(1+t) \int_{B(t)} (C + C|\xi|t)^2 d\xi.$$

Let $M^{2\alpha} = \dfrac{1}{1+t}$. We have for the term of the right hand side

$$\int_{B(t)} (1+|\xi|t)^2 d\xi = \int_0^{2\pi} \int_0^M (1+rt)^2 r dr d\tau$$

$$\leqslant 4\pi \int_0^M (r + r^3 t^2) dr$$

$$\leqslant 4\pi \left(\frac{1}{2} \frac{1}{(1+t)^{\frac{1}{\alpha}}} + \frac{1}{4} \frac{t^2}{(1+t)^{\frac{2}{\alpha}}} \right).$$

Then

$$\frac{d}{dt}[(1+t)^2 \int_{R^2} |\hat{\theta}(\xi,t)|^2 d\xi] \leqslant C \left[\frac{1}{(1+t)^{\frac{1}{\alpha}-1}} + \frac{1}{(1+t)^{\frac{2}{\alpha}-3}} \right],$$

which yields by integrating in time over $[0,t]$,

$$(1+t)^2 \int_{R^2} |\hat{\theta}(\xi,t)|^2 d\xi \leqslant \|\theta_0\|^2 + C \int_0^t \left[\frac{1}{(1+s)^{\frac{1}{\alpha}-1}} + \frac{1}{(1+s)^{\frac{2}{\alpha}-3}} \right] ds. \qquad (3.6.23)$$

Case 1: $0 < \alpha < \dfrac{1}{2}$. In this case, $\dfrac{2}{\alpha} - 2 > \dfrac{1}{\alpha} > 2$ and hence

$$(1+t)^2 \int_{R^2} |\hat{\theta}(\xi,t)|^2 d\xi \leqslant \|\theta_0\|^2 + C \int_0^t [\frac{1}{(1+s)^{\frac{1}{\alpha}-1}} + \frac{1}{(1+s)^{\frac{2}{\alpha}-3}}] ds$$

$$\leqslant \|\theta_0\|^2 + C\left[\frac{1}{\left(2 - \dfrac{1}{\alpha}\right)(1+t)^{\frac{1}{\alpha}-2}} \right.$$

$$\left. + \frac{1}{\left(4 - \dfrac{2}{\alpha}\right)(1+t)^{\frac{2}{\alpha}-4}} - \frac{3}{4 - \dfrac{2}{\alpha}} \right]$$

$$\leqslant \|\theta_0\|^2 + C\left[\frac{1}{(1+t)^{\frac{1}{\alpha}-2}} + \frac{1}{(1+t)^{\frac{2}{\alpha}-4}} + \frac{3}{\dfrac{2}{\alpha} - 4} \right].$$

Noting $\dfrac{1}{\alpha} > 2$, the last term in the right hand of inequality is nonnegative. Then the above inequality gives the following estimate

$$\int_{R^2} |\hat{\theta}(\xi,t)|^2 d\xi \leqslant \frac{\|\theta_0\|^2}{(1+t)^2} + C\left[\frac{1}{(1+t)^{\frac{1}{\alpha}}} + \frac{1}{(1+t)^{\frac{2}{\alpha}-2}} + \frac{3}{\left(\dfrac{2}{\alpha}-4\right)(1+t)^2} \right]$$

$$\leqslant \frac{C}{(1+t)^2}.$$

Case 2: $\alpha = \dfrac{1}{2}$. In this case, $\dfrac{1}{\alpha} - 1 = 1$, $\dfrac{2}{\alpha} - 3 = 1$, and hence

$$\int_{R^2} |\hat{\theta}(\xi,t)|^2 d\xi \leqslant \frac{\|\theta_0\|^2}{(1+t)^2} + C\frac{\ln(1+t)}{(1+t)^2} \leqslant \frac{C}{(1+t)}.$$

Case 3: $\dfrac{1}{2} < \alpha < 1$. In this case, $0 < \dfrac{2}{\alpha} - 2 < \dfrac{1}{\alpha} < 2$, and hence

$$\int_{R^2} |\hat{\theta}(\xi,t)|^2 d\xi \leqslant \frac{\|\theta_0\|^2}{(1+t)^2} + C\left[\frac{1}{\left(2 - \dfrac{1}{\alpha}\right)(1+t)^{\frac{1}{\alpha}}} + \frac{1}{\left(4 - \dfrac{2}{\alpha}\right)(1+t)^{\frac{2}{\alpha}-2}} \right.$$

$$\left. - \frac{3}{\left(4 - \dfrac{2}{\alpha}\right)(1+t)^2} \right]$$

$$\leqslant \frac{\|\theta_0\|^2}{(1+t)^2} + C\left[\frac{1}{\left(2-\frac{1}{\alpha}\right)(1+t)^{\frac{1}{\alpha}}} + \frac{1}{\left(4-\frac{2}{\alpha}\right)(1+t)^{\frac{2}{\alpha}-2}}\right]$$

$$\leqslant \frac{C}{(1+t)^{\frac{2}{\alpha}-2}},$$

where the last term in the right hand of the first inequality is negative and hence dropped.

After obtaining the decay of the temperature, now we consider the decay of the velocity. Taking inner product of (3.6.17) with u in L^2, we obtain

$$\frac{1}{2}\frac{\mathrm{d}}{\mathrm{d}t}\|u\|^2 + \mu_1\|\Delta u\|^2 + 2\mu_0\int_\Omega (\epsilon + |e(u)|^2)^{\frac{p-2}{2}}|e(u)|^2\mathrm{d}x = (\eta\theta, u) - (f(u), u),$$

which gives

$$\frac{1}{2}\frac{\mathrm{d}}{\mathrm{d}t}\|u\|^2 \leqslant (\eta\theta, u) + (-f(u), u).$$

Obviously,

$$(\eta\theta, u) \leqslant \frac{l\,\|u\|^2}{2} + \frac{\|\theta\|^2}{2l}.$$

Combining the restricted condition for $f(u)$, then $(-f(u), u) \leqslant -l\|u\|^2$. Thus

$$\frac{1}{2}\frac{\mathrm{d}}{\mathrm{d}t}\|u\|^2 \leqslant -\frac{l\,\|u\|^2}{2} + \frac{\|\theta\|^2}{2l}.$$

From the decay of the temperature, we know

Case 1: $0 < \alpha < \frac{1}{2}$. In this case,

$$\frac{\mathrm{d}}{\mathrm{d}t}\|u\|^2 + l\,\|u\|^2 \leqslant \frac{\|\theta\|^2}{l} \leqslant \frac{C}{l\,(t+1)^2}.$$

Applying the Gronwall inequality,

$$\|u(t)\|^2 \leqslant \exp^{-lt}\|u_0\|^2 + \frac{C}{l}\int_0^t \exp^{-l(t-s)}(s+1)^{-2}\mathrm{d}s.$$

We divide the integral into two parts,

$$\int_0^t \exp^{-l(t-s)}(s+1)^{-2}ds = \exp^{-lt}\left[\int_0^{\frac{t}{2}} \exp^{ls}(s+1)^{-2}ds \right.$$

$$\left. + \int_{\frac{t}{2}}^t \exp^{ls}(s+1)^{-2}ds\right]$$

$$\leqslant \frac{\exp^{-lt}}{l}\left[(\exp^{\frac{lt}{2}}-1) + \frac{1}{\left(\dfrac{t}{2}+1\right)^2}(\exp^{lt}-\exp^{\frac{lt}{2}})\right]$$

$$\leqslant \frac{\exp^{-lt}}{l}\left[\exp^{\frac{lt}{2}} + \frac{\exp^{lt}}{\left(\dfrac{t}{2}+1\right)^2}\right]$$

$$\leqslant \frac{\exp^{-\frac{lt}{2}}}{l} + \frac{1}{l\left(\dfrac{t}{2}+1\right)^2}.$$

From the above estimates, we have

$$\|u(t)\|^2 \leqslant \exp^{-lt}\|u_0\|^2 + \frac{C}{l^2}\left(\exp^{-\frac{lt}{2}} + \frac{1}{\left(\dfrac{t}{2}+1\right)^2}\right).$$

Case 2: $\alpha = \dfrac{1}{2}$. In this case,

$$\frac{d}{dt}\|u\|^2 + l\|u\|^2 \leqslant \frac{\|\theta\|^2}{l} \leqslant \frac{C}{l(t+1)}.$$

Similarly, we can obtain the following inequality

$$\|u(t)\|^2 \leqslant \exp^{-lt}\|u_0\|^2 + \frac{C}{l}\int_0^t \exp^{-l(t-s)}(s+1)^{-1}ds,$$

and

$$\int_0^t \exp^{-l(t-s)}(s+1)^{-1}ds \leqslant \frac{\exp^{-\frac{lt}{2}}}{l} + \frac{1}{l\left(\dfrac{t}{2}+1\right)},$$

then

$$\|u(t)\|^2 \leqslant \exp^{-lt}\|u_0\|^2 + \frac{C}{l^2}\left(\exp^{-\frac{lt}{2}} + \frac{1}{\left(\dfrac{t}{2}+1\right)}\right).$$

Case 3: $\dfrac{1}{2} < \alpha < 1$. In this case,

$$\frac{\mathrm{d}}{\mathrm{d}t}\|u\|^2 + l\,\|u\|^2 \leqslant \frac{\|\theta\|^2}{l} \leqslant \frac{C}{l\,(t+1)^{\frac{2}{\alpha}-2}}.$$

Similarly, we can obtain the following inequality

$$\|u(t)\|^2 \leqslant \exp^{-lt}\|u_0\|^2 + \frac{C}{l}\int_0^t \exp^{-l(t-s)}(s+1)^{-\left(\frac{2}{\alpha}-2\right)}\mathrm{d}s,$$

and

$$\int_0^t \exp^{-l(t-s)}(s+1)^{-\left(\frac{2}{\alpha}-2\right)}\mathrm{d}s \leqslant \frac{\exp^{-\frac{lt}{2}}}{l} + \frac{1}{l\left(\dfrac{t}{2}+1\right)^{\frac{2}{\alpha}-2}},$$

then

$$\|u(t)\|^2 \leqslant \exp^{-lt}\|u_0\|^2 + \frac{C}{l^2}\left(\exp^{-\frac{lt}{2}} + \frac{1}{\left(\dfrac{t}{2}+1\right)^{\frac{2}{\alpha}-2}}\right).$$

Obviously, if t tends to infinity, the L^2 norm of velocity tends to zero.

Remark 3.6.2 $f(0) = 0,\ f'(u) \geqslant l > 0$ *is a special case satisfying the restricted condition for* $f(u)$.

Remark 3.6.3 *Furthermore, the result of Theorem 3.6.2 can be modified as follows:*
Case 1: $0 < \alpha < \dfrac{1}{2}$,

$$\|u(t)\|^2 \leqslant \exp^{-lt}\|u_0\|^2 + \frac{C}{l^2\left(\dfrac{t}{2}+1\right)^2},$$

$$\|\theta(t)\|^2 \leqslant \frac{C}{(1+t)^2}.$$

Case 2: $\alpha = \dfrac{1}{2}$,

$$\|u(t)\|^2 \leqslant \exp^{-lt}\|u_0\|^2 + \frac{C}{l^2\left(\dfrac{t}{2}+1\right)},$$

$$\|\theta(t)\|^2 \leqslant \frac{C}{(1+t)}.$$

Case 3: $\dfrac{1}{2} < \alpha < 1$,

$$\|u(t)\|^2 \leqslant \exp^{-lt}\|u_0\|^2 + \frac{C}{l^2\left(\dfrac{t}{2}+1\right)^{\frac{2}{\alpha}-2}},$$

$$\|\theta(t)\|^2 \leqslant \frac{C}{(1+t)^{\frac{2}{\alpha}-2}}.$$

Remark 3.6.4 *If $\alpha = 1$, the proof becomes complex, the above method is invalid. In fact, we cannot obtain the decay estimates of temperature similar to inequality (3.6.23), but by other ways we can solve this problem. First, multiplying equation (3.6.20) with $f(t) = (\ln(e+t))^3$, by Plancherel's theorem and detailed calculation, we can obtain*

$$\|\theta(t)\| \leqslant \frac{C}{\ln(e+t)}.$$

Then, multiplying equation (3.6.20) with $f(t) = (1+t)^2$, the corresponding decay estimates can be obtained.

3.7   Boundary value problems

This section introduces the boundary value problems of fractional differential equations by the harmonic extension method. In recent years, many researchers explored the properties of fractional Laplacian and related fractional partial differential equations defined on a domain with boundaries from different points of view. In particular, Ma *et al* investigated the regional fractional Laplacian by the concept of generators of random processes. They obtained some integration by parts formulae and existence and uniqueness of some boundary value problems were obtained, cf. [98—100]. Caffarelli *et al* obtained some important results on boundary value problems and obstacle problems of the "elliptic" equation with fractional Laplacian based on the harmonic extension method, spectral decomposition and the Sobolev trace theorem. The basic idea is to transform the nonlocal problem with fractional Laplacian to a local problem in a higher dimensional space, cf. [31, 32]. Cabre and Tan [30] considered the positive solutions of nonlinear problems with fractional Laplacian. We only make a simple introduction here and the interested readers may refer to the literature above and the references therein.

Consider the case when $\alpha = 1/2$. Let u be a smooth solution of the problem

$$\begin{cases} \Delta u(x,y) = 0 & x \in \mathbf{R}^d, \ y > 0 \\ u(x,0) = g(x), & x \in \mathbf{R}^d. \end{cases} \tag{3.7.1}$$

Letting $T : g \mapsto -u_y(x,0)$, then we have $(T \circ T)(g)(x) = T(-u_y(x,0))(x) = u_{yy}(x,0) = -\Delta_x g(x)$. By integration by part, we know that T is a positive operator, thus $T = (-\Delta)^{1/2}$ and $(-\Delta)^{1/2}g(x) = -u_y(x,0)$. In other words, the operator $(-\Delta)^{1/2}$ coincides with the Dirichlet to Neumann operator in the upper half space of $\mathbf{R}^{d+1}$.

Now, we consider a general fractional Laplacian $(-\Delta)^{\alpha/2}$. Consider the Dirichlet problem

$$\Delta_x u + \frac{a}{y} u_y + u_{yy} = 0 \quad x \in \mathbf{R}^d, \ y > 0 \tag{3.7.2}$$

and

$$u(x,0) = g(x), \quad x \in \mathbf{R}^d, \tag{3.7.3}$$

where $g : \mathbf{R}^d \to \mathbf{R}$ and $u : \mathbf{R}^d \times [0,\infty) \to \mathbf{R}$. The equation (3.7.2) can also be rewritten as

$$\nabla \cdot (y^a \nabla u) = 0, \tag{3.7.4}$$

where $\nabla = (\nabla_x, \nabla_y)$. By taking coordinate transform $z = \left(\dfrac{y}{1-a}\right)^{1-a}$, we have $y^a u_y = u_z$ and (3.7.4) is transformed into a non-divergence form

$$\Delta_x u + z^{\frac{-2a}{1-a}} u_{zz} = 0. \tag{3.7.5}$$

It can be shown that there exists some constant C such that

$$C(-\Delta)^{\alpha/2}g(x) = -u_z(z,0), \quad \alpha = 1 - a. \tag{3.7.6}$$

To this end, we first derive a Poisson formula. Consider the "$n + 1 + a$-dimensional" Laplace equation (3.7.2). When $n > 1 + a$, the fundamental solution at the origin can be expressed as $\Gamma(X) = C_{n+1+a}|X|^{1-n-a}$, where $C_{n+1+a} = n^{\frac{n+1+a}{2}} \Gamma\left(\dfrac{n+1+a}{2} - 1\right)/4$ and $X = (x,y)$. It can be directly checked that Γ is a solution of (3.7.2) when $y \neq 0$ and $\lim_{y \to 0+} y^a u_y = -C\delta_0$. Using the transform $z = \left(\dfrac{y}{1-a}\right)^{1-a}$ yields the fundamental solution

$$\tilde{\Gamma}(x,z) = C_{n+1+a} \frac{1}{\left(|x|^2 + (1-a)^2|z|^{2/(1-a)}\right)^{\frac{n-1+a}{2}}},$$

which solves (3.7.5) when $z \neq 0$ and $u_z(x,z) \to -\delta_0$ as $z \to 0$.

On the other hand, letting $P(x,y) = C_{n,a} y^{1-a}/(|x|^2 + |y|^2)^{\frac{n+1-a}{2}}$, then the solution of (3.7.2)-(3.7.3) can be expressed by the Poisson formula

$$u(X) = \int_{\mathbf{R}^d} P(x - \xi, y) g(\xi) \mathrm{d}\xi.$$

The corresponding Poisson kernel of (3.7.5) can be obtained by

$$\tilde{P}(x, z) = C_{n,a} \frac{z}{\left(|x|^2 + (1-a)^2|z|^{2/(1-a)}\right)^{\frac{n+1-a}{2}}}. \tag{3.7.7}$$

By the Poisson formula, we can compute by definition of the fractional Laplacian

$$
\begin{aligned}
u_z(z, 0) &= \lim_{z \to 0} \frac{u(x, z) - u(x, 0)}{z} \\
&= \lim_{z \to 0} \frac{1}{z} \int_{\mathbf{R}^d} \tilde{P}(x - \xi, z)(g(\xi) - g(x)) \mathrm{d}\xi \\
&= \lim_{z \to 0} \frac{1}{z} \int_{\mathbf{R}^d} \frac{C}{\left(|x - \xi|^2 + (1-a)^2|z|^{2/(1-a)}\right)^{\frac{n+1-a}{2}}} (g(\xi) - g(x)) \mathrm{d}\xi \\
&= C \mathrm{P.V.} \int_{\mathbf{R}^d} \frac{g(\xi) - g(x)}{|x - \xi|^{n+1-a}} \mathrm{d}\xi \\
&= -C(-\Delta)^{\frac{1-a}{2}} g(x),
\end{aligned}
$$

where the limit in the third step exists as long as g is regular enough. On the other hand, direct computation yields $y^a u_y = \left(\dfrac{1}{1-a}\right)^{-a} u_z$. Therefore, (3.7.6) follows.

By employing the extension method, one can obtain some important results similar to the Harnack inequality. Let $u : \mathbf{R}^d \times [0, \infty) \to \mathbf{R}$ be a solution of (3.7.2) such that $\lim_{y \to 0} y^a u_y(x, y) = 0$ for $|x| \leqslant r$, then the extension $\tilde{u}(x, y)$ is a weak solution of (3.7.4) in $B_R = \{(x, y) : |x|^2 + |y|^2 \leqslant R^2\}$, where $\tilde{u}(x, y) = u(x, y)$ for $y \geqslant 0$ and $\tilde{u}(x, y) = u(x, -y)$ for $y < 0$. Indeed, let $h \in C_0^\infty(B_R)$ be a test function and $\varepsilon > 0$, then by (3.7.2) and integration by parts we have

$$
\begin{aligned}
\int_{B_R} \nabla \tilde{u} \cdot \nabla h |y|^a \mathrm{d}X &= \int_{B_R \setminus \{|y| < \varepsilon\}} + \int_{B_R \cap \{|y| < \varepsilon\}} \\
&= \int_{B_R \setminus \{|y| < \varepsilon\}} \nabla \cdot (|y|^a h \nabla \tilde{u}) \mathrm{d}x \\
&\quad + \int_{B_R \cap \{|y| < \varepsilon\}} \nabla \tilde{u} \cdot \nabla h |y|^a \mathrm{d}X \tag{3.7.8} \\
&= \int_{B_R \setminus \{|y| = \varepsilon\}} h \tilde{u}_y(x, \varepsilon) \varepsilon^a \mathrm{d}x \\
&\quad + \int_{B_R \cap \{|y| < \varepsilon\}} \nabla \tilde{u} \cdot \nabla h |y|^a \mathrm{d}X.
\end{aligned}
$$

The second term on the RHS clearly goes to zero since $|y|^a|\nabla u|^2$ is locally integrable and the first term converges to zero if $\varepsilon^a \tilde{u}_y(x,\varepsilon) \to 0$ as $\varepsilon \to 0$. Therefore, $\tilde{u}$ is a weak solution of (3.7.4) in the B_R across $y = 0$ if $\varepsilon^a \tilde{u}_y(x,\varepsilon) \to 0$ as $\varepsilon \to 0$.

We have the following

Theorem 3.7.1　*Let $f : \mathbf{R}^d \to \mathbf{R}$ be a nonnegative function such that $(-\Delta)^s f = 0$ on B_r. Then there exists a constant $C = C(s,d)$ such that*

$$\sup_{B_{r/2}} f \leqslant C \inf_{B_{r/2}} f.$$

Let u be the extension of f that solves (3.7.2). Since f is non-negative, so is u. By reflecting it through the hyperplane $y = 0$, since $(-\Delta)^{\alpha/2} f = 0$, then u is weak solution of (3.7.4). Theorem 3.7.1 follows from [84].

To further expound the idea of harmonic extension in treating the boundary value problems, we continue to consider the nonlinear problem in a smooth domain $D \subset \mathbf{R}^d$

$$\begin{cases} (-\Delta)^{1/2}u = f(u), & \text{in } D, \\ u|_{\partial D} = 0, \text{ and } u > 0 \text{ in } D, \end{cases} \tag{3.7.9}$$

Here $(-\Delta)^{1/2}$ is defined by the eigenvalue problem of the standard Laplacian. Let $\{\lambda_k, \varphi_k\}$ be eigenvalues and corresponding eigenfunction of the problem

$$-\Delta\varphi_k = \lambda_k\varphi_k, \text{ such that } \varphi_k|_{\partial D} = 0 \text{ and } \|\varphi_k\|_{L^2(D)} = 1, \tag{3.7.10}$$

then $(-\Delta)^{1/2}$ is defined by

$$u = \sum_{k=1}^{\infty} c_k\varphi \mapsto (-\Delta)^{1/2}u = \sum_{k=1}^{\infty} c_k\lambda_k^{1/2}\varphi_k,$$

which clearly maps $H_0^1(D)$ to $L^2(D)$. For a given function u defined on D, consider its extension v on the cylindrical region $\mathcal{C} = D \times (0,\infty)$ such that $v = 0$ on $\partial_L \mathcal{C} = \partial D \times (0,\infty)$. Similar to the case of $\mathbf{R}^d$, the fractional Laplacian can be constructed by the extension method. Let v satisfy

$$\begin{cases} \Delta v = 0 \text{ and } v > 0 \text{ in } \mathcal{C}, \\ v = 0, \text{ on } \partial_L \mathcal{C} \text{ and } \dfrac{\partial v}{\partial n} = f(v) \text{ on } D \times \{0\}, \end{cases}$$

then the trace $u = \mathrm{Tr}v$ on $D \times \{0\}$ is a solution of (3.7.9). Indeed, since $\partial_y v$ is still harmonic and vanishes on $\partial_L \mathcal{C}$, it follows that $(-\Delta)^{1/2}u = -v_y(\cdot,0)$.

Let $H_{0,L}^1(\mathcal{C}) = \{v \in H^1(\mathcal{C})|v = 0 \text{ a.e. on } \partial_L\mathcal{C}\}$ equipped with the norm $\|v\| = \left(\int_{\mathcal{C}} |\nabla v|^2 dxdy\right)^{1/2}$ and Tr_D be the trace operator defined by $\mathrm{Tr}_D v =$

$v(\cdot,0)$. It then follows that $\mathrm{Tr}_D v \in H^{1/2}(D)$. We define $\mathcal{V}_0(D) = \{u = \mathrm{Tr}_D v : v \in H^1_{0,L}(\mathcal{C})\}$.

Proposition 3.7.1 *Let* $l(x) = dist(x, \partial D)$, *then*

$$
\mathcal{V}_0(D) = \left\{ u \in H^{1/2}(D) : \int_D \frac{u^2(x)}{l(x)} \mathrm{d}x < \infty \right\}
$$

$$
= \left\{ u \in L^2(D) : u = \sum_{k=1}^\infty b_k \varphi_k \ such\ that \ \sum_{k=1}^\infty b_k^2 \lambda_k^{1/2} < \infty \right\},
$$

(3.7.11)

and is a Banach space under the norm $\|u\|_{\mathcal{V}_0(D)} = \left\{ \|u\|^2_{H^{1/2}(D)} + \int_D \frac{u^2}{l} \mathrm{d}x \right\}^{1/2}$.

Proposition 3.7.2 *Let* $u = \sum_{k=1}^\infty b_k \varphi_k \in \mathcal{V}_0(D)$, *then there exists a unique harmonic extension* $v \in H^1_{0,L}(\mathcal{C})$ *in* $\mathcal{C}$ *of* u *having the expression*

$$
v(x,y) = \sum_{k=1}^\infty b_k \varphi_k(x) \exp\{-\lambda_k^{1/2} y\}, \quad \forall (x,y) \in \mathcal{C}.
$$

Thus the operator $(-\Delta)^{1/2}$ *is given by the Dirichlet-Neumann map*

$$
(-\Delta)^{1/2} u = \frac{\partial v}{\partial n}\Big|_{D \times \{0\}} = \sum_{k=1}^\infty b_k \lambda_k^{1/2} \varphi_k.
$$

These two propositions will be proved in what follows. We first give some properties of the space $H^1_{0,L}(\mathcal{C})$. Let $\mathcal{D}^{1,2}(\mathbf{R}^{d+1}_+)$ be the closure of smooth functions compactly supported on $\overline{\mathbf{R}^{d+1}_+}$ under the norm $\|w\|_{\mathcal{D}^{1,2}(\mathbf{R}^{d+1}_+)} = (\int_{\mathbf{R}^{d+1}_+} |\nabla w|^2 \mathrm{d}x\mathrm{d}y)^{1/2}$. Then, for $w \in \mathcal{D}^{1,2}(\mathbf{R}^{d+1}_+)$, there holds the Sobolev trace inequality

$$
\left(\int_{\mathbf{R}^d} |w(x,0)|^{2d/(d-1)} \mathrm{d}x \right)^{(d-1)/2d} \leqslant C(d) \left(\int_{\mathbf{R}^{d+1}_+} |\nabla w(x,y)|^2 \mathrm{d}x\mathrm{d}y \right)^{1/2}.
$$

(3.7.12)

From [137], there exists an optimal constant $C(d) = (d-1)\sigma_d^{1/d}/2$ as well as $w \in \mathcal{D}^{1,2}(\mathbf{R}^{d+1}_+)$ such that equality in (3.7.12) hold, where σ_d is the Lebesgue measure of the d-dimensional unit sphere.

When $d \geqslant 2$, we let $2^* = \dfrac{2d}{d-1}$. Let $v \in H^1_{0,L}(\mathcal{C})$ and extend it to $\mathbf{R}^{d+1}_+ \backslash \mathcal{C}$ by zero. Then the extended function can be approximated by functions that are compactly supported on $\overline{\mathbf{R}^{d+1}_+}$. It then follows from the Sobolev trace inequality that

$$\left(\int_D |v(x,0)|^{2^*} \mathrm{d}x\right)^{1/2^*} \leqslant C \left(\int_{\mathcal{C}} |\boldsymbol{\nabla} v(x,y)|^2 \mathrm{d}x\mathrm{d}y\right)^{1/2}, \quad d \geqslant 2.$$

It then follows by Hölder inequality that

$$\left(\int_D |v(x,0)|^q \mathrm{d}x\right)^{1/q} \leqslant C \left(\int_{\mathcal{C}} |\boldsymbol{\nabla} v(x,y)|^2 \mathrm{d}x\mathrm{d}y\right)^{1/2},$$

where $1 \leqslant q \leqslant 2^*$ for $d \geqslant 2$ and $1 \leqslant q < \infty$ for $d = 1$. That is, $\mathrm{Tr}_D(H_{0,L}^1(\mathcal{C})) \hookrightarrow L^q(D)$ continuously and the embedding is compact since $\mathrm{Tr}_D(H_{0,L}^1(\mathcal{C})) \hookrightarrow H^{1/2}(D)$ and $H^{1/2}(D) \hookrightarrow L^q(D)$ compactly. Here, $\|\cdot\|_{H^{1/2}(D)}$ is given by

$$\|u\|_{H^{1/2}(D)}^2 = \int_D \int_D \frac{|u(x) - u(\tilde{x})|^2}{|x - \tilde{x}|^{d+1}} \mathrm{d}x\mathrm{d}\tilde{x} + \int_D |u(x)|^2 \mathrm{d}x. \qquad (3.7.13)$$

The space $H_{0,L}^1(\mathcal{C})$ can also be characterized as follows.

Lemma 3.7.1 *There exists some constant C depending on D such that*

$$\int_D \frac{|v(x,0)|^2}{l(x)} \mathrm{d}x \leqslant C \int_{\mathcal{C}} |\boldsymbol{\nabla} v(x,y)|^2 \mathrm{d}x\mathrm{d}y, \quad \forall v \in H_{0,L}^1(\mathcal{C})$$

where $l(x) = dist(x, \partial D)$.

Proof First consider $d = 1$ and $D = (0,1)$. For $x_0 \in (0, 1/2)$, we have

$$v(x_0, 0) = v(t, x_0 - t)|_{t=0}^{x_0} = \int_0^{x_0} (\partial_x v - \partial_y v)(t, x_0 - t)\mathrm{d}t.$$

So that,

$$|v(x_0, 0)|^2 \leqslant x_0 \int_0^{x_0} 2|\boldsymbol{\nabla} v(t, x_0 - t)|^2 \mathrm{d}t.$$

Dividing the equation by x_0, integrating with respect to x_0 on $(0, 1/2)$ and taking the change of variables $x = t, y = x_0 - t$, we finally obtain

$$\int_0^{1/2} \frac{|v(x_0, 0)|^2}{x_0} \mathrm{d}x_0 \leqslant 2 \int_0^{1/2} \mathrm{d}x \int_0^{1/2} |\boldsymbol{\nabla} v|^2 \mathrm{d}y \leqslant 2 \int_{\mathcal{C}} |\boldsymbol{\nabla} v|^2 \mathrm{d}x\mathrm{d}y.$$

When $x_0 \in (1/2, 1)$, the lemma can be proved in a similar manner.

In the case of high spatial dimensions, suppose $D = \{x = (x', x_d) : |x'| < 1, 0 < x_d < 1/2\}$ and $v = 0$ on $\{x_d = 0, |x'| < 1\} \times (0, \infty)$. According to the results from one-dimensional case, we see that as long as $|x'| < 1$

$$\int_0^{1/2} \frac{|v(x,0)|^2}{x_d} \mathrm{d}x_d \leqslant C \int_0^{1/2} \int_0^\infty |\boldsymbol{\nabla} v|^2 \mathrm{d}x_d\mathrm{d}y.$$

Integration with respect to x' then yields

$$\int_D \frac{|v(x,0)|^2}{x_d}\,\mathrm{d}x = \int_D \int_0^{1/2} \frac{|v(x,0)|^2}{x_d}\,\mathrm{d}x'\mathrm{d}x_d \leqslant C \int_{\mathcal{C}} |\nabla v|^2 \mathrm{d}x\mathrm{d}y.$$

The results for a general domain can be derived by boundary flatten skills.

From this lemma, we can indeed prove the first equality in Proposition 3.7.1. Let $u \in H^{1/2}(D)$ satisfy $\int_D \frac{u^2}{l} < \infty$. Let $\tilde{u}$ be the extension of u in all of $\mathbf{R}^d$ by assigning $\tilde{u} = 0$ in $\mathbf{R}^d\backslash D$, then there exists a constant C such that

$$\|\tilde{u}\|_{H^{1/2}(\mathbf{R}^d)}^2 \leqslant C\left\{\|u\|_{H^{1/2}(D)}^2 + \int_D \frac{u^2(x)}{l(x)}\mathrm{d}x\right\} < \infty,$$

where $\|\tilde{u}\|_{H^{1/2}(\mathbf{R}^d)}^2$ is given by (3.7.13) with D being replaced by $\mathbf{R}^d$. Thus $\tilde{u} \in H^{1/2}(\mathbf{R}^d)$ is the trace in $\mathbf{R}^d = \partial\mathbf{R}_+^{d+1}$ of a certain function $\tilde{v} \in H^1(\mathbf{R}_+^{d+1})$. Next, there exists a local bi-Lipschitz maps that maps $\overline{\mathbf{R}_+^{d+1}}$ into $\overline{D \times [0,\infty)}$ being identity on $D \times \{0\}$ and maps $\mathbf{R}^d\backslash D$ into $\partial D \times [0,\infty)$. By composing such a bi-Lipschitz map with the function $\tilde{v}$, we obtain a $H^1_{0,L}(\mathcal{C})$ function, whose trace is u on $D \times \{0\}$. Therefore, the first equation of (3.7.11) is valid.

For a given function $u \in \mathcal{V}_0(D)$, consider the following minimizing problem

$$\inf\left\{\int_{\mathcal{C}} |\nabla v|^2 \mathrm{d}x\mathrm{d}y : v \in H^1_{0,L}(\mathcal{C}), \text{ and } v(\cdot,0) = u \text{ in } D\right\}. \qquad (3.7.14)$$

By definition, the set of v is nonempty. By lower weak semi-continuity and compact embedding $\mathrm{Tr}_D(H^1_{0,L}(\mathcal{C})) \hookrightarrow L^q(D)$, there exists minimizer $v \in H^1_{0,L}(\mathcal{C})$ which is the harmonic extension of u to $\mathcal{C}$ vanishing on $\partial_L\mathcal{C}$. Furthermore, the minimizer is unique. This can be seen from the inequality

$$0 \leqslant J\left(\frac{v_1 - v_2}{2}\right) = \frac{1}{2}J(v_1) + \frac{1}{2}J(v_2) - J\left(\frac{v_1 + v_2}{2}\right) \leqslant 0,$$

where $J(v) = \int_{\mathcal{C}} |\nabla v|^2 \mathrm{d}x\mathrm{d}y$.

To study the relationship between v and u, we denote $v = h(u)$ to be the harmonic extension from u to $\mathcal{C}$ vanishing on $\partial_L\mathcal{C}$. By divergence theorem and Lemma 3.7.1, we know there exists a constant C such that

$$\|u\|_{\mathcal{V}_0(D)} \leqslant C\|h(u)\|_{H^1_{0,L}(\mathcal{C})}, \quad \forall u \in \mathcal{V}_0(D).$$

On the other hand, h is bijective from $\mathcal{V}_0(D)$ to $\mathcal{H}$, the subspace of $H^1_{0,L}(\mathcal{C})$ made of harmonic functions in $H^1_{0,L}(\mathcal{C})$. Moreover, since $\mathcal{V}_0(D)$ and $\mathcal{H}$ are both Banach spaces, the open mapping theorem gives that there exists a constant C such that

$$\|h(u)\|_{H_{0,L}^1(\mathcal{C})} \leqslant C\|u\|_{\mathcal{V}_0(D)}, \quad \forall u \in \mathcal{V}_0(D). \tag{3.7.15}$$

Let $\mathcal{V}_0(D)$ be the dual space of by $\mathcal{V}_0^*(D)$ whose norm is given by

$$\|g\|_{\mathcal{V}_0^*(D)} = \sup_{u \in \mathcal{V}_0(D), \|u\|_{\mathcal{V}_0(D)}=1} \{\langle u, g \rangle\}.$$

Let $\xi \in \mathcal{V}_0(D)$ be a smooth function, then from divergence theorem, it follows

$$\int_{\mathcal{C}} \nabla v \nabla \eta \, dx dy = \int_D \frac{\partial v}{\partial n} \xi dx,$$

which yields by (3.7.15) that

$$\left| \int_D \frac{\partial v}{\partial n} \xi dx \right| \leqslant C\|u\|_{\mathcal{V}_0(D)} \|\xi\|_{\mathcal{V}_0(D)}.$$

Therefore, $\dfrac{\partial v}{\partial n}|_D \in \mathcal{V}_0^*(D)$ and $\left\| \dfrac{\partial h(u)}{\partial n} \right\|_{\mathcal{V}_0^*(D)} \leqslant C\|u\|_{\mathcal{V}_0(D)}$. Thus we have

Lemma 3.7.2   *The operator* $(-\Delta)^{1/2} : u \mapsto \dfrac{\partial v}{\partial n}|_{D \times \{0\}}$ *is a linear bounded mapping from* $\mathcal{V}_0(D)$ *to* $\mathcal{V}_0^*(D)$, *where* $v = h(u) \in H_{0,L}^1(\mathcal{C})$ *is the harmonic extension of* u *in* $\mathcal{C}$ *vanishing on* $\partial_L \mathcal{C}$.

In what follows, we consider the spectral representation of $(-\Delta)^{1/2}$ and the corresponding structure of the space $\mathcal{V}_0(D)$. Let $u \in \mathcal{V}_0(D) \subset L^2(D)$ have the expansion $u = \sum_{k=1}^{\infty} b_k \varphi_k$ and consider

$$v(x, y) = \sum_{k=1}^{\infty} b_k \varphi_k(x) \exp\{-\lambda_k^{1/2} y\}, \quad y > 0.$$

Obviously, $v(x, 0) = u(x)$ and $\Delta v(x, y) = 0$ when $y > 0$. On the other hand,

$$\int_0^{\infty} \int_D |\nabla v|^2 dx dy = \int_0^{\infty} \int_D \{|\nabla_x v|^2 + |\partial_y v|^2\} dx dy$$

$$= 2 \sum_{k=1}^{\infty} b_k^2 \lambda_k \int_0^{\infty} \exp\{-2\lambda_k^{1/2} y\} dy$$

$$= 2 \sum_{k=1}^{\infty} b_k^2 \lambda_k \frac{1}{2\lambda_k^{1/2}} = \sum_{k=1}^{\infty} b_k^2 \lambda_k^{1/2}.$$

This shows that $v \in H^1_{0,L}(\mathcal{C})$ if and only if $\displaystyle\sum_{k=1}^{\infty} b_k^2 \lambda_k^{1/2} < \infty$. Therefore, if this condition holds, then $v \in H^1_{0,L}(\mathcal{C})$ and hence $v = h(u)$. This is the second equality in Proposition 3.7.1.

By direct calculation of $\left. -\dfrac{\partial v}{\partial y}\right|_{y=0}$ yields $(-\Delta)^{1/2} u = \displaystyle\sum_{k=1}^{\infty} b_k \lambda_k^{1/2} \varphi_k \in V_0^*(D)$. Therefore, Proposition 3.7.2 holds.

Next, we consider the inverse of $(-\Delta)^{1/2}$.

Definition 3.7.1 *Let $B : g \mapsto \mathrm{Tr}_D v$ be a map from $V_0^*(D)$ to $V_0(D)$, where v is the unique weak solution of the problem*

$$
\begin{cases}
\Delta v = 0, & in\ \mathcal{C} \\[2mm]
v = 0, & on\ \partial_L \mathcal{C}\ and\ \dfrac{\partial v}{\partial n} = g(x)\ on\ D \times \{0\}.
\end{cases}
\tag{3.7.16}
$$

That is, $v \in H^1_{0,L}(\mathcal{C})$ and satisfies

$$
\int_{\mathcal{C}} \nabla v \nabla \xi \, dx dy = \langle g, \xi(\cdot, 0)\rangle, \quad \forall \xi \in H^1_{0,L}(\mathcal{C}).
\tag{3.7.17}
$$

The existence and uniqueness of the weak solutions follows from the Lax-Milgram theorem by studying the functional in $H^1_{0,L}(\mathcal{C})$:

$$
I(v) = \frac{1}{2} \int_{\mathcal{C}} |\nabla v|^2 dx dy - \langle g, v(\cdot, 0)\rangle, \quad g \in V_0^*(D).
$$

It is obvious that the operator B is the inverse operator of $(-\Delta)^{1/2}$ and $(B \circ B)g = (-\Delta)^{-1} g$. Furthermore, we have

Proposition 3.7.3 $B \circ B|_{L^2(D)} = (-\Delta)^{-1} : L^2(D) \to L^2(D)$ *is a bounded linear operator, where $(-\Delta)^{-1}$ is the inverse Laplacian in D with zero Dirichlet boundary conditions.*

The operator $B : L^2(D) \to L^2(D)$ is self-adjoint. For arbitrary $v_1, v_2 \in H^1_{0,L}(\mathcal{C})$, there holds

$$
\int_{\mathcal{C}} (v_2 \Delta v_1 - v_1 \Delta v_2) dx dy = \int_{D} \left(v_2 \frac{\partial v_1}{\partial n} - v_1 \frac{\partial v_2}{\partial n}\right) dx,
$$

from which it follows

$$
\int_{D} B g_2 \cdot g_1 dx = \int_{D} B g_1 \cdot g_2 dx,
$$

and

$$
\int_{D} v_2(x, 0)(-\Delta)^{-1/2} v_1(x, 0) dx = \int_{D} v_1(x, 0)(-\Delta)^{-1/2} v_2(x, 0) dx.
$$

Taking $\xi = v$ in (3.7.17) and using the compactness of the embedding $\mathrm{Tr}_D(H^1_{0,L}(\mathcal{C})) \hookrightarrow H^{1/2}(D)$, we know B is a positive and compact operator on $L^2(D)$. From the spectral theory of compact self-adjoint operator, all the eigenvalues of B are positive real numbers and the corresponding eigenvectors consists of an orthonormal basis of $L^2(D)$. Furthermore, such eigenvalues and eigenvectors can be expresses in an explicit form. This leads to the following

Proposition 3.7.4   *Let $\{\varphi_k\}$ be an orthonormal basis of $L^2(D)$ with $\{\lambda_K\}$ being the corresponding Dirichlet eigenvalues, forming a spectral decomposition of $-\Delta$ in D with Dirichlet boundary conditions as in (3.7.10). Then for all $k \geqslant 1$, there holds*

$$\begin{cases} (-\Delta)^{-1/2}\varphi_k = \lambda_k^{1/2}\varphi_k, & in\ D \\ \varphi_k = 0, & on\ \partial D. \end{cases} \qquad (3.7.18)$$

In particular, $\{\varphi_k\}$ is also a basis formed by the eigenfunctions of $(-\Delta)^{1/2}$, with eigenvalues $\{\lambda_k^{1/2}\}$.

We end this section by giving a result of the following problem

$$\begin{cases} (-\Delta)^{1/2}u = f(x), & in\ D \\ u = 0, & on\ \partial D, \end{cases} \qquad (3.7.19)$$

where $f \in \mathcal{V}_0^*(D)$ and D is a smooth bounded domain in $\mathbf{R}^d$. By the extension method, the solution of the problem can be represented by $u = \mathrm{Tr}_D v$ for some $v \in H^1_{0,L}$ being the solution of (3.7.16) and $v(x,0) = u \in \mathcal{V}_0(D)$. The following proposition is parallel to the regularity results of $W^{2,p}$ estimates and Schauder estimates in elliptic equations, whose proof is omitted here for simplicity, cf. [30].

Proposition 3.7.5   *Let $\alpha \in (0,1)$, D be a $C^{2,\alpha}$ bounded domain in $\mathbf{R}^d$, $g \in \mathcal{V}_0^*(D)$, $v \in H^1_{0,L}(\mathcal{C})$ be the weak solution of (3.7.16) and $u = \mathrm{Tr}_D v$ be the weak solutions of (3.7.19). Then,*
   (1) if $g \in L^2(D)$, then $u \in H^1_0(D)$,
   (2) if $g \in H^1_0(D)$, then $u \in H^2(D) \cap H^1_0(D)$,
   (3) if $g \in L^\infty(D)$, then $v \in W^{1,q}(D \times (0,R))$ for all $R > 0$ and $1 < q < \infty$.
In particular, $v \in C^\alpha(\overline{\mathcal{C}})$ and $u \in C^\alpha(\overline{D})$,
   (4) if $g \in C^\alpha(\overline{D})$ and $g|_{\partial D} = 0$, then $v \in C^{1,\alpha}(\overline{\mathcal{C}})$ and $u \in C^{1,\alpha}(\overline{D})$, and
   (5) if $g \in C^{1,\alpha}(\overline{D})$ and $g|_{\partial D} = 0$, then $v \in C^{2,\alpha}(\overline{\mathcal{C}})$ and $u \in C^{2,\alpha}(\overline{D})$.

Chapter 4

Numerical Approximations in Fractional Calculus

Recent decades have witnessed a fast growing applications of fractional calculus to diverse scientific and engineering fields regarding anomalous diffusion, constitutive modelling in viscoelasticity, signal processing and control, fluid mechanics, image processing, and researches on soft matter behaviors, to just mention a few. Compared to integer-order calculus, fractional calculus has the capacity of providing a more simple and accurate description of complex mechanical and physical processes featuring history dependency and space nonlocality, and has thus induced the occurrences of a series of fractional differential equations. Although the analytical solutions of some of fractional differential equations are obtainable, yet these solutions are expressed in terms of special functions which are usually difficult for numerical evaluation, and the solutions are even inaccessible for some of fractional nonlinear equations. These naturally lead to a rapid increasing developments of numerical methods for fractional differential equations. Due to the history dependency and space nonlocality of fractional calculus, numerical solution of fractional differential equations usually characterizes extremely high computational cost and memory requirements. Even though a high-performed computer is employed, it is still difficult to perform a long-time or large-domain simulation, whose operations are found to increase exponentially with time. Up to now, the "short memory principle" has been proposed to reduce the computational effort, but this principle, as pointed out by Ford et al, will yield instability in numerical computations for some specific problems, which implies the applicability of the "short memory principle" seems not very appealing. It is therefore an open issue how to successfully implement the long-time simulation for fractional calculus. On the other hand, less are now known about the systematic analyzes on the stability of numerical methods concerning fractional calculus, together with the solution techniques for high-dimensional fractional differential equations, especially for nonlinear equations.

The succeeding three chapters mainly concern numerical methods for solving fractional differential equations. These methods include: (1) finite difference methods based on Euler explicit, Euler implicit, Crank-Nicolson, and predictor-corrector schemes; (2) series approximation methods comprising Adomain decomposition method, variational iteration method, homotopy perturbation method, homotopy analysis method and differential transform method; (3) finite element method; and (4) other methods such as spectral methods, mesh-free methods, etc. The above methods possess their respective merits and drawbacks, and are applicable for problems having different governing equations and initial and/or boundary value conditions. It should be also noted that much less are known about the rigorous theoretical analyses, such as stability and convergence analyses, of some existing methods.

Before embarking upon the numerical methods for fractional differential equations, we first present some typical discretization schemes for fractional derivative (or integral). These schemes are mainly based on the definition of Grünwald-Letnikov fractional derivative (or integral), the numerical discretization of Riemann-Liouville fractional derivative (or integral), the numerical integration formulas, and the extensions of conventional finite difference schemes.

Throughout the chapter, unless stated otherwise, we always assume $f(t)$ a sufficiently smooth function defined on $[a, T]$, along with the notations $t_j = a + jh, f(t_j) = f_j, j = 0, 1, \cdots, [(T - a)/h]$ and

$$b_j^{(\alpha)} = (j + 1)^{1-\alpha} - j^{1-\alpha},$$

where $[x]$ takes the integer part of x, being the maximum integer that does not exceed x.

4.1   Fundamentals of fractional calculus

There have been different types of definitions of fractional derivatives, and different fractional derivatives are usually associated with different discretization schemes and thus with different stability and convergence analyses. It suffices in this section to mention the following three types of fractional derivatives that usually appear in the fractional differential equations of practical interest.

1. Grünwald-Letnikov fractional derivative

Integer-order derivatives can take the form of the limit of backward difference quotient of the corresponding order:

$$\frac{\mathrm{d}f(t)}{\mathrm{d}t} = \lim_{h \to 0} \frac{1}{h}\Big(f(t) - f(t-h)\Big),$$

$$\frac{\mathrm{d}f^2(t)}{\mathrm{d}t^2} = \lim_{h \to 0} \frac{1}{h^2}\Big(f(t) - 2f(t-h) + f(t-2h)\Big),$$

$$\vdots$$

$$\begin{aligned}
\frac{\mathrm{d}^n f(t)}{\mathrm{d}t^n} &= \lim_{h \to 0} \frac{1}{h^n} \sum_{k=0}^{n} (-1)^k \binom{n}{k} f(t-kh) \\
&= \lim_{h \to 0} \frac{1}{h^n} \sum_{k=0}^{\infty} \frac{(-1)^k \Gamma(n+1)}{\Gamma(k+1)\Gamma(n-k+1)} f(t-kh), \quad n \in \mathbb{N},
\end{aligned} \tag{4.1.1}$$

where $\binom{n}{k} = 0$ for $k > n$.

Extending the integer-order derivative above to arbitrary-order derivative, i.e., replacing the differential order n in (4.1.1) by an arbitrary real number α, leads to the standard Grünwald-Letnikov fractional derivative:

$$^{GL}\mathcal{D}^\alpha f(t) = \lim_{h \to 0} \frac{1}{h^\alpha} \sum_{k=0}^{\infty} \frac{(-1)^k \Gamma(\alpha+1)}{\Gamma(k+1)\Gamma(\alpha-k+1)} f(t-kh), \quad \alpha > 0. \tag{4.1.2}$$

Given a function $f(t)$ defined on $[a, T]$ and vanishing for $t < a$, the Grünwald-Letnikov fractional derivative can be written as

$$^{GL}\mathcal{D}^\alpha f(t) = \lim_{h \to 0} \frac{1}{h^\alpha} \sum_{k=0}^{[(t-a)/h]} \omega_k^{(\alpha)} f(t-kh), \quad \alpha > 0, \tag{4.1.3}$$

where $\omega_k^{(\alpha)} = (-1)^k \binom{\alpha}{k} = \dfrac{(-1)^k \Gamma(\alpha+1)}{\Gamma(k+1)\Gamma(\alpha-k+1)}$ is called the Grünwald-Letnikov coefficients.

Also, according to [158], a shifted Grünwald-Letnikov formula is defined as

$$^{GL}\mathcal{D}^\alpha f(t) = \lim_{h \to 0} \frac{1}{h^\alpha} \sum_{k=0}^{[(t-a)/h+p]} \omega_k^{(\alpha)} f(t-(k-p)h), \quad \alpha > 0. \tag{4.1.4}$$

2. Riemann-Liouville fractional integral and derivative

n-th order integral, where n is an positive integer, can be written as

$$_0\mathcal{D}^{-n} f(t) = \frac{1}{\Gamma(n)} \int_0^t (t-\tau)^{n-1} f(\tau)\mathrm{d}\tau, \tag{4.1.5}$$

and replacing the n in (4.1.5) by the arbitrary real number α yields Riemann-Liouville fractional integral, which reads

$$_a\mathcal{D}^{-\alpha} f(t) = \frac{1}{\Gamma(\alpha)} \int_a^t (t-\tau)^{\alpha-1} f(\tau)\mathrm{d}\tau. \tag{4.1.6}$$

By letting $\beta = m - \alpha (m - 1 < \beta \leqslant m)$ for integer m, the Riemann-Liouville fractional derivative of order β takes the form of

$$
\begin{aligned}
{}_a\mathcal{D}^\beta f(t) &= {}_a \mathcal{D}^m \, {}_a\mathcal{D}^{-\alpha} f(t) \\
&= \frac{d^m}{dt^m} \left[\frac{1}{\Gamma(\alpha)} \int_a^t (t - \tau)^{\alpha - 1} f(\tau) d\tau \right].
\end{aligned}
\tag{4.1.7}
$$

In particular, for $a = -\infty$, (4.1.7) is called the Liouville fractional derivative. It is trivial that if $f(t)$ vanishes for $t \leqslant a$, Riemann-Liouville and Liouville fractional derivatives are just the same.

3. Caputo fractional derivative

Caputo fractional derivative is defined by

$$
{}_a^C\mathcal{D}^\alpha f(t) = \begin{cases} \dfrac{1}{\Gamma(m - \alpha)} \displaystyle\int_a^t (t - \tau)^{m - \alpha - 1} f^{(m)}(\tau) d\tau \right], & m - 1 < \alpha < m, \\ f^{(m)}(t), & \alpha = m. \end{cases}
\tag{4.1.8}
$$

where m is a positive integer. Particularly, if $a = 0$, then ${}_a^C\mathcal{D}^\alpha f(t)$ is abbreviated to ${}^C\mathcal{D}^\alpha f(t)$.

4. Relations among three types of fractional derivatives and the essential difference between fractional- and integer-order derivatives

Proposition 4.1.1 [218]   *Let $m - 1 < \alpha \leqslant m, m \in \mathbb{N}, f(t) \in C^m[a, b]$, then it holds that*

$$
{}^{GL}\mathcal{D}^\alpha f(t) = {}_a\mathcal{D}^\alpha f(t).
\tag{4.1.9}
$$

Proposition 4.1.2 [218]   *If, for $m - 1 < \alpha \leqslant m, m \in \mathbb{N}, {}_a\mathcal{D}^\alpha f(t)$ and the $(m - 1)$-th order derivative of $f(t)$ at $t = a$ are both bounded, then*

$$
\begin{aligned}
{}_a^C\mathcal{D}^\alpha f(t) &= {}_a\mathcal{D}^\alpha [f - T_{m-1}[f; a]](t) \\
&= {}_a\mathcal{D}^\alpha f(t) - \sum_{k=0}^{m-1} \frac{f^{(k)}(a)}{\Gamma(k - \alpha + 1)} (t - a)^{k - \alpha},
\end{aligned}
\tag{4.1.10}
$$

where $T_{m-1}[f; a]$ is the $(m - 1)$-th order Taylor expansion of f:

$$
T_{m-1}[f; a] = \sum_{k=0}^{m-1} \frac{(t - a)^k}{k!} f^{(k)}(a).
$$

Remark 4.1.1   *1. Riemann-Liouville and Grünwald-Letnikov fractional derivatives are equivalent under a condition that is easy to satisfy for many practical problems, and we thus allow this equivalence without further explicit statement.*

2. Caputo and Riemann-Liouville fractional derivatives are equivalent if $f^{(k)}(a) = 0, k = 0, 1, \cdots, m - 1$.

3. *If the conditions of Proposition 4.1.2 hold, then the relation among three types of derivatives afore-mentioned will be given by:*

$$^{GL}\mathcal{D}^\alpha f(t) = \sum_{k=0}^{m-1} \frac{f^{(k)}(a)}{\Gamma(k-\alpha+1)}(t-a)^{k-\alpha} + {}_a^C\mathcal{D}^\alpha f(t) = {}_a\mathcal{D}^\alpha f(t). \quad (4.1.11)$$

4. *The essential difference between fractional differential operator and its integer-order counterpart is: the former is a non-local operator whereas the latter is a local one. The integral nature of fractional derivatives or integral underlies the very nonlocality of our interest.*

4.2   G-Algorithms for Riemann-Liouville fractional derivative

From the definition of Grünwald-Letnikov definition (4.1.3), a simple but effective approach to approximate the Riemann-Liouville fractional derivative $\mathcal{D}^\alpha f(t)$ is to remove the limit symbol in the definition of Grünwald-Letnikov fractional derivative, thereby leading to a discretization scheme in form of truncated series. We call the resulting scheme the Grünwald-Letnikov approximation scheme. The scheme is commonly used to evaluate the Riemann-Liouville fractional derivative because of the equivalence between the derivative and Grünwald-Letnikov fractional derivative, and is one of the numerical methods that have ever been utilized to approximate fractional derivative (or integral) in researches of early period (see Chapter 7 in [179] and §8.2 in [176]). The Grünwald-Letnikov approximation scheme can be given by

$$_a\mathcal{D}_t^\alpha f(t) \approx h^{-\alpha} \sum_{k=0}^{[(t-a)/h]} \omega_k^{(\alpha)} f(t-kh) := \left({}_a\mathcal{D}^\alpha f(t)\right)_{GL}. \quad (4.2.1)$$

Letting $f(a) = 0$, taking $h = \dfrac{t-a}{N}$ and using the relation

$$\omega_j^{(\alpha)} = (-1)^j \binom{\alpha}{j} = \binom{j-\alpha-1}{j} = \frac{\Gamma(j-\alpha)}{\Gamma(-\alpha)\Gamma(j+1)}, \quad (4.2.2)$$

we obtain the following detailed approximation scheme

$$_a\mathcal{D}_t^\alpha f(t) \approx \frac{\left(\dfrac{t-a}{N}\right)^{-\alpha}}{\Gamma(-\alpha)} \sum_{j=0}^{N-1} \frac{\Gamma(j-\alpha)}{\Gamma(j+1)} f\left(t - j\left(\frac{t-a}{N}\right)\right). \quad (4.2.3)$$

In particular, for $a = 0$, the scheme amounts to

$$_0\mathcal{D}_t^\alpha f(t) \approx \frac{t^{-\alpha} N^\alpha}{\Gamma(-\alpha)} \sum_{j=0}^{N-1} \frac{\Gamma(j-\alpha)}{\Gamma(j+1)} f\left(t - \frac{jt}{N}\right), \quad (4.2.4)$$

which we call the "G1-algorithm".

G1-algorithm can be written in a compact form as

$$\left({}_a\mathcal{D}_t^\alpha f(t_n) \right)_{G1} = h^{-\alpha} \sum_{k=0}^{n} \omega_k^{(\alpha)} f_{n-k}, \qquad (4.2.5)$$

which is also called the "fractional backward difference quotient" approximation scheme.

Similarly, using the definition of shifted Grünwald-Letnikov fractional derivative (4.1.4) produces the approximation scheme below:

$$_a\mathcal{D}_t^\alpha f(t) \approx h^{-\alpha} \sum_{k=0}^{[(t-a)/h+p]} \omega_k^{(\alpha)} f(t-(k-p)h) := \left({}_a\mathcal{D}^\alpha f(t) \right)_{G_{S(p)}}, \qquad (4.2.6)$$

and we call it the shifted Grünwald approximation scheme, abbreviated to "$G_{S(p)}$-algorithm".

Generally, for the non-negative integer p, the $G_{S(p)}$-algorithm can be represented by:

$$\left({}_a\mathcal{D}_t^\alpha f(t_n) \right)_{G_{S(p)}} = h^{-\alpha} \sum_{k=0}^{[(t-a)/h]+p} \omega_k^{(\alpha)} f_{n-k+p}. \qquad (4.2.7)$$

The Grünwald-Letnikov coefficients above $\omega_j^{(\alpha)} = (-1)^j \binom{\alpha}{j}$ are actually the Taylor expansion coefficients of generating function $\omega(z) = (1-z)^\alpha$, and these coefficients can be secured using the following recursion relations:

$$\omega_0^{(\alpha)} = 1, \quad \omega_j^{(\alpha)} = \left(1 - \frac{\alpha+1}{j} \right) \omega_{j-1}^{(\alpha)}, \quad j = 1, 2, \cdots. \qquad (4.2.8)$$

In addition, Oldham and Spanier [176] presented in 1974 the approximation schemes given by

$$_a\mathcal{D}_t^{-1} f(t) = \lim_{h \to 0} h \sum_{j=0}^{[\frac{t-a}{h}-\frac{1}{2}]} f\left(t - \left(j + \frac{1}{2} \right) h \right) \qquad (4.2.9)$$

$$_a\mathcal{D}_t^{1} f(t) = \lim_{h \to 0} h^{-1} \sum_{j=0}^{[\frac{t-a}{h}+\frac{1}{2}]} (-1)^j f\left(t - \left(j - \frac{1}{2} \right) h \right). \qquad (4.2.10)$$

These schemes feature fast convergence, from which one can derive an improved Grünwald-Letnikov fractional derivative defined by (i.e., letting $p =$

$\alpha/2$ in Eq.(4.1.4))

$$_aD_t^\alpha f(t) = \lim_{h \to 0} \frac{h^{-\alpha}}{\Gamma(-\alpha)} \sum_{j=0}^{[(t-a)/h+\alpha/2]} \frac{\Gamma(j-\alpha)}{\Gamma(j+1)} f\left(t - \left(j - \frac{1}{2}\alpha\right)h\right). \quad (4.2.11)$$

Letting $a = 0$ rewrites (4.2.11) to

$$_0D_t^\alpha f(t) = \lim_{h \to 0} \frac{h^{-\alpha}}{\Gamma(-\alpha)} \sum_{j=0}^{[t/h+\alpha/2]} \frac{\Gamma(j-\alpha)}{\Gamma(j+1)} f\left(t - \left(j - \frac{1}{2}\alpha\right)h\right). \quad (4.2.12)$$

Removal of the limit operation in above equation produces the "fractional central difference quotient" approximation scheme, which is usually called the "G2-algorithm". This scheme needs the functional values at the non-grid points and thus requires function interpolation. For instance, a three-point interpolating formula reads:

$$\begin{aligned} f\left(t - \left(j - \frac{1}{2}\alpha\right)h\right) &\approx \left(\frac{\alpha}{4} + \frac{\alpha^2}{8}\right) f(t - (j-1)h) \\ &+ \left(1 - \frac{\alpha^2}{4}\right) f(t - jh) \\ &+ \left(\frac{\alpha^2}{8} - \frac{\alpha}{4}\right) f(t - (j+1)h), \end{aligned} \quad (4.2.13)$$

then the corresponding G2-algorithm can be presented by:

$$\begin{aligned} \left(_aD_t^\alpha f(t_n)\right)_{G2} &= h^{-\alpha} \sum_{j=0}^{n-1} w_j^{(\alpha)} \left(f_{n-j} + \frac{1}{4}\alpha\left(f_{n-j+1} - f_{n-j-1}\right)\right. \\ &\left. + \frac{1}{8}\alpha^2\left(f_{n-j+1} - 2f_{n-j} + f_{n-j-1}\right)\right). \end{aligned} \quad (4.2.14)$$

Remark 4.2.1   *G1-,G2- and G_S- algorithms are all developed from the definition of Grünwald-Letnikov fractional derivative or integral, and they can thus be used to approximate either the fractional derivative ($\alpha \geqslant 0$) or fractional integral ($\alpha \leqslant 0$).*

Theorem 4.2.1 [158]   *Suppose $f(t) \in L_1(R)$ and $f \in \wp^{\alpha+1}(R)$, and let*

$$A_h f(t) = h^{-\alpha} \sum_{k=0}^{\infty} w_k^{(\alpha)} f(t - (k-p)h), \quad (4.2.15)$$

where p is non-negative real number, and $Af(t) = {}_\infty D^\alpha f(t)$ be the Liouville fractional derivative (namely, the Riemann-Liouville fractional derivative with $a = -\infty$, see (4.1.7)), then, as $h \longrightarrow 0$,

$$A_h f(t) = Af(t) + O(h), \quad t \in R. \quad (4.2.16)$$

Proof Let $\hat{f}(k) = \mathcal{F}\{f(t); k\} = \int_{-\infty}^{\infty} e^{ikt} f(t) dt$ be the Fourier transform of $f(t)$, then $\mathcal{F}\{f(t-h); k\} = e^{ikh} \hat{f}(k)$.

For arbitrary complex number z and $\alpha > 0$ it holds that

$$(1-z)^{\alpha} = \sum_{k=0}^{\infty}(-1)^k \binom{\alpha}{k} z^k = \sum_{k=0}^{\infty} \omega_k^{(\alpha)} z^k, \qquad (4.2.17)$$

thus the Fourier transform of (4.2.15) yields

$$\mathcal{F}\{A_h f(t); k\} = h^{-\alpha} \sum_{m=0}^{\infty} \omega_k^{(\alpha)} e^{ik(m-p)h} \hat{f}(k)$$

$$= h^{-\alpha} e^{-ikph} \hat{f}(k) \sum_{m=0}^{\infty} \omega_k^{(\alpha)} e^{ikmh} \qquad (4.2.18)$$

$$= h^{-\alpha} e^{-ikph} \hat{f}(k)(1 - e^{ikh})$$

$$= (-ik)^{\alpha} \phi(-ikh) \hat{f}(k),$$

where

$$\phi(z) = \left(\frac{1 - e^{-z}}{z}\right)^{\alpha} e^{zp} = 1 + \left(p - \frac{\alpha}{2}\right) z + O(|z|^2). \qquad (4.2.19)$$

It is straightforward to see that there exists some $c > 0$, such that

$$|\phi(-ix) - 1| \leqslant c|x|, \quad \forall x \in R.$$

So that we have

$$\mathcal{F}\{A_h f(t); k\} = (-ik)^{\alpha} \hat{f}(k) + (-ik)^{\alpha} \hat{f}(k)[\phi(-ikh) - 1]$$

$$= \mathcal{F}\{A f(t); k\} + \hat{\varphi}(h, k), \qquad (4.2.20)$$

where $\hat{\varphi}(h, k) = (-ik)^{\alpha}[\phi(-ikh) - 1]\hat{f}(k)$, and $|\hat{\varphi}(h, k)| \leqslant |k|^{\alpha} c|hk||\hat{f}(k)|$. Since $f(t) \in L_1(R)$, and $f \in \wp^{\alpha+1}(R)$, we see that

$$I = \int_{-\infty}^{\infty}(1 + |k|)^{\alpha+1}|\hat{f}(k)| dk < \infty.$$

Accordingly, we finally obtain

$$|\varphi(h, x)| = \left|\frac{1}{2\pi i}\int_{-\infty}^{\infty} e^{-ikx}\hat{\varphi}(h, k) dk\right| \leqslant Ich.$$

Remark 4.2.2 *1. From (4.2.19), it can been seen that for $p = \alpha/2$, the error of A_h takes its minimum and a second-order accuracy is accordingly achieved, but note that interpolation should be used to derive the functional values at non-grid points. The corresponding approximation scheme is (4.2.11), namely, the G2-algorithm.*

2. *Avoiding interpolation can simplify the computation. This can be done by letting $t_n - (k - p)h$ be the grid points, where we need to find an optimal non-negative integer p, such that $|p - \alpha/2|$ is minimum. It is obvious that for $0 < \alpha \leqslant 1$, $p = 0$ is acceptable; while for $1 < \alpha \leqslant 2$, $p = 1$ is optimal.*

3. *If $f(t) = 0$ for $t \leqslant 0$, then $A_h f(t)$ is comprised of a finite number of terms, and $Af(t)$ is equivalent to Riemann-Liouville fractional derivative. This indicates that when $f(t)$ is sufficiently smooth at $t = a$ and $f(a) = 0$, G-algorithms can achieve first-order accuracy, which leads to the following conclusion:*

Corollary 4.2.1 [179, 218] *Suppose $f \in C^n[a, T]$, $\alpha \geqslant 0$, $N = (T - a)/h \in \mathbb{N}$, then the finite Grünwald-Letnikov fractional differential operator*

$$\left({}_a\mathcal{D}^\alpha f(t) \right)_{G_{s(p)}} = h^{-\alpha} \sum_{k=0}^{N+P} \omega_k^{(\alpha)} f(t - (k - p)h) \tag{4.2.21}$$

is the first-order approximation of Riemann-Liouville fractional differential operator ${}_a\mathcal{D}_t^\alpha$ if $f(a) = 0$, namely,

$$\left({}_a\mathcal{D}^\alpha f(t) \right)_{G_{s(p)}} = \mathcal{D}_t^\alpha f(t) + O(h) \Leftrightarrow f(a) = 0.$$

Otherwise, if $f(a) \neq 0$, then it holds that

$$\left(\mathcal{D}^\alpha f(t) \right)_{G_{s(p)}} = \mathcal{D}_t^\alpha f(t) + O(h) + O(f(a)).$$

Next we use $G_{s(p)}$−algorithm to evaluate the fractional derivative of $sin(x)$ at $x = 1$. By using the properties of fractional calculus [179], the explicit expressions of the fractional derivative are given by

$$\mathcal{D}^\alpha sint = \begin{cases} t^{1-\alpha} \sum_{i=0}^{\infty} \dfrac{(-1)^i t^{2i}}{\Gamma(2i + 2 - \alpha)}, & 0 < \alpha < 1; \\[3mm] t^{2-\alpha} \sum_{i=0}^{\infty} \dfrac{(-1)^{i+1} t^{2i+1}}{\Gamma(2i + 4 - \alpha)}, & 1 < \alpha < 2. \end{cases} \tag{4.2.22}$$

The errors of $G_{s(p)}$ are tabulated in Table 4.2.1. It can be observed that the convergence of the algorithm is of the first-order, which originates from the fact that the errors halve when the step h halves. Moreover, through comparing the errors, we see that taking $p = 0$ and $p = 1$ turns out to be sensible and optimal when $\alpha = 0.2$ and $\alpha = 1.6$, respectively. This conclusion is in accordance with the discussions given in the remarks mentioned above.

Table 4.2.1   Absolute errors from $G_{s(p)}$-algorithm for different
approximation step h

h	$\alpha = 0.2$			$\alpha = 1.6$		
	$p = 0$	$p = 1$	$p = 2$	$p = 0$	$p = 1$	$p = 2$
0.1	-0.0032	0.0243	0.0412	0.0890	-0.0214	-0.1249
0.05	-0.0016	0.0132	0.0254	0.0439	-0.0108	-0.0637
0.01	$-3.1650\mathrm{e}{-004}$	0.0028	0.0058	0.0087	-0.0022	-0.0130
0.005	$-1.5826\mathrm{e}{-004}$	0.0014	0.0030	0.0043	-0.0011	-0.0065
0.001	$-3.1653\mathrm{e}{-005}$	$2.8447\mathrm{e}{-004}$	$5.9954\mathrm{e}{-004}$	$8.6735\mathrm{e}{-004}$	$-2.1676\mathrm{e}{-004}$	-0.0013
0.0005	$-1.5827\mathrm{e}{-005}$	$1.4234\mathrm{e}{-004}$	$3.0024\mathrm{e}{-004}$	$4.3362\mathrm{e}{-004}$	$-1.0838\mathrm{e}{-004}$	$-6.5022\mathrm{e}{-004}$

4.3   D-Algorithm for Riemann-Liouville fractional derivative

In 1997, Kai Diethelm [65] presented the numerical integration formulas (see [64]) for finite-part integrals in order to approximate the fractional integral and derivative.

Lemma 4.3.1 [179, 218]   *Given $m - 1 < \alpha < m, m \in \mathbb{N}, \alpha \notin \mathbb{N}$ and $f(t) \in C^m[0, T]$, Reimann-Liouville fractional derivative can be expressed in terms of Hadamard finite-part integral:*

$$\mathcal{D}^\alpha f(t) = \frac{1}{\Gamma(-\alpha)} \int_0^t \frac{f(\tau)}{(t - \tau)^{\alpha+1}} \mathrm{d}\tau. \tag{4.3.1}$$

Similarly, Caputo fractional derivative can also be represented by a Hadamard finite-part:

$$^C\mathcal{D}^\alpha f(t) = \frac{1}{\Gamma(-\alpha)} \int_0^t \frac{f(\tau) - T_{m-1}[f; 0](\tau)}{(t - \tau)^{\alpha+1}} \mathrm{d}\tau. \tag{4.3.2}$$

To discretize (4.3.1), transform the variable interval from $[0, t]$ to $[0, 1]$, and select equispaced grid points $t_j = jh$. We thus have the Reimann-Liouville fractional derivative (4.3.1) written by

$$_0\mathcal{D}_t^\alpha f(t_n) = \frac{t_n^{-\alpha}}{\Gamma(-\alpha)} \int_0^1 \frac{f(t_n - t_n\xi)}{\xi^{\alpha+1}} \mathrm{d}\xi = \frac{t_n^{-\alpha}}{\Gamma(-\alpha)} \int_0^1 \frac{g_n(\xi)}{\xi^{\alpha+1}} \mathrm{d}\xi, \tag{4.3.3}$$

where $g_n(\xi) = f(t_n - t_n\xi)$.

So far, the numerical approximation of Reimann-Liouville fractional derivative has been transformed to the approximation of Hadamard finite-part integral

$$\int_0^1 \frac{g(\xi)}{\xi^{\alpha+1}} \mathrm{d}\xi.$$

The compound integration quadrature formula [64] given by Diethellm can be employed in numerical evaluation of the above integral, whose procedure can be described as: (1) Partition the integration interval $[0,1]$ as $0 = x_0 < x_1 < \cdots < x_n = 1$; (2) Interpolate the integrand g: construct the piecewise interpolating polynomial of degree d, i.e. $\tilde{g}_d$, each fragment of which is the interpolating function on the sub-interval $[x_{l-1}, x_l](l = 1, 2, \cdots, n)$ with respect to $d+1$ equispaced points $x_{l-1} + \dfrac{\mu}{d}(x_l - x_{l-1})$, $\mu = 0, 1, \cdots, d$; (3) Exactly compute the weighted integration of $\tilde{g}_d$ with weight function $\xi^{-\alpha-1}$. Ultimately, we obtain the numerical integration formula:

$$Q_n[g] = \int_0^1 \frac{\tilde{g}_d(\xi)}{\xi^{\alpha+1}} d\xi,$$

which depends on n, d, α and selection of grid points.

In particular, taking $x_k = k/j$, $k = 0, 1, \cdots, j$ and making piecewise linear interpolation, i.e. letting $d = 1$, yields

$$Q_n[g] \approx \sum_{k=0}^n w_{k,n} g(k/n)$$

where

$$
w_{k,n} = \frac{n^\alpha}{\alpha(1-\alpha)}
\begin{cases}
-1, & k = 0 \\
2k^{1-\alpha} - (k-1)^{1-\alpha} - (k+1)^{1-\alpha}, & 1 \leqslant k \leqslant n-1 \\
(\alpha-1)n^{-\alpha} - (n-1)^{1-\alpha} + n^{1-\alpha}, & k = n
\end{cases}
$$
$$
= \frac{n^\alpha}{\alpha(1-\alpha)}
\begin{cases}
-1, & k = 0 \\
b_{k-1}^{(\alpha)} - b_k^{(\alpha)}, & 1 \leqslant k \leqslant n-1 \\
(\alpha-1)n^{-\alpha} + b_{n-1}^{(\alpha)}, & k = n.
\end{cases}
\tag{4.3.4}
$$

With the weighting factors $w_{k,n}$ just derived, the approximation scheme of the Reimann-Liouville fractional derivative is given by:

$$\mathcal{D}^\alpha f(t_n) \approx \frac{t_n^{-\alpha}}{\Gamma(-\alpha)} \sum_{k=0}^n w_{k,n} f(t_n - kh) := \left(\mathcal{D}^\alpha f(t_n)\right)_D. \tag{4.3.5}$$

We call this scheme the "D-algorithm".

Theorem 4.3.1   *If $\alpha \in (0,2), \alpha \neq 1, f(t) \in C^2[0,T], t_n = nh \in [0,T]$, then there exists α-dependent constant $c_\alpha > 0$, such that the truncated error of D-algorithm satisfies*

$$\left|\mathcal{D}^\alpha f(t_n) - \left(\mathcal{D}^\alpha f(t_n)\right)_D\right| \leqslant c_\alpha \|f''\|_\infty h^{2-\alpha}. \tag{4.3.6}$$

See the proof in [66] (Theorem 2.3) and [65] (lemma 2.2).

We still consider the evaluation of fractional derivative of $sin(x)$ at $x = 1$ but using the D-algorithm discussed here. The errors are given in Table 4.3.1. The algorithm is proved to be convergent since for different α, the error reduces with the decreasing step. To see the convergence rate, we consider the variation of error with the step h, which is shown in Fig. 4.3.1, where only logarithmic coordinates are considered, and where the dotted and solid lines denote the numerical results and line $y = (2 - \alpha)x$, respectively.

From the figure, one can see that the error variation with approximation step under logarithmic coordinates is linear and parallel to the line $y = (2 - \alpha)x$, which imply that the algorithm can achieve a $(2 - \alpha)$-order convergent rate, namely, $|\mathcal{D}^\alpha f(t_n) - \left(\mathcal{D}^\alpha f(t_n)\right)_D| = O(h^{2-\alpha})$.

Table 4.3.1   Absolute errors from D-algorithm for different approximation step h

h	$\alpha = 0.01$	$\alpha = 0.5$	$\alpha = 1.5$
0.01	4.8999e-007	1.9018e-004	0.0693
0.002	2.4486e-008	1.7370e-005	0.0310
0.001	6.6534e-009	6.1709e-006	0.0219
0.0002	3.1609e-010	5.5547e-007	0.0098
0.0001	8.4463e-011	1.9668e-007	0.0069

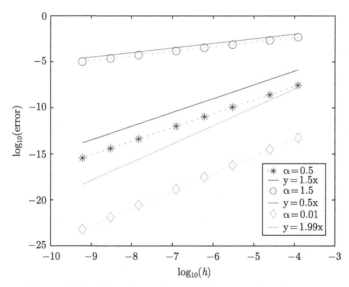

Figure 4.3.1   Variation of error with approximation step.

Analogously, we can also apply the ideas behind D-algorithm to Caputo fractional derivative by using the relation between the derivative and the

Riemann-Liouville fractional derivative. Incidentally, the D-algorithm can also be exploited to the case $\alpha < 0$, and there, it is actually the R2-algorithm having second-order accuracy, which shall be presented in the next section.

4.4   R-Algorithms for Riemann-Liouville fractional integral

The key point in numerically computing fractional integral equations, especially linear and non-linear Abel integral equations as well as integro-differential equations having fractional integral operators, is to find an effective numerical approximation of Riemann-Liouville integral. In what follows, we detail an approximation scheme proposed by Lubich (see also the paper [145] of Lubich and his collaborators).

Riemann-Liouville fractional integral is defined by (let $q < 0$)

$$
\begin{aligned}
{}_0\mathcal{J}_t^{-q} f(t) = {}_0\mathcal{D}_t^q f(t) &= \frac{1}{\Gamma(-q)} \int_0^t \frac{f(\tau)}{(t-\tau)^{q+1}} \mathrm{d}\tau \\
&= \frac{1}{\Gamma(-q)} \int_0^t \frac{f(t-\tau)}{\tau^{q+1}} \mathrm{d}\tau.
\end{aligned}
\tag{4.4.1}
$$

Applying different integration quadrature formulas will lead to different numerical approximation schemes for Riemann-Liouville fractional integral, e.g. the applications of different compound integration quadrature formulas. Note that (4.4.1) can be written by

$$
{}_0\mathcal{J}_t^{-q} f(t) = {}_0\mathcal{D}_t^q f(t) = \frac{1}{\Gamma(-q)} \sum_{j=0}^{[t/h]} \int_{t_j}^{t_{j+1}} \frac{f(t-\tau)}{\tau^{q+1}} \mathrm{d}\tau.
\tag{4.4.2}
$$

Accordingly, numerical approximation of Riemann-Liouvile fractional integral has been transformed to the approximation of (4.4.2). We call this type of approximation schemes the "R-algorithms".

Taking compound rectangle quadrature will produce

$$
\begin{aligned}
{}_0\mathcal{J}_t^{-q} f(t_n) = {}_0\mathcal{D}_t^q f(t_n) &= \frac{1}{\Gamma(-q)} \sum_{j=0}^{n-1} \int_{t_j}^{t_{j+1}} \frac{f(t_n-\tau)}{\tau^{q+1}} \mathrm{d}\tau \\
&\approx \frac{1}{\Gamma(-q)} \sum_{j=0}^{n-1} f(t_n - t_{j+1}) \int_{t_j}^{t_{j+1}} \frac{1}{\tau^{q+1}} \mathrm{d}\tau \\
&= \frac{h^{-q}}{\Gamma(1-q)} \sum_{j=0}^{n-1} [(j+1)^{-q} - j^{-q}] f(t_n - t_{j+1}).
\end{aligned}
\tag{4.4.3}
$$

With considering the above approximation scheme, the so-called "R0-algorithm" for approximating Riemann-Liouville fractional integral can be presented by:

$$\left({}_0\mathcal{J}_t^{-q}f(t_n)\right)_{R0} = \left({}_0\mathcal{D}_t^q f(t_n)\right)_{R0} =: \frac{h^{-q}}{\Gamma(1-q)}\sum_{j=0}^{n-1} b_j^{(1+q)} f_{n-j-1}. \quad (4.4.4)$$

Alternatively, taking compound trapezoidal quadrature will yield

$$\int_{t_j}^{t_{j+1}} \frac{f(t-\tau)}{\tau^{q+1}}d\tau \approx \frac{f(t-t_j)+f(t-t_{j+1})}{2}\int_{t_j}^{t_{j+1}} \frac{d\tau}{\tau^{q+1}}$$
$$= \frac{f(t-t_j)+f(t-t_{j+1})}{-2q}(t_{j+1}^{-q} - t_j^{-q}), \quad (4.4.5)$$

from which and (4.4.2), we derive the "R1-algorithm":

$$\left({}_0\mathcal{J}_t^{-q}f(t)\right)_{R1} = \left({}_0\mathcal{D}_t^q f(t)\right)_{R1} = \frac{h^{-q}}{2\Gamma(1-q)}\sum_{j=0}^{n-1} b_j^{(1+q)}\left(f_{n-j}+f_{n-j-1}\right). \quad (4.4.6)$$

Furthermore, if using linear interpolation for integrand f, namely,

$$\int_{t_j}^{t_{j+1}} \frac{f(t-\tau)}{\tau^{q+1}}d\tau$$
$$\approx \int_{t_j}^{t_{j+1}} \frac{(1+j-h\tau)f(t-t_j)+(h\tau-j)f(t-t_{j+1})}{\tau^{q+1}}d\tau, \quad (4.4.7)$$

one will obtain the "R2-algorithm" which is given by:

$$\left({}_0\mathcal{J}_t^{-q}f(t_n)\right)_{R2} = \left({}_0\mathcal{D}_t^q f(t_n)\right)_{R2}$$
$$= \frac{h^{-q}}{\Gamma(1-q)}\sum_{j=0}^{n-1}\left\{b_j^{(1+q)}\frac{(j+1)f_{n-j}-jf_{n-j-1}}{-q} + b_j^{(q)}\frac{f_{n-j-1}-f_{n-j}}{1-q}\right\}.$$
$$(4.4.8)$$

Remark 4.4.1   *1. Integral terms in (4.4.2) can also be approximated by other high-order quadrature rules; in other words, the integrand $f(t-\tau)$ can be approximated by other interpolating formulas such as piecewise quadratic interpolation.*

2. R-algorithms can still take unequispaced grid points.

3. The accuracy of R-algorithms is of the one- , $(1-q)$-, and two- orders for R0-, R1-, and R2- algorithms, respectively. For details please see [72] and the lemmas 5.2.1 and 5.2.2 in the succeeding chapter.

Now use R-algorithms to evaluate the Riemann-Liouville fractional integral $_0\mathcal{J}_t^\alpha f(t)$ of the Mittag-Leffler function $f(t) = E_{2,1}(-t^2)$ at $t = 1$. The Mittag-Leffler function is defined in form of series

$$E_{\alpha,\beta}(z) := \sum_{n=0}^{\infty} \frac{z^n}{\Gamma(\alpha n + \beta)}, \quad z \in \mathbb{C}, \beta > 0, \tag{4.4.9}$$

and using the basic properties of fractional integral can obtain the explicit expression of the fractional integral, i.e.

$$_0\mathcal{J}_t^\alpha E_{\mu,1}(-t^\mu) = t^\alpha E_{\mu,1+\alpha}(-t^\mu). \tag{4.4.10}$$

Table 4.4.1 gives the absolute errors from the R-algorithms. It can be seen that the error reduces as the step h decreases which indicates the con-

Table 4.4.1   Absolute errors from R-algorithms for different approximation step h

h	R0		R1		R2	
	$\alpha = 0.05$	$\alpha = 0.8$	$\alpha = 0.05$	$\alpha = 0.8$	$\alpha = 0.05$	$\alpha = 0.8$
0.01	0.0073	0.0027	0.0031	1.3139e-005	1.7314e-006	7.2609e-006
0.005	0.0036	0.0014	0.0015	4.3150e-006	4.5968e-007	1.8159e-006
0.001	6.9375e-004	2.7149e-004	2.7918e-004	2.9340e-007	2.0747e-008	7.2659e-008
0.0005	3.4217e-004	1.3569e-004	1.3487e-004	8.9663e-008	5.4268e-009	1.8166e-008
0.0001	6.6357e-005	2.7125e-005	2.4892e-005	5.5004e-009	2.3813e-010	7.2668e-010

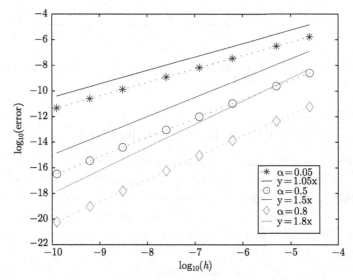

Figure 4.4.1   Variations of errors from R0- and R2- algorithms with approximation step.

vergence of the algorithms. The variations of errors from R-algorithms with
approximation steps are shown in Figs. 4.4.1 and 4.4.2 where only logarithmic
coordinates are considered. From Fig. 4.4.1, we see that for different α, the
error variations of R0- and R2- algorithms with approximation step are linear
and parallel to line $y = x$ and line $y = 2x$, respectively, which implies the
accuracy of the two algorithms are of the first- and second- orders separately.
Additionally, Fig. 4.4.2 shows the variation of error from R1-algorithm with
approximation step is parallel to the line $y = (1 + \alpha)x$, which corresponds to
a $(1 + \alpha)$-order accuracy of the R1-algorithm.

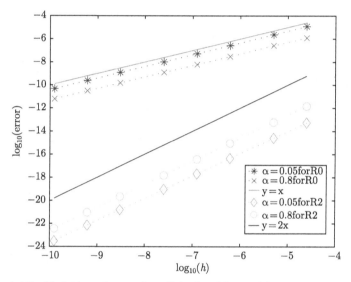

Figure 4.4.2   Variation of error from R1- algorithms with approximation step.

4.5   L-Algorithms for fractional derivative

R-algorithms are intended for approximating Riemann-Liouvile fractional integral. Extending the ideas behind R-algorithms to the approximation for
fractional derivative leads to the "L-algorithms" which will be elaborated in
this section. The fundamental principle behind the L-algorithms is to numerically differentiate the derivative of f, e. g. f' or f'', which appears in
integrand.

For $0 \leqslant \alpha < 1$, the Caputo fractional derivative can be approximated by:

$$
\begin{aligned}
^{C}\mathcal{D}_t^{\alpha} f(t_n) &= \frac{1}{\Gamma(1-\alpha)} \int_0^{t_n} \frac{f'(\tau)\mathrm{d}\tau}{(t_n - \tau)^{\alpha}} \\
&= \frac{1}{\Gamma(1-\alpha)} \sum_{j=0}^{n-1} \int_{t_j}^{t_{j+1}} \frac{f'(t_n - \tau)\mathrm{d}\tau}{\tau^{\alpha}}
\end{aligned}
$$

$$
\begin{aligned}
&\approx \frac{1}{\Gamma(1-\alpha)} \sum_{j=0}^{n-1} \frac{f(t_n - t_j) - f(t_n - t_{j+1})}{h} \int_{t_j}^{t_{j+1}} \tau^{-\alpha}\mathrm{d}\tau \\
&= \frac{h^{-\alpha}}{\Gamma(2-\alpha)} \sum_{j=0}^{n-1} (f_{n-j} - f_{n-j-1})[(j+1)^{1-\alpha} - j^{1-\alpha}].
\end{aligned}
\tag{4.5.1}
$$

We call this approximation scheme the 'L1-algorithm' which can be written in a compact form:

$$
\left(^{C}\mathcal{D}_t^{\alpha} f(t_n) \right)_{L1} = \frac{h^{-\alpha}}{\Gamma(2-\alpha)} \sum_{j=0}^{n-1} b_j^{(\alpha)} (f_{n-j} - f_{n-j-1}).
\tag{4.5.2}
$$

For $1 \leqslant \alpha < 2$, the Caputo fractional derivative is defined by

$$
\begin{aligned}
^{C}\mathcal{D}_t^{\alpha} f(t_n) &= \frac{1}{\Gamma(2-\alpha)} \int_0^{t_n} \frac{f''(\tau)\mathrm{d}\tau}{(t_n - \tau)^{\alpha}} \\
&= \frac{1}{\Gamma(2-\alpha)} \sum_{j=0}^{n-1} \int_{t_j}^{t_{j+1}} \frac{f''(t_n - \tau)\mathrm{d}\tau}{\tau^{\alpha-1}},
\end{aligned}
\tag{4.5.3}
$$

and using the second-order central difference quotient approximation for f'', i.e.,

$$
\begin{aligned}
\int_{t_j}^{t_{j+1}} \frac{f''(t_n - \tau)}{\tau^{\alpha-1}}\mathrm{d}\tau &\approx \frac{f(t_n - t_{j-1}) - 2f(t_n - t_j) + f(t_n - t_{j+1})}{h^2} \int_{t_j}^{t_{j+1}} \frac{\mathrm{d}\tau}{\tau^{\alpha-1}} \\
&= \frac{h^{-\alpha}}{2-\alpha} \left(f_{n-j+1} - 2f_{n-j} + f_{n-j-1} \right)[(j+1)^{2-\alpha} - j^{2-\alpha}],
\end{aligned}
\tag{4.5.4}
$$

leads to the so-called 'L2-algorithm' given by:

$$
\left(^{C}\mathcal{D}_t^{\alpha} f(t_n) \right)_{L2} =: \frac{h^{-\alpha}}{\Gamma(3-\alpha)} \sum_{j=0}^{n-1} b_j^{(\alpha-1)} \left(f_{n-j+1} - 2f_{n-j} + f_{n-j-1} \right).
\tag{4.5.5}
$$

In a similar manner, it is straightforward to deduce the algorithms corresponding to cases $2 \leqslant \alpha < 3, 3 \leqslant \alpha < 4, \cdots$ (L3-algorithm, L4-algorithm, etc.)

Additionally, owing to the relation (4.1.10) between Riemann-Liouville and Caputo fractional derivatives, the L-algorithms can also be applied to the approximation of Riemann-Liouville fractional derivative. In fact, from (4.1.10), one can learn that the only difference between the L-algorithms for Riemann-Liouville fractional derivative and those for Caputo derivative is the first several terms. It is interesting to see that using L1-algorithms to approximate Riemann-Liouville derivative will yields completely the same approximation scheme as that derived from D-algorithm despite a totally different deductions. This coincidence can be clearly seen in (4.6.5) and (4.6.10). In this connection, we see that the convergence rate of L-algorithms are the same as that of D-algorithm which has been analysed in the preceding section.

4.6   General form of fractional difference quotient approximations

All the approximation schemes mentioned in previous sections for fractional integral and derivative can be uniformly written by

$$\mathcal{D}^\alpha f(t_n) \approx h^{-\alpha} \sum_{j=0}^{N} c_{n,j}^{(\alpha)} f_j, \tag{4.6.1}$$

where the weighting coefficients $c_{n,j}^{(\alpha)}$ depend on n, j, α, but are independent of f. Below is a list of weighting coefficients determined by different approximation schemes:

G1-algorithm

$$c_{n,j}^{(\alpha)} = \begin{cases} \omega_{n-j}^{(\alpha)} = \dfrac{\Gamma(n-j-\alpha)}{\Gamma(-\alpha)\Gamma(n-j+1)}, & 0 \leqslant j \leqslant n; \\ 0, & others. \end{cases} \tag{4.6.2}$$

$G_{s(p)}$-algorithm (p is a positive integer)

$$c_{n,j}^{(\alpha)} = \begin{cases} \omega_{n-j+p}^{(\alpha)} = \dfrac{\Gamma(n-j+p-\alpha)}{\Gamma(-\alpha)\Gamma(n-j+p+1)}, & 0 \leqslant j \leqslant n+p; \\ 0, & others. \end{cases} \tag{4.6.3}$$

G2-algorithm

$$c_{n,j}^{(\alpha)} = \begin{cases} \dfrac{\Gamma(n-1-\alpha)}{\Gamma(-\alpha)\Gamma(n)}\left(\dfrac{\alpha^2}{8} - \dfrac{\alpha}{4}\right), & j = 0; \\[2mm] \dfrac{\Gamma(n-2-\alpha)}{\Gamma(-\alpha)\Gamma(n)}\left[(1 - \dfrac{\alpha}{2} + \dfrac{\alpha^2}{8})n + \dfrac{\alpha^2}{8} - \dfrac{\alpha}{4} - 2\right], & j = 1; \\[2mm] \dfrac{\Gamma(n-j-1-\alpha)}{\Gamma(-\alpha)\Gamma(n-j+2)}\Big\{(n-j)^2 - \dfrac{(n-j)\alpha}{2}(\alpha+3) \\[1mm] \quad + (\alpha+1)\left(\dfrac{\alpha^3}{8} + \dfrac{\alpha^2}{2} - 1\right)\Big\}, & 2 \leqslant j \leqslant n-1; \\[2mm] -\dfrac{\alpha^3}{8} - \dfrac{\alpha^2}{2} + 1, & j = n; \\[2mm] \dfrac{\alpha^2 + 2\alpha}{8}, & j = n+1; \\[2mm] 0, & others. \end{cases} \tag{4.6.4}$$

D-algorithm

$$c_{n,j}^{(\alpha)} = \dfrac{1}{\Gamma(2-\alpha)}\begin{cases} (1-\alpha)n^{-\alpha} - b_{n-1}^{(\alpha)}, & j = 0; \\ b_{n-j}^{(\alpha)} - b_{n-j-1}^{(\alpha)}, & 1 \leqslant j \leqslant n-1; \\ 1, & j = n; \\ 0, & others. \end{cases} \tag{4.6.5}$$

R0-algorithm

$$c_{n,j}^{(\alpha)} = \dfrac{1}{\Gamma(1-\alpha)}\begin{cases} b_{n-j-1}^{(1+\alpha)}, & 0 \leqslant j \leqslant n-1; \\ 0, & others. \end{cases} \tag{4.6.6}$$

R1-algorithm

$$c_{n,j}^{(\alpha)} = \dfrac{1}{2\Gamma(1-\alpha)}\begin{cases} b_{n-1}^{(1+\alpha)}, & j = 0; \\ b_{n-j}^{(1+\alpha)} + b_{n-j-1}^{(1+\alpha)}, & 1 \leqslant j \leqslant n-1; \\ 1, & j = n; \\ 0, & others. \end{cases} \tag{4.6.7}$$

R2-algorithm

$$c_{n,j}^{(\alpha)} = \dfrac{1}{\Gamma(2-\alpha)}\begin{cases} (1-\alpha)n^{-\alpha} - b_{n-1}^{(\alpha)}, & j = 0; \\ b_{n-j}^{(\alpha)} - b_{n-j-1}^{(\alpha)}, & 1 \leqslant j \leqslant n-1; \\ 1, & j = n; \\ 0, & others. \end{cases} \tag{4.6.8}$$

L1-algorithm (Caputo fractional derivative)

$$c_{n,j}^{(\alpha)} = \dfrac{1}{\Gamma(2-\alpha)}\begin{cases} -b_{n-1}^{(\alpha)}, & j = 0; \\ b_{n-j}^{(\alpha)} - b_{n-j-1}^{(\alpha)}, & 1 \leqslant j \leqslant n-1; \\ 1, & j = n; \\ 0, & others. \end{cases} \tag{4.6.9}$$

L1-algorithm (Riemann-Liouville fractional derivative)

$$
c_{n,j}^{(\alpha)} = \frac{1}{\Gamma(2-\alpha)}
\begin{cases}
(1-\alpha)n^{-\alpha} - b_{n-1}^{(\alpha)}, & j = 0; \\
b_{n-j}^{(\alpha)} - b_{n-j-1}^{(\alpha)}, & 1 \leqslant j \leqslant n-1; \\
1, & j = n; \\
0, & others.
\end{cases}
\tag{4.6.10}
$$

L2-algorithm (Caputo fractional derivative)

$$
c_{n,j}^{(\alpha)} = \frac{1}{\Gamma(3-\alpha)}
\begin{cases}
b_{n-1}^{(\alpha-1)}, & j = 0; \\
-2b_{n-1}^{(\alpha-1)} + b_{n-2}^{(\alpha-1)}, & j = 1; \\
b_{n-j+1}^{(\alpha-1)} - 2b_{n-j}^{(\alpha-1)} + b_{n-j-1}^{(\alpha-1)}, & 2 \leqslant j \leqslant n-1; \\
2^{2-\alpha} - 3 & j = n \\
1, & j = n+1; \\
0, & others.
\end{cases}
\tag{4.6.11}
$$

Remark 4.6.1 *1. The positive and negative α in the general form (4.6.1) correspond to the approximations of fractional derivative and integral, respectively. For G-algorithms, α can be either positive or negative; for R-algorithms, α is limited to a negative number; and for L-algorithms, α should take a positive number.*

2. D-algorithm, R2-algorithm and L1-algorithm (Riemann-Liouville type) have the same weighting coefficients, although the deductions of these coefficients and the approximating objectives are both different. Moreover, D-algorithm is only suitable for uniform grid while R- and L- algorithms can be extended to unequisapced grid points.

3. The general form (4.6.1) generally takes $N = n$ for G1-, D-, R2-, and L1- algorithms; $N = n - 1$ for R0-algorithm; $N = n + 1$ for G2- and L2-algorithms; and $N = n + p$ for $G_{s(p)}-$algorithm.

4.7   Extensions of integer-Order numerical differentiation and integration

4.7.1   Extensions of backward and central difference quotient schemes

In analogy with deriving the G1- and G2- algorithms from extending integer-order derivative to fractional derivatives, we can also directly derive the fractional difference schemes from extending the difference quotient schemes of the integer-order derivative, namely, backward and central difference schemes, to their fractional-order counterpart.

We first consider the shift operator E^h and difference operators $\nabla_h, \triangle_h, \delta_h$ (for backward, forward and central difference, respectively), where $h \in \mathbb{R}$.

Imposing these operators on $u(t), t \in \mathbb{R}$ yields

$$\begin{cases} E^h u(t) = u(t+h), \\ \nabla_h u(t) = u(t) - u(t-h), \\ \triangle_h u(t) = u(t+h) - u(t), \\ \delta_h u(t) = u(t+h/2) - u(t-h/2). \end{cases} \tag{4.7.1}$$

Obviously, shift operator E^h possesses the following properties:

$$E^{\sigma+\tau} = E^\sigma E^\tau, \quad \sigma, \tau \in \mathbb{R}, \tag{4.7.2}$$

together with the relation:

$$\nabla_h = I - E^{-h}, \quad \triangle_h = E^h - I, \quad \delta_h = E^{h/2} - E^{-h/2}. \tag{4.7.3}$$

By using the above notations, the backward and central difference quotient of first-order derivative can be represented by:

$$u'(t) = D^1 u(t) = \frac{u(t) - u(t-h)}{h} + O(h) = \frac{[\nabla_h u(t)]}{h} + O(h),$$

$$u'(t) = D^1 u(t) = \frac{u(t+h/2) - u(t-h/2)}{h} + O(h^2) = \frac{\delta_h u(t)}{h} + O(h^2).$$

Here and hereafter, we assume $u(t)$ sufficiently smooth. These approximation schemes can be generalized to the approximations of high-order derivatives $u^{(n)}(t) = D^n u(t), \ n \in \mathbb{N}$:

$$D^n u(t) = \frac{[\nabla_h^n u(t)]}{h^n} + O(h) = h^{-n}(I - E^{-h})^n u(t) + O(h)$$

and

$$D^n u(t) = \frac{\delta_h^n u(t)}{h^n} + O(h^2) = h^{-n}(E^{h/2} - E^{-h/2})^n u(t) + O(h^2).$$

Here, we assume $h > 0$. The power operation of the difference operators ∇_h^n, δ_h^n can be determined by binomial expansion:

$$\nabla_h^n = \sum_{j=0}^{n} (-1)^j \binom{n}{j} E^{-jh},$$

$$\delta_h^n = \sum_{j=0}^{n} (-1)^j \binom{n}{j} E^{(n-j)h/2} E^{-jh/2} = \sum_{j=0}^{n} (-1)^j \binom{n}{j} E^{(n/2-j)h},$$

from which we further have:

$$h^{-n} \sum_{j=0}^{n} (-1)^j \binom{n}{j} u(t-jh) = D^n u(t) + O(h), \tag{4.7.4}$$

$$h^{-n} \sum_{j=0}^{n} (-1)^j \binom{n}{j} u(t + (n/2 - j)h) = D^n u(t) + O(h^2). \qquad (4.7.5)$$

Now we extend the above schemes to the fractional-order cases:

$$\nabla_h^\alpha = \sum_{j=0}^{\infty} (-1)^j \binom{\alpha}{j} E^{-jh},$$

$$\delta_h^\alpha = \sum_{j=0}^{\infty} (-1)^j \binom{\alpha}{j} E^{(\alpha/2 - j)h}.$$

Note that the schemes above possess the similar series forms to the following expansions (replace E^{-h} by z and the series are convergent when $|z| < 1$):

$$(1-z)^\alpha = \sum_{j=0}^{\infty} (-1)^j \binom{\alpha}{j} z^j = \sum_{j=0}^{\infty} (-1)^j w_j^{(\alpha)} z^j,$$

$$(z^{-1/2} - z^{1/2})^\alpha = z^{-\alpha/2} \sum_{j=0}^{\infty} (-1)^j \binom{\alpha}{j} z^j = \sum_{j=0}^{\infty} w_j^{(\alpha)} z^{j-\alpha/2}.$$

From the power of the difference operators, we derive again the Grünwald-Letnikov approximation scheme:

$$h^{-\alpha} \nabla_h^\alpha u(t) = h^{-\alpha} \sum_{j=0}^{\infty} w_j^{(\alpha)} u(t - jh) = {}_0^{GL} \mathcal{D}^\alpha u(t) + O(h),$$

as well as the fractional central difference scheme:

$$h^{-\alpha} \delta_h^\alpha u(t) = h^{-\alpha} \sum_{j=0}^{\infty} w_j^{(\alpha)} u(t - (j - \alpha/2)h) = {}_0^{GL} \mathcal{D}^\alpha u(t) + O(h^2).$$

If $u(t)$ vanishes for $t \leqslant 0$, then we have

$$h^{-\alpha} \nabla_h^\alpha u(t) = h^{-\alpha} \sum_{j=0}^{[t/h]} w_j^{(\alpha)} u(t - jh) = {}_0^{GL} \mathcal{D}^\alpha u(t) + O(h),$$

$$h^{-\alpha} \delta_h^\alpha u(t) = h^{-\alpha} \sum_{j=0}^{[t/h+\alpha/2]} w_j^{(\alpha)} u(t - (j - \alpha/2)h) = {}_0^{GL} \mathcal{D}^\alpha u(t) + O(h^2).$$

The above two schemes are associated with schemes (4.2.1) and (4.2.11), respectively.

Remark 4.7.1 *1. The forward difference scheme*

$$h^{-n}\triangle^n u(t) = D^n u(t) + O(t),$$

is not very suitable to be extended to the approximation of fractional derivative.

2. If $u(t)$ at $t = 0$ fails to smoothly extend to the negative semi-axis, then the approximation schemes will give inaccurate predictions. It is usually required $u(t) = 0, \forall t \leqslant 0$.

4.7.2   Extension of interpolation-type integration quadrature formulas

Classical interpolation-type integration quadrature formulas approximate the integrand using the interpolation polynomial. We can extend this idea to approximating the fractional calculus. Without loss of generality, we consider the approximation of Riemann-Liouville fractional integral using compound trapezoidal quadrature rule.

Let $t_j = a + jh, f(t_j) = f_j (j = 0, 1, \cdots)$ with step h and write the Riemann-Liouville fractional integral of $f(t)$ in form of

$$
\begin{aligned}
\left(_a\mathcal{J}_t^\alpha f(t)\right)(t_n) &= \frac{1}{\Gamma(\alpha)} \int_a^{t_n} (t_n - \tau)^{\alpha-1} f(\tau) d\tau \\
&= \frac{1}{\Gamma(\alpha)} \sum_{j=0}^{n-1} \int_{t_j}^{t_{j+1}} (t_n - \tau)^{\alpha-1} f(\tau) d\tau.
\end{aligned}
\tag{4.7.6}
$$

The $f(t)$ in integrand can be replaced by interpolation polynomials of different orders. For instance, using first-order Newton interpolation leads to

$$
\begin{aligned}
&\left(_a\mathcal{J}_t^\alpha f(t)\right)(t_n) \\
&= \frac{1}{\Gamma(\alpha)} \sum_{j=0}^{n-1} \int_{t_j}^{t_{j+1}} (t_n - \tau)^{\alpha-1} \left[f(t_j) + \frac{f(t_{j+1}) - f(t_j)}{h}(\tau - t_j) \right] d\tau \\
&= \frac{h^\alpha}{\Gamma(1+\alpha)} \sum_{j=0}^{n-1} b_{n-j-1}^{(1-\alpha)} f_j \\
&\quad + \frac{h^\alpha}{\Gamma(1+\alpha)} \sum_{j=0}^{n-1} (f_{j+1} - f_j) \left[\frac{b_{n-j-1}^{(-\alpha)}}{1+\alpha} - (n-j-1)^\alpha \right] \\
&= h^\alpha \sum_{j=0}^{n} \bar{c}_{j,n} f_j
\end{aligned}
\tag{4.7.7}
$$

where $\bar{c}_{j,n} = c_{j,n}^{(-\alpha)}$. The $c_{j,n}^{(-\alpha)}$ is actually given by (4.6.8), which indicate that the approximation scheme (4.7.7) is the same as the foregoing R2-algorithm.

Particularly, for $\alpha = 1$, (4.7.7) reduces to compound trapezoidal quadrature rule; while for $\alpha = 0$, (4.7.7) reduces to $\left({}_a\mathcal{J}_t^\alpha f(t) \right)(t_n) = f_n$.

Similarly, using higher-order interpolation polynomials to approximate $f(t)$ can produce integration quadrature formulas of higher accuracy. For example, using second-order Newton interpolation polynomial will lead to an extension of compound Simpson quadrature rule.

4.7.3   Extension of linear multi-step method: Lubich fractional linear multi-step method

We first review some fundamentals of linear multi-step method for solving first-order integral equation, and then extend the ideas to fractional linear multi-step method.

Consider the following integral equation

$$y(t) = \mathcal{J}u(t) = \int_0^t u(\tau)\mathrm{d}\tau. \tag{4.7.8}$$

Denote $t_k = kh$, $y_k \approx y(t_k)(k = 0, 1, 2, \cdots)$, let $u_k = \begin{cases} u_h(kh), & k \geqslant 0 \\ 0, & k < 0 \end{cases}$,

and denote by z the backward shifting operator $z = E^{-h}$:

$$zu_n = u_{n-1}, \quad z^k u_n = u_{n-k}.$$

The generic linear multi-step method for solving integral equation (4.7.8) can be written by

$$\alpha_p y_n + \alpha_{p-1} y_{n-1} + \cdots + \alpha_0 y_{n-p} = h(\beta_p u_n + \beta_{p-1} u_{n-1} + \cdots + \beta_0 u_{n-p}), \tag{4.7.9}$$

where the coefficients α_k, β_k are prescribed. Considering a polynomial of degree p

$$\rho(z) = \alpha_p + \alpha_{p-1}z + \cdots + \alpha_0 z^p, \quad \sigma(z) = \beta_p + \beta_{p-1}z + \cdots + \beta_0 z^p,$$

and letting

$$\omega(z) = \frac{\rho(z)}{\sigma(z)},$$

we call $\omega^{-1}(z)$ is the generating function of linear multi-step method (4.7.9) for first-order integral equation (4.7.8). As a result, the linear multi-step method (4.7.9) can be further represented as

$$\rho(z)y_n = h\sigma(z)u_n,$$

or

$$y_n = h\omega^{-1}(z)u_n \approx \mathcal{J}u(nh), \qquad (4.7.10)$$

which is briefly called the (ρ, σ) linear multi-step method. Expanding the generating function $\omega^{-1}(z)$ in Taylor series yields

$$\omega^{-1}(z) = \omega_0 + \omega_1 z + \omega_2 z^2 + \cdots,$$

and accordingly the linear multi-step method (4.7.9) or (4.7.10) can be given by

$$y_n = h \sum_{j=0}^{\infty} \omega_j u_{n-j} \approx \mathcal{J}u(nh). \qquad (4.7.11)$$

Since $u_k = 0$ for $k < 0$, (4.7.11) can be reduced to

$$y_n = h \sum_{j=0}^{n} \omega_j u_{n-j} \approx \mathcal{J}u(nh). \qquad (4.7.12)$$

Similarly, using

$$u_n = h^{-1}\omega(z)y_n \approx y'(nh) \qquad (4.7.13)$$

can define the generating function $\omega(z)$ of linear multi-step method for first-order differential equation.

The generating function of p-order backward multi-step method for first-order differential equation is [146]

$$\omega(z) = \sum_{k=0}^{p} \omega_k z^k = \sum_{k=1}^{p} \frac{1}{k}(1-z)^k := W_p(z). \qquad (4.7.14)$$

Extending the ideas of approximating first-order integral operator $\mathcal{J}u(t)$ to the approximation of fractional integral operator leads to

$$\mathcal{J}^\alpha u(t) \approx h^\alpha(\omega(z))^{-\alpha}u(t) = h^\alpha \sum_{j=0}^{[t/h]} \omega_j^{(-\alpha)} u(t-jh) \qquad (4.7.15)$$

or

$$^{RL}\mathcal{D}_t^\alpha y(t) = \mathcal{J}^{-\alpha}y(t) = h^{-\alpha}(\omega(z))^\alpha \approx h^{-\alpha} \sum_{j=0}^{[t/h]} \omega_j^{(\alpha)} y(t-jh), \qquad (4.7.16)$$

where the coefficients $\omega_j^{(\beta)}$, $j = 0, 1, 2, \cdots$, are the Taylor expansion coefficients of a given generating function, namely

$$\omega_0^{(\beta)} + \omega_1^{(\beta)} z + \omega_2^{(\beta)} z^2 + \cdots = \omega^{(\beta)}(z). \qquad (4.7.17)$$

The cases $\beta < 0$ and $\beta > 0$ respectively represent the generating functions related to Riemman-Liouville fractional integral and differential operators, and they can be obtained by using the β-th power of the generating function (4.7.14):

$$\omega^{(\beta)}(z) = \left(\omega(z)\right)^{\beta}. \tag{4.7.18}$$

Generating functions vary with the order p. Using generating function (4.7.14), one can derive the one- to four- order generating functions of fractional linear multi-step method [146]:

$$W_1^{(\beta)}(z) = \left(W_1(z)\right)^{\beta} = (1 - z)^{\beta},$$

$$W_2^{(\beta)}(z) = \left(W_2(z)\right)^{\beta} = \left(\frac{3}{2} - 2z + \frac{1}{2}z^2\right)^{\beta},$$

$$W_3^{(\beta)}(z) = \left(W_3(z)\right)^{\beta} = \left(\frac{11}{6} - 3z + \frac{3}{2}z^2 - \frac{1}{3}z^3\right)^{\beta},$$

$$W_4^{(\beta)}(z) = \left(W_4(z)\right)^{\beta} = \left(\frac{25}{12} - 4z + 3z^2 - \frac{1}{3}z^3 + \frac{1}{4}z^4\right)^{\beta},$$

$$W_5^{(\beta)}(z) = \left(W_5(z)\right)^{\beta} = \left(\frac{137}{60} - 5z + 5z^2 - \frac{10}{3}z^3 + \frac{5}{4}z^4 - \frac{1}{5}z^5\right)^{\beta},$$

$$W_6^{(\beta)}(z) = \left(W_6(z)\right)^{\beta} = \left(\frac{147}{60} - 6z + \frac{15}{2}z^2 - \frac{20}{3}z^3 + \frac{15}{4}z^4 - \frac{6}{5}z^5 + \frac{1}{6}z^6\right)^{\beta}.$$

In fact,the G1-algorithm in Section 4.2 can be seen as a type of linear multi-step method whose coefficients $w_j^{(\alpha)} = (-1)^j \binom{\alpha}{j} (j = 0, 1, 2, \cdots)$ is the Taylor expansion coefficients of generating function $W_1^{(\alpha)}(z) = (1 - z)^{\alpha}$.

If only using (4.7.16) in approximation, Lubich has proved that this approximation possesses the accuracy of the order $O(h^{\nu}) + O(h^p)$ for $f(t) = t^{\nu-1}$, where $\nu > 0$ and p is the order of the corresponding multi-step method $(2 \sim 6)$. It should be noted that for a fixed ν, even for a larger p, the error is still limited to the order $O(h^{\nu})$. To achieve a higher accuracy, Lubich presented in 1986 a technique which added a correction term in approximation scheme [45]. The approximation scheme can be given by

$$\mathcal{D}_t^{\beta} f(t_n) \approx h^{-\beta} \sum_{j=0}^{n} w_{n-j}^{(\beta)} f(t_j) + h^{-\beta} \sum_{j=0}^{s} \varpi_{n,j} f(t_j). \tag{4.7.19}$$

Addition of the correction term is able to remove the error term $O(h^\nu)$, thereby leading the final error to $O(h^p)$. The correction term coefficients $\varpi_{n,j}$ in (4.7.19) can be derived as follows. Taking

$$\mathcal{A} = \{\gamma = k + l\beta, 0 \geqslant k, l = 0, 1, 2, \cdots; \gamma \leqslant p - 1\},$$

letting s be the number of elements in set $\mathcal{A}$ minus one, and substituting $f(t) = t^q, q \in \mathcal{A}$ in (4.7.19), then lead to a linear system with unknowns ϖ_{nj}:

$$\sum_{j=0}^{s} \varpi_{n,j} j^q = \frac{\Gamma(a+1)}{\Gamma(1-\beta+q)} n^{q-\beta} - \sum_{j=1}^{n} \omega_{n-j}^{(\beta)} j^q, q \in \mathcal{A}. \qquad (4.7.20)$$

Theorem 4.7.1 *Suppose that $f(t)$ is defined on $[0,T]$ and is sufficiently differentiable, that the coefficients ω_j^β are given by (4.7.18) and (4.7.14), and that $\varpi_{n,j}$ is determined by linear system (4.7.20), then it holds that*

$$h^{-\beta} \sum_{j=0}^{n} \omega_{n-j}^{(\beta)} f(t_j) + h^{-\beta} \sum_{j=0}^{s} \varpi_{n,j} f(t_j), -\mathcal{D}_t^\beta f(t_n) = O(h^p), \qquad (4.7.21)$$

where $t_n \in [0,T], \omega_k^{(\beta)} = O(k^{\beta-1}), \varpi_{n,j} = O(n^{\beta-1})$. See the proof in [145].

Remark 4.7.2 *1. For each grid point t_n, one needs to solve the linear system (4.7.20) to derive a group of correction term coefficients $\varpi_{n,j}$; the system matrix is invariable (having Vondermonde structure) even if different sets of grid points are considered; and the constants in the right-hand side of the linear system vary with the selection of grid points.*

2. The property of the system matrix of the linear system (4.7.20) is intimately linked with α. The condition number of the matrix would turn very large for some α.

3. The high accuracy is achieved by this approximation scheme at the expense of increasing computation effort, where the computations of coefficients $\omega_j^{(\alpha)}$ and $\varpi_{n,j}$ become complicated. As an antidote to the problem, Lubich and his collaborator suggested using fast Fourier transform in computing these coefficients.

4.8   Applications of other approximation techniques

4.8.1   Approximations of fractional integral and derivative of periodic function using fourier expansion

For a periodic function with period $2L$, using Fourier expansion writes $f(t)$ which is defined on $[-L, L]$ in form of a trigonometric series:

$$f(t) = \frac{a_0}{2} + \sum_{n=1}^{\infty} (a_n \cos \frac{n\pi}{L} t + b_n \sin \frac{n\pi}{L} t) \qquad (4.8.1)$$

where

$$
\begin{cases}
a_n = \dfrac{1}{L} \displaystyle\int_{-L}^{L} f(t) \cos \dfrac{n\pi t}{L} dt, & n = 0, 1, 2, \cdots \\[4mm]
b_n = \dfrac{1}{L} \displaystyle\int_{-L}^{L} f(t) \sin \dfrac{n\pi t}{L} dt, & n = 1, 2, 3, \cdots
\end{cases} \tag{4.8.2}
$$

Note that the integer-order derivatives of sine and cosine functions take the form:

$$
\frac{d^k}{dt^k}[\sin at] = a^k \sin\left(at + \frac{k\pi}{2}\right), \quad \frac{d^k}{dt^k}[\cos at] = a^k \cos\left(at + \frac{k\pi}{2}\right). \tag{4.8.3}
$$

It can be proved from Cauchy integral formula that for a Riemann-Liouville fractional derivative (i. e. k is a real number), the (4.8.3) still holds [215]. So that, letting k equals a real number α and using (4.8.3) give the fractional derivative of $f(t)$:

$$
\begin{aligned}
&_a\mathcal{D}_t^\alpha f(t) \\
&= \frac{a_0}{\Gamma(1-\alpha)} t^{-\alpha} + \sum_{n=1}^{\infty} \left(\frac{n\pi}{L}\right)^\alpha \left[a_n \cos\left(\frac{n\pi}{L}t + \frac{\alpha\pi}{2}\right) + b_n \sin\left(\frac{n\pi}{L}t + \frac{\alpha\pi}{2}\right)\right].
\end{aligned} \tag{4.8.4}
$$

4.8.2   Short memory principle

For $t \gg a$, the approximation scheme (4.6.1) will have a very large number of adding terms. While for a large t, and with certain assumption, the "history" contribution of $f(t)$ in the neighbourhood of $t = a$ to the final approximating value can be ignored. This is what is called "short memory principle", that is, to only consider the functional values in the "latest past" $[t - L, t]$ where L is called the "memory length":

$$
_a\mathcal{D}_t^\alpha f(t) \approx {}_{t-L}\mathcal{D}_t^\alpha f(t), \quad t > a + L. \tag{4.8.5}
$$

In other words, based on short memory principle (4.6.1), replace the original fractional derivative defined on $[a, t]$ by a new fractional derivative defined on $[t - L, t]$, and then apply difference approximation scheme (4.6.1) to the new derivative. In this fashion, the number of adding terms will not exceed $\left[\dfrac{L}{h}\right]$. It should be noticed that this kind of simplification reduces the computation effort at the expense of a bit loss in accuracy.

If $|f(t)| \leqslant M, \forall a \leqslant t \leqslant b$, which is actually easy to satisfy in many practical problems, then based on

$$
_a\mathcal{D}_t^\alpha f(t) = \frac{1}{\Gamma(-\alpha)} \int_a^t \frac{f(\tau)d\tau}{(t-\tau)^{\alpha+1}}, \quad \alpha \neq 0, 1, 2, \cdots,
$$

the error estimation induced by the short memory principle is

$$\Delta(t) = |_a\mathcal{D}_t^\alpha f(t) - _{t-L}\mathcal{D}_t^\alpha f(t)| \leqslant \frac{ML^{-\alpha}}{|\Gamma(1-\alpha)|}. \tag{4.8.6}$$

This inequality enables the selection of suitable memory length L according to the accuracy required. If $L \leqslant \left(\dfrac{M}{\varepsilon|\Gamma(1-\alpha)|}\right)^{1/\alpha}$, then $|\Delta(t)| \leqslant \varepsilon$.

Alternatively, Ford and Simpson investigated the nonlinear fractional differential equation, analyzed the fixed memory principle, and presented the nested mesh scheme. In their studies, the variable step computation is permitted and a better approximation is achieved at a reasonable computational cost [86, 87].

Chapter 5

Numerical Methods for the Fractional Ordinary Differential Equations

This chapter presents the numerical methods for fractional ordinary differential equations. It is noted that many researchers have studied the numerical solution of Abel-Volterra integral equations of the first kind and the second kind. These equations are also termed fractional integral equations in which Riemann-Liouville fractional integral are considered (see the references [97] and [91]). On the other hand, research on numerical methods for fractional derivative equations merely commences in recent decades. Here we investigate the numerical solution of fractional ordinary differential equation and fractional integral equation based on the approximation schemes for fractional derivative and fractional integral, respectively.

5.1 Solution of fractional linear differential equation

Firstly, consider the linear fractional ordinary differential equation stated in the general form as [179]

$$a_m \mathcal{D}_t^{\beta_m} y(t) + a_{m-1} \mathcal{D}_t^{\beta_{m-1}} y(t) + \cdots + a_1 \mathcal{D}_t^{\beta_1} y(t) + a_0 \mathcal{D}_t^{\beta_0} y(t) = u(t) \quad (5.1.1)$$

where $u(t)$ can be a function and/or its fractional derivatives of different orders, and it is assumed $\beta_m > \beta_{m-1} > \cdots > \beta_1 > \beta_0$.

Supposing a zero initial-value condition of function $y(t)$ and applying Laplace transform lead to

$$G(s) = \frac{Y(s)}{U(s)} = \frac{1}{a_m s^{\beta_m} + a_{m-1} s^{\beta_{m-1}} + \cdots + a_1 s^{\beta_1} + a_0 s^{\beta_0}}$$

where $G(s)$ is also called the fractional transfer function. The exact solution is presented in [179], but it appears hard-to-implement in computer programming. We thus here discuss other numerical solution techniques instead.

Generally, fractional difference quotient formula (see its general formula (4.6.1)) is employed in numerical solution. The resulting numerical solution of the differential equation (5.1.1) can be directly deduced as

$$y_N = \cfrac{1}{\displaystyle\sum_{i=0}^{m} \cfrac{a_i c_{n,N}}{h^{\beta_i}}} \left(u(t_n) - \sum_{i=0}^{m} \frac{a_i}{h^{\beta_i}} \sum_{j=0}^{N-1} c_{n,j}^{(\beta_i)} y_j \right).$$

It should be noted that when using the generalized-form fractional difference quotient formula (4.6.1), special attention should be paid to the applicability of different approximation schemes and sometimes modifications needs to be made. The afore-mentioned numerical method is based on the direct discretization of the fractional derivative or integral and thus belongs to the direct methods. Another group of methods are called indirect methods which through variable substitution equivalently transform the fractional differential equation to a series of fractional differential equations [67, 71, 77]. This group of method can also solve the nonlinear multi-order fractional differential equations. Other types of methods have also be developed. Podlubny [179] presented some numerical methods but with the absence of the corresponding error analyses. After transforming the multi-order fractional differential equation to a set of fractional differential equations, Kiethelm [67, 71] introduced the linear multi-step method and the predictor-corrector method to handle the transformed equations and the related stability and convergence analyses were given. The Poisson transform method is applied by Ali to solve the linear multi-order fractional integro-differential equation [8]. The differential transform method is applied by Erturk et al to solve the fractional multi-order differential equation. There are also some other methods including Adomian decomposition method [60, 80] and separate variable method [61].

5.2   Solution of the general fractional differential equations

Consider the fractional ordinary differential equation as follows

$$\frac{\partial^\alpha y(t)}{\partial t^\alpha} = f(t, y(t)), \quad t \in [0, T], \tag{5.2.1}$$

where $\alpha > 0, m = [\alpha] + 1$, and fractional derivative operator $\dfrac{\partial^\alpha y(t)}{\partial t^\alpha}$ belongs to Caputo type or to Riemann-Liouville type.

Proper initial-value conditions should be added to guarantee the solution existence and uniqueness (see the proof of the solution existence and uniqueness in [70]). The initial-value condition in Caputo sense is

$$\mathcal{D}^k y(0) = y_0^{(k)}, k = 0, 1, \cdots, m - 1. \tag{5.2.2}$$

while

$$\mathcal{D}^{\alpha-k}y(0) = y_0^{(k)}, k = 1, 2, \cdots, m, \tag{5.2.3}$$

in Riemann-Liouville sense.

Since the initial-value condition for Riemann-Liouville sense is written in a fractional derivative form, the physical meaning is not clear; whereas the condition for Caputo sense is given in classical integer-order derivative having explicit physical meaning. Therefore, the remaining part of this section centers on the numerical methods for Caputo-type fractional ordinary differential equations.

Note that, under certain continuity condition, the Grünwald-Letnikov fractional derivative is equivalent to Riemann-Liouville fractional derivative. Particularly, for the homogeneous initial-value conditions $y_0^{(k)} = 0$, these two derivatives take the equivalent form as the Caputo fractional derivative.

Initial-value problem (5.2.1)+(5.2.2) can be transformed to [70]

$$y(t) = \sum_{k=0}^{m-1} \frac{t^k}{k!}y_0^{(k)} + \frac{1}{\Gamma(\alpha)} \int_0^t (t-\tau)^{\alpha-1}f(\tau, y(\tau))\mathrm{d}\tau. \tag{5.2.4}$$

i.e.

$$y(t) = \sum_{k=0}^{m-1} \frac{t^k}{k!}y_0^{(k)} + \mathcal{J}^\alpha f(t, y(t)). \tag{5.2.5}$$

Many numerical methods developed for the integer-order ordinary differential equations can be extended to the solution of equations of fractional order. However, the nonlocality of the fractional derivative induces significant differences in solution process of fractional equations compared to solving integer-order equations. Numerical methods are mainly classified to the direct methods and indirect methods. The finite difference which is established directly on the original equation (5.2.1) is so called the direct method. The fractional derivative $^C\mathcal{D}^\alpha$ given in different expression (weak or strong integral kernel) lead to different approximation schemes. Using the relation between the Caputo fractional derivative and Riemann-Liouville/ Grünwald-Letnikov fractional derivative, we can use the G-, D- and L- algorithms, and fractional linear multi-step method mentioned in Chapter 4. Meanwhile transform (5.2.1) to Volterra integral equation (5.2.4), and then use the numerical methods originally developed for Volterra problems, particularly the numerical integration formulas, we can get the so called indirect method. The R-algorithm discussed in Chapter 4 and the predictor-corrector method to be introduced both belong to this group.

Mesh the computational domain: take equispaced grid points $t_j = jh, j = 0, 1, \cdots, [T/h]$, and denote $y_n \approx y(t_n), f_n = f(x_n, y_n)$.

5.2.1   Direct method

Firstly, consider the fractional differential equation with homogeneous initial-value condition

$$\begin{cases} \dfrac{\partial^\alpha y(t)}{\partial t^\alpha} = f(t, y(t)), & t \in [0, T], \\ y^{(k)}(0) = 0, k = 0, 1, \cdots, m-1. \end{cases} \tag{5.2.6}$$

By the homogeneous initial conditions, we have

$$\frac{\partial^\alpha y(t)}{\partial t^\alpha} = {}^C\mathcal{D}^\alpha y(t) = {}^{RL}\mathcal{D}^\alpha y(t) = {}^{GL}\mathcal{D}^\alpha y(t).$$

Applying the general fractional difference quotient approximation formula (4.6.1), we obtain

$$h^{-\alpha} \sum_{j=0}^{N} c_{n,j}^{(\alpha)} y_j = f(t_n, y_n), \quad n = 0, 1, \cdots, [t/h]. \tag{5.2.7}$$

As a result, the numerical solution can be determined by

$$y_N = \frac{h^\alpha}{c_{n,N}^{(\alpha)}} f(t_n, y_n) - \frac{1}{c_{n,N}^{(\alpha)}} \sum_{j=1}^{N-1} c_{n,j}^{(\alpha)} y_j, n = 1, \cdots, [T/h], \tag{5.2.8}$$

where $N = n$ (for G1- algorithm, D- algorithm, L1- algorithm and linear multi-step method) or $N = n+1$ (for G2- algorithm and L2- algorithm).

Note 1. When $N = n+1$ (for G2- algorithm and L2- algorithm), it can be point-wise explicitly computed according to (5.2.8). Notice that L2- algorithm can only be used for the case $1 < \alpha \leqslant 2$. In addition, the L2-algorithm lacks the systematic theoretical analyses, especially the stability analysis.

2. When $N = n$ (for G1- algorithm, D- algorithm, L1- algorithm and linear multi-step method), if f is linear, it can be point-wise computed according to (5.2.8). Also, if the initial-value condition is nonhomogeneous, it can be transformed to a homogeneous problem [218]:

(i) Caputo-type problem

$$y(t) = \sum_{k=0}^{m-1} y^{(k)}(0) t^k + z(t) \tag{5.2.9}$$

(ii) Riemann-Liouville-type problem

$$y(t) = \sum_{k=1}^{m} \mathcal{D}^{\alpha-k} y(0) t^{\alpha-k} + z(t) \qquad (5.2.10)$$

The original nonhomogeneous problem is finally transformed to a homogeneous problem in terms of the new variable $z(t)$. Then the transformed problem can be solved using finite difference scheme (5.2.8). For a nonlinear initial-value condition, it is required to solve nonlinear equation or linear equations.

3. Problem with nonlinear nonhomogeneous initial-value condition

(i) G1-algorithm

The correction term [218] should be added to the finite difference scheme (5.2.8) of the Caputo fractional ordinary differential equation. When $0 < \alpha \leqslant 1$, the correction term is

$$y_n = h^\alpha f(t_n, y_n) - \sum_{k=1}^{n} \omega_k^{(\alpha)} y_{n-k} - \left(\frac{n^{-\alpha}}{\Gamma(n-\alpha)} - \sum_{j=0}^{n} \omega_j^{(\alpha)} \right) y_0, \qquad (5.2.11)$$

where $n = 1, \cdots, [T/h]$.

(ii) D-algorithm

Based on the relation between the Caputo derivative and Riemann-Liouville derivative (see Proposition 4.1.2)

$$^C\mathcal{D}^\alpha y(t) = \mathcal{D}^\alpha y(t) - \mathcal{D}^\alpha T_{m-1}[y; 0](t) \qquad (5.2.12)$$

where

$$T_{m-1}[y; a] = \sum_{k=0}^{m-1} \frac{t^k}{k!} y^{(k)}(0),$$

and using D-algorithm, we obtain

$$h^{-\alpha} \sum_{j=0}^{n} c_{n,j} y_j - \sum_{k=0}^{m-1} \frac{t_n^k}{k!} y^{(k)}(0) = f(t_n, y_n). \qquad (5.2.13)$$

It follows

$$y_n = h^\alpha f(t_n, y_n) + h^\alpha \sum_{k=0}^{m-1} \frac{t_n^k}{k!} y^{(k)}(0) - \sum_{j=0}^{n-1} c_{n,j} y_j, \qquad (5.2.14)$$

where the coefficient $c_{n,j}$ is defined by (4.6.5). Letting $\alpha = 1$, we obtain the classical simplest backward difference scheme for the first-order differential equation. However, the approximation theory for this method is still not perfect, and two important problems have not yet been well resolved: the solvability of equation (5.2.14) and the error analysis.

Diethelm [65, 68] discussed the two problems for the special case $0 < \alpha < 1$, $f(t, y) = \mu y + q(x)$. Equation (5.2.23) becomes

$$(_0\mathcal{D}_t^\alpha[y(t) - y(0)])(x) = \beta y(x) + f(x), 0 < x < 1, \beta \leqslant 0. \tag{5.2.15}$$

The numerical error is proved to be $O(h^{2-\alpha})$ when $y(t) \in C^2[0, T]$. Diethelm and Walz [74] further obtained the asymptotic expansion of y_n

$$y_n = y(x_n) + \sum_{l=2}^{M_1} a_l n^{l-\alpha} + \sum_{j=1}^{M_2} b_j n^{-2j} + O(x^{-\lambda_M})(n \to \infty) \tag{5.2.16}$$

where the nature numbers M_1, M_2 are defined by the smoothness of function $f(x)$ and $y(x)$. Constants $a_k(k = 2, \cdots, M_1)$ and $b_j(j = 1, \cdots, M_2)$ depend on $k - \alpha$, $2j$ and $M = \min[\alpha - M_1, 2M_2]$. An extrapolation method for the numerical solution of equation (5.2.15) is illustrated by this asymptotic estimation formula (5.2.16).

(iii) Linear multi-step method

Fractional linear multi-step method is firstly proposed by Lubich [143-148] and Hairer, Schlichte [110].

Based on the relation (5.2.12) and the approximation scheme (4.7.19), the $p \in \{1, 2, \cdots, 6\}$ order Lubich fractional linear multi-step method for solving the Caputo fractional differential equation is stated as

$$h^{-\alpha} \sum_{j=0}^{n} w_{n-j}^{(\alpha)} y_j + h^{-\alpha} \sum_{j=0}^{s} \varpi_{nj} y_j - \mathcal{D}^\alpha T_{m-1}[y; 0](t_n)$$
$$= f(t_n, y_n), \quad n = 1, \cdots, N. \tag{5.2.17}$$

It can be written as

$$y_n = h^\alpha f(t_n, y_n) + h^\alpha \mathcal{D}^\alpha T_{m-1}[y; 0](t_n)$$
$$- \sum_{j=1}^{n} w_{n-j}^{(\alpha)} y_j - \sum_{j=0}^{s} \varpi_{nj} y_j, \quad n = 1, \cdots, N \tag{5.2.18}$$

where coefficient $w_k^{(\alpha)}$ is generated by function

$$w^\alpha(z) = \left(\sum_{k=1}^{p} \frac{1}{k}(1 - z)^k \right)^\alpha, \tag{5.2.19}$$

and the start weight ϖ_{mj} can be obtained by the following equations

$$\sum_{j=0}^{s} \varpi_{nj} j^q = \frac{\Gamma(1 + q)}{\Gamma(1 + q - \alpha)} n^{q-\alpha} - \sum_{j=1}^{n} w_{n-j}^{(\alpha)} j^q.$$

The details can be seen in Chapter 4. There exists small $\varepsilon > 0$, such that the error of the approximation scheme (5.2.18) on arbitrary grid point is $O(h^{p-\varepsilon})$, and $\omega_k = O(k^{\alpha-1})$.

(iv) L-algorithm

Shkhanukov [200] firstly applied the difference method to solve the following Dirichlet problem

$$\begin{cases} Ly \equiv \dfrac{\mathrm{d}}{\mathrm{d}x}\left[k(x)\dfrac{\mathrm{d}}{\mathrm{d}x}y(x)\right] - r(x){}_0\mathcal{D}_x^\alpha y(x) - q(x)y(x) = -f(x), \quad 0 < x < 1, \\ y(0) = y(1) = 0; \quad k(x) \geqslant c_0 > 0, r(x) \geqslant 0, q(x) \geqslant 0, \end{cases}$$

$$(5.2.20)$$

where $0 < \alpha < 1$, ${}_0\mathcal{D}_x^\alpha$ is Riemann-Liouville fractional derivative. His method is based on the approximation of fractional derivative

$$_0\mathcal{D}_x^\alpha y(x_i) = \frac{1}{\Gamma(2-\alpha)}\sum_{k=1}^{i}(x_{i-k+1}^{1-\alpha} - x_{i-k}^{1-\alpha})y_{\bar{x}k} \tag{5.2.21}$$

where $y_{\bar{x}k} = \dfrac{y(x_k) - y(x_{k-1})}{x_k - x_{k-1}}$ is the first order forward difference quotient of $y(x_k)$. The above formula is L1-algorithm (4.5.2). Here, we still take the uniform grid points $\{x_j = jh : \ j = 0, 1 \cdots, N-1\}$, where $h = 1/N$ is the step length. Applying the approximation formula, Shkhanukov obtained the difference scheme of the problem (5.2.20) and further proved its stability and convergency. Using difference approximation (5.2.21), Shkhanukov presented the difference scheme of the fractional partial differential equation with initial-boundary values as

$$\begin{cases} \mathcal{D}_t^\alpha u(x,t) = \dfrac{\partial^2 u(x,t)}{\partial x^2} + f(x,t), \ 0 < x < 10 < t < T, \\ u(0,t) = u(1,t) = 0, \ \ 0 \leqslant t \leqslant T; \\ u(x,0) = 0, \ \ \mathcal{D}_t^\alpha u(x,t)|_{t=0} = 0, \ \ 0 \leqslant x \leqslant 1, \end{cases}$$

$$(5.2.22)$$

and also derived the stability and convergency of the difference scheme on a uniform grid.

5.2.2   Indirect method

Linear multi-step method

We still consider the following fractional ordinary differential equation

$$\begin{cases} {}^C\mathcal{D}^\alpha y(t) = f(t, y(t)), \quad t \in [0, T], \\ y^{((k))}(0) = b_k, k = 0, 1, \cdots, m-1, \end{cases} \tag{5.2.23}$$

where $\alpha > 0, m = [\alpha] + 1$. Caputo fractional derivative operator $^C\mathcal{D}^\alpha$ is taken due to the fact that the initial-value condition related to this type of operator possesses clear physical meaning.

It is easily noted that the fractional differential equation (5.2.23) can be transformed to Abel-Volterra integral equation

$$y(t) = T_{m-1}[y; 0](t) + \mathcal{J}^\alpha f(t, y(t)), \tag{5.2.24}$$

where

$$T_{m-1}[y; 0](t) = \sum_{k=0}^{m-1} \frac{t^k}{k!} b_k$$

$$\mathcal{J}^\alpha f(t, y(t)) = \frac{1}{\Gamma(\alpha)} \int_0^t (t - \tau)^{\alpha-1} f(\tau, y(\tau)) d\tau.$$

Applying the $p \in \{1, 2, \cdots, 6\}$ order Lubich fractional linear multi-step method to the above equation yields

$$\begin{aligned}
y_n =\, & T_{m-1}[y; 0](t_n) + h^\alpha \sum_{j=0}^n w_{n-j}^{(-\alpha)} f(t_j, y_j) \\
& + h^\alpha \sum_{j=0}^s \varpi_{nj} f(t_j, y_j), \quad m = 1, \cdots, N,
\end{aligned} \tag{5.2.25}$$

where the convolution coefficient $w_k^{(-\alpha)}$ is given by function

$$w^{-\alpha}(z) = \left(\sum_{k=1}^p \frac{1}{k} (1 - z)^k \right)^{-\alpha}. \tag{5.2.26}$$

and the starting weight ϖ_{nj} is obtained by the following equations

$$\sum_{j=0}^s \varpi_{nj} j^q = \frac{\Gamma(1+q)}{\Gamma(1+q+\alpha)} n^{q+\alpha} - \sum_{j=1}^n w_{n-j}^{(-\alpha)} j^q.$$

R-Algorithm

Diethelm and Freed [73] considered the following nonlinear fractional differential equation:

$$(_0\mathcal{D}_t^\alpha [y(t) - y(0)])(x) = f[x, y(x)](0 < x < 1; 0 < \alpha < 1) \tag{5.2.27}$$

whose equivalent form is the second kind of Volterra integral equation

$$y(x) = y(0) + \frac{1}{\Gamma(\alpha)} \int_0^x \frac{f[t, y(t)]}{(x - t)^{1-\alpha}} dt. \tag{5.2.28}$$

The integral in the above equation can be seen as weighted integral with the weighting function $(t_{n+1} - t)^{\alpha-1}$. Take the nodes $t_j (j = 0, 1, \cdots, n+1)$ and apply the trapezoidal quadrature formula to solve problem. This method is the R-algorithm mentioned in Chapter 4.

R-algorithm is frequently used in the predictor-corrector scheme, therefore, we will analyze its specific applications in the following section.

Fractional predictor-corrector method

We still consider the fractional ordinary differential equation (5.2.23).

Mesh the computational domain: take the uniform grid points $t_j = jh(h = T/N)$ and denote $y_j = y_h(t_j) \approx y(t_j), f_j = f(x_j, y_j), j = 0, 1, \cdots, N$. (5.2.23) amounts to the Abel-Volterra integral equation (5.2.24), i.e.

$$y(t) = \sum_{k=0}^{m-1} \frac{t^k}{k!} b_k + J^\alpha f(t, y(t)). \tag{5.2.29}$$

The first term on the right-hand of (5.2.29) is totally determined by the initial-value condition and is thus a known quantity. The second term is the Riemann-Liouville integral of function f which can be approximated by R-algorithm previously mentioned. Using relatively accurate R2-algorithm leads to

$$y_h(t_{n+1}) = \sum_{k=0}^{m-1} \frac{t_{n+1}^k}{k!} b_k + h^\alpha \sum_{j=0}^{n+1} a_{j,n+1} f(t_j, y_h(t_j)). \tag{5.2.30}$$

where the coefficients are

$$a_{j,n} = \frac{1}{\Gamma(2+\alpha)} \begin{cases} (1+\alpha)n^\alpha - n^{1+\alpha} + (n-1)^{1+\alpha}, & j = 0; \\ (n-j+1)^{1+\alpha} - 2(n-j)^{1+\alpha} \\ + (n-j-1)^{1+\alpha}, & 1 \leqslant j \leqslant n-1; \\ 1, & j = n. \end{cases} \tag{5.2.31}$$

The difference approximation scheme (5.2.30) is called the Adams-Moulton method.

For this method, since both sides of the equation include the unknown variables $y_h(t_{n+1})$ and due to the non-linearity of f, it is often difficult to derive $y_h(t_{n+1})$. Therefore iteration procedure is usually employed. To achieve a better approximate solution, substitute a predicted value $y_h(t_{n+1})$ into the right-hand of (5.2.30).

Let $y_h^p(t_{n+1})$ be the predicted value, which can be obtained by some simple method (explicit form). For instance, use relatively inaccurate R0-algorithm to derive the predicted value:

$$y_h^p(t_{n+1}) = \sum_{k=0}^{m-1} \frac{t_{n+1}^k}{k!} b_k + h^\alpha \sum_{j=0}^{n} b_{j,n+1} f(t_j, y_h(t_j)). \tag{5.2.32}$$

The formula above is called the fractional Euler method or fractional Adams-Bashforth method, where

$$b_{j,n} = \frac{(n-j)^\alpha - (n-j-1)^\alpha}{\Gamma(1+\alpha)}.$$

Replacing $y_h(t_{n+1})$ in the right-hand of (5.2.30) by (5.2.32) gives

$$\begin{aligned} y_h(t_{n+1}) = &\sum_{k=0}^{m-1} \frac{t_{n+1}^k}{k!} b_k + \frac{h^\alpha}{\Gamma(2+\alpha)} f(t_{n+1}, y_h^p(t_{n+1})) \\ &+ h^\alpha \sum_{j=0}^{n} a_{j,n+1} f(t_j, y_h(t_j)). \end{aligned} \tag{5.2.33}$$

The method determined by (5.2.32) and (5.2.33) is called fractional Adams-Bashforth-Moulton method.

The computing process of the fractional Adams-Bashforth-Moulton method mainly includes four steps:

(1) Predict: predict $y^p(t_{n+1})$ from (5.2.32);

(2) Evaluate: compute $f(t_{n+1}, y_{n+1}^p)$;

(3) Correct: correct $y(t_{n+1})$ by (5.2.33);

(4) Evaluate: compute $f(t_{n+1}, y_h(t_{n+1}))$ to prepare for the next loop iteration.

Hence, it is more common to call this method predictor-corrector scheme, or PECE (Predict, Evaluate, Correct, Evaluate) method.

Lemma 5.2.1 [72] *Suppose* $g(t) \in C^1[0,T]$, *then*

$$\left| \mathcal{J}^\alpha g(t_n) - h^\alpha \sum_{j=0}^{n-1} b_{j,n} g(t_j) \right| \leqslant \frac{1}{\Gamma(1+\alpha)} \|g'\|_\infty t_n^\alpha h. \tag{5.2.34}$$

Lemma 5.2.2 [72] *Suppose* $g(t) \in C^2[0,T]$, *then there exists a constant* C_α *dependent on* α *such that*

$$\left| \mathcal{J}^\alpha g(t_n) - h^\alpha \sum_{j=0}^{n} a_{j,n} g(t_j) \right| \leqslant C_\alpha \|g''\|_\infty t_n^\alpha h^2. \tag{5.2.35}$$

Theorem 5.2.1 [72] *Suppose* $\alpha > 0, y(t)$ *is sufficiently smooth,* $^C\mathcal{D}^\alpha y(t) \in C^2[0,T]$ *and function* $f(t,y)$ *satisfy the Lipschitz condition with respect to the second variable, namely,*

$$f(t, y_1) - f(t, y_2) \leqslant L|y_1 - y_2|, \tag{5.2.36}$$

then the error of the correct scheme (5.2.32)+(5.2.33) satisfies

$$\max_{0 \leqslant j \leqslant N} |y(t_j) - y_h(t_j)| = \begin{cases} O(h^2), & \alpha \geqslant 1; \\ O(h^{1+\alpha}), & \alpha < 1, \end{cases} \tag{5.2.37}$$

i.e.

$$\max_{0 \leqslant j \leqslant N} |y(t_j) - y_h(t_j)| = O(h^p), \tag{5.2.38}$$

where $p = \min\{2, 1 + \alpha\}, N = [T/h]$.

Proof It will be shown that for arbitrary $j = 0, 1, \cdots, N$ and sufficiently small h, there exists a constant C such that

$$|y(t_j) - y_h(t_j)| \leqslant Ch^p. \tag{5.2.39}$$

Since the initial-value condition is given, the above inequality holds for $j = 0$. Assume (5.2.39) holds for $j = 0, 1, \cdots, k$, now we prove the inequality also holds for $j = k + 1$.

Firstly, we observe the error of predicted value y_{k+1}^P. From (5.2.29) and (5.2.32), it follows

$$\begin{aligned}
&|y(t_{k+1} - y_{k+1}^P)| \\
&= |[\mathcal{J}^\alpha f(t, y(t))]_{t=t_{k+1}} - h^\alpha \sum_{j=0}^{k} b_{j,k+1} f(t_j, y_j)| \\
&\leqslant |[\mathcal{J}^{\alpha\,C} \mathcal{D}^\alpha y(t)]_{t=t_{k+1}} - h^\alpha \sum_{j=0}^{k} b_{j,k+1} [^C \mathcal{D}^\alpha y(t)]_{t=t_{k+1}}| \\
&\quad + h^\alpha \sum_{j=0}^{k} b_{j,k+1} |f(t_j, y(t_j)) - f(t_j, y_j)| \\
&\leqslant c_1 t_{k+1}^\alpha h + c_2 t_{k+1}^\alpha h^p,
\end{aligned} \tag{5.2.40}$$

where c_1, c_2 are constants depending on α.

Then, we attempt to obtain the error of the corrected value. By (5.2.29) and (5.2.33), we have

$$\begin{aligned}
|y(t_{k+1} - y_{k+1})| &= |[\mathcal{J}^\alpha f(t, y(t))]_{t=t_{k+1}} - h^\alpha a_{k+1,k+1} f(t_{k+1}, y_{k+1}^P) \\
&\quad - h^\alpha \sum_{j=0}^{k} a_{j,k+1} f(t_j, y_j)| \\
&\leqslant |[\mathcal{J}^{\alpha\,C} \mathcal{D}^\alpha y(t)]_{t=t_{k+1}} - h^\alpha \sum_{j=0}^{k+1} a_{j,k+1} [^C \mathcal{D}^\alpha y(t)]_{t=t_{k+1}}|
\end{aligned}$$

$$+h^\alpha \sum_{j=0}^{k} a_{j,k+1}|f(t_j, y(t_j)) - f(t_j, y_j)|$$

$$+h^\alpha a_{k+1,k+1}|f(t_{k+1}, y(t_{k+1})) - f(t_{k+1}, y_{k+1}^P)| \tag{5.2.41}$$

$$\leqslant c_3 t_{k+1}^\alpha h^2 + c h^\alpha h^p \sum_{j=0}^{k} a_{j,k+1} + c' h^\alpha a_{k+1,k+1}(h + h^p)$$

$$\leqslant c_3 t_{k+1}^\alpha h^2 + c_4 h^{1+\alpha} + c_5 t_{k+1}^\alpha h^p \leqslant C h^p.$$

The error is obtained under a relatively strict condition ($^C\mathcal{D}^\alpha y(t) \in C^2[0,T]$). But for some smooth function $y(t)$, its fractional derivative $^C\mathcal{D}^\alpha y(t)$ is likely non-smooth. Diethelm et al also gave the error evaluations under some other conditions [72].

The convergence evaluation given below is derived for a smooth $y(t)$.

Theorem 5.2.2   *Let $0 < \alpha \leqslant 1, y(t) \in C^2[0,T]$, and $f(t,y)$ satisfy the Lipschitz condition with respect to the second variable (5.2.36), then*

$$|y(t_j) - y_h(t_j)| = C t_j^{\alpha-1} \times \begin{cases} h^{1+\alpha}, & 0 < \alpha \leqslant 1/2; \\ h^{2-\alpha} & 1/2 < \alpha < 1, \end{cases} \tag{5.2.42}$$

where C is a constant independent of j and h.
    See the proof in [72].

Note: 1.   Compared to the integer order derivative, the fractional order derivative is non-local operator. It means that the computation of fractional derivative on each point depends on not only the date in the neighbourhood of the present instant, but also the data in the whole history. This property can describe the physical phenomenon having memory features, but leads to some troublesome in numerical computation. The time complexity of the present method is $O(N^2)$ (while the complexity is only $O(N)$ when solving integer-order problem), where N is the number of the computational points. Short memory principle [179] can be employed to lower the complexity at the expense of the accuracy and stability for some problems. Nest memory concept can also be a suitable choice since its complexity is reduced to merely $O(NlogN)$ when retaining the original accuracy [87].

2.   The stability analysis of method is equivalent to that of classical Adams—Bashforth—Moutton scheme. One of the methods to improve the stability is the so-called $P(EC)^m E$ algorithm, which corrects m times for each calculation. The stability of the method is improved at the expense of increasing the correcting iteration number while keeping the convergence and complexity unchanged.

3. Richardson extrapolation method can be used to improve the accuracy of the method. Compute the value $u_{i,2N}^n$ on a grid two times denser than the original one in each time step, and then take the Richardson extrapolation value $2u_{2i,2N}^n - u_i^n$ as the new value $\bar{u}_i^n$. Thus the spatial convergence order is increased to $O(h^2)$.

4. The algorithm idea can be generalized and applied to the unequis-paced grid case, in which the weighting factors in the predictor and corrector formulas needs some adjustments but the Richardson extrapolation method will fail.

Chapter 6

Numerical Methods for Fractional Partial Differential Equations

Finite difference methods and series approximation methods (mainly including Adomian decomposition method and variational iteration method) are the dominant numerical methods for solving fractional partial differential equations. The corresponding theoretically analyzing methods include Fourier methods, energy estimation, matrix eigenvalue method and mathematical induction. There still exist other types of methods but with either somewhat weaker applicability or absence of relatively sound theoretical analyses. The developments of the correlative numerical methods can be briefly reviewed as follows. From the end of 20th century, Gorenflo et al have published a series of papers [93-95, 149, 150] regarding the finite difference schemes for solving time, space, and time-space fractional diffusion equations. These schemes are formulated by using the equivalence of the Riemann-Liouville and Grünwald-Letnikov fractional derivatives, and are further interpreted as the discrete random walk models in terms of time, space and time-space levels. To guarantee the stability of the schemes, shifted Grünwald-Letnikov approximation schemes are constructed in place of the standard schemes. It should be also noted these finite difference schemes can be easily extended to solution of the generic fractional partial differential equations. In studying the saltwater intrusion into aquifer systems, Liu et al presented the "Method of Lines" [137] which transforms the fractional partial differential equation to a system of fractional ordinary differential equations. Their approach takes the backward difference scheme with variable-order and variable-step, and has been widely accepted and extensively used to solution of space fractional partial differential equations. In 2004, Meerschaert and Tadjeran [157] presented the finite difference scheme for space advection-dispersion equation with variable coefficients, together with its error analysis. Afterwards, Tadjeran et al, in 2006, derived a temporally second-order, spatially first-order accurate and unconditionally stable finite difference scheme by combining Grünwald-Letnikov formula with

Crank-Nicolson method. Space extrapolation was used to increase the spatial convergence rate to the second-order, and the proposed finite difference scheme is further applied to solving other types of space fractional partial differential equations [75, 158, 202, 208]. It is worth noting that the finite difference schemes mentioned above are all based on the Grünwald-Letnikov approximation schemes. In addition to that, L-algorithms can also be utilized to construct the finite difference schemes [46, 81, 137, 199, 224], yet in the absence of rigorous theoretical analyses for most of schemes.

For solving time fractional partial differential equations, there have been two leading finite difference schemesone is based on G-algorithms [92, 226] and the other on L-algorithms [129].

For solving time-space fractional equations, G- and L- algorithms are usually combined to form the finite difference schemes [140, 141, 231].

Besides, special care has still been taken of solving high-dimensional problems. Meerschaert et al [156] have presented finite difference scheme for solving two-dimensional fractional diffusion equations with variable coefficients based on alternating direction implicit method, along with the stability and convergence analyses. Chen and Liu [43] considered the two-dimensional fractional advection-diffusion equation and proposed the alternating directional Euler method. Matrix eigenvalue method is employed to analyze the stability of the method and the Richardson extrapolation to increase the accuracy to the second-order. Liu has investigated the two- and three- dimensional fractional advection-diffusion equations in the dissertation, where several modified alternating direction methods are developed and where the Richardson extrapolation is considered as well.

Apart from the foregoing numerical methods, a finite element scheme given by Roop in 2006 is used for solving space fractional differential equations [193]. In 2005, Adomian decomposition method has been applied to solving time-space fractional telegraph equation [166] as well as fractional diffusion-wave equations [6]. In 2006, Rawashdeh [187] combined collocation method with polynomial spline in solving a type of fractional integral equation, but in the absence of the numerical analysis. In 2007, Zhang used the finite element method in his dissertation to solve fractional partial differential equation, which achieves high-order approximation accuracy. Additionally, Lin and Xu [134] applied spectral method for solving time fractional diffusion equation.

In principle, for fractional partial differential equations, the researches on their numerical methods are still on the early stage, and the corresponding theoretical analyses and the potential improvements on the methods seems somewhat inadequate. Emphatically, the nonlocality of fractional differential and integral operators leads to high computational cost as well as large memory requirements in the solution of fractional partial differential equations.

Finite difference schemes for fractional partial differential equations generally originate from the Grünwald-Letnikov approximation schemes, and the approximation accuracy is of the first-order for most cases. At present, in the literature regarding numerical methods, Fourier analysis, eigenvalue methods, mathematical induction and energy methods are commonly used to prove the stability and convergence of the numerical methods. In what follows, we elaborate on the basics of some typical finite difference schemes.

On the other hand, fractional diffusion equations are widely utilized to model problems in physics [161], finance [96] and hydrology [19,20]. In particular, fractional advection-diffusion equations are claimed to better simulate the solute transport process which characterizes long-tail phenomenon.

In addition, Liu et al [138] considered the time fractional advection-diffusion equation, and derived the fundamental solution using Mellin and Laplace transforms. This fundament solution is a Fox function comprised of a probability density function and of a complete error function. Huang and Liu [116] further derived the fundamental solutions for problems in $n-$dimensional space and half space. They also obtained the analytical solution of time-space fractional advection-diffusion equations [115].

In the succeeding three sections, we respectively introduce the finite difference schemes [43, 129, 141, 157, 199, 226] for solving space, time and time-space fractional advection-diffusion equations. Consider the following equation with variable coefficients:

$$\frac{\partial^\alpha u(x,t)}{\partial t^\alpha} = -v(x,t)\mathcal{D}_x^\beta u(x,t) + d(x,t)\mathcal{D}_x^\gamma u(x,t) + f(x,t),$$
$$0 < t \leqslant T, L < x < R, \tag{6.0.1}$$

where $0 < \alpha, \beta \leqslant 1, 1 < \gamma \leqslant 2$, and $v, d \geqslant 0$ (i.e. the fluid moves from the left to the right). $\frac{\partial^\alpha u(x,t)}{\partial t^\alpha} = {}^C\mathcal{D}_t^\alpha u(x,t)$ and $\mathcal{D}_x^\mu u(x,t)$ are the Caputo time fractional derivative and the Riemann-Liouville space fractional derivative, respectively. The existence and uniqueness of the (6.0.1) can be seen in [82].

6.1　Space fractional advection-diffusion equation

Consider the space fractional advection-diffusion equation below (namely, letting $\alpha, \beta = 1, v = v(x), d = d(x)$ in (6.0.1)):

$$\frac{\partial u(x,t)}{\partial t} = -v(x)\frac{\partial u(x,t)}{\partial x} + d(x)\mathcal{D}_x^\gamma u(x,t) + f(x,t),$$
$$0 < t \leqslant T, L < x < R, \tag{6.1.1}$$

with initial-boundary value conditions:

$$u(x,t=0) = \psi(x), L < x < R;$$
$$u(x=L,t) = 0, u(x=R,t) = 0. \tag{6.1.2}$$

Space fractional derivative can be discretized by using G- or L- algorithms, thus resulting in the finite difference schemes based these two type of algorithms. Without loss of generality, we consider the finite difference scheme based on G-algorithm.

The first-order temporal and spatial derivatives in (6.1.1) can be approximated by first-order difference quotient, while the space fractional derivative is discretized by using the G-algorithm, i.e., using the equivalence of Riemann-Liouville and Grünwald-Letnikov fractional derivatives. Meerschaert et al [157] has proven that explicit, implicit and C-N finite difference schemes based on standard Grünwald-Letnikov approximation schemes are instable. They suggested shifted Grünwald-Letnikov approximation schemes in place of the original schemes. Since $1 < \gamma \leqslant 2$, the optimal shift number p should be 1 (see Theorem 4.2.1 and its remark). We thus have the following approximation:

$$\mathcal{D}_x^\gamma u(x,t) \approx h^{-\gamma} \sum_{k=0}^{[x-L/h]} \omega_k^{(\gamma)} u(x - (k-1)h, t). \tag{6.1.3}$$

We denote $0 \leqslant t_n = n\tau \leqslant T, x_i = L + ih, h = (R-L)/M, i = 0, 1, \cdots, M$. $u_i^n \approx u(x_i, t_n), v_i = v(x_i), d_i = d(x_i), f_i^n = f(x_i, t_n)$.

Theorem 6.1.1 [157] *The implicit finite difference scheme for solving space fractional advection-diffusion equation (6.1.1)*

$$\frac{u_i^{n+1} - u_i^n}{\tau} = -v_i\frac{u_i^{n+1} - u_{i-1}^{n+1}}{h} + \frac{d_i}{h^\gamma}\sum_{k=0}^{i+1}\omega_k^{(\gamma)}u_{i-k+1}^{n+1} + f_i^{n+1}, \tag{6.1.4}$$

which is based on modified Grünwald-Letnikov approximation scheme (6.1.3), is continuous, unconditionally stable and thus convergent.

Proof Considering the boundary condition $(u(L,t) = 0)$ and using the Collorary 4.2.1, we see that the approximation accuracy of the shifted Grünwald-Letnikov approximation scheme (6.1.3) can reach $O(h)$. So that the accuracy of the scheme (6.1.4) is $O(h) + O(\tau)$, i. e., the scheme is continuous.

Denote $E_i = v_i\tau/h$, $B_i = d_i\tau/h^\gamma$, then the scheme (6.1.4) can be represented as

$$u_i^{n+1} - u_i^n = -E_i(u_i^{n+1} - u_{i-1}^{n+1}) + B_i \sum_{k=0}^{i+1} \omega_k^{(\gamma)} u_{i-k+1}^{n+1} + \tau f_i^{n+1}, \quad (6.1.5)$$

or

$$-B_i\omega_0^{(\gamma)}u_{i+1}^{n+1} + (1 + E_i - B_i\omega_1^{(\gamma)})u_i^{n+1} - (E_i + B_i\omega_2^{(\gamma)})u_{i-1}^{n+1}$$
$$-B_i \sum_{k=3}^{i+1} \omega_k^{(\gamma)} u_{i-k+1}^{n+1} = u_i^n + \tau f_i^{n+1}. \quad (6.1.6)$$

Consideration of the column vector notation

$$\underline{U}^{n+1} = [u_1^{n+1}, u_2^{n+1}, \cdots, u_{M-1}^{n+1}]^T,$$

$$\underline{F}^{n+1} = [f_1^{n+1}, f_2^{n+1}, \cdots, f_{M-1}^{n+1}]^T,$$

rewrites (6.1.6) in a matrix-vector form, i. e., $\underline{A}\underline{U}^{n+1} = \underline{U}^n + \tau\underline{F}^{n+1}$, where $\underline{A} = [A_{i,j}]$ is coefficient matrix, the entry of which $A_{i,j}$ is defined by (note that $u_0^{n+1} = u_M^{n+1} = 0$)

$$A_{i,j} = \begin{cases} 0, & j \geqslant i+2, \\ -B_i\omega_0^{(\gamma)} & j = i+1, \\ 1 + E_i - B_i\omega_1^{(\gamma)} & j = i, \\ -E_i - B_i\omega_2^{(\gamma)} & j = i-1, \\ -B_i\omega_{i-j+1}^{(\gamma)} & j \leqslant i-1 \end{cases} \quad (6.1.7)$$

for $i, j = 1, 2, \cdots, M-1$. Let λ be the eignevalue of the matrix $\underline{A}$, $\underline{X}$ be the corresponding eigenvector and thus we have $\underline{A}\underline{X} = \lambda\underline{X}$. Find some i, such that $\|x_i\| = \max\{|x_j| : j = 1, \cdots, M-1\}$. So that from $\sum_{j=1}^{M-1} A_{i,j}x_j = \lambda x_i$, we have

$$\lambda = A_{i,i} + \sum_{j=1, j\neq i}^{M-1} A_{i,j}\frac{x_j}{x_i}, \quad (6.1.8)$$

substituting (6.1.7) into (6.1.8) leads to

$$\lambda = 1 + E_i - B_i\omega_1^{(\gamma)} - B_i\omega_0^{(\gamma)}\frac{x_{i+1}}{x_i} - (E_i + B_i\omega_2^{(\gamma)})\frac{x_{i-1}}{x_i} - B_i\sum_{j=1}^{i-2}\omega_{i-j+1}^{(\gamma)}\frac{x_j}{x_i}$$
$$= 1 + E_i(1 - x_{i-1}/x_i) - B_i\left[\omega_1^{(\gamma)} + \sum_{j=1, j\neq i}^{i+1}\omega_{i-j+1}^{(\gamma)}\frac{x_j}{x_i}\right]. \quad (6.1.9)$$

It follows from $\sum_{k=0}^{\infty} w_k^{(\gamma)} = 0, 1 < \gamma \leqslant 2$, that there only exists one negative

Grunwald weighting coefficient $w_1^{(\gamma)} = -\gamma$, along with $-w_1^{(\gamma)} \geqslant \sum_{k=1, k \neq 1}^{j} w_k^{(\gamma)}$,

$j = 0, 1, 2, \cdots$. Moreover, since $|x_j/x_i| \leqslant 1$ and $w_j^{(\gamma)} \geqslant 0, j = 2, 3, 4, \cdots$, it thus holds that

$$\sum_{j=1, j \neq i}^{i+1} w_{i-j+1}^{(\gamma)} |x_j/x_i| \leqslant \sum_{j=1, j \neq i}^{i+1} w_{i-j+1}^{(\gamma)} \leqslant -w_1^{(\gamma)}.$$

From the above inequalities, we have

$$w_1^{(\gamma)} + \sum_{j=1, j \neq i}^{i+1} w_{i-j+1}^{(\gamma)} \|x_j/x_i\| \leqslant 0.$$

From the fact that the parameters B_i, E_i are both non-negative real numbers, one can see that the eigenvalues of coefficient matrix $\underline{A}$ satisfy $\|\lambda\| \geqslant 1$. It follows that the coefficient matrix is invertible and that the eigenvalues η of the inverse matrix $\underline{A}^{-1}$ satisfy $\|\eta\| \leqslant 1$ (namely, the spectral radius of $\underline{A}^{-1}$ does not exceed one). By letting the error of $\underline{U}^k$ be $\underline{\varepsilon}^k$, the recursive formula of error reads $\underline{\varepsilon}^1 = \underline{A}^{-1} \underline{\varepsilon}^0$. It is straightforward to see that $\|\underline{\varepsilon}^1\| \leqslant \|\underline{\varepsilon}^0\|$ and thus the approximation scheme (6.1.4) is unconditionally stable. The convergence of the scheme can be proved by using the Lax equivalence theorem.

Remark 6.1.1   *1. Local truncated error of the scheme (6.1.4) is $O(\tau)+O(h)$.*

2. For time-dependent coefficients, namely, $v = v(x,t) \geqslant 0, d = d(x,t) \geqslant 0$, the conclusion of the theorem still holds.

3. The theorem is still valid for other types of right boundary value condition, e. g.

$$u(R,t) = b_R(t), \quad or \quad u(R,t) + \nu \frac{\partial u}{\partial t} u(R,t) = \phi(t), \nu \geqslant 0.$$

4. For $\gamma = 2$, the finite difference scheme (6.1.4) reduces to the classical second-order central difference quotient which is used for approximating second derivative. In such case, the shifted Grünwald-Letnikov scheme (6.1.3) reduces to the classical central quotient: (i.e., $w_0^{(2)} = 1, w_1^{(2)} = -2, w_2^{(2)} = 1, w_4^{(2)} = w_4^{(2)} = 0$)

$$\frac{\partial^2 u(x_i, t_n)}{\partial x^2} \approx \frac{u_{i+1}^n - 2u_i^n + u_{i-1}^n}{h^2}.$$

5. The scheme (6.1.4) can be extended to solving other types of equations.

Meerschaert et al applied the scheme to solving two-dimensional [156] and two-sided [158] space fractional partial differential equations. They proved that the resulting explicit finite difference scheme is conditionally stable, while the implicit one is unconditionally stable. Moreover, these stable conditions can be seen as the extensions of those of the explicit finite difference schemes that are used for solving classical parabolic and hyperbolic equations.

6. Applications of other types of finite difference schemes based on shifted Grünwald-Letnikov approximation.

(a) Advection term can be discretized by central difference quotient as is shown in Lax-Wendroff scheme (conditionally stable) [202]. (b) Weighted average methods. The corresponding finite difference scheme for (6.1.1) is given by [208].

$$\frac{\partial u(x,t)}{\partial t}\Big|_{(x_j,t_{n+\frac{1}{2}})} = (1-\lambda)\Big[-v(x)\frac{\partial u(x,t)}{\partial x} + d(x)\mathcal{D}_x^\gamma u(x,t)\Big]_{(x_j,t_{n+1})}$$

$$+\lambda\Big[-v(x)\frac{\partial u(x,t)}{\partial x} + d(x)\mathcal{D}_x^\gamma u(x,t)\Big]_{(x_j,t_n)} + f(x_j,t_{n+\frac{1}{2}}),$$

$$(6.1.10)$$

where $0 \leqslant \lambda \leqslant 1$ *is the weighting coefficient. Second-order central difference quotient, first-order backward difference quotient and shifted Grünwald-Letnikov approximation are employed to discretize the time derivative, first-order space derivative and space fractional derivative, respectively. In particular, for* $\lambda = 1/2$, *the scheme (6.1.10) is called fractional Crank-Nicholson scheme. Similar to the proof of the theorem, it can also be proved that the scheme (6.1.10) is stable and convergent [209]. Furthermore, using Richardson extrapolation can increase both the spatial and temporal accuracy to the second-order. The weighted average methods can still be used to solve two-sided space fractional advection-diffusion equation [75].*

7. Applications of the finite difference schemes based on L-algorithms.

L-algorithms can also formulate the fractional Euler schemes for approximating space fractional derivative, but most of which have not been given stability and convergence analyses. For instance, the L2-algorithm has been used to discretize Riemann-Liouville fractional derivative [81], the two-sided space fractional derivative [137] and Riesz space fractional derivative [46, 224]. Besides, Shen [199] has given a theoretical analysis to L1-algorithm-based finite difference scheme for fractional diffusion equation having Caputo fractional derivative.

6.2   Time fractional partial differential equation

Time fractional diffusion equations are widely considered in physical applications, which can describe the anomalous transport processes with long-time memory.

Consider the following time fractional diffusion equation:

$$\frac{\partial u(x,t)}{\partial t} = K_\mu \mathcal{D}_t^{1-\mu} \frac{\partial^2 u(x,t)}{\partial x^2}, 0 \leqslant x \leqslant L, t > 0, \tag{6.2.1}$$

with the initial-boundary value conditions:

$$u(x, t = 0) = g(x), 0 \leqslant x \leqslant L. \tag{6.2.2}$$

$$u(x = 0, t) = \varphi(x), \quad u(x = L, t) = \varphi_2(x) \tag{6.2.3}$$

Before proceeding further, we denote $x_i = ih, i = 0, 1, \cdots, M; h = L/N;$
$t_k = k\tau, k = 0, 1, \cdots, M; \tau = T/M.$

6.2.1   Finite difference schemes

Respectively using first-order forward difference quotient and second-order central difference quotient for approximating the temporal first-order derivative and spatial second-order derivative in (6.2.1) leads to the Forward time and centered space method (FTCS method) :

$$\frac{\partial u}{\partial t} u(x_j, t_k) = \frac{[u]_j^{k+1} - [u]_j^k}{\tau} + O(\tau), \tag{6.2.4}$$

$$\frac{\partial^2 u}{\partial x^2} u(x_j, t_k) = \frac{[u]_{j-1}^k - 2[u]_j^k + [u]_{j+1}^k}{(h)^2} + O(h^2). \tag{6.2.5}$$

Substituting the above schemes into the (6.2.1) leads to

$$\frac{[u]_j^{k+1} - [u]_j^k}{\tau} = K_\mu \mathcal{D}_t^{1-\mu} \frac{[u]_{j-1}^k - 2[u]_j^k + [u]_{j+1}^k}{(h)^2} + T(x,t), \tag{6.2.6}$$

where the truncated error is $T(x,t)$.

To discretize the fractional derivative, high-order linear multi-step method (4.7.16) is considered, namely

$$\mathcal{D}_t^{1-\mu} f(t) = \bar{h}^{-(1-\mu)} \sum_{j=0}^{[t/\bar{h}]} \omega_j^{(1-\mu)} f(t - j\bar{h}) + O(\bar{h}^p), \tag{6.2.7}$$

where $\bar{h} = \tau$ is the approximation step, and the coefficient $\omega_j^{(\alpha)}$ is derived from the corresponding generating function $W_p^{(\alpha)}(z)$ (the functions for $p = 1, 2, \ldots, 6$ have been given in Chapter 4). For $p - 1$, $W_1^{(\alpha)} = (1 - z)^\mu$, which corresponds to G1-algorithm, also called fractional first-order backward difference formula (BDF1 for short). For $p = 2$, $W_2^{(\alpha)} = \left(\dfrac{3}{2} - 2x + \dfrac{1}{2}x^2 \right)^\alpha$ is called fractional second-order backward difference formula (BDF2).

Substituting all the approximation formulas afore-mentioned in the equation (6.2.1) and omitting the truncated error yield

$$u_j^{k+1} = u_j^k + S_\mu \sum_{m=0}^{k} w_m^{(1-\mu)} \left(u_{j-1}^{k-m} - 2u_j^{k-m} + u_{j+1}^{k-m} \right), \tag{6.2.8}$$

where$S_\mu = K_\mu \dfrac{\tau^\mu}{h^2}$.

6.2.2   Stability analysis: Fourier-von Neumann method

Let $u_j^k = \zeta_k e^{iqjh}$ where q is the wave number, and substitute them into (6.2.8) to obtain

$$\zeta_{k+1} = \zeta_k - 4S_\mu \sin^2 \left(\frac{qh}{2} \right) \sum_{m=0}^{k} w_m^{(1-\mu)} \zeta_{k-m}, \tag{6.2.9}$$

which is the discrete form of the following fractional differential equation:

$$\frac{d\psi(t)}{dt} = -4C \sin^2 \left(\frac{qh}{2} \right) \mathcal{D}_t^{1-\mu} \psi(t), \tag{6.2.10}$$

where $C = S_\mu \tau^\mu$ and the solution of which can be represented in terms of Mittag-Leffler function [179]. Let

$$\zeta_{k+1} = \xi \zeta_k, \tag{6.2.11}$$

assume $\xi = \xi(q)$ to be time-independent, and then substitute ξ into (6.2.9) to obtain

$$\xi = 1 - 4S_\mu \sin^2 \left(\frac{qh}{2} \right) \sum_{m=0}^{k} w_m^{(1-\mu)} \xi^{-m}. \tag{6.2.12}$$

If there exists some q such that $|\xi| > 1$, then the finite difference scheme is instable.

Considering the extreme case $\xi = -1$, we have

$$S_\mu \sin^2 \left(\frac{qh}{2} \right) \leqslant \frac{1/2}{\displaystyle\sum_{m=0}^{k} (-1)^m w_m^{(1-\mu)}} \equiv \bar{S}_{\mu,k}. \tag{6.2.13}$$

The right-hand side of inequality (6.2.13) weakly depends on the iteration number k. Let $\bar{S}_\mu = \lim_{k \to \infty} \bar{S}_{\mu,k}$, which can be determined by inequality (6.2.13) as well as the generating function $W_p^{(\beta)}(z) = (1-z)^\beta = \sum_{m=0}^{k} \omega_m^{(\beta)} z^m$ (let $z = -1, \beta = 1 - \mu$ and see the definition of generating function in Chapter 4).

Hence, the prerequisite for a stable scheme is to satisfy (sufficient condition)

$$S_\mu \sin^2 \left(\frac{qh}{2} \right) \leqslant \bar{S}_\mu = \frac{1}{2W_p^{(1-\mu)}(-1)}. \tag{6.2.14}$$

The reference [226] conclude through numerical investigation that the above inequality is still the necessary condition for a stable scheme. We thus have obtained the sufficient and necessary condition for a stable finite difference scheme (6.2.6):

$$S_\mu \leqslant \frac{\bar{S}_\mu}{\sin^2 \left(\dfrac{qh}{2} \right)}. \tag{6.2.15}$$

Noting this, we can see that the scheme is stable if

$$S_\mu = K_\mu \frac{\tau^\mu}{(h)^2} \leqslant \bar{S}_\mu. \tag{6.2.16}$$

6.2.3   Error analysis

From (6.2.6), we see that the truncated error term is

$$T(x,t) = \frac{[u]_j^{k+1} - [u]_j^k}{\tau} - K_\mu \mathcal{D}_t^{1-\mu} \frac{[u]_{j-1}^k - 2[u]_j^k + [u]_{j+1}^k}{h^2}. \tag{6.2.17}$$

Since

$$\frac{[u]_j^{k+1} - [u]_j^k}{\tau} = u_t + \frac{1}{2} u_{tt} \tau + O(\tau)^2, \tag{6.2.18}$$

and

$$\begin{aligned} &\mathcal{D}_t^{1-\mu} \left([u]_{j-1}^k - 2[u]_j^k + [u]_{j+1}^k \right) \\ &= \frac{1}{\bar{h}^{1-\mu}} \sum_{m=0}^{k} \omega_m^{(1-\mu)} \left(u_{xx} + \frac{1}{12} u_{xxx}(h)^2 + \cdots \right) + O(\bar{h}^p), \end{aligned} \tag{6.2.19}$$

we have

$$\begin{aligned} T(x,t) &= O(\bar{h}^p) + \frac{1}{2} u_{tt} \tau - \frac{K_\mu h^2}{12} \mathcal{D}_t^{1-\mu} u_{xxxx} + \cdots \\ &= O(\bar{h}^p) + O(\tau) + O(h^2). \end{aligned} \tag{6.2.20}$$

Hence, (i) assuming the initial-boundary conditions of u are consistent (also assumed for the classical FTCS method) and (ii) assuming that u is sufficiently smooth in the neighbourhood of $t = 0$ (namely, the prerequisite of linear multi-step method (6.2.7)), then FTCS method is unconditionally continuous, namely,

$$T(x,t) \longrightarrow 0 \quad \text{as } \bar{h}, \tau, h \to 0.$$

Remark 6.2.1 *1. In particular, for $p = 1$, $W_1^{(\alpha)} = (1-z)^\alpha$, so that $\bar{S}_\mu =$*
$\dfrac{1}{2^{2-\mu}}$; *for $p = 2$, $W_2^{(\alpha)}(z) = \left(\dfrac{3}{2} - 2z + \dfrac{1}{2}z^2\right)^\alpha$, which leads to $\bar{S}_\mu = \dfrac{1}{2^{3/2-\mu}}$.*

2. Note that for $\mu < 1$, $\dfrac{1}{2^{3/2-\mu}} \leqslant \dfrac{1}{2^{2-\mu}}$, which implies that the stability of BDF2($p = 2$) method is somewhat lower than that of BDF1($p = 1$) method.

3. In practical computations, we usually let $\bar{h} = \tau$. It follows from (6.2.20) that the higher-order linear multi-step method ($p = 2$) for fractional derivative cannot authentically improve the accuracy of FTCS method. In terms of stability, as mentioned above, the stability of the higher-order method ($p > 1$) is worse than that of the lower-order method ($p = 1$). Noting this, we usually let $p = 1$, i. e., use the FTCS method based on G1-algorithm.

4. Global error analysis can be found in [43], where the implicit finite difference scheme is given, together with its stability and convergence analyses.

5. FTCS method based on L1-algorithm

Caputo time fractional derivative $\dfrac{\partial^\alpha u(x,t)}{\partial t^\alpha} = {}^C\mathcal{D}_t^\alpha u(x,t)$ can be discretized by L1-algorithm (4.5.2):

$$\frac{\partial^\alpha u(x_i, t_{k+1})}{\partial t^\alpha} = \frac{\tau^{-\alpha}}{\Gamma(2-\alpha)} \sum_{j=0}^{k} b_j^{(\alpha)} (u(x_i, t_{k-j+1}) - u(x_i, t_{k-j})) + O(\tau) \quad (6.2.21)$$

where $b_j^{(\alpha)} = (j+1)^{1-\alpha} - j^{1-\alpha}$.

Using first-order backward difference quotient and second-order central difference quotient to respectively approximate first-order time derivative and second-order space derivative yield [129]

$$(1+2\rho)u_j^{k+1} - \rho u_{j+1}^{k+1} - \rho u_{j-1}^{k+1} = (1+2\rho)u_j^k - \rho u_{j+1}^k$$

$$-\rho u_{j-1}^k + \rho \frac{\mu}{(k+1)^{1-\mu}} \Delta_h^2 u_j^0 + \rho \sum_{l=0}^{k-1} b_{k-l}^{(1-\mu)} (\Delta_h^2 u_j^{l+1} - \Delta_h^2 u_j^l), \qquad (6.2.22)$$

where $\rho = \dfrac{K_\mu \tau^\mu}{h^2 \Gamma 1 + \mu}, \Delta_h^2 u_j^k = u_{j+1}^k - 2u_j^k + u_{j-1}^k$.

Langlands and Henry gave a brief but not very rigorous stability and convergence analyses for this implicit finite difference scheme. They derived the *local truncated error being $O((\tau)^{2-\mu}) + O((h)^2)$, but without further giving the global error.*

6.3   Time-space fractional partial differential equation

Consider the fractional advection-diffusion equation (6.0.1) with variable coefficients, of which the initial-boundary conditions are

$$u(x, t = 0) = g(x), 0 \leqslant x \leqslant L. \tag{6.3.1}$$

$$u(x = 0, t) = 0, \quad u(x = L, t) = \varphi(t). \tag{6.3.2}$$

6.3.1   Finite difference schemes

Take time-space grid with time step τ and space step h and let $x_i = ih, i = 0, 1, \cdots, N, h = L/N; t_k = k\tau, k = 0, 1, \cdots, M; \tau = T/M$.

Using L1-algorithm (4.5.2) to approximate Caputo time fractional derivative $\dfrac{\partial^\alpha u(x,t)}{\partial t^\alpha} = {}^C\mathcal{D}_t^\alpha u(x,t)$ produces:

$$\frac{\partial^\alpha u(x_i, t_{k+1})}{\partial t^\alpha} = \frac{\tau^{-\alpha}}{\Gamma(2-\alpha)} \sum_{j=0}^{k} b_j^{(\alpha)} (u(x_i, t_{k-j+1}) - u(x_i, t_{k-j})) + O(\tau) \tag{6.3.3}$$

where $b_j^{(\alpha)} = (j+1)^{1-\alpha} - j^{1-\alpha}$ and we write $b_j^{(\alpha)} = b_j$.

Riemann-Liouville space fractional derivative is discretized by G-algorithm. According to the remarks below Theorem 4.2.1, since $0 < \beta \leqslant 1, 1 < \gamma \leqslant 2$, when using G-algorithm to evaluate $\mathcal{D}_x^\beta u(x,t)$ and $\mathcal{D}_x^\gamma u(x,t)$, the optimal shift number should take $p = 0$ and $p = 1$, respectively. That is, to use G1- and $G_{S(1)}-$ algorithms for discretization:

$$\mathcal{D}_x^\beta u(x_i, t_{k+1}) = h^{-\beta} \sum_{j=0}^{i} \omega_j^{(\beta)} u(x_i - jh, t_{k+1}) + O(h)), \tag{6.3.4}$$

$$\mathcal{D}_x^\gamma u(x, t) = h^{-\gamma} \sum_{j=0}^{i} \omega_j^{(\gamma)} u(x_i - (j-1)h, t_{k+1}) + O(h)), \tag{6.3.5}$$

where $\omega_j^\mu = (-1)^j \dfrac{\mu(\mu-1)\cdots(\mu-l+1)}{j!}$. So that, we have the following implicit finite difference scheme:

$$\sum_{j=0}^{k} b_j(u_i^{k-j+1} - u_i^{k-j}) = -r_{i,k+1}^{(1)} \sum_{l=0}^{i} \omega_l^{(\beta)} u_{i-l}^{k+1} + r_{i,k+1}^{(2)} \sum_{l=0}^{i+1} \omega_l^{(\gamma)} u_{i+1-l}^{k+1} + \bar{f}_i^{k+1},$$

$$\tag{6.3.6}$$

which can be further rearranged to

$$u_i^{k+1} + r_{i,k+1}^{(1)} \sum_{l=0}^{i} \omega_l^{(\beta)} u_{i-l}^{k+1} - r_{i,k+1}^{(2)} \sum_{l=0}^{i+1} \omega_l^{(\gamma)} u_{i+1-l}^{k+1} = \sum_{j=0}^{k-1}(b_j - b_{j+1})u_i^{k-j}$$
$$+ b_k u_i^0 + \bar{f}_i^{k+1}, \ i = 1, 2, \cdots, M; k = 0, 1, \cdots, N, \tag{6.3.7}$$

where

$$v_i^k = v(ih, k\tau), \quad d_i^k = d(ih, k\tau), \quad r_{i,k}^{(1)} = \frac{v_i^k \tau^\alpha \Gamma(2-\alpha)}{h^\beta},$$

$$r_{i,k}^{(2)} = \frac{d_i^k \tau^\alpha \Gamma(2-\alpha)}{h^\gamma}, \quad f_i^k = f(ih, k\tau), \quad \bar{f}_i^k = \tau^\alpha \Gamma(2-\alpha) f_i^k.$$

Similarly, we can also derive the following explicit scheme:

$$u_i^{k+1} = b_k u_i^0 + \sum_{j=0}^{k-1}(b_j - b_{j+1})u_i^{k-j} - r_{i,k+1}^{(1)} \sum_{l=0}^{i} \omega_l^{(\beta)} u_{i-l}^k$$
$$+ r_{i,k+1}^{(2)} \sum_{l=0}^{i+1} \omega_l^{(\gamma)} u_{i+1-l}^k + \bar{f}_i^{k+1}, \ i = 1, 2, \cdots, M; k = 0, 1, \cdots, N, \tag{6.3.8}$$

with the initial-boundary conditions:

$$u_i^0 = g(ih), \ u_0^k = 0, \quad u_M^k = \varphi(k\tau), i = 0, 1, \cdots, M, \ k = 0, 1, \cdots, N. \tag{6.3.9}$$

Lemma 6.3.1   *The coefficients* $b_j, \omega_j^{(\beta)}, \omega_j^{(\gamma)}$ *satisfy*

$$b_0 = 1, \quad b_j > 0, \quad b_{j+1} > b_j, \ j = 0, 1, 2, \cdots$$

$$\omega_0^{(\beta)} = 1, \quad \omega_1^{(\beta)} = -\beta, \quad \omega_j^{(\beta)} < 0(j > 1),$$

$$\sum_{j=0}^{\infty} \omega_j^{(\beta)} = 0, \quad \forall K, \sum_{j=0}^{K} \omega_j^{(\beta)} > 0;$$

$$\omega_0^{(\gamma)} = 1, \quad \omega_1^{(\gamma)} = -\gamma, \quad \omega_j^{(\gamma)} > 0(j \neq 1),$$

$$\sum_{j=0}^{\infty} \omega_j^{(\gamma)} = 0, \quad \forall K, \sum_{j=0}^{K} \omega_j^{(\gamma)} < 0.$$

   For convenience of analysis, we assume that v, d are constants irrespective of x, t, and denote $r_{i,k}^{(m)} = r_m, m = 1, 2$. In fact, this assumption will not affect the analyses of the stability and convergence of the scheme.

6.3.2   Stability and convergence analysis

1. Implicit finite difference scheme and its stability

Define two difference operators L_1 and L_2 as

$$L_1 u_i^{k+1} = u_i^{k+1} + r_1 \sum_{l=0}^{i} \omega_l^{(\beta)} u_{i-l}^{k+1} - r_2 \sum_{l=0}^{i+1} \omega_l^{(\gamma)} u_{i+1-l}^{k+1} \qquad (6.3.10)$$

$$L_2 u_i^k = b_k u_i^0 + \sum_{j=0}^{k-1} (b_j - b_{j+1}) u_i^{k-j}, \qquad (6.3.11)$$

then the implicit finite difference scheme (6.3.6) can be written by

$$L_1 u_i^{k+1} = L_2 u_i^k + \bar{f}_i^{k+1}. \qquad (6.3.12)$$

Let $\tilde{u}_i^j$ be the approximant derived from the finite difference schemes (6.3.7) and (6.3.9), and $\varepsilon_i^j = \tilde{u}_i^j - u_i^j$ be the numerical error, such that

$$L_1 \varepsilon_i^{k+1} = L_2 \varepsilon_i^k, \qquad (6.3.13)$$

and $E^k = (\varepsilon_1^k, \varepsilon_2^k, \cdots, \varepsilon_{M-1}^k)^T$ be the error vector.

Theorem 6.3.1   *The numerical errors induced by initial-value conditions in finite difference schemes (6.3.7) and (6.3.9) satisfy*

$$\|E^{k+1}\|_\infty \leqslant \|E^0\|_\infty, \ k = 0, 1, 2, \cdots \qquad (6.3.14)$$

namely, the schemes are unconditionally stable.

Proof Consider the mathematical deduction.

When $k = 0$, letting $|\varepsilon_l^1| = \max_{1 \leqslant i \leqslant M} |\varepsilon_i^1|$ and from the lemma 6.3.1, we have

$$\begin{aligned}
|\varepsilon_l^1| &\leqslant \left(1 + r_1 \sum_{j=0}^{l} \omega_j^{(\beta)} - r_2 \sum_{j=0}^{l+1} \omega_j^{(\gamma)} \right) |\varepsilon_l^1| \\
&\leqslant |\varepsilon_l^1| + r_1 \sum_{j=0}^{l} \omega_j^{(\beta)} |\varepsilon_{l-j}^1| - r_2 \sum_{j=0}^{l+1} \omega_j^{(\gamma)} |\varepsilon_{l+1-j}^1| \\
&= (1 + r_1 + r_2\gamma)|\varepsilon_l^1| + r_1 \sum_{j=0}^{l} \omega_j^{(\beta)} |\varepsilon_{l-j}^1| - r_2 \sum_{j=0, j\neq 1}^{l+1} \omega_j^{(\gamma)} |\varepsilon_{l+1-j}^1| \\
&\leqslant |\varepsilon_l^1 + r_1 \sum_{j=0}^{l} \omega_j^{(\beta)} \varepsilon_{l-j}^1 - r_2 \sum_{j=0}^{l+1} \omega_j^{(\gamma)} \varepsilon_{l+1-j}^1| \\
&= |L_1 \varepsilon_l^1| = |L_2 \varepsilon_l^0| = |\varepsilon_l^0| \leqslant \|E^0\|_\infty.
\end{aligned} \qquad (6.3.15)$$

Thus, $\|E^1\|_\infty \leqslant \|E^0\|_\infty$.

Now assuming $\|E^j\|_\infty \leqslant \|E^0\|_\infty$, $j = 1, 2, \cdots, k$, and letting $|\varepsilon_l^{k+1}| = \max\limits_{1\leqslant i\leqslant M-1} |\varepsilon_i^{k+1}|$, it follows from the lemma 6.3.1 that

$$
\begin{aligned}
|\varepsilon_l^{k+1}| &\leqslant \left(1 + r_1\sum_{j=0}^{l}\omega_j^{(\beta)} - r_2\sum_{j=0}^{l+1}\omega_j^{(\gamma)}\right)|\varepsilon_l^{k+1}| \\
&\leqslant |\varepsilon_l^{k+1}| + r_1\sum_{j=0}^{l}\omega_j^{(\beta)}|\varepsilon_{l-j}^{k+1}| - r_2\sum_{j=0}^{l+1}\omega_j^{(\gamma)}|\varepsilon_{l+1-j}^{k+1}| \\
&= (1 + r_1 + r_2\gamma)|\varepsilon_l^{k+1}| + r_1\sum_{j=0}^{l}\omega_j^{(\beta)}|\varepsilon_{l-j}^{k+1}| - r_2\sum_{j=0,j\neq1}^{l+1}\omega_j^{(\gamma)}|\varepsilon_{l+1-j}^{k+1}| \\
&\leqslant |\varepsilon_l^{k+1} + r_1\sum_{j=0}^{l}\omega_j^{(\beta)}\varepsilon_{l-j}^{k+1} - r_2\sum_{j=0}^{l+1}\omega_j^{(\gamma)}\varepsilon_{l+1-j}^{k+1}| \\
&= |L_1\varepsilon_l^{k+1}| = |L_2\varepsilon_l^{k}|,
\end{aligned}
\tag{6.3.16}
$$

Thus,

$$
\begin{aligned}
\|E^{k+1}\|_\infty &\leqslant |L_2\varepsilon_l^{k}| = |b_k\varepsilon_l^0 + \sum_{j=0}^{k-1}(b_j - b_{j+1})\varepsilon_l^{k-j}| \\
&\leqslant (b_k + \sum_{j=0}^{k-1}(b_j - b_{j+1})\|E^0\|_\infty = \|E^0\|_\infty.
\end{aligned}
\tag{6.3.17}
$$

That is, the implicit finite difference scheme is unconditionally stable for arbitrary initial value conditions.

2. Convergence of the implicit finite difference scheme

Let $u(x_i, t_k)(i = 1, 2, \cdots, M-1; k = 1, 2, \cdots, N)$ be the exact solution of equations (6.0.1),(6.3.1) and (6.3.2)) at grid points, define the error between exact and numerical solutions by $\eta_i^k = u(x_i, t_k) - u_i^k, i, k = 1, 2, \cdots$, and denote $Y^k = (\eta_1^k, \eta_2^k, \cdots, \eta_{M-1}^k)^T$. Obviously, $Y^0 = 0$. It follows from equations (6.3.3)$\sim$(6.3.5) that the error satisfy

$$
\begin{cases}
L_1\eta_i^{k+1} = L_2\eta_i^k + R_i^{k+1}, \\
\eta_i^0 = 0,
\end{cases}
i = 1, 2, \cdots, M-1; k = 0, 1, 2, \cdots, N-1, \tag{6.3.18}
$$

where $|R_i^k| \leqslant C\tau^\alpha(\tau + h)$ derived from (6.3.3)$\sim$(6.3.5).

Lemma 6.3.2   *The errors between exact solutions and numerical solutions derived from implicit finite difference scheme (6.3.7) and(6.3.9) satisfy*

$$
\|Y^{k+1}\|_\infty \leqslant Cb_k^{-1}(\tau^{1+\alpha} + \tau^\alpha h), k = 1, 2, \cdots, n.
\tag{6.3.19}
$$

Proof Consider the mathematical deduction.

When $k = 0$, letting $\|Y^1\|_\infty = |\eta_l^1| = \max\limits_{1 \leqslant i \leqslant M-1} |\eta_i^1|$, and from the lemma 6.3.1, we have

$$
\begin{aligned}
|\eta_l^1| &\leqslant \left(1 + r_1 \sum_{j=0}^{l} \omega_j^{(\beta)} - r_2 \sum_{j=0}^{l+1} \omega_j^{(\gamma)}\right) |\eta_l^1| \\
&\leqslant |\eta_l^1| + r_1 \sum_{j=0}^{l} \omega_j^{(\beta)} |\eta_{l-j}^1| - r_2 \sum_{j=0}^{l+1} \omega_j^{(\gamma)} |\eta_{l+1-j}^1| \\
&= (1 + r_1 + r_2\gamma)|\eta_l^1| + r_1 \sum_{j=0}^{l} \omega_j^{(\beta)} |\eta_{l-j}^1| - r_2 \sum_{j=0,j\neq 1}^{l+1} \omega_j^{(\gamma)} |\eta_{l+1-j}^1| \\
&\leqslant |\eta_l^1 + r_1 \sum_{j=0}^{l} \omega_j^{(\beta)} \eta_{l-j}^1 - r_2 \sum_{j=0}^{l+1} \omega_j^{(\gamma)} \eta_{l+1-j}^1| \\
&= |L_1 \eta_l^1| = |L_2 \eta_l^0 + C\tau^\alpha(\tau + h)| = |\eta_l^0 + C\tau^\alpha(\tau + h)| \leqslant C\tau^\alpha(\tau + h).
\end{aligned}
\tag{6.3.20}
$$

Thus, $\|Y^1\|_\infty \leqslant Cb_0^{-1}\tau^\alpha(\tau + h)$.

Now assuming $\|Y^j\|_\infty \leqslant Cb_{j-1}^{-1}\tau^\alpha(\tau + h)$, $j = 1, 2, \cdots, k$. and letting $|\eta_l^{k+1}| = \max\limits_{1 \leqslant i \leqslant M-1} |\eta_i^{k+1}|$, it follows from lemma 6.3.1 that $b_k^{-1} \geqslant b_j^{-1}(j = 0, 1, \cdots, k)$, which leads to

$$
\|Y^j\|_\infty \leqslant Cb_k^{-1}\tau^\alpha(\tau + h), \ j = 1, 2, \cdots, k.
$$

Similarly, use $\eta_l^0 = 0$ to obtain

$$
\begin{aligned}
|\eta_l^{k+1}| &\leqslant \left(1 + r_1 \sum_{j=0}^{l} \omega_j^{(\beta)} - r_2 \sum_{j=0}^{l+1} \omega_j^{(\gamma)}\right) |\eta_l^{k+1}| \\
&\leqslant |\eta_l^{k+1}| + r_1 \sum_{j=0}^{l} \omega_j^{(\beta)} |\eta_{l-j}^{k+1}| - r_2 \sum_{j=0}^{l+1} \omega_j^{(\gamma)} |\eta_{l+1-j}^{k+1}| \\
&= (1 + r_1 + r_2\gamma)|\eta_l^{k+1}| + r_1 \sum_{j=0}^{l} \omega_j^{(\beta)} |\eta_{l-j}^{k+1}| - r_2 \sum_{j=0,j\neq 1}^{l+1} \omega_j^{(\gamma)} |\eta_{l+1-j}^{k+1}| \\
&\leqslant |\eta_l^{k+1} + r_1 \sum_{j=0}^{l} \omega_j^{(\beta)} \eta_{l-j}^{k+1} - r_2 \sum_{j=0}^{l+1} \omega_j^{(\gamma)} \eta_{l+1-j}^{k+1}| \\
&= |L_1 \eta_l^{k+1}| = |L_2 \eta_l^k + C\tau^\alpha(\tau + h)| \\
&= |b_k \eta_l^0 + \sum_{j=0}^{k-1} (b_j - b_{j+1})\eta_l^{k-j} + C\tau^\alpha(\tau + h)| \\
&\leqslant \sum_{j=0}^{k-1} (b_j - b_{j+1})\eta_l^{k-j} + C\tau^\alpha(\tau + h) \\
&\leqslant (b_k + \sum_{j=0}^{k-1} (b_j - b_{j+1})b_k^{-1} C\tau^\alpha(\tau + h) = Cb_k^{-1}\tau^\alpha(\tau + h).
\end{aligned}
\tag{6.3.21}
$$

That is, $\|Y^{k+1}\|_\infty \leqslant Cb_k^{-1}\tau^\alpha(\tau + h)$.

Theorem 6.3.2   *The errors between the exact solutions and numerical solutions derived from implicit finite difference schemes (6.3.7) and (6.3.9) satisfy*

$$|u_i^k - u(x_i, t_k)| \leqslant \bar{C}(\tau + h), i = 1, 2, \cdots, M - 1; k = 1, 2, \cdots, N.$$

That is, the schemes are convergent.

Proof   Since

$$\lim_{k\to\infty} \frac{b_k^{-1}}{k^\alpha} = \lim_{k\to\infty} \frac{k^{-\alpha}}{(k+1)^{1-\alpha} - k^{1-\alpha}} = \lim_{k\to\infty} \frac{k^{-1}}{(1 + \frac{k}{l})^{1-\alpha} - 1} \qquad (6.3.22)$$
$$= \lim_{k\to\infty} \frac{k^{-1}}{(1 - \alpha)k^{-1}} = \frac{1}{1 - \alpha},$$

there exists a constant $\tilde{C}$ such that

$$b_k^{-1} \leqslant \tilde{C}k^\alpha. \qquad (6.3.23)$$

As $k\tau \leqslant T$ is finite, it follows from the proof of the preceding lemma that

$$|u_i^k - u(x_i, t_k)| \leqslant |\eta_i^k| \leqslant Cb_k^{-1}\tau^\alpha(\tau + h) \leqslant C\tilde{C}k^\alpha\tau^\alpha(\tau + h) \leqslant \bar{C}(\tau + h).$$

That is, the schemes are convergent.

  3. Convergence of the explicit finite difference scheme

  Similarly, let $\tilde{u}_i^j$ be the approximant derived from the finite difference schemes (6.3.8) and (6.3.9), and $\varepsilon_i^j = \tilde{u}_i^j - u_i^j$ be the numerical error such that

$$\varepsilon_i^{k+1} = b_k\varepsilon_i^0 + \sum_{j=0}^{k-1}(b_j - b_{j+1})\varepsilon_i^{k-j} - r_1\sum_{l=0}^{i}\omega_l^{(\beta)}\varepsilon_{i-l}^k + r_2\sum_{l=0}^{i+1}\omega_l^{(\gamma)}\varepsilon_{i+1-l}^k. \ (6.3.24)$$

Here, $k = 0, 1, \cdots, N - 1; i = 1, 2, \cdots, M - 1$ and $E^k = (\varepsilon_1^k, \varepsilon_2^k, \cdots, \varepsilon_{M-1}^k)^T$ be the error vector.

Theorem 6.3.3   *If*

$$r_1 + r_2\beta < 2 - 2^{1-\alpha} = 1 - b_1, \qquad (6.3.25)$$

then the errors induced by initial-value conditions in explicit finite difference schemes (6.3.8) and (6.3.9) satisfy

$$\|E^{k+1}\|_\infty \leqslant \|E^0\|_\infty, \ k = 0, 1, 2, \cdots \qquad (6.3.26)$$

That is, the schemes are conditionally stable for arbitrary initial-value conditions.

Proof Consider the mathematical deduction.

When $k = 0$, letting $|\varepsilon_l^1| = \max\limits_{1 \leqslant i \leqslant M} |\varepsilon_i^1|$ and from the lemma 6.3.1 and the condition (6.3.25), we have

$$b_0 - r_1 - r_2\beta > b_0 - 1 + b_1 = b_1 > 0. \tag{6.3.27}$$

Thus,

$$
\begin{aligned}
|\varepsilon_l^1| &\leqslant b_0|\varepsilon_l^0| - r_1\sum_{j=0}^{l} \omega_j^{(\beta)}|\varepsilon_{l-j}^0| - r_2\sum_{j=0}^{l+1} \omega_j^{(\gamma)}|\varepsilon_{l+1-j}^0| \\
&= \left(1 - r_1\sum_{j=0}^{l} \omega_j^{(\beta)} + r_2\sum_{j=0}^{l+1} \omega_j^{(\gamma)}\right) \|E^0\|_\infty \leqslant \|E^0\|_\infty.
\end{aligned}
\tag{6.3.28}
$$

Further we have $\|E^1\|_\infty \leqslant \|E^0\|_\infty$.

Now assuming $\|E^j\|_\infty \leqslant \|E^0\|_\infty$, $j = 1, 2, \cdots, k$ and letting $|\varepsilon_l^{k+1}| = \max\limits_{1 \leqslant i \leqslant M-1} |\varepsilon_i^{k+1}|$, it follows from (6.3.25) that

$$b_0 - b_1 - r_1 - r_2\beta > b_0 - b_1 - 1 + b_1 = 0.$$

Using lemma 6.3.1 yields

$$
\begin{aligned}
|\varepsilon_l^{k+1}| &\leqslant b_k|\varepsilon_l^0| + \sum_{j=0}^{k-1}(b_j - b_{j+1})\varepsilon_i^{k-j} - r_1\sum_{j=0}^{l} \omega_j^{(\beta)}\varepsilon_{l-j}^0 + r_2\sum_{j=0}^{l+1} \omega_j^{(\gamma)}\varepsilon_{i+1-l}^0 \\
&\leqslant \left(b_k + \sum_{j=0}^{k-1}(b_j - b_{j+1}) - r_1\sum_{j=0}^{l} \omega_j^{(\beta)} + r_2\sum_{j=0}^{l+1} \omega_j^{(\gamma)}\right) \|E^0\|_\infty \\
&\leqslant \|E^0\|_\infty.
\end{aligned}
\tag{6.3.29}
$$

Hence, $\|E^{k+1}\|_\infty \leqslant \|E^0\|_\infty$. That is, the explicit finite difference schemes are conditionally stable for arbitrary initial-value conditions.

Remark 6.3.1 *When the equation coefficients v, b are expressed in form of the function of x, t, the stable condition is replaced by*

$$\lambda = \max_{\substack{1 \leqslant i \leqslant M-1 \\ 1 \leqslant k \leqslant N}} \left[(r_{i,k}^{(1)} + r_{i,k}^{(2)}\beta)\right] - 2 - 2^{1-\alpha} < 0.$$

4. Convergence of the explicit finite difference scheme

Let $u(x_i, t_k)(i = 1, 2, \cdots, M - 1; k = 1, 2, \cdots, N)$ be the exact solutions of the equations (6.0.1), (6.3.1) and (6.3.2)) at grid points, define the errors between exact and numerical solutions as $\eta_i^k = u(x_i, t_k) - u_i^k, i, k = 1, 2, \cdots,$

and denote $Y^k = (\eta_1^k, \eta_2^k, \cdots, \eta_{M-1}^k)^T$. Obviously, $Y^0 = 0$. The errors satisfy the following equation:

$$
\begin{cases}
\eta_i^{k+1} = b_k \eta_i^0 + \displaystyle\sum_{j=0}^{k-1}(b_j - b_{j+1})\eta_i^{k-j} - r_1 \sum_{l=0}^{i} \omega_l^{(\beta)} \eta_{i-l}^k \\
\qquad\quad + r_2 \displaystyle\sum_{l=0}^{i+1} \omega_l^{(\gamma)} \eta_{i+1-l}^k + R_i^{k+1} \\
\eta_i^0 = 0, \qquad i = 1, 2, \cdots, M-1; \ k = 0, 1, 2, \cdots, N-1,
\end{cases}
\tag{6.3.30}
$$

where $|R_i^k| \leqslant C\tau^\alpha(\tau + h)$.

Lemma 6.3.3 *If the condition (6.3.25) holds, then the errors of the exact solutions and the numerical solutions derived from the explicit finite difference schemes (6.3.8) and (6.3.9) satisfy*

$$
\|Y^{k+1}\|_\infty \leqslant Cb_k^{-1}(\tau^{1+\alpha} + \tau^\alpha h), k = 1, 2, \cdots, n.
\tag{6.3.31}
$$

Proof Consider the mathematical deduction.

First give the proof of the case $k = 0$. Letting $\|Y^1\|_\infty = |\eta_l^1| = \displaystyle\max_{1\leqslant i\leqslant M-1} |\eta_i^1|$, we have

$$
|\varepsilon_l^1| = |R_l^1| \leqslant C\tau^\alpha(\tau + h).
\tag{6.3.32}
$$

Thus, $\|Y^1\|_\infty \leqslant Cb_0^{-1}\tau^\alpha(\tau + h)$.

Now suppose $\|Y^j\|_\infty \leqslant Cb_{j-1}^{-1}\tau^\alpha(\tau+h)$, $j = 1, 2, \cdots, k$. Letting $|\eta_l^{k+1}| = \displaystyle\max_{1\leqslant i\leqslant M-1} |\eta_i^{k+1}|$ and considering $b_k^{-1} \geqslant b_j^{-1}(j = 0, 1, \cdots, k)$ and (6.3.25), lead to

$$
|\eta_l^{k+1}| \leqslant b_k|\eta_l^0| + \sum_{j=0}^{k-1}(b_j - b_{j+1})\eta_i^{k-j} - r_1 \sum_{j=0}^{l} \omega_j^{(\beta)} \eta_{l-j}^k
$$
$$
+ r_2 \sum_{j=0}^{l+1} \omega_j^{(\gamma)} \eta_{i+1-l}^k + |R_l^{k+1}|
$$
$$
\leqslant \left(1 + \sum_{j=0}^{k-1}(b_j - b_{j+1}) - r_1 \sum_{j=0}^{l} \omega_j^{(\beta)} + r_2 \sum_{j=0}^{l+1} \omega_j^{(\gamma)} + b_k\right) b_k^{-1}C\tau^\alpha(\tau + h)
$$
$$
\leqslant b_k^{-1}C\tau^\alpha(\tau + h).
\tag{6.3.33}
$$

It further follows from (6.3.23) that $\|Y^{k+1}\|_\infty \leqslant Cb_k^{-1}\tau^\alpha(\tau+h) \leqslant \bar{C}(k\tau)^\alpha(\tau+h)$.

Since $k\tau \leqslant T$ is finite, we have the following theorem.

Theorem 6.3.4 *If the condition (6.3.25) holds, then the explicit schemes (6.3.8) and (6.3.9) are convergentand the errors satisfy*

$$|u_i^k - u(x_i, t_k)| \leqslant C(\tau + h), i = 1, 2, \cdots, M-1; k = 1, 2, \cdots, N.$$

That is, the schemes are convergent.

Remark 6.3.2 *It has been proven that the convergence of implicit and explicit schemes are both of the order $O(\tau + h)$, where $O(\tau)$ is the error order of the L1-algorithm and $O(h)$ is that of D-algorithm. But Langlands and Henry [129] proved, $u(t)$ can be expressed in terms of the following Taylor expansion:*

$$u(t) = u(0) + tu'(0) + \int_0^t u''(t-s)ds, \tag{6.3.34}$$

from which we see that the accuracy of L1-algorithm (6.3.3) can reach $O(\tau^{2-\alpha})$, which is higher than $O(\tau)$ (it should be noted that, the numerical results show, even though $u(t)$ has no Taylor expansion (6.3.34), L1-algorithm can still achieve the accuracy of the order $O(\tau^{2-\alpha})$). In such case, it can be shown that the convergence of both implicit and explicit finite difference schemes should be of the order $O(\tau^{2-\alpha} + h)$.

6.4 Numerical methods for non-linear fractional partial differential equations

6.4.1 Adomina decomposition method

In the 1980s, G. Adomian presented a decomposition method for deriving the semi-analytical solutions of the non-linear differential equations. This method gives the semi-analytical solutions in form of series and can be applied to solving mathematical, physical, linear, and non-linear ordinary/partial differential equations [4, 5]. During recent years, the method has been used to solve fractional differential equations. Momani et al applied Adomian decomposition method for solving non-linear fractional ordinary differential system as well as multi-term fractional linear ordinary differential equation [4,5,168]. Ray and Beral used the method to solve fractional Bayley-Trvk equation [188]. Jafari and Daftardar-Gjji solved the fractional non-linear two-point boundary-value problem [117]. Momani [167] derived the solution of space-time fractional telegraph equation having specific initial-boundary value conditions. Al-khaled and Momani have solved fractional diffusion-wave equation [6]. Odibat and Momani employed the modified Adomian decomposition method, namely, the matrix method, to solve space-time fractional diffusion-wave equation [174].

Consider the following generic time fractional partial differential equaiton:

$$^{C}\mathcal{D}_t^\alpha u(x,t) + \mathcal{L}u(x,t) + \mathcal{N}u(x,t) = g(x,t), \ m-1 < \alpha \leqslant m, \quad (6.4.1)$$

which is equivalent to

$$u(x,t) = \sum_{k=0}^{m-1} \frac{\partial^k u(x,0)}{\partial t^k} \frac{t^k}{k!} + \mathcal{J}^\alpha g(x,t) - \mathcal{J}^\alpha[\mathcal{L}u(x,t) + \mathcal{N}u(x,t)]. \quad (6.4.2)$$

Let

$$u(x,t) = \sum_{n=0}^{\infty} u_n(x,t), \quad (6.4.3)$$

$$\mathcal{N}u(x,t) = \sum_{n=0}^{\infty} A_n, \quad (6.4.4)$$

where A_n is the so-called Adomian polynomials. Substituting (6.4.3) and (6.4.4) into (6.4.2) gives

$$\sum_{n=0}^{\infty} u_n(x,t) = \sum_{k=0}^{m-1} \frac{\partial^k u(x,0)}{\partial t^k} \frac{t^k}{k!} + \mathcal{J}^\alpha g(x,t)$$
$$- \mathcal{J}^\alpha\Big[\mathcal{L}\Big(\sum_{n=0}^{\infty} u_n(x,t)\Big) + \sum_{n=0}^{\infty} A_n\Big]. \quad (6.4.5)$$

Then iteratively solving the equation via the following basic relations yields:

$$\begin{cases} u_0(x,t) = \sum_{k=0}^{m-1} \frac{\partial^k u(x,0)}{\partial t^k} \frac{t^k}{k!} + \mathcal{J}^\alpha g(x,t) \\ u_1(x,t) = -\mathcal{J}^\alpha(\mathcal{L}u_0 + A_0), \\ u_2(x,t) = -\mathcal{J}^\alpha(\mathcal{L}u_1 + A_1), \\ \vdots \\ u_{n+1}(x,t) = -\mathcal{J}^\alpha(\mathcal{L}u_n + A_n), \\ \vdots \end{cases} \quad (6.4.6)$$

Adomian polynomial A_n can be derived from

$$\begin{cases} v = \sum_{i=0}^{\infty} \lambda^i u_i, \\ \mathcal{N}(v) = \mathcal{N}\Big(\sum_{i=0}^{\infty} \lambda^i u_i\Big) = \sum_{n=0}^{\infty} \lambda^n A_n. \end{cases} \quad (6.4.7)$$

Differentiating the above equation with respect to λ for n times leads to the general form of the Adomian polynomial:

$$A_n = \frac{1}{n!}\frac{d^n}{d\lambda^n}\left[\mathcal{N}\left(\sum_{i=0}^{\infty}\lambda^i u_i\right)\right]_{\lambda=0}. \tag{6.4.8}$$

Hence, the solution of equation (6.4.1) can be given by

$$u(x,t) = \lim_{N\to\infty}\left(\sum_{n=0}^{N-1}u_n(x,t)\right). \tag{6.4.9}$$

Adomian decomposition method possesses the similar convergence to that of Taylor series, and its convergence and truncated error analyses can be found in [1,45]. In addition, Adomian decomposition method can also be used for space-time fractional reaction-diffusion equation with variable coefficients [225] and for other non-linear equations [117].

The merits of Adomian decomposition method is that, in addition to avoiding the discretization of equation, the method gives semi-analytical solutions that are fast convergent to exact solution, and enjoys low computational cost. The method also embraces a broad field of applications. Nevertheless, the method usually requires the fractional integral of a given function, which may not be easily obtained sometimes.

6.4.2   Variational iteration method

Variational iteration method [112] is somewhat similar to the Adomian decomposition method. It is originally developed for quantum mechanics and is subsequently applied to solving non-linear equations.

Consider the following generic time fractional partial differential equation

$${}^{C}\mathcal{D}_t^{\alpha}u(x,t) = f(u, u_x, u_{xx}) + g(x,t), \tag{6.4.10}$$

where f is a non-linear function, g is the source term, $m-1 < \alpha \leqslant m$, and the initial-boundary value conditions read:

for $0 < \alpha \leqslant 1$,

$$\begin{cases} u(x,0) = \varphi_1(x), \\ u(x,t) \to 0, \ as \ |x| \to \infty; \end{cases} \tag{6.4.11}$$

while for $1 < \alpha \leqslant 2$,

$$\begin{cases} u(x,0) = \varphi_1(x), \quad \partial_t u(x,0) = \varphi_2(x), \\ u(x,t) \to 0, \ as \ |x| \to \infty. \end{cases} \tag{6.4.12}$$

The modified functional of equation (6.4.10) is

$$u_{k+1}(x,t) = u_k(x,t) + \int_0^t \lambda(\xi)({}^{C}\mathcal{D}_{\xi}^{\alpha}u(x,\xi)$$
$$- f(\tilde{u}_k, (\tilde{u}_k)_\tau, (\tilde{u}_k)_{\tau\tau}) - g(\tau,\xi))d\xi \tag{6.4.13}$$

where u_k is the k-th approximation solution and $\tilde{u}_k$ is the constrain variation satisfying $\delta\tilde{u}_k = 0$. λ is the generalized Lagrange multiplier which can be derived by imposing variation operation on both sides of (6.4.13):

$$
\begin{aligned}
\delta u_{k+1}(x,t) &= \delta u_k(x,t) + \delta \int_0^t \lambda(\xi) \\
&\quad \cdot \left({}^C\mathcal{D}_\xi^\alpha u(x,\xi) - f(\tilde{u}_k, (\tilde{u}_k)_x, (\tilde{u}_k)_{xx}) - g(x,\xi) \right) d\xi \quad (6.4.14) \\
&= \delta u_k(x,t) + \delta \int_0^t \lambda(\xi)\left({}^C\mathcal{D}_\xi^\alpha u(x,\xi) - g(x,\xi) \right) d\xi \\
&= 0.
\end{aligned}
$$

It follows that

$$
\lambda = -1,\ for\ m = 1; \qquad \lambda = \xi - t,\ for\ m = 2.
$$

In practical computations, we usually use integration by part to first extract the λ from the integrand, and then by comparing the coefficients in the resulting equation and realizing the arbitrariness of $u_n(t)$, we obtain the value of λ.

Ultimately, we derive the following variation iteration scheme:
when $m = 1$,

$$
\begin{cases}
u_{k+1}(x,t) = u_k(x,t) - \displaystyle\int_0^t \left({}^C\mathcal{D}_\xi^\alpha u_k(x,\xi) \right. \\
\qquad\qquad \left. - f(u_k, (u_k)_x, (u_k)_{xx}) + g(x,\xi) \right) d\xi, \\
u_0(x,t) = \varphi_1(x);
\end{cases}
\qquad (6.4.15)
$$

when $m = 2$

$$
\begin{cases}
u_{k+1}(x,t) = u_k(x,t) - \displaystyle\int_0^t (\xi - t)\left({}^C\mathcal{D}_\xi^\alpha u_k(x,\xi) \right. \\
\qquad\qquad \left. - f(u_k, (u_k)_x, (u_k)_{xx}) + g(x,\xi) \right) d\xi, \\
u_0(x,t) = \varphi_1(x) + t\varphi_2(x).
\end{cases}
\qquad (6.4.16)
$$

The solution of (6.4.10) is given by

$$
u(x,t) = \lim_{k\to\infty} u_k(x,t). \qquad (6.4.17)
$$

It can be seen that, the basic principle behind the variational iteration method is to first derive the Lagrange multiplier via variation principle, and then to rapidly obtain the approximation solution by arbitrarily selecting initial iteration value u_0. The advantage of the method over Adomian decomposition method is to avoid obtaining Adomian polynomial, whereas, similar to Adomian decomposition method, the variational iteration method needs the fractional derivative of a given function, which may not be obtainable sometimes. The comparison of these two methods for fractional differential equations can be found in [170, 174].

Bibliography

[1] K. Abbapio and Y. Cherruault. Convergence of Adomian's method applied to differential equations. *Comput. Math. Appl.*, 28:103–109, 1994.

[2] M. Abramowitz and I. Stegun. *Handbook of mathematical functions.* Dover, New York: 1972.

[3] R.A. Adams and J.J.F. Fournier. *Sobolev spaces, 2nd edition.* Academic Press, Amsterdam: 2003.

[4] G. Adomian. A review of the decomposition method in applied mathematics. *J. Math. Anal. Appl.*, 135:501–544, 1988.

[5] G. Adomian. *Solving Frontier Problems of Physics: The Decomposition Method.* Kluwer Academic Publisher, Boston: 1994.

[6] K. AI-Khaled and S. Momani. Approximate Solution for a fractional Diffusion-Wave Equation Using the Dcomposition Method. *Appl. Math. Comput.*, 165:473–483, 2005.

[7] A.I. Akhiezer, V.G. Yakhtar, and S.V. Peletminskii. *Spin waves.* North-Holland, Amsterdam: 1968.

[8] I. Ali, V. Kiryakov, and S.L. Kalla. Solutions of fractional multi-order integral and differential equations using a Poisson-type transform. *J. Math. Anal. Appl.*, 269(1):172–199, 2002.

[9] S. Alinhac and P. Gerard. *Pseudo-differentiels et theoreme de Nash-Moser.* EDP Sciences, Paris, 1991.

[10] F. Alouges and A. Soyeur. On global weak solutions for Landau-Lifshitz equations: Existence and nonuniqueness. *Nonl. Anal. TMA*, 19(11).

[11] T. Atanackovic, S. Pilipovic, and D. Zorica. Existence and calculation of the solution to the time distributed order diffusion equation. *Phys. Scr.*, T136:014012, 2009.

[12] T. Atanackovic, S. Pilipovic, and D. Zorica. Time distributed-order diffusive-wave equation. I. Volterra-type equation. *Proc. R. Soc. A*, 465:1869–1891, 2009.

[13] J.P. Aubin. Un theoreme de compacite. *C. R. Acad. Sci.*, 256:5042–5044, 1963.

[14] R.L. Bagley and P.J. Torvic. On the existence of the order domain and the solution of distributed order equations, I. *Int. J. Appl. Math.*, 2:865–882, 2000.

[15] R.L. Bagley and P.J. Torvic. On the existence of the order domain and the solution of distributed order equations, II. *Int. J. Appl. Math.*, 2:965–987, 2000.

[16] D. Baleanu, K. Biethelm, E. Scalas, and J. Trujillo. *Fractional Calculus, Models and Numerical Methods.* World Scientific, Singapore, 2012.

[17] E. Barkai, R. Metzler, and J. Klafter. From continuous time random walks to the fractional Fokker-Planck equation. *Phys. Rev. E*, 61:132–138, 2000.

[18] J.T. Beale, T. Kato, and A. Majda. Remarks on breakdown of smooth solutions for the three-dimensional Euler equations. Comm. *Math. Phys.*, 94:61–66, 1984.

[19] D.A. Benson, S.W. Wheatcraft, and M.M. Meerschaer. Application of fractional advectiondispersion equation. *Water Resource. Res.*, 36(6):1403–1412, 2000.

[20] D.A. Benson, S.W. Wheatcraft, and M.M. Meerschaer. Application of fractional advectiondispersion equation. *Water Resource. Res.*, 36(6):1413–1423, 2000.

[21] J. Bergh and J. Löofströom. *Interpolation spaces.* Springer-Verlag, Berlin Heidelberg, 1976.

[22] L.C. Berselli. Vanishing viscosity limit and long time behavior for 2D quasi-geostrophic equations. *Indian Univ. Math. J.*, 51(4):905–930, 2002.

[23] H. Bessaih and F. Flandoli. Weak attractor for a dissipative Euler equation. *J. Dyn. Diff. Eqs.*, 12(4):713–732, 2000.

[24] P. Biler, T. Funaki, and W.A. Woyczynski. Fractal Burgers equations. *J. Differential Equations*, 148:9–46, 1998.

[25] G.W. Scott Blair. The role of psychophysics in rheology. *J. Colloid Sciences*, 2:21–32, 1947.

[26] G.W. Scott Blair. *Measurements of mind and matter.* Dennis Dobson, London, 1950.

[27] F. Bloom and W. Hao. Regularization of a non-Newtonian system in an unbounded channel: Existence and uniqueness of solutions. *Nonl. Anal.*, 44:281–309, 2001.

[28] B.O'Shaugnessy and I. Procaccia. Analytical Solutions for Diffusion on Fractal Objects. *Phys. Rev. Lett.*, 54(5):455–458, 1985.

[29] J. Bourgain. Fourier restriction phenomena for certain lattice subsets and applications to nonlinear evolution equations, Part I: Schrödinger equations; Part II: the KdV equation. *Geom. Funct. Anal.*, 3:107–156; 209–262, 1993.

[30] X. Cabré and J. Tan. Positive solutions of nonlinear problems involving the square root of the Laplacian. *Adv. Math.*, 224(5):2052–2093, 2010.

[31] L. Caffarelli, S. Salsa, and L. Silvestre. Regularity estimates for the solution and the free boundary of the obstacle problem for the fractional Laplacian. *Invent. Math.*, 171(2):425–461, 2008.

[32] L. Caffarelli and L. Silvestre. An entension problem related to the fractional Laplacian. *Commun. Partial Differential Equation.*, 32(8):1245–1260, 2007.

[33] L. Cafferelli, R. Kohn, and L. Nirenberg. Partial regularity of suitable weak solutions of the Navier-Stokes equations. *Comm. Pure Appl. Math.*, 35:771–831, 1982.

[34] L. Cafferelli and A. Vasseur. Drift diffusion equations with fractional diffusion and the quasi-geostrophic equation. *Ann. Math.*, 171(3):1903–1930, 2010.

[35] M. Caputo. Linear model of dissipation whose Q is almost frequency independent–II. *Geophys. J. R. Astr. Soc.*, 13:529–539, 1967.

[36] M. Caputo. *Elasticità e Dissipazione*. Zanichelli, Bologna, 1969.

[37] M. Caputo. Mean fractional-order-derivatives differential equations and filters. *Annali dell'Universita di Ferrara*, 41(1):73–84, 1995.

[38] A. Carpinteri and F. Mainardi. *Fractals and fractional calculus in continuum mechanics*. Springer-Verlag, Vienna-New York, 1997.

[39] D. Chae. On the regularity conditions for the dissipative quasi-geostrophic equations. *SIAM J. Math. Anal.*, 37:1649–1656, 2006.

[40] D. Chae and J. Lee. Global well-posedness in the super-critical dissipative quasigeostrophic equations. *Commun. Math. Phys.*, 223:297–311, 2003.

[41] N.H. Chand and K. Uhlebeck. Schrödinger maps. Commun. *Pure Appl. Math.*, 53(5):590–602, 2000.

[42] A.V. Chechkin, R. Goreno, I.M. Sokolov, and V.Y. Gonchar. Distributed order time fractional diffusion equation. *Fract. Calc. Appl. Anal.*, 6:259–279, 2003.

[43] C. Chen, F. Liu, and K. Burrage. Finite difference methods and a fourier analysis for the fractional reaction-subdiffusion equation. *Appl. Math. Comput.*, 198:754–769, 2008.

[44] Y. Chen. The weak solutions to the evolution problems of harmonic maps. *Math. Z.*, 201:69–74, 1989.

[45] Y. Cherrualt. Convergence of Adomian's method. *Kybernetes*, 18:31–38, 1989.

[46] M. Ciesielski and J. Leszczynski. Numerical treatment of an initial-boundary value problem for fractional partial differential equations. *Signal Processing*, 86:2619–2631, 2006.

[47] I. Cimrak. *On the Landau-Lifshitz equation of ferromagnetism*. PhD Thesis, Ghent University, 2005.

[48] R. Coifman and Y. Meyer. *Au delá des opérateurs psedodifférentieles*. Astérisque **57**, Société Mathématique de France, 1978.

[49] A. Compte. Continuous time random walks on moving uids. *Phys. Rev. E*, 55:6821–6831, 1997.

[50] A. Compte and M.O. Caceres. Fractional Dynamics in Random Velocity Fields. *Phys. Rev. Lett.*, 81:3140–3143, 1998.

[51] A. Compte, R. Metzler, and J. Camacho. Biased continuous time random walks between parallel plates. *Phys. Rev. E*, 56:1445–1454, 1997.

[52] P. Constantin, D. Cordoba, and J. Wu. On the critical dissipative quasi-geostrophic equation. *Indiana Univ. Math. J.*, 50:97–107, 2001.

[53] P. Constantin, A.J. Majda, and E. Tabak. Formation of strong fronts in the 2-D quasigeostrophic thermal active scalar. *Nonlinearity*, 7:1495–1533, 1994.

[54] P. Constantin and J. Wu. Behavior of solutions of 2D quasi-geostrophic equations. *SIAM J. Math. Anal.*, 30(5):937–948, 1999.

[55] P. Constantin and J. Wu. Regularity of Hölder continuous solutions of the supercritical quasi-geostrophic equation. *Ann. Inst. H. Poincaré Anal. Non Linéaire*, 25(6):1103–1110, 2008.

[56] P. Constantin and J. Wu. Hölder continuity of solutions of supercritical dissipative hydrodynamic transport equations. *Ann. Inst. H. Poincaré Anal. Non Linéaire*, 26(1):159–180, 2009.

[57] A. Cordoba and D. Cordoba. A pointwise estimate for fractionary derivatives with applications to P.D.E. *Proc. Natl. Acad. Sci.*, 100(26):15136–15317, 2003.

[58] A. Cordoba and D. Cordoba. A maximum principle applied to quasi-geostrophic equations. *Commum. Math. Phys.*, 249:511–528, 2004.

[59] C.X.Miao. *Harmonic analysis and partial differential equations (second edition)*. Science Press, 2004.

[60] V. Daftardar-Gejji and H. Jafar. Solving a multi-order fractional differential equation using Adomian decomposition. *Appl. Math. Comput.*, 189:541–548, 2007.

[61] V. Daftardar-Gejji and S.Bhalekara. Boundary value problems for multi-term fractional differential equations. *J. Math. Anal. Appl.*, 345(2):754–765, 2008.

[62] G. David and J.L. journé. A boundedness criterion for generalized Calderon-Zygmund operators. *Ann. Math.*, 120:371–397, 1984.

[63] A. DeSimone, R.V. Kohn, S. Müller, and F. Otto. A reduced theory for thin-film micromagnetics. *Comm. Pure Appl. Math.*, 55:1408–1460, 2002.

[64] K. Diethelm. *Numerical approximation for finite-part integrals with generalized compound quadrature formulae*. Hildesheimer Informatikberichte, 1995.

[65] K. Diethelm. An algorithm for the numerical solution of differential equations of fractional order. *Electron. Trans. Numer. Anal.*, 5:1–6, 1997.

[66] K. Diethelm. Generalized compound quadrature formulae for finite-part integrals. *IMA J. Numer. Anal.*, 17:479–493, 1997

[67] K. Diethelm. Efficient Solution of Multi-Term Fractional Differential Equations Using $P(EC)^m E$ Methods. *Computing*, 71:305–319, 2003.

[68] K. Diethelm. *Fractional Differential Equations, Theory and Numerical Treatment*. TU Braunschweig, Braunschweig, 2003.

[69] K. Diethelm. *The Analysis of Fractional Differential Equations–An Application-Oriented Exposition Using Differential Operator of Caputo Type*. Springer-Verlag, Berlin and Heidelberg, 2010.

[70] K. Diethelm and N.J. Ford. Analysis of fractional differential equations. *J. Math. Anal. Appl.*, 265(2):229–248, 2002.

[71] K. Diethelm and N.J. Ford. Multi-order fractional differential equations and their numerical solution. *Applied Mathematics and Computation*, 154:621–640, 2004.

[72] K. Diethelm, N.J. Ford, and A.D. Freed. Detailed error analysis for a fractional Adams method. *Numerical Algorithms*, 36(1):31–52, 2004.

[73] K. Diethelm and A. Freed. On the solution of nonlinear fractional-order differential equations used in the modelling of viscoplasticity, in "Scientific Computing in Chemical Engineering II–Computational Fluid Dynamics, Reaction Engineering, and Molecular Properties, F.Keil, W.Mackens, H.Vob and J.Werther(eds)". *Springer*, pages 217–224, 1999.

[74] K. Diethelm and G. Walz. Numerical solution of fractional order differential eqeuations by extrapolation. *Numer. Algorit.*, 16(3-4):231–253, 1997.

[75] Z. Ding, A. Xiao, and M. Li. Weighted finite difference methods for a class of space fractional partial differential equations with variable coeffcients. *J. Comput. Appl. Math.*, 233:1905–1914, 2010.

[76] C.R. Doering, J.D. Gibbon, and C.D. Levermore. Weak and strong solutions of the complex Ginzburg-Landau equation. *Phys. D.*, 71:285–318, 1994.

[77] J.T. Edwards, N.J. Ford, and C.A.Simpson. The numerical solution of linear multiterm fractional differential equations: systems of equations. *J. Comput. Appl. Math.*, 148(2):401–418, 2002.

[78] A.M.A. El-sayed. Multivalued fractional equations of fractional order. *Appl. Math. Comput.*, 49:1–11, 1994.

[79] A.M.A. El-sayed. Fractional order evolution equations. *J. Frac. Calculus*, 7:89–100, 1995.

[80] A.M.A. El-Sayed, M.M. Saleh, and E.A.A. Ziada. Analytical and numerical solution of multi-term nonlinear differential equations of arbitrary orders. *J. Appl. Math. Comput.*, 33:375–388, 2010.

[81] V.E. Lynch *et. al.* Numerical methods for the solution of partial differential equations of fractional order. *J. Comput. Phy.*, 192:406–421, 2003.

[82] V.J. Ervin and J.P. Roop. Variational solution of the fractional advection dispersion equation on bounded domains in R2. *Numer. Methods Partial Differential Equations*, 23(2):256–281, 2007.

[83] E. Fabes, C. Kenig, and R. Serapioni. The local regularity of solutions of degenerate elliptic equations. *Commun. Partial Differential Equations*, 7(1):77–116, 1982.

[84] A.D. Fokker. Die mittlere energie rotierender elektrischer Dipole im Strahlungsfeld. *Annalen der Physik*, 43:810–820, 1914.

[85] G. Folland. *Introduction to Partial differential equations, 2nd edition.*

[86] N.J. Ford and A.C. Simpson. *Numerical approaches to the solution of some fractional differential equations, Numerical Analysis Report.* Manchester Centre for Computational Mathematics, Manchester, 2003.

[87] N.J. Ford and A.C. Simpson. *The numerical solution of fractional differential equations: speed versus accuracy, Numerical Analysis Report.* Manchester Centre for Computational Mathematics, Manchester, 2003.

[88] A.N. Gerasimov. A generalization of linear laws of deformation and its application to inner friction poblems. *Prikl. Mat. Mekh.*, 12:251–259, 1948.

[89] D. Gilbarg and N.S. Trudinger. *Elliptic partial differential equations of second order, 2nd edition.* Springer-Verlag, 1983.

[90] T.L. Gilbert. A Lagrangian formulation of gyromagnetic equation of the magnetization field. *Phys. Rev.*, 100:1243–1255, 1955.

[91] R. Gorenflo. Fractional calculus: Some numerical methods, in: "Carpinteri A. and Mainardi F.(Eds.), Fractals and Fractional Calculus in Continuum Mechanics, vol.378 of CISM Coursed and Lectures". *Springer-Verlag*, pages 277–290, 1997.

[92] R. Gorenflo and E.A. Abdel-Rehim. Convergence of the Grünwald-Letnikov scheme for time-fractional diffusion. J. Comput. *Appl. Math.*, 205:871–881, 2007.

[93] R. Gorenflo, G.D. Fabritiis, and F. Mainardi. Discrete random walk models for symmetric Levy-Feller diffusion Processes. *Physcica A*, 269:79–89, 1999.

[94] R. Gorenflo and F. Mainardi. Random walk models for space-fractional diffusion processes. *Fract. Calc. Appl. AnaL.*, 1(2):167–191, 1998.

[95] R. Gorenflo, F. Mainardi, D. Moretti, G. Pagnini, and P. Paradisi. Diserete Random Walk Models for Space-Time Fractional Diffusion. *Chemical Physics*, 284:521–541, 2002.

[96] R. Gorenflo, E. Scalas, and F. Mainardi. Fractional calculus and continuous-time finance. *Phys. A*, 284:376–384, 2000.

[97] R. Gorenflo and S. Vessela. *Abel Integral Equations, Vol. 146, Lect. Math. Note.* Springer-Verlag, Berlin, 1991.

[98] Q. Guan. Integration by parts formula for regional fractional Laplacian. *Commun. Math. Phys.*, 266(2):289–329, 2006.

[99] Q. Guan and Z. Ma. Boundary problems for fractional Laplacian. *Stoch. Dynam.*, 5(3):385–424, 2005.

[100] Q. Guan and Z. Ma. Reflected symmetric α-stable processes and regional fractional Laplacian. *Probab. Th. Rel. Fields*, 134(4):649–694, 2006.

[101] B. Guo. *Nonlinear evolution equations (in Chineae)*. Shanghai Science and Technology Press, Shanghai, 1995.

[102] B. Guo and S. Ding. *Landau-Lifshitz equations*. World Scientific, Singapore, 2008.

[103] B. Guo, Y. Han, and J. Xin. Existence of the global smooth solution to the period boundary value problem of fractional nonlinear Schrödinger equation. *Appl. Math. Comput.*, 204:468–477, 2008.

[104] B. Guo and M. Hong. The Landau-Lifshitz equation for the ferromagnetic spin chian and harmonic maps. *Calc. Var.*, 1:311–334, 1993.

[105] B. Guo, H.Y. Huang, and M.R. Jiang. *Ginzburg-Landau equation*. Science Press, Beijing, 2002.

[106] B. Guo and Z. Huo. Global well-posedness for the fractional nonlinear Schrodinger equation. *Commun. Partial Differential Equations*, 36(2):247–255, 2011.

[107] B. Guo and X.K. Pu. *Infinite Dimensional Random Dynamical system*. Beihang University Press, Beijing, 2009.

[108] B. Guo and M. Zeng. Solutions for the fractional Landau-Lifshitz equation. *J. Math. Anal. Appl.*, 361:131–138, 2009.

[109] C. Guo, J. Zhang, and B. Guo. Existence, uniqueness and decay of solution for fractional Boussinesq approximation. *Acta Math. Sci. Ser. B Engl. Ed.*, to appear.

[110] E. Hairer, Ch. Lubich, and M. Schlichte. Fast numerical solution of nonlinear volterra convolution equations. *SIAM J. Sci. Statist. Comput.*, 6(3):532–541, 1985.

[111] Y.S. Han. *Modern harmonic analysis method and its application*. Science Press, 1999.

[112] J. He. Variational iteration method.Some recent results and new interpretations. *J. Comput. Appl. Math.*, 207:3–17, 2007.

[113] L. Hormander. *The analysis of linear partial differential operators, Vol.1.* Springer, 1983.

[114] L. Hormander. *The analysis of linear partial differential operators, Vol.3-4.* Springer, 1985.

[115] F. Huang and F. Liu. The space-time fractional diffusion equation with Caputo derivatives. *J. Appl. Math. Comp.*, 19(1-2):179–190, 2005.

[116] F. Huang and F. Liu. The time fractional diffusion equation and the advection-dispersion equation. *ANZIAM. J.*, 46:317–330, 2005.

[117] H. Jafari and V. Daftardar-Gejji. Solving linear and non-linear fractional diffusion and wave equations by Adomian decomposition. *Appl. Math. Comput.*, 180(2):488–497, 2006.

[118] L.S. Jiang and Y.Z. Chen. *Lectures on Mathematical Physical Equations.* Higher Education Press, 1986.

[119] N. Ju. Existence and uniqueness of the solution to the dissipative 2D quasi-geostrophic equations in the Sobolev space. *Commum. Math. Phys.*, 252:365–376, 2004.

[120] N. Ju. The maximum principle and the global attractor for the dissipative 2D quasigeostrophic equations. *Commum. Math. Phys.*, 255, 2005.

[121] T. Kato. *Quasi-linear equations of evolution, with applications to partial differential equations, Lect. Note Math., Vol. 448.* Springer-Verlag, Berlin, 1975.

[122] C. Kenig, G. Ponce, and L. Vega. The Cauchy problem for the Korteweg-de Vries equation in Sobolev spaces of negative indices. *Duke Math. J.*, 71:1–21, 1993.

[123] C. Kenig, G. Ponce, and L. Vega. Well-posedness and scattering results for the generalized Korteweg-de Vries equation via the contraction principle. *Comm. Pure Appl. Math.*, 46(4):453–620, 1993.

[124] C. Kenig, G. Ponce, and L. Vega. A bilinear estimate with applications to the KdV equation. *J. Amer. Math. Soc.*, 9:573–603, 1996.

[125] A. Kiselev, F. Nazarov, and A. Volberg. Global well-posedness for the critical 2D dissipative quasi-geostrophic equation. *Invent. Math.*, 167:445–453, 2007.

[126] J. Klafter, A. Blumen, and M.F. Shlesinger. Stochastic pathway to anomalous diffusion. *Phys. Rev. A*, 35:3081–3085, 1987.

[127] J.J. Kohn and L. Nirenberg. An algebra of pseudo-differential operators. *Comm. Pure Appl. Math.*, 18:269–305, 1965.

[128] L.D. Landau and E.M. Lifshitz. On the theory of the dispersion of magnetic permeability in ferromagnetic bodies. *Phys. Z. Sowj.*, 8:153–169, 1935.

[129] T.A.M. Langlands and B.I. Henry. The accuracy and stability of an implicit solution method for the fractional diffusion equation. *J. Comput. Phys.*, 205:719–736, 2005.

[130] N. Laskin. Fractional Schrödinger equations. *Phys. Rev. E*, 66(5):056108, 2002.

[131] E.H. Lieb. Gaussian kernels have only Gaussian maximizer. *Invent. Math.*, 102:179–208, 1990.

[132] E.M. Lifshitz and L.P. Pitaevsky. *Stochastical physics, Landau course on theoretical physics, Vol.9.* Pergamon Press, Oxford-New York, 1980.

[133] F. Lin and C. Wang. *The analysis of harmonic maps and their heat flows.* World Scientific, 2008.

[134] Y. Lina and C. Xu. Finite difference/spectral approximations for the time-fractional diffusion equation. *J. Comput. Phys.*, 225(2):1533–1552, 2007.

[135] J.L. Lions. *Quelques methodes de resolution des problemes aux limites non lineaires.* Dunod Gauthier-Villars, Paris, 1969.

[136] P.L. Lions. The concentration-compactness principle in the calculus of variations, the limit case ii. *Rev. Mat. Iberoamericana*, 1:45–121, 1985.

[137] F. Liu, V. Anh, and I. Turner. Error analysis of an explicit finite difference approximation for the space fractional diffusion equation with insulated ends. *ANZIAM J.*, 46(E):C871– 887, 2005.

[138] F. Liu, V. Anh, I. Turner, and P. Zhuang. Time fractional advection-dispersion equation. *J. Appl. Math. Comp.*, 13(1-2):233–246, 2003.

[139] F. Liu, P. Zhang, V. Anh, and I. Turner. A fractional-order implicit difference approximation for the space-time fractional diffusion equation. *ANZIAM J.*, 47:48–68, 2006.

[140] F. Liu, P. Zhuang, V. Anh, and I. Turner. A fractional-order implicit difference approximation for the space-time fractional diffusioon equation. *ANZIAM J.*, 47(E):48–68, 2006.

[141] F. Liu, P. Zhuang, V. Anh, I. Turner, and K. Burrage. Stability and convergence of the difference methods for the space-time fractional advection-diffusion equation. *Appl. Math. Comput.*, 191(E):12–20, 2007.

[142] C.F. Lorenzo and T.T. Hartley. Variable order and distributed order fractional differential operators. *Nonl. Dym.*, 29:57–98, 2002.

[143] Ch. Lubich. On the stability of linear multistep methods for Volterra convolution equations. *IMA J. Numer. Anal.*, 3(4):439–465, 1983.

[144] Ch. Lubich. Fractional linear multistep methods for Abel-Volterra integral equations of the second kind. *IMA J. Numer. Anal.*, 45(172):463–469, 1985.

[145] Ch. Lubich. Discretized fractional calculus. *SIAM J. Math. Anal.*, 17:704–719, 1986.

[146] Ch. Lubich. A stability analysis of convolution quadratures for Abel-Volterra integral equations. *IMA J. Numer. Anal.*, 6(1):87–101, 1986.

[147] Ch. Lubich. Fractional linear multistep methods for Abel-Volterra integral equations of the first kind. *IMA J. Numer. Anal.*, 7(1):97–106, 1987.

[148] Ch. Lubich. Convolution quadrature and discretized operational calculus. I. *Numer. Math.*, 52(2):129–145, 1988.

[149] F. Mainardi and R. Goren o. Approximation of Levy-Feller diffusion by random walk. *J. Anal. Appl.*, 18:231–246, 1999.

[150] F. Mainardi, R.Goren o, and Yu. Luehko. Wright function as scale-invariant solutions of the diffusion-wave equation. *J. Comp. Appl. Math.*, 118:175–191, 2000.

[151] A. Majda. *Compressiblefluidflow and systems of conservation laws in seberal space variables*. Springer, New York, 1984.

[152] A.J. Majda and A.L. Bertozzi. *Vorticity and incompressibleflow*. Cambridge University Press, 2002.

[153] B.B. Mandelbrot. *The fractional geometry of nature*. Freeman, New York, 1983.

[154] B.B. Mandelbrot and J.W. van Ness. Fractional Brownian motions, fractional noises and applications. *SIAM Rev.*, 10:422–437, 1968.

[155] F. Maniardi. *Applications of fractional calculus in mechanics, in P. Rusev and V. Kiryakova, Transform Methods and special functions, Verna '96.* SCT Publishers, Singapore, 1997.

[156] M.M. Meerschaert, H.P. Scheffler, and C. Tadjeran. Finite difference methods for twodimensional fractional dispersion equation. *J. Comp. Phy.*, 211:249–261, 2006.

[157] M.M. Meerschaert and C. Tadjeran. Finite difference approximations for fractional advection-dispersionflow equations. *J. Comput. Appl. Math.*, 172:65–77, 2004.

[158] M.M. Meerschaert and C. Tadjeran. Finite difference approximations for two-sided space fractional partial differential equations. *Appl. Numer. Math.*, 56:80–90, 2006.

[159] R. Metzler, E. Barkai, and J. Klafter. Anomalous Diffusion and Relaxation Close to Thermal Equilibrium: A Fractional Fokker-Planck Equation Approach. *Phys. Rev. Lett.*, 82:3563–3567, 1999.

[160] R. Metzler, E. Barkai, and J. Klafter. Deriving fractional Fokker-Planck equations from a generalised master equation. *Europhys. Lett.*, 46:431–436, 1999.

[161] R. Metzler and J.Klafter. The restaurant at the random walk: recent developments in the description of anomalous transport by fractional dynamics. *J. Phys. A*, 37:R161–R208, 2004.

[162] R. Metzler and J. Klafter. The random walk's guide to anomalous diffusion: a fractional dynamics approach. *Phys. Rep.*, 339:1–77, 2000.

[163] N. Meyers and J. Serrin. $H = W$. *Proc. Nat. Acad. Sci. U.S.A.*, 51:1055–1056, 1964.

[164] C. Miao, B. Yuan, and B. Zhang. Well-posedness of the Cauchy problem for the fractional power dissipative equations. *Nonl. Anal.*, 68:461–484, 2008.

[165] K.S. Miller and B. Ross. *An introduction to the fractional calculus and fractional differential equations.* John Wiley & Sons, New York, 1993.

[166] S. Momani. Analytic and Approximate Solutions of the Space-and Time-Fractional Telegraph Equations. *Appl. Math. Comput.*, 170:1126–1134, 2005.

[167] S. Momani. Analytic and approximate solutions of the space- and time-fractional telegraph equations. *Applied Mathematics and Computation*, 170(2):1126–1134, 2005.

[168] S. Momani and K. Al-Khaled. Numerical Solutions for Systems of Fractional Differential Equations by the Decomposition Method. *J. Appl. Math. Comp.*, 162:1351–1365, 2005.

[169] S. Momani and Z. Odibat. Analytical solution of a time-fractional Navier-Stokes equation by Adomian decomposition method. *Appl. Math. Comput.*, 177:488–494, 2006.

[170] S. Momania and Z. Odibatb. Numerical comparison of methods for solving linear differential equations of fractional order. *Chaos, Solitons & Fractals*, 31(5):1248–1255, 2007.

[171] M. Naber. Distributed order fractional sub-diffusion. *Fractals*, 12(23):23–32, 2004.

[172] M. Naber. Time fractional Schrödinger equation. *J. Math. Phys.*, 45(8):3339–3352, 2004.

[173] A. Nahmod, A. Stefanov, and K. Uhlenbeck. On Schrödinger maps. *Commun. Pure Appl. Math.*, 56(1):114–151, 2003.

[174] Z. Odibat and S. Momani. Numerical methods for nonlinear partial differential equations of fractional order. *Appl. Math. Model.*, 32:28–39, 2008.

[175] K.B. Oldham and J. Spanier. *The fractional calculus*. Academic Press, New York, 1974.

[176] K.B. Oldham and J. spanier. *The fractional calculus: Theory and applications of differentiation and integration to arbitrary order*. Academic press, 1974.

[177] A. Pazy. *Semigroups of linear operators and applications to partial differential equations*. Springer-Verlag, New York, 1983.

[178] M. Planck. Ueber einen Satz der statistichen Dynamik und eine Erweiterung in der Quantumtheorie. *Sitzungberichte der Preussischen Akadademie der Wissenschaften*, pages 324–341, 1917.

[179] I. Podlubny. *Fractional differential equations: An introduction to fractional derivatives, fractional differential equations, to methods of their solution and some of their applications*. Academic Press, New York, 1999.

[180] J. Prüss. *Evolutionary integral equations and applications*. Birkhauser Verlag, Basel, Switzerland, 1993.

[181] X. Pu and B. Guo. Existence and decay of solutions to the two-dimensional fractional quasigeostrophic equation. *J. Math. Phys.*, 51(8):083101–15, 2010.

[182] X. Pu and B. Guo. Global Weak solutions to the 1-D fractional Landau-Lifshitz equation. *Discret. Contin. Dyn. Syst. Ser. B*, 14(1):199–207, 2010.

[183] X. Pu and B. Guo. The fractional Landau-Lifshitz-Gilbert equation and the heat ow of harmonic maps. *Calc. Var.*, 42:1–19, 2011.

[184] X. Pu and B. Guo. Well-posedness and dynamics for the fractional ginzburg-landau equation. *Applicable Analysis*, 92(2):318–334, 2013.

[185] X. Pu and B. Guo. Well-posedness for the fractional Landau-Lifshitz equation without Gilbert damping. *Calc. Var.*, 46:441–460, 2013.

[186] M.Y. Qi. *Linear partial differential operator on Introduction, Volume I (in Chinese)*. Science Press, Beijing, 1986.

[187] E.A. Rawashdeh. Numerical solution of fractional integro-differential equations by collocation method. *Appl. Math. Comput.*, 176:1–6, 2006.

[188] S.S. Ray and R.K. Bera. Analytical Solution of the Bagley-Torvik Equation by Adomian Decomposition Method. *J. Appl. Math.Comp.*, 168:398–410, 2005.

[189] M. Reed and N. Simon. *Methods of modern mathematical physics*. Academic Press, New York, 1975.

[190] S. Resnick. *Dynamical problems in nonlinear advective partial differential equations*. Ph.D. Thesis, University of Chicago, 1995.

[191] H. Risken. *The Fokker-Planck equation*. Springer, Berlin, 1989.

[192] J. Robinson. *Inffnite-dimensional dynamical systems: an introduction to dissipative parabolic PDEs and the theory of global attractors*. Cambridge Univ. Press, Cambridge, 2001.

[193] J.P. Roop. Computational aspects of FEM approximation of frac- tional advection dispersion equations on bounded domain in R^2. *J. Comp. Appl. Math.*, 193:243–268, 2006.

[194] Yu.A. Rossikhin and M.V. Shitikova. Applications of fractional calculus to dynamic problems of linear and nonlinear hereditary mechanics of solids. *Appl. Mech. Rev.*, 50(1):15–67, 1997.

[195] S.G. Samko, A.A. Kilbas, and O.I. Marichev. *Fractional integrals and derivatives: theory and applications*. Gordon and Breach Science, New York, 1993.

[196] H. Scher, M.F. Shlesinger, and J.T. Bendler. Time-Scale Invariance in Transport and Relaxation. *Phys. Today*, 44(1):26–34, 1991.

[197] M. Schonbek and T. Schonbek. Asymptotic behavior to dissipative quasigeostrophic ows. *SIAM J. Math. Anal.*, 35:357–375, 2003.

[198] G.R. Sell. Global attractors for the three-dimensional Navier-Stokes equations. *J. Dyn. Diff. Eqs.*, 8:1–33, 1996.

[199] S. Shen and F. Liu. Numerical solution of the space fractional Fokker-Planck equation. *J. Comput. Appl. Math.*, 166:209–219, 2004.

[200] M.K. Shkhanukov. On the convergence of difference schemes for differential equations with a fractional derivative. *Dekl. Akad. Nauk*, 348(6):746–748, 1996.

[201] J. Simon. Compact sets in space $L^p(0; T; B)$. *Ann. Math. Pura. Appl.*, 146:65–96, 1987.

[202] E. Sousa. Finite difference approximations for a fractional advection diffusion problem. *J. Comput. Phys.*, 228:4038–4054, 2009.

[203] E.M. Stein. *Singular integrals and differentiability properties of functions*. Princeton University Press, Princeton, 1970.

[204] E.M. Stein. *Harmonic Analysis*. Princeton University Press, Princeton, 1993.

[205] E.M. Stein and R. Shakarchi. *Fourier Analysis: An introduction*. Princeton University Press, Princeton, 2003.

[206] E.M. Stein and G. Weiss. *Introduction to Fourier Analysis on Euclidean Space*. Princeton University Press, Princeton, 1971.

[207] R.S. Strichart. Multipliers in fractional Sobolev spaces. *J. Math. Mech.*, 16:1031–1060, 1967.

[208] L. Su, W. Wang, and Q. Xu. Finite difference methods for fractional dispersion equations. *Appl. Math. Comput.*, 216:3329–3334, 2010.

[209] C. TadjeranMark, M.M. Meerschaert, and H.P. Scheffler. A second-order accurate numerical approximation for the fractional diffusion equation. *J. Comp. Phy.*, 213:205–213, 2006.

[210] V.E. Tarasov. Fractional Heisenberg equation. *Phys. Lett. A*, 372:2984–2988, 2008.

[211] V.E. Tarasov and G.M. Zaslavsky. Fractional Ginzburg-Landau equation for fractal media. *Phys. A*, 354:249–261, 2005.

[212] M. Taylor. *Partial Differential Equations, Vol. II.* Springer-Verlag, 1996.

[213] R. Temam. *Infinite dimensional dynamical systems in mechanics and physics (2nd edition).* Springer-Verlag, 1998.

[214] H. Triebel. *Theory of function sapces.* Birkhuser, 1983.

[215] C-C. Tseng, S-C. Pei, and S-C. Hsia. Computation of fractional derivatives using Fourier transform and digital FIR differentiator. *Signal Processing*, 80:151–159, 2000.

[216] T.Tao. Multilinear weighted convolution of L^2 functions, and applications to nonlinear dispersive equation. *Amer.J. Math.*, 123:839–908, 2001.

[217] N.G. van Kampen. *Stochastic processes in physics and chemistry.* North-Holland, Amersterdam, 1981.

[218] M. Weilbee. *Efficient Numerical Methods for Fractional Differetnial Equations and their Analytical Background.* Technical University of Braunschweig, 2005.

[219] H. Weyl. Bemerkungen Zum Bergriff des Differentialquotienten gebrochener Ordnung. *Vierteliahresschr. Naturforsh. Ges. Zurich*, 62:296–302, 1917.

[220] J. Wu. Inviscid limit and regularity estimates for the solutions of the 2-D dissipative quasi-geostrophic equations. *Indiana Univ. Math. J.*, 46:1113–1124, 1997.

[221] J. Wu. The quasi-geostrophic equation and its two regularizations. *Comm. Partial Differential Equations*, 27(5-6):1161–1181, 2002.

[222] J. Wu. Generalized MHD equations. *J. Differential Equations*, 195:284–312, 2003.

[223] J. Wu. Solutions of the 2D quasi-geostrophic equation in Hölder spaces. *Nonl. Anal.*, 62:579–594, 2005.

[224] Q. Yang, F. Liu, and I. Turner. Numerical methods for fractional partial differential equations with Riesz space fractional derivatives. *Appl. Math. Model.*, 34:200–218, 2010.

[225] Q. Yu, F. Liu, V. Anh, and I. Turner. Solving linear and non-linear space-time fractional reaction-diffusion equations by the Adomian decomposition method. *Int. J. Numer. Meth. Engng*, 74:138–158, 2008.

[226] S.B. Yuste and L. Acedo. An explicit finite difference method and a new Von Neumann-type stablity analysis for fractional diffusion equations. *SIAM J. Numer. Anal.*, 42(5):1862–1874, 2005.

[227] Z. Zhai. Well-posedness for fractional Navier-Stokes equations in critical spaces close to $\dot{B}_{\infty,\infty}^{-(2\beta-1)}(R^n)$. *Dynamics of PDE*, 7(1):25–44, 2010.

[228] M.Q. Zhou. *Lectures on Harmonic Analysis (in Chinese)*. Peking University Press, Beijing, 1999.

[229] Y. Zhou. Regularity criteria for the generalized viscous MHD equations. *Ann. I. H. Poincaré*, 24:491–505, 2007.

[230] Y. Zhou, B. Guo, and S. Tan. Existence and uniqueness of smooth solution for system of ferromagnetic chain. *Sci. China, Ser. A*, 34:257–266, 1991.

[231] P. Zhuang, F. Liu, I. Turner, and V. Anh. Numerical Treatment for the Fractional Fokker- Planck Equation. *ANZIAM J.*, 48(E):759–774, 2007.

[232] W.P. Ziemer. *Weakly differentiable functions*. Springer-Verlag, Heidelberg, 1989.

Printed in the United States
By Bookmasters